T0220888

FOREST MENSURATION

FOREST MENSURATION

Fifth Edition

JOHN A. KERSHAW, JR.
MARK J. DUCEY
THOMAS W. BEERS
BERTRAM HUSCH

WILEY Blackwell

This edition first published 2017 © 1972, 1983, 1993, 2003, 2017 by John Wiley & Sons, Ltd

Registered Office
John Wiley & Sons, Ltd, The Atrium, Southern Gate, Chichester, West Sussex, PO19 8SQ, UK

Editorial Offices
9600 Garsington Road, Oxford, OX4 2DQ, UK
The Atrium, Southern Gate, Chichester, West Sussex, PO19 8SQ, UK
111 River Street, Hoboken, NJ 07030-5774, USA

For details of our global editorial offices, for customer services and for information about how to apply for permission to reuse the copyright material in this book please see our website at www.wiley.com/wiley-blackwell.

Library of Congress Cataloging-in-Publication Data

Names: Kershaw, John A., Jr., 1962– author. | Ducey, Mark J., author. | Beers, Thomas W., author. |
    Husch, Bertram, 1923– author.
Title: Forest mensuration / by John A Kershaw, Jr., Mark J Ducey, Thomas W Beers, Dr. Bertram Husch.
Description: Fifth edition. | Chichester, UK ; Hoboken, NJ : John Wiley & Sons, 2017. |
    Includes bibliographical references and index.
Identifiers: LCCN 2016036142| ISBN 9781118902035 (cloth) | ISBN 9781118902004 (epub)
Subjects: LCSH: Forests and forestry–Measurement. | Forest surveys.
Classification: LCC SD555 .K47 2017 | DDC 634.9–dc23
LC record available at https://lccn.loc.gov/2016036142

A catalogue record for this book is available from the British Library.

Wiley also publishes its books in a variety of electronic formats. Some content that appears in print may not be available in electronic books.

Set in 10/12pt Times by SPi Global, Pondicherry, India

1   2017

CONTENTS

PREFACE

It was not without some trepidation that the two lead authors undertook the revision of this text. The 12 years since the last revision has seen many substantial changes to the field of forest mensuration and has marked a passing of a generation of mensurationists as reflected in the change of authorship of this text. The authors attempted to reflect this change while preserving the classic coverage that this text has been known for.

We first want to mark the passing of Dr. Bertram Husch, who served as lead author on this text through four editions and over 40 years. The first edition of this text, released in 1963, marked the first comprehensive coverage of the field of forest mensuration from a statistical perspective. We also acknowledge the passing of several other great mensurationists since the publication of the last edition, many of whom are cited throughout this text: Dr. Walter Bitterlich, Dr. Lew Grosenbaugh, Dr. Al Stage, Dr. Benno Hesske, Dr. George Furnival, Dr. Boris Zeide, Dr. Paul Van Deusen, and Mr. Bill Carr. They dedicated their lives to forest mensuration, advanced our knowledge of its principles and applications, and contributed much to the education of the current generation of mensurationists.

Forest mensuration is a living science, and one that continues to advance and grow as we build our understanding and as society's needs and expectations of forests change. We have attempted to reflect those changes in this new edition. The fourth edition saw a major reorganization of the materials and the introduction of nontimber vegetation measurement and carbon estimation. This edition builds upon those changes. We moved the planning of a forest inventory from late in the book upfront to Chapter 1 and place a greater emphasis on estimation throughout. The coverage of nontimber vegetation and carbon accounting has expanded to reflect the current emphasis on these factors in forest management and forest inventory. These factors are integrated throughout the text rather than covered in separate chapters. We have developed an expanded chapter on sampling and estimating dead and down woody debris (Chapter 12) and added a new chapter on the use of remote sensing in forest inventory (Chapter 13). We have developed worked examples for all of the sample designs and have provided the base data so that instructors and students can work through examples in their classes. We have maintained the use of both Imperial units and SI units throughout the text. One criticism of the fourth edition was the deletion of the tables of variable probability sampling factors that appeared in the third edition; we have included these tables in this edition and appreciated the feedback from our colleagues.

The authors acknowledge the help of many friends and colleagues during the preparation of this revision. Drs. David Larsen, Peter Marshall, Robert Froese, Tzeng Yih Lam, Kim Iles, and Andrew Robinson commented on the structure of the book early during its development, and Tom Lynch provided especially valuable comments on a late draft. Dr. Jeffrey Gove commented on many of the datasets and examples used throughout the book. Dr. Jim Chamberlain provided data on black cohosh used in Chapter 3. R. Andy Colter provided LiDAR data from the USDA Forest Service, and Jamie Perkins assisted with the optical imagery in Chapter 13. Dr. Joel Hartter, Russell Congalton, Forrest Stevens, and Michael Palace provided outstanding insights on the role of remote sensing through the Communities and Forests in Oregon (CAFOR) project. Tzeng Lam and Andrew Robinson provided critical feedback on several chapters. Laird van Damme provided feedback on the structure of the book from a practitioner's perspective. Ethan Belair carefully proofread the final versions of the chapters, and Julia Smith and Madison Poe helped with page proofs. Finally, we want to thank our families for their patience, encouragement, tolerance, and occasional distractions as this project progressed and protracted far longer than we led them to believe at the start.

1

INTRODUCTION

In the first widely available book on *Forest Mensuration* in North America, Henry S. Graves (1906) wrote: "Forest mensuration deals with the determination of the volume of logs, trees, and stands, and with the study of increment and yield." *The Dictionary of Forestry* (Helms, 1998) states that "Forest mensuration is the determination of dimensions, form, weight, growth, volume, and age of trees, individually or collectively, and of the dimensions of their products." This definition is essentially a paraphrase of the 1906 definition given by Henry S. Graves. Although some foresters feel this definition is still adequate, this text considers that mensuration should embrace new measurement problems that have arisen or have been recognized as the horizons of forestry have expanded.

If we accept the challenge of a broader scope, we must ask: "To what degree should mensuration be concerned with measurement problems of wildlife management, recreation, watershed management, and the other aspects of multiple–use forestry?" One might argue that it is unrealistic to imagine that forest mensuration can take as its domain such a diverse group of subjects. The objection becomes irrelevant if we recognize forest mensuration, not as a collection of specific techniques, but as a subject of study that provides principles applicable to a wide range of measurement problems. We view the measurement and quantification of all aspects of forest vegetation as within the domain of forest mensuration. Moreover, many ideas, approaches, and techniques have been developed within the context of traditional forest mensuration that have broad applicability for forest ecology, wildlife habitat, recreation, and watershed management. This book, in addition to a treatment of the traditional product-oriented measurement problems of forestry, will also provide a unified foundation of principles for solving measurement problems in other aspects of forestry.

During the latter half of the twentieth century, the application of statistical theory and the use of computers, electronics, and lasers wrought a revolution in the solution of forest

Forest Mensuration, Fifth Edition. John A. Kershaw, Jr., Mark J. Ducey, Thomas W. Beers and Bertram Husch.
© 2017 John Wiley & Sons, Ltd. Published 2017 by John Wiley & Sons, Ltd.

measurement problems. Consequently, mensurationists must have a degree of competence in their use as well as in basic mathematics and statistics. Knowledge of calculus is also desirable. In addition, familiarity with systems analysis and operational research, approaches to problem solving that depend on model building and techniques that include simulation and mathematical programming, will also be valuable, especially in advanced and more sophisticated treatments of forest mensurational problems. We do not presume that all readers of this text will have such a deep and broad background, and have tried to present forest mensuration in a way that is accessible to new students but provides a comprehensive overview of the possibilities of the field.

1.1. ROLE OF FOREST MENSURATION IN FOREST MANAGEMENT

Forest mensuration is one of the cornerstones in the foundation of forestry. Forestry in the broadest sense is a management activity involving forest land, the plants and animals on the land, and humans as they use the land. Much of the forest land in North America and in other parts of the world is under active forest management. In many jurisdictions, foresters are required to complete detailed long-term forest management plans, especially on public lands. These plans require foresters to make detailed predictions about the growth and yield of forest resources, and how harvesting and other forest management activities influence the flow of timber and other resources. Based on the outputs from these models, forest managers make decisions about where, when, how, and how much forest land should be treated. Elsewhere, management planning may reflect shorter time horizons, but the decisions are no less critical. Good forest management decisions require good tools to analyze the impacts of management activities on the quantities and flows of the various forest resources and on the state of the forest itself. These tools require good models and, ultimately, these models require good data. The acquisition of this data is the subject of this book.

Foresters are faced with many decisions in the management of a forest. The following questions are examples of the problems that must be solved for a particular forest:

1. What silvicultural treatment will result in best regeneration and growth?
2. What species is most suitable for reforestation?
3. Is there sufficient timber to supply a forest industry and for an economical harvesting operation?
4. What is the value of the timber and land?
5. What is the recreational potential?
6. What is the wildlife potential?
7. What is the status of biodiversity on the area?
8. What is the status of the forest as a carbon sink?

A forester needs information to answer these and countless other questions and to make intelligent decisions or recommendations to a client. This information often is needed in quantifiable terms. In most situations, the axiom holds, "You can't efficiently make, manage, or study anything you don't locate and measure." At the same time, resources for measurement are usually limited, so information must be acquired efficiently. In this sense, forest mensuration is the application of measurement principles to obtain quantifiable information for forest management decision-making.

To summarize, *forest mensuration* is concerned with obtaining information about forest resources and conditions. The ultimate objective of forest mensuration is to provide quantitative information about the forest and its resources that will allow making reasonable decisions on its destiny, use, and management.

1.2. FOREST MENSURATION AS A TOOL FOR MONITORING FORESTS

To many, a forest, if not affected by cutting, fire, or some other calamity, is a stable, unchanging entity. Actually, a forest is a dynamic system that is continuously changing. Although this may not be evident over a short term, such as a few years, change is always present: some trees increase their dimensions, others die, and new trees germinate and enter the forest. Consequently, the information obtained about the status of a forest area at a given time is only valid for a length of time that depends on the vegetation itself, and on environmental and external pressures affecting the forest. This means that the mensurational information regarding the forest must be updated periodically by monitoring procedures so that the appropriate management and policy decisions may be taken.

Throughout the twentieth century, the demand on forest resources increased worldwide (Westoby, 1987). In the opening decades of the twenty-first century, this increase is expected to continue. During the last 40 years, not only have the demands for timber increased, foresters also have been required to manage for other resources including wildlife habitat, water quality, recreational opportunities, and biodiversity. An increase in the public awareness of the influence of human activities on the environment has resulted in the development of a number of forest certification procedures to ensure that forest management activities are sustainable both economically and environmentally. These procedures require forest managers to document and monitor the impacts of forest management activities on a wide range of forest resources, not just timber.

Monitoring must consider changes in composition, structure, size and health of forests (Max et al., 1996). To be effective, monitoring must be comprehensive. In these situations, foresters must increase the scope of their inventories, and the models they use, to include information on multiple aspects of forest structure, not just timber-producing trees. To be cost effective, forest managers will be required to design and implement new sampling strategies and measurement procedures to meet the demand for increased information.

In the early 1900s, as professional forestry was beginning in North America, the need to use the quantitative tools forest mensuration and forest inventory offered to monitor forest resources was quickly recognized (Bates and Zon, 1922). Zon (1910), in one of the first attempts to assess global forest resources, recognized the need for systematic monitoring of forest resources on both a national and global basis. The Scandinavian countries (Norway in 1919, Finland in 1920, and Sweden in 1923) were the first countries to implement systematic national forest inventories (NFIs) based on modern statistical principles (Tokola, 2006; Tomppo, 2006). The United Kingdom began NFIs in 1924 and the United States in 1930. The Food and Agricultural Organization of the United Nations began compiling global assessments of forest resources in 1947. Many other European and Asian countries began NFIs in the 1960s and 1970s. In the late 1980s and early 1990s, many countries, including the United States and Sweden, redesigned their NFIs, adopting a systematic-grid-based sample design and consistent plot designs and remeasurement intervals. A number of countries around the world have redesigned or adopted similar protocols for their NFIs. Although most NFIs began with a focus on timber and related resources, nearly all are actively broadening to address the full range of economic, conservation, and environmental challenges faced by regional and national policymakers. Most countries have also integrated remote sensing, aerial photography, and/or LiDAR data into the analyses of ground-based data.

While the various NFIs differ by the specific measurements made, plot sizes and layouts, grid intensities, and remeasurement intervals, the newer designs make it easier to compare estimates across countries (Liknes et al., 2013). The information collected has many important uses including

1. helping policymakers at national and regional levels to formulate good forest policy, and to assess the sustainability of current and past policy;

2. enabling land managers to devise better management plans and to assess effects of current and past management practices on the land;

3. serving as a starting point for scientific investigations in a variety of areas that involve changes in forest ecosystems over time;

4. formulating business plans that will be both economically and ecologically sustainable over time;

5. keeping the public informed about the health and sustainability of a nation's forests; and

6. providing consistent and reliable reporting statistics to demonstrate compliance with various treaty obligations including conservation and biodiversity commitments and carbon offset accounting.

Most countries provide online access to their NFI data and extensive online report generating capabilities.

1.3. RELEVANCE OF FOREST MENSURATION FOR ECOLOGY AND NONTIMBER RESOURCES

Ecologists, conservation biologists, and wildlife managers, like forest managers, require quantitative information to make informed decisions. Sometimes, these decisions are about choices of management actions, and sometimes about choices between competing scientific hypotheses; but either type of choice depends on high-quality data that are collected and analyzed in a cost-effective way. While the emphasis of this text is the measurement, sampling, and estimation of the tree component of the forest vegetation, the basic principles of mensuration can be applied to a range of forest structures and attributes, and we have included some coverage of measurements for these characteristics. Estimation of species diversity, abundance, biomass, and carbon content utilize similar techniques for measurement and sampling, whether the focus is on trees, shrubs, herbs, or grasses. Of course, the types of measurements made, the tools used, the size, shape, and type of sample plots, and the sample intensity will vary depending upon the size of individuals, the spatial heterogeneity, and the estimates required (Bonham, 2013).

Ecological studies require measurements that are effective, accurate, and precise (Ford, 2000). Many of the stand structure parameters useful to foresters for making management decisions (e.g., density, basal area, volume, biomass, and crown cover) are useful parameters for estimating nontimber resources and ecological indices. Wildlife managers often rely on measures of stand structure to assess habitat quality and suitability. The need for accurate measurements and robust sample designs are just as important for these applications as they are for timber management.

An unfortunate consequence of the increasing channelization and specialization of knowledge in the twentieth and twenty-first century has been the divergence of scholarship and teaching between those working in areas of forestry most closely related to timber management, and those working in areas that have more conventionally been the domain of programs in ecology and conservation biology. Our view is that the two groups have much to learn from each other. Recent decades have seen dramatic advances in sampling techniques and forest measurements designed with timber applications in mind, but many of these advances are not widely known even though they have direct application to other challenges in forestry, forest conservation, and ecology in general. Conversely, some foresters whose training is narrowly focused on timber production have found the emerging science and practice in areas related to biodiversity or carbon accounting a bit mysterious, even though the basic measurement principles—and even some of the techniques—are quite similar to those they already know. An integrated treatment of these areas of forest mensuration will help both scientists and professionals approach forests with greater technical versatility and a broader outlook.

1.4. DESIGN AND PLANNING OF INVENTORIES

A *forest inventory* is the procedure for obtaining information on the quantity, quality, and condition of the forest resource, associated vegetation and components, and many of the characteristics of the land area on which the forest is located. The goals of a forest inventory depend on the management and policy questions it is designed to support. Most forest inventories have been and will continue to be focused on timber estimation. The need for information on forest health, water, soils, recreation, wildlife and scenic values, and other nonwood values has stimulated the development of integrated or multiresource inventories (Schreuder et al., 1993; Lightner et al., 2001). Interest in forests as a carbon sink has generated considerable interest in estimating forest biomass and carbon content (Satoo and Madgwick, 1982; Brown, 2002; Smith et al., 2006; Pearson et al., 2007). Development and execution of multiresource inventories often requires cooperation among many specialists in pertinent fields.

1.4.1. Timber Estimation

A complete forest inventory for timber evaluation provides the following information: estimates of the area, description of topography, ownership patterns, accessibility, transportation facilities, estimates of timber quantity and quality, and estimates of growth and drain. The emphasis placed on specific elements will differ depending upon the purpose of the inventory (Husch, 1971). For example, if the purpose of an inventory is for the preparation of a harvesting plan, major emphasis would be put on a description of topography, determination of accessibility and transportation facilities, and estimation of timber quantity.

Other aspects would be given little emphasis or eliminated. If the purpose of an inventory is for the preparation of a management plan, major emphasis would be put on timber quantity, growth, and harvest levels with lesser detail on other aspects.

Information is obtained in a forest inventory for timber evaluation by measuring and assessing the trees and various characteristics of the land. The information may be obtained from measurements made on the ground or from remotely sensed data (aerial photographs, satellite imagery, or LiDAR*). When measurements are made on all trees in a forest, the inventory is a complete or 100% inventory. When measurements are made for a sample of the forest, it is a sampling inventory. The terms *cruise* in North America and *enumeration* in other English-speaking areas are frequently used instead of inventory.

In executing a forest inventory for timber evaluation, even one based on a fairly small sample, it is impossible to measure directly quantities such as volume or weight of standing trees. Consequently, a relationship is established between directly measurable tree or stand characteristics (e.g., dbh, height) and the desired quantity. This may be done as follows:

1. Make detailed field measurements of trees or stands and compute the desired quantities from these measurements. For example, one might make detailed diameter measurements at determined heights along the stem of standing trees and determine volumes by formulas or graphical methods (Section 6.1).

2. Estimate the desired quantities in trees or stands by utilizing relationships previously derived from other trees or stands. For example, one might measure tree or stand characteristics such as dbh, height, and form, and determine the corresponding volume or weight from an equation or a table (Section 6.2).

Timber inventories can estimate gross and net volumes. An estimate of *gross volume* shows the volume of wood, usually without bark, based on the exterior measurements of the trees: dbh, height and, form, without any deductions for defects. An estimate of *net volume* reduces the gross volume due to defects such as rot, diseased portions, and stem irregularities. The reduction for defect can be determined in two ways: (1) at the time of measurement of each tree, the external defects can be noted, internal defects estimated (or directly determined by boring) and recorded, and the loss in volume estimated per tree; and (2) by applying a cull factor expressed as a percent that will reduce the gross volume of the inventory to net volume. This cull percent can be determined directly by carrying out a study that requires the felling of a sample of trees and determining their gross and net volumes to determine cull percentages. Frequently cull percentages are based on previous experience in the kind of timber being inventoried. Inventories can be designed to estimate net volumes or quantities by product classes such as poles, pulpwood, sawtimber, and veneer logs. Of course, this requires specifying the criteria to be used at the time of measurement of sample trees during fieldwork.

Other factors in estimating net usable quantity of timber depend not on the defects in the trees themselves but on several external causes. Impacts of factors such as accessibility, legal restrictions (protected areas, silvicultural requirements, and harvesting restrictions), nonutilizable species, minimum quality characteristics and sizes of logs, and losses due to breakage in logging and transport are required to obtain an estimate of the usable net volume.

---

*Light Detection and Ranging, Section 13.1.4.

1.4.2. Nontimber Estimation

As Hassan et al. (1996) have pointed out, "the need to look into non-wood benefits of forestry is ... becoming an increasingly important component of forest management planning in view of the fact that the contribution of forestry goes beyond wood production." The parameters measured in multiresource inventories vary, depending on the information required. These parameters may include timber estimates together with data on nontimber vegetation and/or other forest characteristics, such as biodiversity, forest health, scenic values, recreation resources, water quality, wildlife habitat, carbon storage, and ownership (USDA, 1992).

Multiresource inventories require considerations in the design and implementation that generally are not required when designing timber inventories. Tree species have large woody structures that persist throughout the year, making identification and quantification relatively easy in most seasons. Many nontree species, on the other hand, have ephemeral structures that may only be present on a site for a relatively short period during the growing season. Nontree vegetation assessments require fieldwork to coincide with periods in which ephemeral species are actively growing and identifiable on the site. Different species also vary in the timing of their growth, with some species present early in the growing season and others later. Estimates of total vegetation diversity may require multiple visits to the same sites (Rosenzweig, 1995; Li et al., 2008). While many of the measurement and sampling techniques used for tree species can be adapted to nontree species, different assessment methods are required depending on whether presence, counts, biomass, or cover estimates are required. Sample designs will also need to be modified since the spatial variability of nontree species is often much finer than for tree species.

Assessment of wildlife poses similar considerations. Wildlife maybe directly assessed or the forest structure might be assessed for either signs of use or its suitability for use by particular species. Direct assessment of wildlife populations is beyond the scope of this text, and the reader is referred to textbooks specializing in wildlife population estimation (Buckland et al., 2001, 2004; Braun, 2005). Estimation of suitability often relies on models that predict habitat suitability for a given species based on forest vegetation attributes found to be correlated with wildlife use or abundance (Morrison et al., 1992). Often, variables used in habitat suitability models are the same as those estimated in timber inventories; however, some require additional measurements such as canopy cover, browse availability, presence of snags, cavities, or down woody debris, and nontree vegetation density.

Water quality, soils, and forest health have specialized variables of interest that may require additional measurements other than those commonly made in timber inventories, though models relating forest structure to these variables are often used (Anhold et al., 1996; Rheinhardt et al., 2012; Fernandes et al., 2014). Recreation value and scenic values are similarly estimated using forest structure attributes and models predicting recreation and scenic values based on user preference surveys (Ribe, 1989).

1.4.3. Inventory Planning

An important step in designing a forest inventory is the development of a comprehensive plan before initiating work. Such a plan ensures that all facets of the inventory, including the data to be collected, financial and logistical support, and compilation procedures, are thought through before the inventory begins. The student is referred to Husch (1971) for a discussion of inventory planning.

The following checklist includes all, or almost all, items that should be considered in planning a forest inventory. The items do not always have the same importance, and are not needed in all plans.

1. Purpose of the inventory
   a. Timber and nontimber parameters to be estimated
   b. How and by whom the information will be used
2. Background information
   a. Past surveys, reports, maps, photographs, and so on
   b. Individual or organization supporting the inventory
   c. Funds available
3. Description of area
   a. Location
   b. Size
   c. Terrain, accessibility, transport facilities
   d. General character of forest
4. Information required in final report
   a. Tables and graphs
   b. Maps, mosaics, or other pictorial material
   c. Narrative report
5. Inventory design
   a. Estimation of area (from aerial or orthographic photographs, maps or field measurements)
   b. Determination of timber quantity (e.g., volume tables, units of volume, etc.)
   c. Methods for estimation of nontimber parameters (e.g., nontimber forest products, regeneration, understory vegetation, dead wood, soil, water, scenic and recreational values)
   d. Size and shape of fixed-size sampling units
   e. Probability sampling
      (i) Simple random sampling
      (ii) Systematic sampling
      (iii) Stratified sampling
      (iv) Multistage sampling
      (v) Double sampling
      (vi) Sampling with varying probability
   f. Nonrandom or selective sampling
   g. Setting precision for inventory
   h. Sampling intensity to meet required precision
   i. Times and costs for all phases of work
6. Procedures for interpretation of photo, satellite imagery, or other remote sensing material
   a. Location and establishment of sampling units
   b. Determination of current stand information, including instructions on measurement of appropriate tree and stand characteristics coordinated with fieldwork

    c. Determination of insect damage, forest cover types, forest fuel types, area, and so on, coordinated with fieldwork

    d. Personnel

    e. Instruments

    f. Recording of information

    g. Quality control

    h. Data conversion and editing

7. Procedures for fieldwork

    a. Crew organization

    b. Logistical support and transportation

    c. Location and establishment of sampling units

    d. Determination of current stand information, including instructions on measurement of trees and sample units coordinated with photo interpretation

    e. Determination of growth, regeneration, insect damage, mortality, forest cover types, forest fuel types, area, and so on, coordinated with photo interpretation

    f. Instruments

    g. Recording of observations

    h. Quality control

    i. Data conversion and editing

8. Compilation and calculation procedures for reduction of remote sensing and field measurements

    a. Conversion of remote sensing material or field measurements to desired expressions of quantity

    b. Calculation of sampling errors

    c. Specific methods and computer programs to use

    d. Description of all phases from handling of raw data to final results, including programs

9. Final report

    a. Outline

    b. Estimated time to prepare

    c. Personnel responsible for preparation

    d. Method of reproduction

    e. Number of copies

    f. Distribution

10. Maintenance

    a. Storage and retrieval of data

    b. Plans for updating inventory

The decision to conduct an inventory depends on the need for information. An inventory is an information-gathering process that provides one of the bases for rational decisions. These decisions may be required for many reasons such as the purchase or sale of timber, for preparation of timber-harvesting plans, in forest and wildlife management, for obtaining a loan, and so on. If the purpose is clarified, one can foresee how the information will be used and know the relative emphasis to put on the different elements of the inventory.

As a first step in planning an inventory, one should obtain the requisite background information and prepare a description of the area. One should then decide on the information required from the inventory and prepare the outlines of tables that will appear in the final report. Table outlines should include titles, column headings, class limits, measurement units, and other categories needed to indicate the inventory results. These table outlines should be prepared before detailed planning begins because the inventory design will depend on the information required for the final report. Area information is usually shown by such categories as land use class, forest-type or condition class, and ownership. Timber quantities are usually given in stand-and-stock tables. A *stand table* gives number of trees by species, dbh, and height classes. A *stock table* gives volumes or weights according to similar classifications (see Chapter 9 for details on construction of stand and stock tables). Stand and stock tables may be on a per unit area basis (per acre or hectare) or for the total forest area and may be prepared for forest types or other classifications (e.g., compartments, watersheds). Since a forest is a living, changing complex, the inventory plan should consider the inclusion of estimates of growth and drain. For a discussion of tree and stand growth, see Chapter 14.

1.4.4. Forest Inventory Design

There must be wide latitude in designing an inventory to meet the variety of forest vegetation, topographic, economic, and transportation conditions that may be encountered. The required forest inventory information can be obtained by observations and measurements in the field and on aerial photographs, satellite images, and other remote sensing sources. The most useful and practical approach is typically to use remote sensing materials for forest classification or stratification, mapping, and area determination, and to employ ground work for detailed information about forest conditions, timber quantities and qualities, and additional nontimber characteristics. One can design a forest inventory utilizing only fieldwork, but it is usually less efficient unless the area under consideration is quite small. If aerial photographs or satellite imagery are available, they should be used. Under some circumstances, it is possible to design a forest inventory based entirely on photographic or remotely sensed interpretations and measurements; however, this method will provide only rough approximations of timber species and the quantity, quality, and sizes present, and information on the nontimber parameters; ground-based reference data are almost always needed to evaluate the accuracy of the remote sensing work, and to provide estimates of those quantities that cannot easily be observed from an aircraft or satellite (uses of remotely sensed data are discussed in Chapter 13).

The funds available and the cost of an inventory will strongly influence the design chosen. The main factors that will affect costs are the type of information required, the standard of precision chosen, the total size of the area to be surveyed, and the minimum size of the unit area for which estimates are required. General information is relatively inexpensive, but the costs increase, as more details are required. The standard of precision chosen greatly influences costs; costs increase since increases in precision usually require more intensive sampling. Costs per unit area will decrease as the size of the inventory area increases. If independent estimates are required for subdivisions of a large forest area, this will also raise costs. Descriptions of basic sampling designs applicable to forest inventory are given in Chapter 10.

1.4.5. Inventory Fieldwork

The size and organization of field crews will vary with sampling procedures, the observations or measurements required, forest conditions, and tradition. For timber inventory work in temperate zone forests, crews of one or two workers are widely used. If additional non-timber information is required, additional members may be needed. Small-sized crews have proved satisfactory in these areas because roads are usually abundant (access by vehicle is good), forest travel is relatively easy (little brush cutting is required), and well-trained technicians are generally available (they require little supervision). For inventory work in tropical forests, large crews are widely used because roads are generally sparse (access by vehicle is poor), forest travel is difficult (workers are required to cut vegetation), and few trained technicians are available (they require considerable supervision). Whatever the crew size or the assigned tasks, specific, clear instructions should be given to each crewmember.

A basic rule of forest inventory is to prepare complete written instructions before field-work starts. To minimize later changes, instructions should be tested before operations begin. Instructions should be clear and specific enough so that individual judgment by field crews on where and how to take measurements is eliminated. For example, field crews should not be permitted to subjectively choose the position of a sampling unit or to move it to a more accessible position or more "typical" stand. The aim should be to standardize all work to obtain uniform quality and the best possible reliability of the measurements, regardless of which individual does the work. The occurrence of mistakes or nonrandom errors must be eliminated or, at least, held to a minimum.

It is best to settle on a standard set of instruments. The use of several kinds of instruments to make the same type of measurements should be avoided. To minimize errors, the instruments should be periodically checked to see if they are in adjustment.

Decisions should be made on the precision required for each of the measurements. Thus, tree diameters may be stipulated to the nearest centimeter, inch, tenth of an inch, and so on, and heights to the nearest foot, meter, one-half log, and so on. If estimates of timber quantity are to be made by quality or product classes, then field instructions must include tree measurement specifications of minimum dbh and upper stem diameter, section lengths, and total usable length and criteria such as crook, scars, catfaces, evidence of rot, and so forth on individual sections and for the entire tree.

Inventory field operations must include a checking or quality control procedure. Whether plots, strips, or points are used, a certain percentage of the sampling units should be remeasured. The results of the remeasurements should be compared to the original measurement to see if the work meets the required standards. If it does not, remedial steps should be taken.

There is no standard form or system for recording field observations. The manner in which measurements are recorded depends in large part on the way they will be processed. Regardless of whether data will be recorded on paper forms or entered directly into electronic data recorders, consistent forms should be developed for recording information and made available to all crews. The forms should be developed such that they reflect the logical order of measurements and instructions to crews. Electronic data recorders can greatly speed up processing of forest inventory data since paper-based data do not have to be subsequently entered into a computer. Electronic data recorders should have internal data quality checks so that measurement or data entry errors can be identified and corrected in the field. Electronic data recorders often require more forethought in the design of data entry forms since much field time can be lost if a user must scroll or page through several forms during field measurement procedures.

Plans for calculation and compilation of photo-interpretation data and field measurements should be made before fieldwork begins. It is illogical to postpone consideration of these things until data are obtained because foreknowledge of the calculation and compilation procedures can influence collection of data. For example, if volume tables will be used to determine tree volumes, there would be little point in measuring diameters to a tenth of a centimeter in the field if the volume tables give volumes by full centimeter dbh classes. The statistical formulas or computer programs to be used for estimating means, totals, and standard errors should be selected early in the planning process. It is worthwhile to check the formulas with simulated data to verify the applicability of the program or to work out the optimum sequence of computational steps.

2

PRINCIPLES OF MEASUREMENT

Knowledge is to a large extent the result of acquisition and systematic accumulation of observations, or measurements, of concrete objects and natural phenomena. Thus, measurement is a basic requirement for the extension of knowledge. In its broadest sense, *measurement* consists of the assignment of numbers to measurable properties. Ellis (1966) gives this definition: "Measurement is the assignment of numerals to things according to any determinative, non-degenerate rule." "*Determinative*" means that the same numerals, or range of numerals, are always assigned to the same things under the same conditions. "*Nondegenerate*" allows for the possibility of assigning different numerals to different things or to the same things under different conditions. This definition implies that we have a scale that allows us to use a rule and that each scale inherently has a different rule that must be adhered to in representing a property by a numerical quantity.

The numbering system in general use throughout the world is the *decimal system*. This can probably be attributed to the fact that human beings have 10 fingers. But the decimal system is merely one of many possible numbering systems that could be utilized. In fact, there are examples of other numbering systems used by earlier civilizations (e.g., the vigesimal system based on 20, utilized by the Mayas, and the sexagesimal system of the Babylonians, based on 60). Our own system of measuring time and angles in minutes and seconds comes from the sexagesimal system. Systems to other bases, such as the duodecimal system, based on 12 (which seems to have lingered on in the use of dozen and gross) may also have been used. For a discussion of the history of number theory, the student is referred to Ore (1988). With the development of the electronic computer, interest has been revived in numbering systems using bases other than 10. Of primary interest is the binary system because electronic digital computers that use two basic states have been found most practical.

Forest Mensuration, Fifth Edition. John A. Kershaw, Jr., Mark J. Ducey, Thomas W. Beers and Bertram Husch.
© 2017 John Wiley & Sons, Ltd. Published 2017 by John Wiley & Sons, Ltd.

2.1. SCALES OF MEASUREMENT

Table 2.1 shows a classification of different kinds of scales and their applications according to Stevens (1946). The four scales of measurement are nominal, ordinal, interval, and ratio. A very important feature of this table is the column "permissible statistics." Measurements made on a *ratio scale* permit the use of all the types of statistics shown for all the scales. Those made on an *interval scale* can utilize the statistics in this scale as well as the preceding scales, but not those on a ratio scale. Similarly, measurements on an *ordinal scale* can utilize statistics permitted for this scale as well as those on a nominal scale. Finally, measurements made on a *nominal scale* can utilize only these statistics shown for this scale and not any of those on the following scales.

The *nominal scale* is used for numbering objects for identification (e.g., numbering of tree species or forest types in a stand map), and, as the name suggests, is essentially a numeric representation of an object's name or type (e.g., the assignment of code numbers to species). Each member of the class is assigned the same numeral. The order in which classes are recorded has no importance. Indeed, the order could be changed without changing the meaning.

The *ordinal scale* is used to express degree, quality, or position in a series, such as first, second, and third. In a scale of this type, the successive intervals on the scale are not necessarily equal. Examples of ordinal scales used in forestry measurements are lumber grades, log grades, piece product grades, Christmas tree grades, nursery stock grades, and site quality classes. The order in which classes or grades are arranged on an ordinal scale has an intrinsic meaning. Classes are arranged in order of increasing or decreasing qualitative rank, so that the position on the scale affords an idea of comparative rank. The continuum of the variable consists of the range between the limits of the established ranks or grades. As many ranks or grades can be established as are deemed suitable. An attempt may be made to have each grade or rank occupy an equal interval of the continuum; however, this will rarely be achieved since ranks are subjectively defined with no assurance of equal increments between ranks.

The *interval scale* includes a series of graduations marked off at uniform intervals from a reference point of fixed magnitude. There is no absolute reference point or true origin for this scale. The origin is arbitrarily chosen. The Celsius temperature scale is a good example of an interval scale. The zero point on the scale (originally defined in terms of the freezing point of water) is arbitrary, but each degree on the scale marks off equal intervals of temperature.

The *ratio scale* is similar to the interval scale in that there is equality of intervals between successive points on the scale; however, an absolute zero of origin is always present or implied. Where the Celsius scale for temperature is an interval scale, the Kelvin scale is a ratio scale; absolute zero on the Kelvin scale is the minimum possible temperature for matter. Ratio scales are the most commonly employed, and the most versatile in that all types of statistical measures are applicable. It is convenient to consider ratio scales as fundamental and derived. Such things as frequency, length, weight, and time intervals represent *fundamental scales*. Such things as stand volume per hectare, stand density, and stand growth per unit of time represent *derived scales*. (These are derived scales in that the values on the scale are functions of two or more fundamental values.)

TABLE 2.1. Classification of Scales of Measurement

Scale	Basic Operation	Mathematical Group Structure	Permissible Statistics	Examples
Nominal	Determination of equality (numbering and counting)	Permutation group $X' = f(X)$, where $f(X)$ means any one to one substitution	Number of cases Mode Contingency correlation	• Numbering of forest stand types • Assignment of code numbers to species in studying stand composition
Ordinal	Determination of greater or less (ranking)	Isotonic group $X' = f(X)$, where $f(X)$ means any increasing monotonic function	Median Percentiles Order correlation	• Lumber grading • Tree and log grading • Site class estimation
Interval	Determination of the equality of intervals or of differences (numerical magnitude of quantity, arbitrary origin)	Linear group $X' = aX + b$, where $a > 0$	Mean Standard deviation Correlation coefficient	• Fahrenheit temperature • Calendar time • Available soil moisture • Relative humidity
Ratio	Determination of the equality of ratios (numerical magnitude of quantity, absolute origin)	Similarity group $X' = cX$, where $c > 0$	Geometric mean Harmonic mean Coefficient of variation	• Length of objects • Frequency of items • Time intervals • Volumes • Weights • Absolute temperature • Absolute humidity

Columns 2, 3, 4, and 5 are cumulative in that all characteristics listed opposite a particular scale are additive to those above it. In the column that records the group structure of each scale are listed the mathematical transformations that leave the scale invariant. Thus, any numeral X can be replaced by another numeral X', where X' is the function of X, $f(X)$, listed in this column. The criterion for the appropriateness of a statistic is invariance under the transformations in column 3. Thus, the case that stands at the median of a distribution maintains its position under all transformations that preserve order (isotonic group), but an item located at the mean remains at the mean only under transformations as restricted as those of the linear group. The ratio expressed by the coefficient of variation remains invariant only under the similarity transformation (multiplication by a constant). The rank-order correlation coefficient is usually considered appropriate to the ordinal scale, although the lack of a requirement for equal intervals between successive ranks really invalidates this statistic (Stevens, 1946).

2.2. UNITS OF MEASUREMENT

To describe a physical quantity, one must establish a unit of measure and determine the number of times the unit occurs in the quantity. Thus, if an object has a length of 3 m, the meter has been taken as the unit of length, and the length dimension of the object contains three of these standard units.

The *fundamental units* in mechanics are measures of length, mass, and time. These are regarded as independent and fundamental variables of nature, although scientists have chosen them arbitrarily. Other fundamental units have been established for thermal, electrical, and illumination quantity measurement.

Derived units are expressed in terms of fundamental units or in units derived from fundamental units. Derived units include ones for the measurement of area (square feet or meters, acres, hectares), volume (cubic feet or meters), velocity (miles per hour, meters per second), force (kilogram-force), and so on. Derived units are often expressed in formula form. For example, the area of a rectangle is defined by the function

$$\text{Area} = W \cdot L$$

where W and L are fundamental units of length.

Physical quantities such as length, mass, and time are called *scalar quantities* or *scalars*. Physical quantities that require an additional specification of direction for their complete definition are called *vector quantities* or *vectors*.

2.3. SYSTEMS OF MEASUREMENT

There are two methods of establishing measurement units. We may select an arbitrary unit for each type of quantity to be measured or we may select fundamental units and formulate from them a consistent system of derived units. The first method was employed extensively in our early history. For example, units for measuring the length of cloth, the height of a horse, or land distances were all different. Reference units were objects such as the width of a barleycorn, the length of a man's foot, the length of a man's forearm (a cubit), and so on; however, these primitive units often lacked uniformity. Vestiges of this system still exist, particularly in English-speaking countries (foot, yard, pound, etc.), although the units still in wide use are now uniform.

The second method of establishing a system of units is illustrated by the International System of Units or SI (abbreviated from the French "Le Système International d'Unités"). In this system, an arbitrary set of units has been chosen that is uniformly applicable to the measurement of any object. Moreover, there is a logical, consistent, and uniform relationship between the basic units and their subdivisions.

2.3.1. International System of Units (Metric System)

This system of weights and measures, commonly referred to as the metric system, was originally formulated by the French Academy of Sciences in 1790 (Alder, 2003). The system was adopted in France in 1799 and made compulsory in 1840. In 1875, the International Metric Convention, which was established by treaty, furnished physical standards of length and mass to the 17 member nations. The General Conference on Weights and Measures (referred to as CGPM from the French "Conférence Générale des Poids et Mesures") is an international organization established under the Convention. This organization meets periodically. The 11th meeting of the CGPM in 1960 adopted the name "International System of Units" with the international abbreviation SI (from the French "Le Système International d'Unités"). This is now the accepted form of the metric system. The CGPM controls the International Bureau of Weights and Measures, which is headquartered at Sèvres, near Paris, and maintains the physical standards of units. The United States Bureau of Standards represents the United States on the CGPM and maintains its standards of measure. For a concise description of the SI and its use, see USMA (2007).

The SI has been adopted by most of the technologically developed countries of the world. Although conversion in Great Britain and the United States has met with resistance, a gradual changeover is taking place. In 1866, the United States Congress enacted legislation authorizing, but not mandating, the use of SI units in the country. Then, in 1975, the Congress enacted the Metric Conversion Act and established the U.S. Metric Board to coordinate voluntary conversion to SI. In 1982, the Office of Metric Programs replaced the Metric Board. To date, no states have enacted legislation mandating the adoption of SI units. Nevertheless, globalization of the world economy has put pressure on the United States to convert to SI.

The SI considers three classes of units: (1) base units, (2) derived units, and (3) supplementary units. There are seven *base units*, which by convention are considered dimensionally independent. These are the meter, kilogram, second, ampere, Kelvin, mole, and candela. Combining base units according to algebraic statements that relate the corresponding quantities forms the derived units. The *supplementary units* are those that the CGPM established without stating whether they are base or derived units.

Here is a list of dimensions measured by these base units, along with definitions of the units. The conventional symbol for each unit is shown in parentheses.

1. *Length—meter or metre* (m). The meter is the length of the path traveled by light in a vacuum during a time interval of 1/299,792,458 of a second.
2. *Mass*—kilogram* (kg). It is equal to the mass of the international prototype standard, a cylinder of platinum-iridium alloy preserved in a vault in Sèvres, France.
3. *Time—second* (s). The second has been calculated by atomic standards to be 9,192,631,770 periods of vibration of the radiation emitted at a specific wavelength by an atom of cesium-133.
4. *Electric current—ampere* (A). The ampere is the current in a pair of equally long, parallel, straight wires (in a vacuum and 1 m apart) that produces a force of 2×10^{-7} Newtons between the wires for each meter of their length.
5. *Temperature—Kelvin* (K). The Kelvin is 1/273.15 of the thermodynamic temperature of the triple point of water. The temperature 0 K is called absolute zero. The Kelvin degree is the same size as the Celsius degree (also called centigrade). The freezing point of water (0 °C) and the boiling point of water (100 °C) correspond

---

*The term "weight" is commonly used for mass although this is, strictly speaking, incorrect. Weight of a body means the force caused by gravity, acting on a mass, which varies in time and space and which differs according to location on the earth. Since it is important to know whether mass or force is being measured, the SI has established two units: the kilogram for mass and the Newton for force.

to 273.15 and 373.15 K, respectively. On the Fahrenheit scale, 1.8° are equal to 1.0°C or 1.0 K. The freezing point of water on the Fahrenheit scale is 32°F.

6. *Amount of substance—mole* (mol). The mole is the base unit used to specify quantity of chemical elements or compounds. It is the amount of substance of a system that contains as many elementary entities as there are atoms in 0.012 kg of carbon-12. When the mole is used, the elementary entities must be qualified. They may be atoms, molecules, ions, electrons, or other particles or specified groups of such particles.

7. *Luminous intensity—candela* (cd). The candela is 1/600,000 of the intensity, in the perpendicular direction, of 1 m² of a black body radiator at the temperature at which platinum solidifies (2045 K) under a pressure of 101,325 N/m².

At present, there are two supplementary units, the radian and the steradian. The radian (rad) is the plane angle between two radii of a circle that cuts off on the circumference an arc equal to the radius. It is 57.29578 degrees for every circle. The steradian (sr) is the solid angle at the center of a sphere subtending a section on the surface equal in area to the square of the radius of the sphere. Derived units are expressed algebraically in terms of the base units by means of mathematical symbols of multiplication and division. Some examples of derived units are shown in Table 2.2. There are a number of widely used units that are not part of SI. These units, which the International committee on Weights and Measures recognized in 1969, are shown in Table 2.3. To form decimal multiples of SI units, the prefixes and symbols shown in Table 2.4 are used.

2.3.2. Imperial System

The imperial system of weights and measures is often referred as the English system because of its predominant use in English-speaking countries; however, the imperial system is a combination of traditional English measurement standards, such as the rod, chain, acre, and bushel, with continental measurement standards, such as the foot, inch, and pound,

TABLE 2.2. Examples of SI Derived Units

Quantity	SI Unit for the Quantity	Symbol
Area	Square meter	m^2
Volume	Cubic meter (the liter, 0.001 cubic meter, is not an SI unit although commonly used to measure fluid volume)	m^3
Specific volume	Cubic meter per kilogram	m^3/kg
Force	Newton ($1\,N = 1\,kg \cdot m \cdot s^{-2}$)	N
Pressure	Pascal ($1\,Pa = 1\,N \cdot m^{-2}$)	Pa
Work	Joule ($1\,J = 1\,N \cdot m$)	J
Power	Watt ($1\,W = 1\,J \cdot s^{-1}$)	W
Speed	Meter per second	m/s
Acceleration	(Meter per second) per second	m/s^2
Voltage	Volt ($1\,V = 1\,W \cdot A^{-1}$)	V
Electric resistance	Ohm ($1 = 1\,V \cdot A$)	Ω
Concentration (amount of substance)	Mole per cubic meter	mol/m^3

TABLE 2.3. Supplemental Units Recognized by SI

Unit	Symbol	Equivalence in SI Units
Minute	min	$1\,min = 60\,s$
Hour	h	$1\,h = 60\,min = 3{,}600\,s$
Day	d	$1\,d = 24\,h = 86{,}400\,s$
Degree (angular)	°	$1° = (\pi/180)\,rad$
Minute (angular)	′	$1' = (1/60)° = (\pi/10{,}800)rad$
Second (angular)	″	$1'' = (1/60)' = (\pi/648{,}000)rad$
Liter	L	$1\,L = 1\,dm^3 = 10^{-3}m^3$
Metric ton	t	$1\,t = 10^3\,kg$

TABLE 2.4. Prefixed and Symbols Used in SI

Prefix	Symbol	Factor								
Yotta	Y	10^{24}	1 000	000	000	000	000	000	000	000
Zetta	Z	10^{21}	1 000	000	000	000	000	000	000	
Exa	E	10^{18}	1 000	000	000	000	000	000		
Peta	P	10^{15}	1 000	000	000	000	000			
Tera	T	10^{12}	1 000	000	000	000				
Giga	G	10^9	1 000	000	000					
Mega	M	10^6	1 000	000						
Kilo	k	10^3	1 000							
Hecto	h	10^2	100							
Deca	da	10^1	10							
Deci	d	10^{-1}	0.1							
Centi	c	10^{-2}	0.01							
Milli	m	10^{-3}	0.001							
Micro	μ	10^{-6}	0.000	001						
Nano	n	10^{-9}	0.000	000	001					
Pico	p	10^{-12}	0.000	000	000	001				
Femto	f	10^{-15}	0.000	000	000	000	001			
Atto	a	10^{-18}	0.000	000	000	000	000	001		
Zepto	z	10^{-21}	0.000	000	000	000	000	000	001	
Yocto	y	10^{-24}	0.000	000	000	000	000	000	000	001

brought to the United Kingdom by the Normans (Irwin, 1960). The imperial system of weights and measures is still widely used in the United States but it is gradually being supplanted by the more logical and consistent SI units. This will be a long procedure since resistance due to custom and tradition is strong, and the conversion is currently voluntary. The United Kingdom has taken a more vigorous stance and conversion to SI is proceeding more rapidly.

The units of the imperial system still commonly used in the United States are practically the same as those employed in the American colonies prior to 1776. The names of these units are generally the same as those of the British imperial system; however, their values differ slightly.

In the imperial system, the fundamental length is the yard. The British yard is the distance between two lines on a bronze bar kept in the Standards Office, Westminster, London. Originally, the United States yard was based on a prototype bar, but in 1893 the United States yard was redefined as 3600/3937 m = 0.9144018 m. The British yard, on the other hand, was 3600/3937.0113 m = 0.914399 m. The foot is defined as the third part of a yard, and the inch as a 12th part of a foot. In 1959, it was agreed by Canada, United States, New Zealand, United Kingdom, South Africa, and Australia that they would adopt the value of 1 yard = 0.9144 m. In the United Kingdom, these new values were used only in scientific work, while the older, slightly different values were used for other measurements.

In the British imperial system, the unit of mass, the avoirdupois pound, is the mass of a certain cylinder of platinum in the possession of the British government (1 lb = 0.45359243 kg). As with the traditional yard, the United States pound and the British pound were not exactly equal. The 1959 agreement of English-speaking countries established a new value of the 1 lb = 0.45359237 kg. In the imperial system, the second is the base unit of time, as in SI.

Secondary units in the imperial system also have different values in the United States and the United Kingdom. For example, the U.S. gallon is defined as 231 in.[3] The British gallon is defined as 277.42 in.[3] There are other secondary units used in English-speaking countries that are as arbitrary as the gallon, but some are derived from the fundamental units. For example, the unit of work in British and American engineering practice has been the foot-pound.

2.3.3. Conversions Between Systems

The SI has been almost universally adopted by the scientific community, even in the United States; however, common and commercial measurements routinely use imperial units, even in countries that have adopted SI units (e.g., Canada and the United Kingdom). In forestry, use of imperial units is still widespread. Common units of volume such as the board foot and cord are widely used. Much dimensional lumber is still milled to common imperial standards (2 by 4, 2 by 6, etc.), and many building codes refer to these standards. Most historical forestry data are in imperial units, and many of the volume tables, growth models, and other quantitative tools commonly used by foresters are in imperial units. As a result, it is common to need conversions from one system of measurement to another. Table A.1 shows unit conversions for length, area, volume, and mass in imperial and SI units.

Conversion between fundament units applied to measurements of dimensions such as individual tree diameter and height are straight forward and only require the application of the appropriate conversion factor found in Table A.1. Conversion between derived units may be more complicated, especially if comparisons are desired between results collected in imperial units to results collected in SI units. Even conversion of relatively simple units, such as volume, is complicated by the fact that standard points for measurement of individual tree dimensions, such as diameter, vary between the two systems (and may vary between countries using the same system). An example of these complications is illustrated in Honer et al.'s (1983) conversion of imperial volume tables to SI volume tables for common timber species in eastern Canada.

The large economies of the United States and the United Kingdom have resulted in foresters around the world needing to be conversant in both systems of measurements. Globalization of forest products trade and the predominant use of SI in scientific literature requires foresters (particularly those from the United States) to become more conversant in the use of SI units. In this text, we present measurements and derivation of sampling units in both measurement systems, examples are presented for both systems, though many of the expanded examples are primarily SI.

2.4. VARIABLES

A *variable* is a characteristic that may assume any given value or set of values. A *variate* is the value of a specific variable. Some variables are *continuous* in that they are capable of exhibiting every possible value within a given range. For example, height, weight, and volume are continuous variables. Other variables are *discontinuous* or *discrete* in that they only have values that jump from one number or position to the next. Counts of the number of employees in a company, of trees in a stand, or of deer per unit area are examples of discrete variables. Numbers of individuals observed in classes may be expressed as percentages of the total in all classes.

Data pertaining to continuous variables are obtained by measuring using interval and ratio scales. Data pertaining to discrete variables are obtained using nominal and ordinal scales. The process of measuring according to nominal and ordinal scales consists of counting frequencies of specified events. Discrete variables describe these events. The general term *event* can refer to a discrete physical standard, such as a tree that exists as a tangible object occupying space, or to an occurrence that cannot be thought of as spatial, such as a timber sale. In either case, the measurement consists of defining the variable and then counting the number of occurrences. There is no choice for the unit of measurement—frequency is the only permissible numerical value. A discrete variable can thus be characterized as a class or series of classes of defined characteristics with no possible intermediate classes or values. It is frequently convenient to assign a code number or letter to each class of a discrete variable. It is important to be aware that the code numbers have no intrinsic meaning but are merely identifying labels. No meaningful mathematical operations can be performed on such code numbers (Table 2.1).

A class established for convenience in continuous-type measurements should not be confused with a discrete variable. Classes for continuous variables (e.g., measurements of tree diameters) are often established to facilitate handling of data in computation. Frequencies may then be assigned to these classes. These frequencies represent the occurrence or recurrence of certain measurements of a continuous variable placed in a group or class of defined limits for convenience. At times, it may not be clear as to whether a discrete or continuous variable is being measured. For example, in counting the number of trees per unit area (acre or hectare), the interval, one tree, is so small and the number of trees so large that analyses are made on the frequencies as though they described a continuous variable. This has become customary and may be considered permissible as long as the true nature of the variable is understood.

2.5. PRECISION, ACCURACY, AND BIAS

The terms *precision* and *accuracy* are frequently used interchangeably in nontechnical parlance and often with varying meaning in technical usage. In this text, they will have two distinct meanings. *Precision* as used here (and generally accepted in forest

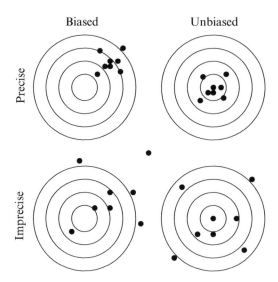

FIG. 2.1. Precision, bias, and accuracy of a target shooter. The target's bull's eye is analogous to the unknown true population parameter, and the holes represent parameter estimates based on different samples. The goal is accuracy, which is the precise, unbiased target (Shiver and Borders, 1996).

mensuration) means the degree of agreement in a series of measurements. The term is also used to describe the resolving power of a measuring instrument or the smallest unit in observing a measurement. In this sense, the more decimal places used in the measurement the more precise the measurement. *Accuracy*, on the other hand, is the closeness of a measurement to the true value. *Bias* refers to the systematic errors that may result from faulty measurement procedures, instrumental errors, flaws in the sampling procedure, errors in the computations, mistakes in recording, and so forth. Bias generally affects accuracy and not precision. Of course, the ultimate objective is to obtain precise, accurate measurements.

In sampling, accuracy refers to the size of the deviation of a sample estimate from the true population value. Precision, expressed as a standard deviation, refers to the deviation of sample values about their mean, which, if biased, does not correspond to the true population value. It is possible to have a very precise estimate in that the deviations from the sample mean are very small; yet, at the same time, the estimate may not be accurate if it differs from the true value due to bias (Fig. 2.1). For example, one might carefully measure a tree diameter repeatedly to the nearest millimeter, with a caliper that reads about 5 mm low. The results of this series of measurements are precise because there is little variation between readings, but they are biased and inaccurate because of faulty adjustment of the instrument. The relation between accuracy A, bias B, and precision P can be expressed as: $A^2 = P^2 - B^2$. This indicates that, if we reduce B^2 to zero, accuracy equals precision (Fig. 2.1). The target's bull's eye is analogous to the unknown true population parameter, and the holes represent parameter estimates based on different samples. The goal is accuracy, which is the precise, unbiased target (Shiver and Borders, 1996).

2.6. SIGNIFICANT DIGITS AND ROUNDING OFF

A significant digit is any digit denoting the true size of the unit at its specific location in the overall number. The term *significant* as used here should not be confused with its use in reference to statistical significance. The significant figures in a number are the digits reading from left to right beginning with the first nonzero digit and ending with the last digit written, which may be a zero. The numbers 25, 2.5, 0.25, and 0.025 all have two significant digits, the 2 and the 5. The numbers 25.0, 0.250, and 0.0250 all have three significant figures, the 2, 5, and 0. When one or more zeros occur immediately to the left of the decimal position and there is no digit to the right of the point, the number of significant digits may be in doubt. Thus, the number 2500 may have two, three, or four significant digits, depending on whether one or both zeros denote an actual measurement or have been used to round off a number and indicate the position of the decimal point. Thus, zero can be a significant figure if used to show the quantity in the position it occupies and not merely to denote a decimal place. A convention sometimes used to indicate the last significant digit is to place a dot above it. Thus, 5,12$\dot{1}$,000 indicates four significant digits and 5,121,$\dot{0}$00 indicates five significant digits. Another method to indicate significant digits is to first divide a number into two factors, one of them being a power of 10. A number such as 150,000,000 could be written as 1.5×10^8 or in some other form such as 15×10^7. A convention frequently used is to show the significant figures in the first factor and to use one nonzero digit to the left of the decimal point. Thus, the numbers 156,000,000, 15.60, and 0.001560 (all with three significant figures) would be written as 1.56×10^8, 1.56×10, and 1.56×10^{-3}.

If a number has a significant zero to the right of the decimal place following a nonzero number, it should not be omitted. For example, 1.560 indicates four significant digits including the last zero to the right. To drop it would reduce the precision of the number. Zeros when used to locate a decimal place are not significant. In the number 0.001560, only the last four digits, 1, 5, 6, and 0 are significant; the first two to the right of the decimal place are not.

When the units used for a measurement are changed, it may change the number of decimal places but not the number of significant digits. Thus, a weight of 355.62 g has five significant figures, as does the same weight expressed as 0.35562 kg, although the number of decimal places has increased. This emphasizes the importance of specifying the number of significant digits in a measurement rather than simply the number of decimal places.

2.6.1. Significant Digits in Measurements

The numbers used in mensuration can be considered as arising as pure numbers, from direct measurements, and from computations involving pure numbers and values from direct measurements. *Pure numbers* can be the result of a count in which a number is exact or they can be the result of some definition. Examples of pure numbers are the number of sides on a square, the value of π, or the number of meters in a kilometer.

Values of *direct measurements* are obtained by reading a measuring instrument (e.g., measuring a length with a ruler). The numerical values obtained in this way are approximations in contrast to pure numbers. The number of significant digits used indicates the precision of the approximation. For example, measurement of a length could be taken to the nearest 1, 10th, or 100th of a foot, and recorded as 8, 7.6, or 7.60. Each of these measurements implies an increasing standard of precision. A length of 8 ft means a length closer to 8 than to 7 or 9 ft. The value of 8 can be considered to lie between 7.5 and 8.5. Similarly, a length of 7.6 means a measurement whose value is closer to 7.6 ft than to 7.5 or 7.7. The value of 7.6 lies between 7.5500…01 and 7.6499…99 or, conventionally, 7.55 and 7.65. In the measurement 7.60, the last digit is significant and the measurement implies a greater precision. The value 7.60 means the actual value lies anywhere between 7.59500…01 and 7.60499…99, or, conventionally, 7.595 and 7.605.

Recording more significant digits than were observed is incorrect. Thus, a length measurement of 8 ft taken to the nearest foot should not be recorded as 8.0 ft since this may mislead the reader into thinking the measurement is more precise than it actually is. On the other hand, one should not omit significant zeros in decimals. For example, one should write 112.0 instead of 112 if the zero is significant.

Since the precision of the final results is limited by the precision of the original data, it is necessary to consider the number of significant digits to take and record in original measurements. Keep in mind that using greater precision than needed is a waste of time and money. A few suggestions follow:

1. Do not try to make measurements to a greater precision (more significant digits) than can be reliably indicated by the measuring process or instrument. For example, it would be illogical to try to measure the height of a standing tree to the nearest 10th of a foot with an Abney level.

2. The precision needed in original data may be influenced by how large a difference is important in comparing results. Thus, if the results of a series of silvicultural treatments are to be compared in terms of volume growth response to the nearest 10th of a cubic meter, then there would be no need to estimate volumes more exactly than the nearest 10th of a cubic meter.

3. The variation in a population sampled and the size of the sample influences the precision chosen for the original measurement. If the population varies greatly or if there are few observations in the sample, then high measurement precision may be worthwhile.

4. Understand how the data are ultimately going to be used. If diameter measurements are going to be lumped into 2 in. classes for a stand and stock table, measurement of diameter to the nearest 2 in. with a Biltmore stick might be appropriate; however, if the trees are part of a permanent sample point and are going to be periodically remeasured to estimate tree growth, measurements to the nearest 0.1 cm using a diameter tape might be required.

2.6.2. Rounding Off

When dealing with the numerical value of a measurement in the usual decimal notation, it is often necessary to round off to fewer significant digits than originally shown. Rounding off can be done by deleting unwanted digits to the right of the decimal point (the fractional part of a number) and by substituting zeros for those to the left of the decimal place (the integer part). Three cases can arise:

1. If the deleted or replaced digits represent less than one-half unit in the last required place, no further change is required.

2. If the deleted or replaced digits represent more than one-half unit in the last required place, then this significant figure is raised by one. (Note that, if the significant figure in the last required place is 9, it is changed to zero and the digit preceding it is increased by one.)

3. If the deleted or replaced digits represent exactly one-half unit in the last required place, a recommended convention is to raise the last digit by one if it is odd but let it stand if it is even. Thus, 31.45 would be rounded to 31.4 but 31.55 would be 31.6. Some authorities differ on how to treat extra digits in the rounding process. For example, to round 31.451 to three significant digits, should the 1 be discarded and then the even-odd rule applied leading to 31.4, or should it be retained leading to 31.5 as the answer? We favor the latter but also note that if this kind of rounding is critical to an important decision, then higher measurement precision is probably needed.

Here are a few examples:

Number	Four Significant Figures	Three Significant Figures	Two Significant Figures
4.6495	4.650	4.65	4.6
93.65001	93.65	93.6	94
567,851	567,900	568,000	57,000
0.99687	0.9969	0.997	1.0

2.6.3. Significant Digits in Arithmetic Operations

In arithmetic operations involving measurements, where figures are only approximations, the question of how many significant digits there are in the result becomes important. In multiplication and division, the factor with the fewest significant figures limits the number of significant digits in the product or quotient. Thus, in multiplying a numerical measurement with five significant figures by another with three significant figures, only the first three figures of the product will be trustworthy, although there may be up to eight digits in the product. For example, if the measurement 895.67 and 35.9 are multiplied, the product is 32,154.553. Only the first three figures in the product—3, 2, and 1—are significant. The number 895.67 represents a measurement between 895.665 and 895.675. The number 35.9 represents a measurement between 35.85 and 35.95. The product of the four possible limiting combinations will differ in all except the first three digits:

$$(895.665)(35.85) = 32,109.59025$$
$$(895.665)(35.95) = 32,199.15675$$
$$(895.675)(35.85) = 32,109.94875$$
$$(895.675)(35.95) = 32,199.51625$$

Therefore, the first three figures are the only reliable ones in the product and the result of $895.67 \times 35.9 = 32,100$. Similarly, in dividing a measurement with eight significant digits by a measurement with three significant figures, the quotient will have only three significant figures.

If a measurement is to be multiplied or divided by an exact number or a factor that is known to any desired number of significant digits, a slightly different situation occurs. For example, a total weight could be estimated as the product of a mean weight having five significant digits times 55. The 55 is an exact number and could also be validly written as 55.000. The product would thus still have five significant digits. It may be helpful to remember that multiplication is merely repetitive addition and the 55, in this case, means that a measurement is added exactly 55 times. Similarly, if the 55 objects had been weighed as a group, to five significant digits, dividing by 55 would give a mean weight to five significant figures. In these cases, the number in the measurement controls the significant digits. Another case occurs if a measurement is to be multiplied or divided by a factor such as π or e (base of Naperian logarithms) that are known to any number of significant figures. The number of significant digits in π or e should be made to agree with the number in the measurement before the operation of multiplication or division so that there is no loss in precision.

A good rule when completing a series of multiplications or divisions is to keep one more digit in the product or quotient than occurs in the shorter of the two factors. This minimizes rounding-off errors in calculations involving a series of operations. At the end of the calculations, the final answer can be rounded off to the proper number of significant figures. Most computer spreadsheet packages do this automatically; however, most rounding functions in these packages round to a specified number of decimal places rather than to a number of significant figures (Iles, 2003).

In addition and subtraction, the position of the decimal points will affect the number of significant digits in a result. It is necessary to align numbers according to their decimal places in order to carry out these operations. The statement that measurements can be added or subtracted when significant digits coincide at some place to the left or right of the decimal point can be used as a primary guide. Also, the number of significant digits in an answer can never be greater than those in the largest of the numbers, but may be fewer. As one example, measurements of 134.023 and 1.5 can be added or subtracted as shown below since significant digits coincide at some place:

$$
\begin{array}{r}
134.023 \\
+ \quad 1.5 \\
\hline
135.523
\end{array}
$$

The sum has only four significant digits and should be expressed as 135.5. The last two significant figures of 134.023 cannot be used since there is no information in the smaller measurement for coinciding positions. Measurements should be taken to uniform standards of significant figures or decimal places to avoid discarding a portion of a measurement, as we did in the case of the last two digits of 134.023.

Another example to consider is the addition of a series of measurements, where the final total may have more figures than any of the individual measurements. The number of significant digits in the total will not exceed the number in the largest measurement. Consider the 11 measurements shown here:

Measurement	Range	
845.6	845.55	845.65
805.8	805.75	805.85
999.6	999.55	999.65
963.4	963.35	963.45
897.6	897.55	897.65
903.1	903.05	903.15
986.9	986.85	986.95
876.3	876.25	876.35
863.2	863.15	863.25
931.2	931.15	931.25
998.1	998.05	998.15
10,070.8	10,070.25	10,071.35

The total, 10,070.8, contains six digits, but only the first four are significant. Each measurement can be thought of as an estimate within a range, as shown in the two right-hand columns. The sum of the lesser values is 10,070.25 and that of the larger is 10,071.35. The total value of the sum of the 11 measurements can fall anywhere within these limits. The significant figures are 1, 0, 0, and 7. Beyond this, the digits are unreliable.

2.7. DATA SUMMARY AND PRESENTATION

When recording, summarizing, and presenting numerical information, the following general precepts should be observed. If text or tables are presented as computer files, the name of the program and the file name should be mentioned. Specifying the version of the program is important since a file prepared with a new version of a program may not be readable by an older version of the program.

The use of points and commas in the recording and presentation of numbers varies in different countries. In the United States and English Canada, a decimal is indicated by a point and commas separate groups of figures for hundreds, thousands, and so on. For example, a figure such as five hundred fifty thousand, three hundred twenty-five, and fifteen hundredths would be written as 550,325.15. In some other countries, the use of points and commas is exactly reversed. Under this system, the number would be recorded as 550.325,15. In international dealings and reports, the system of numerical presentation should be specified.

Similarly, in the United States and Canada, the term billions is used but not in some other countries. For example, in the United States, a number could be specified as five billion, three hundred thirty million, two hundred thirty-four thousand, one hundred fifteen, and written as 5,330,234,115. This same number in some other countries would then be described as five thousand three hundred thirty million, two hundred thirty-four thousand, one hundred fifteen.

2.7.1. Tables

Tables are useful for the recording and presentation of data. The following points should be observed in their use:

1. All tables should have a brief but descriptive title. In general, the title should state what is being shown, where it came from, and when it was measured.
2. If several tables are presented, they should be given numbers. If information from the table is presented in the text, the table number should be mentioned.
3. The type or types of units in the table should be shown as headings in the columns or rows.
4. The headings of columns or rows should be brief. Legends at the bottom of tables can be used to clarify the meaning of concise headings.
5. If classes are used for summarization of numerical information, the midpoints should be shown. Class limits should be explained in the text or as a footnote to the table.

2.7.2. Graphic Presentation

Charts and *graphs* are pictorial or diagrammatic representations of relationships between two or more variables. They are used for three purposes: illustration or communication, analysis of relationships between variables, and as a graphical means of carrying out arithmetical operations or computations. For a detailed discussion of graphic techniques in forest mensuration, see Husch (1963); and for a more general overview of graphical techniques, see Cleveland (1994) or Tufte (1990).

A *graph* is a diagram on a plane surface illustrating the relationship between variables. Since paper presents a plane surface with two usable dimensions, length and width, it is convenient to depict a two-variable relationship by assigning a variable to each of the two dimensions. It is possible to depict a third variable with a more complex form, but to illustrate or solve functions of more than three variables is impossible for practical purposes.

While most graphs are now prepared using computer software, hand-drawn graphs can be facilitated by using different forms of graph paper. There are many types of graph paper available, including numerous highly specialized designs. The three most important kinds used in forest mensuration are rectilinear or rectangular coordinate, logarithmic, and semilogarithmic graph papers. On *rectilinear paper*, both the abscissa and ordinate axes are divided using uniform scales and a series of equidistant points. *Semilogarithmic paper* has one of its axes divided using a uniform scale, but a logarithmic scale divides the other axis. *Logarithmic paper* has both axes divided with logarithmic scales.

The relationship between a dependent and independent variable of the type $Y=f(X)$ can be graphically represented on rectilinear or the logarithmic papers by a line. Conventionally, the *abscissa* or *X-axis* is assigned to the independent variable and the *ordinate* or *Y-axis* to the dependent variable. The line representing the relationship, whether straight or curved, is given the general name *curve*. A relationship involving one dependent and two independent variables requires three dimensions and is represented by a *surface*.

A curve may be plotted from a given mathematical function where the relationship between two variables has been defined by a specific equation. Common mathematical functions used in forest mensuration are the straight line, parabola, and exponential. Empirical curves are developed by plotting paired observations of two variables on graph paper and placing a line to best show the general relationship. The relationship and curve can be determined by regression analysis (see Section 3.8) or by fitting a freehand curve.

The following points should be observed when constructing figures:

1. All figures should have a brief but descriptive title. As with tables, the title should state what is being measured, where it came from, and when the data were collected.
2. If several figures are presented, each figure should be given a number. If the figure is referred to in the text, the figure number should be mentioned.
3. Each axis should include an appropriate range of values for the data being shown. The range should be inclusive of all data and not too broad to distort the relationship being shown. It was customary to always include the origin (0,0) with broken lines on the axis, although this has fallen out of common usage.
4. Each axis should be labeled clearly stating what values are being shown and include their units.
5. If multiple sources of data are being shown on the figure, appropriate symbols should be used to distinguish between the data and a legend included identifying each symbol.
6. If multiple curves are presented, different line types should be used and a legend included identifying each line.

For a complete discussion on the visual presentation of data, see Cleveland (1994) or Tufte (1990).

2.7.3. Class Limits

Measurements of a variable can be recorded as falling within limits of size classes. This may be done for assignment to a frequency table and subsequent statistical calculations, or for recording and presentation of a large number of measurements and results in a tabular form. The use of size classes for indicating tree diameters and heights, quantities of timber, and so on, is a commonplace requirement in forest mensuration. The following, taken from Husch (1963), indicates the main points to keep in mind when using size classes.

In establishing classes, the total range of size is divided into suitable intervals. Classes for measurements of continuous variables must be of equal size if analysis methods applicable to continuous type data are to be used. If unequal size classes are established, only analysis methods for discrete variables are possible. Considering continuous-type data, no universal rules are possible to indicate the number and size of classes to employ. It is important to bear in mind that the series of measurements will probably be analyzed by statistical procedures, and the requirements for these procedures and the interpretation of their results will influence the chosen class interval. There are a number of guides to keep in mind when using classes:

1. The chosen class size should result in a sufficient number of classes to exhibit the frequency pattern associated with the type of data. For example, if tree diameters were measured to the nearest 0.1-in., and diameter classes of 0.1 in. were then chosen, there would be an excessive number of classes with small frequencies in

each. The frequency pattern would not be clear, and the advantages of establishing classes would not be realized. If, on the other hand, 5-in. classes were used, there would only be a few classes with large frequencies in each, again not very revealing. The use of 1-in. classes would most likely be more appropriate.

2. The value of any item in a class must be assumed to be at the midpoint of the class; therefore, classes must not be so large as to cause appreciable differences between actual measurements and assumed midpoint values. The limits of an interval, of course, are equidistant from the midpoint.

3. The precision of the defined class limits will depend on the precision to which the measurements are taken. The actual limits are the limits of implied accuracy when recording a measurement. For example, tree height measurements may have been taken to the nearest foot and then classified by using a 10-ft class interval, as shown below. The actual class limits are presented together with their approximations and midpoints. (The approximate class limits and midpoints, as shown, are used to simplify presentation and allow computational work.)

Actual Height Class Limits (ft)	Approximate Height Class Limits (ft)	Approximate Midpoint (ft)
55.500…01–65.499…9	55–65	60
65.500…01–75.499…9	65–75	70
75.500…01–85.499…9	75–85	80

4. Measurements that fall exactly at a class boundary, such as 65, should be assigned to either the upper or lower class on a random basis since it is just as likely that the true value is above as below the boundary value; however, it is generally acceptable to consistently assign boundary values to one of the classes (e.g., 65.0 could always be assigned to the 70 class and 75.0 to the 80 class).

5. It is acceptable to establish any desired class interval if these principles are maintained, and if there is proper recognition of midpoints.

2.8. FUNDAMENTAL MEASUREMENTS

In this section, we describe the most important fundamental and derived measurements used in forest mensuration: linear, time, weight, area, and volume. Other measurements such as for thermal, electrical, and illumination quantities, for velocity, force, and so on may be required at times and reference should be made to texts on these specific topics.

2.8.1. Linear Measurements

Linear or *length measurement* consists of determining the length of a line from one point to another. Since the configuration of objects varies, a length might be the straight-line distance between two points on an object or the curved or irregular line distance between two points on an object. An example of the latter is the periphery of the cross section of a tree stem.

Length measurements can be made directly or indirectly. *Direct length measurement* is accomplished by placing a prototype standard of a defined unit beside the object to be measured. The number of units between terminals is the length. The application of a foot or meter rule is an example of a direct measurement. *Indirect linear measurement* is accomplished by employing geometry or trigonometry, or by using knowledge of the speed of sound or light.

The varied tasks of length measurement in forest mensuration include determination of the diameter or height of a tree, the length of a log, the width of a tree crown's image on an aerial photograph, the length of the boundary of a tract of land, and so on. These length measurements are covered in Chapters 4 and 5.

2.8.2. Time Measurements

Time is utilized in forestry (1) to denote position in a continuum of time at which some event took place, (2) to measure duration of a given event, and (3) to determine speed or rate at which an event or physical change occurred. An important time measurement in forest mensuration is the determination of tree and stand age (Sections 5.1 and 8.1).

Accurate measurement of time by establishing time standards poses difficult technological problems. Any measure of time is ultimately based on counting cycles of some regularly recurring phenomenon and accurately measuring fractions of a cycle. The ultimate standard for time is provided by the natural frequencies of vibration in atoms and molecules. Therefore, the fundamental time unit that has been internationally accepted is the atomic second (Section 2.3.1).

Several types of studies in forestry involve the relation of time with forest measurements. These include time series and motion and time studies. In studying a time series, the primary object is to measure and analyze the chronological variation in the value of a variable. For example, one may study variation in air temperature over a period of time or fluctuations in the quantity of pulpwood produced. Motion and time studies consist of studying a complete process and dividing it into its fundamental steps. The length of time necessary to carry out the individual steps is then observed. Attempts are then made to eliminate superfluous steps or motions and to minimize the work and time of the essential steps by improving the human contribution, by changing machines, or by rearranging the sequence of the steps. For a detailed overview of these types of studies, see Niebel (1992); and for applications related to forest mensuration, see Köpf (1976) and Miyata et al. (1981).

2.8.3. Weight Measurements

The force of attraction that the earth exerts on a body—that is, the pull of gravity on it—is called the *weight* of a body. Weight is often used as a measure of mass; however, the two are not the same. Mass is the measure of the amount of matter present in a body and thus has the same value at different locations (including zero gravity). Weight varies depending on location of the body in the earth's gravitational field (or the gravitational field of some other astronomical body). Since the gravitational effect varies from place to place on the earth, the weight of a given mass also varies. The distinction between weight and mass is confused by the use of the same units of measure—the gram, the kilogram, and the pound.

A decision to use weight as a measure of quantity depends on these factors:

1. *Physical characteristics of the substance.* The volume of material that occurs as irregular pieces, such as pulpwood, wood chips, coal, soil, seed fertilizer, and other solids, can be measured by filling a space or container of known volume; however, the ratio of air space to solid material will vary with the shape, arrangement, and compaction of pieces. This can seriously affect the accuracy and reliability of volume measurement.

2. *Logicalness of weight as an expression of quantity.* Weight may be the most useful and logical expression of quantity. For example, the weight measurement of pulpwood is advantageous because it can be done rapidly and accurately, and because the derived product, pulp, is expressed in weight. Indeed, when weight is the ultimate expression of quantity, it is logical to apply it consistently from the beginning. (Since transportation charges for forest products are based primarily on weight, weight becomes an even more logical expression of quantity.) Furthermore, a material may be composed of several components. For example, a tree consists of roots, stem, and crown; each part is composed of additional units. The ratio of any component to the whole, such as bark to wood in the merchantable stem, can best be expressed in terms of weight. When the entire tree is of interest (roots, branches, etc.), weight is the most logical measure of quantity.

3. *Feasibility of weighing.* The substance in question must be physically separable from material not relevant to the measurement or weighing is not feasible. In addition, a weighing device must be available. For example, it might be desirable to determine the weight of the merchantable stem of a standing tree. But unless one felled the tree, separated the merchantable stem, and had a weighing machine available, it would be impossible to directly determine weight.

4. *Relative cost of weighing.* Weight might be a better expression of quantity than volume, but costs of weighing might be greater than costs of volume estimates, and vice versa. The final decision is influenced by the value of the material to be measured. When the value of material is high, the costlier procedure may be better if it is more accurate.

2.8.4. Area Measurements

Area is the measure of the size of a surface region, usually expressed in units that are the square of linear units: for example, square feet or square meters. In elementary geometry, formulas for the areas of simple plane figures and the surface area of simple solids are derived from the linear dimensions of these figures. Examples are given in Tables A.2 and A.3. The area of irregular figures, plane, or solid can be computed by the use of coordinates and integral calculus (Section 4.4). Also, the areas of simple plane figures can be obtained, or closely approximated, by the use of dot grids, line transects, or planimeters. *Geographic information systems* (Aronoff, 1989) provide a number of tools to determine areas based on computational implementations of these manual methods.

The most common area determinations in forest mensuration are the basal and surface area of trees (Section 5.2) and land areas (Sections 4.3 and 4.4). Common conversion factors for different units of area are given in Table A.1B.

2.8.5. Volume Measurements

Volume is the measure of the solid content or capacity, usually expressed in units that are cubes of linear units, such as cubic meters and cubic feet, or in units of dry and liquid measure, such as bushels, gallons, and liters. Conversion factors for volume units in the imperial and SI systems are given in Table A.1C.

3

BASIC STATISTICAL CONCEPTS

The range of statistical applications in forest mensuration is immense. It is impossible to treat all aspects of statistics in a single chapter. The aim of this chapter is simply to provide an overview of statistical concepts utilized in forest mensuration and data analysis and to provide a brief summary of the theory for the fundamental statistical parameters frequently encountered in forest mensuration. The main purpose of a chapter on statistics in this text is to explain basic concepts and present basic calculation methods. Understanding the statistical concepts in this chapter will give students a foundation for understanding the more advanced procedures presented in the remaining chapters. The practice of forest mensuration requires a much more complete understanding of statistics than can be presented in a text on mensuration. The application of statistical methods in forest mensuration requires an understanding of statistics that must be covered in specialized texts and in courses on statistics. For a broad overview of statistical methods, the reader is referred to Zar (2009); and for a more detailed review of statistics with applications in forestry, to Prodan (1968a) and Freese (1974).

3.1. DESCRIPTIVE STATISTICS

As defined in Chapter 2, measurement is the assignment of numbers to measurable properties. Collections of numbers are referred to as *data*. Raw, unordered data are generally not very useful to the forester for making management decisions. The raw data can be organized, for example, in a frequency table (see Section 2.7.1), for better understanding of what they show; however, most data, to be truly useful, need to be summarized into a set of concise, descriptive statistics.

Forest Mensuration, Fifth Edition. John A. Kershaw, Jr., Mark J. Ducey, Thomas W. Beers and Bertram Husch.
© 2017 John Wiley & Sons, Ltd. Published 2017 by John Wiley & Sons, Ltd.

The two primary objectives of statistical methods are (Freese, 1974) (1) estimation of population parameters and (2) testing of hypotheses about these parameters. While statistical hypothesis testing plays a fundamental role in the estimation of parameters, and in assessing differences between populations, treatments, and methods, the focus of this text is on estimation. Statistical hypothesis testing is not covered in detail here. For a comprehensive introduction to the classical approach to statistical hypothesis testing, the reader is referred to Zar (2009); and for critical reviews of statistical hypothesis testing, see Hagen (1997), Johnson (1999), Eberhardt (2003), and Läärä (2009).

3.1.1. Population

A *population* in a statistical sense is a collection of elements that belong to a defined group. The *elements* may be individuals or a collection of individuals that may differ in their individual characteristics, but they all belong to the same category. The population elements should be defined at the same scale at which information is desired. For example, if we desire to know average diameter of trees in a given area, then our population elements are the individual trees within that area; however, if we want to know average volume per unit area, then our population elements are typically collections of individual trees, usually on a plot of some predetermined size, within the area of interest. A *census* is a complete enumeration of the desired information about every element in the population. If the information is collected without error, then the population parameters can be computed exactly.

A population in which there is a limited number of physical elements or occurrences, such as the number of trees in a stand, is called *finite*. An *infinite* population has an unlimited number of elements. An infinite population is present when there are unlimited physical objects or occurrences or when individuals are selected from a population and are replaced before the next selection.

3.1.2. Sample

In many situations, the population is infinite or so large that, for all practical purposes, it is not feasible to enumerate or measure every element in the population. In this situation, a subset of individuals is observed, and this subset is used to *estimate* the population parameters. This subset is commonly referred to as a *sample*. The individual items in a sample are termed *sampling units*. It is not information about the sample that is of primary interest, but rather about the population. The assumption is made that the information about the sample also holds for the population from which it was selected. Thus, inferences about the population are made from the sample information.

Consider the following example. It is desired to know the density in stems per hectare of trees on a 25-ha (62 acre) forested area. The area is divided into one hundred 50 m by 50 m (164 ft × 164 ft) plots. The stems/ha for each plot is shown in Figure 3.1. In this case, the population is composed of these one hundred 50 m by 50 m plots. Both the selection of units from the population for inclusion in the sample and projection of sample estimates for the entire population are facilitated by defining the population in the same units as those selected in the sample (Shiver and Borders, 1996). The *sampling frame* is a list of all the sampling units in the population. In the example illustrated in Figure 3.1, the sampling frame is the 100 row–column pairs that define the 50 m by 50 m plots: {1,1}, {1,2}, {1,3}, ..., {10,8}, {10,9}, and {10,10}.

800	848	960	980	824	728	1112	1028	988	800
1084	704	776	820	932	916	1304	788	884	908
904	828	912	968	924	1004	872	948	784	796
1136	1296	1068	804	936	940	772	816	812	684
1412	1456	1476	1148	816	808	936	844	980	820
1440	1536	1472	1376	968	736	1044	1004	876	720
1360	1372	1336	1040	996	1044	760	1124	932	1120
1376	1208	1036	940	924	928	876	768	808	848
1384	1216	796	912	748	1064	876	816	808	876
968	868	728	1024	892	912	772	560	996	824

FIG. 3.1. A 25-hectare example forest divided into one hundred 50 m by 50 m plots. The numbers in each plot represent the stems/hectare for that plot.

Every sample has a sampling frame. In the example of Figure 3.1, the sampling frame is easy to define since the population is composed of only 100 sampling units. However, in many situations, the sampling frame may be unknown or difficult to list prior to completion of the sample. For example, if the average stem diameter for a forested area is desired, then the sampling units are the individual trees. For even a small, forested area, the sampling frame might consist of several thousand individual trees. It would be impractical to attempt to list all of these individual trees and even more impractical to try to label these trees in the field. The sample frame and how we select individuals from that frame determines the probability that an individual will be selected and its subsequent weight in the sample; therefore, it is important to be aware that the sampling frame exists despite the lack of an actual list (Shiver and Borders, 1996).

Sampling unit selection can be made with or without replacement. When sampling units are *selected with replacement*, each sampling unit may be selected more than once. For example, the 100 plots represented in Figure 3.1 could be identified on small pieces of paper and placed in a hat. A sampling unit would be selected by choosing one of the slips of paper from the hat. The plot number would be noted and the paper returned to the hat for potential reselection. When sampling units are *selected without replacement*, each sampling unit may only be selected once. In this case, once a slip of paper is removed from the hat, it is not placed back in the hat for possible reselection.

Sample size refers to the number of sampling units selected from the population and *sample intensity* refers to the proportion of the population sampled (Zar, 2009). For any given sample size, there may exist many different possible combinations of sampling units that could be selected from the population. If sampling with replacement, the number of possible sampling unit combinations is (Shiryayev, 1984)

$$C = \frac{(N+n-1)!}{n!(N-1)!}$$ (3.1)

where $n=$ the sample size (i.e., the number of sampling units included in the sample)

$N=$ the population size (i.e., the total number of sampling units)

and ! denotes the factorial expansion $(r! = r \cdot (r-1) \cdot (r-2) \cdots 3 \cdot 2 \cdot 1)$. If sampling without replacement, the number of sampling unit combinations is

$$C = \frac{N!}{n!(N-n)!}$$ (3.2)

where n and N are as above. For example, consider a sample of size 20 selected from the population shown in Figure 3.1. When sampling with replacement, the number of potential combinations of sampling units from this population would be

$$C = \frac{(100+20-1)!}{20!(100-1)!}$$
$$= \frac{119!}{20! \cdot 99!}$$
$$= \frac{119 \cdot 118 \cdot 117 \cdots 3 \cdot 2 \cdot 1}{(20 \cdot 19 \cdot 18 \cdots 3 \cdot 2 \cdot 1)(99 \cdot 98 \cdot 97 \cdots 3 \cdot 2 \cdot 1)}$$
$$= \frac{119 \cdot 118 \cdot 117 \cdots 102 \cdot 101 \cdot 100}{20 \cdot 19 \cdot 18 \cdots 3 \cdot 2 \cdot 1}$$
$$= \frac{5.97 \cdot 10^{40}}{2.43 \cdot 10^{18}}$$
$$= 2.46 \cdot 10^{22}$$

and when sampling without replacement, the number of possible combinations of sampling units would be

$$C = \frac{100!}{20!(100-20)!}$$
$$= \frac{100!}{20! \cdot 80!}$$
$$= \frac{100 \cdot 99 \cdot 98 \cdots 3 \cdot 2 \cdot 1}{(20 \cdot 19 \cdot 18 \cdots 3 \cdot 2 \cdot 1)(80 \cdot 79 \cdot 78 \cdots 3 \cdot 2 \cdot 1)}$$
$$= \frac{100 \cdot 99 \cdot 98 \cdots 83 \cdot 82 \cdot 81}{20 \cdot 19 \cdot 18 \cdots 3 \cdot 2 \cdot 1}$$
$$= \frac{1.30 \cdot 10^{39}}{2.43 \cdot 10^{18}}$$
$$= 5.36 \cdot 10^{20}$$

So even relatively small populations, such as shown in Figure 3.1, can have many possible sampling unit combinations and each of these combinations may result in different parameter estimates. Methods of selecting samples from a population are discussed in Chapter 10.* The variation in parameter estimates that arise because of sampling is referred to as *sampling error* and will be discussed in detail below.

3.1.3. Statistics

Almost every population can be described or characterized by a set of parameters. Because these parameters describe certain qualities of the population, they are frequently referred to as *descriptive statistics*. In statistics, Greek letters are often used to symbolize population parameters. A population parameter is a quantitative characteristic describing the population, such as an arithmetic mean or variance. A population mean is symbolized by the Greek letter μ and the population variance as σ^2. A quantitative characteristic describing a sample is called *a statistic* and is written in the Roman alphabet. Thus, the mean of a sample is $\bar{x}$ and its variance is s^2. In summary, statistics based on samples are used to estimate population parameters.

3.2. FREQUENCY DISTRIBUTIONS

A series of measurements can be consolidated into more manageable form by grouping measurements into classes (as described in Section 2.7.1) in a *frequency table*. The total list of measurements can then be assigned one by one to the appropriate classes, resulting in an ordered table showing the number of measurements falling into individual classes. The frequency data can then be shown graphically in a *histogram, frequency polygon*, or *frequency curve*. A histogram is a bar chart prepared by plotting frequency, either in number or percent, over size class. (Figure 3.2*a* is an example of a histogram developed from the 100 plots shown in Figure 3.1.) A series of rectangles of equal width is formed. Connecting the plotted points representing frequencies over size classes with straight lines forms a frequency polygon (solid line in Fig. 3.2*b*). If, instead of connecting each point, a smooth line is fitted to represent the general trend, a frequency curve is formed (dashed line in Fig. 3.2*b*). A cumulative frequency curve is prepared by plotting the cumulative frequency numbers or percentages over size.

There are a number of typical frequency distributions that have been discovered and are used in statistical work. Mathematicians have devised numerous theoretical distributions, some of which are of great practical value in applied work because actually observed measurements have been found to fit or approximate them. The properties of the theoretical distributions can then be used in explaining relationships in observed data. A number of these distributions are presented throughout various chapters as they relate to topics being discussed. In this chapter, we only present the normal or Gaussian distribution because of its importance to statistics and to the distributions arising from random samples of populations.

---

*Chapter 10 presents sampling designs used in forest inventory for estimating forest vegetation parameters but the reader should understand that the designs described may be applied for sampling any kind of population.

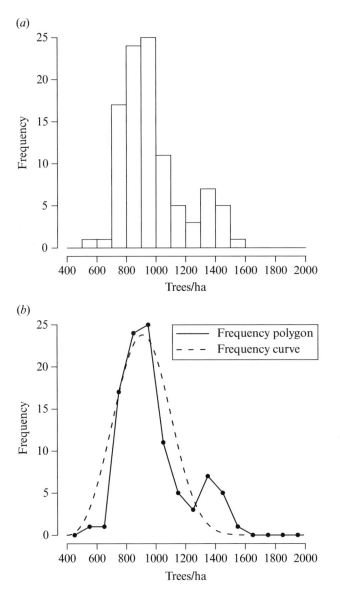

FIG. 3.2. Distribution of trees/ha for the 100 sample plots from the example forest shown in Fig. 3.1: (*a*) histogram and (*b*) frequency polygon and frequency curve.

Normal Distribution. The normal or Gaussian distribution is the typical bell-shaped distribution that most people commonly associate with statistics (Fig. 3.3). The normal distribution is defined by two parameters: the mean, μ; and the standard deviation, σ. As shown in Figure 3.3*a*, the distribution is symmetric about the mean, and the mean "locates" the population along the size axis. The standard deviation describes the spread of individuals about the mean and is a measure of the average distance of individuals from the mean. As the standard deviation increases, the proportion of individuals near the mean decreases and

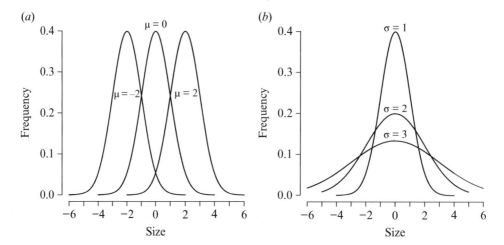

FIG. 3.3. The normal distribution: (*a*) influence of changes in the value of the mean, *μ*, and (*b*) influence of changes in the value of the standard deviation, *σ*.

the proportion farther from the mean increases (Fig. 3.3*b*). While very few forest popula-tions are actually well represented by the normal distribution, it is still a theoretical distri-bution of great practical value in sampling. The importance of the normal distribution for describing the distribution of sample means is developed further in the next sections.

3.3. MEASURES OF CENTRAL TENDENCY

Frequency tables and graphs reveal an overall picture of the range and pattern of a series of measurements, but they lack the conciseness and explicitness that is conveyed by an *average*. An average is a numerical quantity that can be used to represent a group of individual values. The individual values tend to cluster about an average, which, consequently, is often referred to as a *measure of central tendency*. There are several measures of central tendency: arithmetic mean, harmonic mean, geometric mean, quadratic mean, median, and the mode. The one to use should be appropriate for the data and for what it is supposed to show.

3.3.1. Arithmetic Mean

The most common measure of central tendency is the *arithmetic mean* or *arithmetic average*. The *population mean*, commonly designated as *μ*, is the arithmetic average of all possible sampling units in the population:

$$\mu = \frac{\sum_{i=1}^{N} X_i}{N} \tag{3.3}$$

where μ = the population mean

X_i = the *i*th observation in the population

N = total number of items in the population

Calculation of μ requires that every individual within the population be measured without error. Most populations of interest in forest mensuration are very large and essentially can be considered infinite. It is a rare situation when μ is ever actually known.

The arithmetic mean of a sample of measurements drawn from a population is

$$\bar{x} = \frac{\sum_{i=1}^{n} X_i}{n} \tag{3.4}$$

where $\bar{x}$ = the sample mean

  X_i = the ith observation in the sample

  n = the sample size

The statistic $\bar{x}$ is an unbiased estimator of the population parameter μ. For any given sample of size n, $\bar{x}$ is the sample estimate of μ. Using the example forest shown in Figure 3.1, the population mean is calculated as

$$\mu = \frac{\sum_{i=1}^{100} X_i}{100}$$

$$= \frac{968 + 1384 + 1360 + 1440 + \cdots + 684 + 796 + 908 + 800}{100}$$

$$= \frac{97{,}092}{100} = 970.92 = 971 \, \text{trees} / \text{ha}$$

A sample of size 20 is generated by randomly selecting row and column numbers: {4,5}, {9,7}, {7,7}, {2,1}, {8,1}, {5,4}, {8,4}, {5,5}, {9,9}, {5,1}, {7,1},{5,3}, {5,7}, {8,2}, {4,9}, {1,10}, {9,2}, {7,5}, {7,8}, and {6,10}. The corresponding trees/ha for these plots are: 996, 1304, 772, 1384, 904, 1376, 968, 968, 796, 1440, 1136, 1472, 1044, 828, 932, 824, 704, 936, 816, and 820. From eq. (3.4), the sample mean associated with this sample is:

$$\bar{x} = \frac{\sum_{i=1}^{20} X_i}{20}$$

$$= \frac{996 + 1304 + 772 + 1384 + \cdots + 704 + 936 + 816 + 820}{20}$$

$$= \frac{20{,}420}{20} = 1{,}021 \, \text{trees} / \text{ha}$$

This sample of 20 plots yields an estimate of the population mean. Other samples of size 20 would yield other estimates of the population mean.

3.3.2. Quadratic Mean

The *quadratic mean* $\bar{x}_Q$ is the square root of the average squared value:

$$\bar{x}_Q = \sqrt{\frac{\sum_{i=1}^{n} x_i^2}{n}} \tag{3.5}$$

The quadratic mean finds its greatest use in the computation of the standard deviation (Section 3.4.1). It is also useful in determining the diameter of the tree of average basal area (Section 8.3.1).

3.3.3. Harmonic and Geometric Means

The harmonic mean is the reciprocal of the arithmetic average of the reciprocals of a series of measurements. This is the appropriate average to use when dealing with rates. An estimate of this population parameter is given by the statistic $\bar{x}_H$:

$$\bar{x}_H = \frac{n}{(1/X_1)+(1/X_2)+\cdots+(1/X_{n-1})+(1/X_n)} = \frac{n}{\sum\limits_{i=1}^{n}(1/X_i)} \tag{3.6}$$

The geometric mean is used when averaging quantities drawn from a series of measurements that follow a geometric progression or the exponential law. The geometric mean is the nth root of the product of n values. The logarithm of the geometric mean is equal to the arithmetic mean of the logarithms of the individual values. An estimate of the population geometric mean is given by $\bar{x}_G$

$$\bar{x}_G = \sqrt[n]{(X_1)(X_2)\cdots(X_{n-1})(X_n)} = \sqrt[n]{\prod_{i=1}^{n}X_i}$$

$$\log(\bar{x}_G) = \frac{[\log X_1 + \log X_2 + \cdots + \log X_{n-1} + \log X_n]}{n} = \frac{\sum\limits_{i=1}^{n}(\log X_i)}{n} \tag{3.7}$$

In forest mensuration, the geometric mean is used to estimate average diameter when tree cross sections are elliptical and the major and minor axes (diameters) of the cross section are measured (Section 5.2).

3.3.4. Median and Mode

The *median* is the middle value when the data are ordered from smallest to largest. It is the value that divides a frequency distribution into an equal number of items above and below. It differs from the arithmetic mean in being an average of position rather than of magnitude of the values. The *mode* is the most frequent value observed in the data. It is the most commonly occurring value or class in a frequency distribution. For a detailed description of the median and mode and their associated statistical properties, refer to Zar (2009).

3.4. MEASURES OF DISPERSION

An average gives a single value that characterizes a distribution of measurements. However, it reveals nothing regarding the spread or range of the values on which it is based. Such information is provided by *a measure of dispersion*. Because individuals within a population are not all identical in terms of size or value, the population is said to have variation. Measures of dispersion define the amount by which individuals in a set of measurements vary from the central tendency. The commonly used measures of dispersion are the range, mean deviation, and the variance or standard deviation. The *range* is the interval between the largest and smallest value in a series. It is a measure of minor significance since it only indicates the difference between extreme values at the ends of a distribution. The *mean deviation* is the average amount by which the individual values in a series of measurements will deviate from the arithmetic mean. To calculate this measure, the arithmetic mean is first obtained. Then the absolute difference (regardless of sign) between each item and the mean is calculated. These differences are summed and divided by the total number to obtain the mean deviation.

3.4.1. Variance and Standard Deviation

The most important measure of dispersion is the variance. *Variance* is the average squared deviation of individuals from the mean. The *population variance* is calculated from

$$\sigma^2 = \frac{\sum_{i=1}^{N}(X_i - \mu)^2}{N} \tag{3.8}$$

where σ^2 = the population variance
μ = the population mean
X_i = the ith observation in the population
N = the total population size

As with μ, the calculation of σ^2 requires that every individual within the population be measured. The variance of a sample selected from a population is calculated using

$$s^2 = \frac{\sum_{i=1}^{n}(X_i - \bar{x})^2}{n-1} \tag{3.9}$$

where s^2 = the sample variance
$\bar{x}$ = the sample mean (the sample estimator of μ)
X_i = the ith observation in the sample
n = the sample size

Like the sample mean, $\bar{x}$, s^2 is an estimator of σ^2. However, unlike the sample estimator for the mean (eq. (3.4)), the formula for the sample estimator of variance is not identical to the formula for the population. The denominator in the sample estimator is $n-1$ rather than n. The $n-1$ is referred to as *degrees of freedom* (d.f.) and reflects the fact that the sample mean $\bar{x}$, is estimated from the data. Use of $n-1$ makes the estimator an unbiased estimator of σ^2. An alternative formula for s^2, which is algebraically equivalent, but computationally simpler, is

$$s^2 = \frac{\sum\limits_{i=1}^{n} X_i^2 - \dfrac{\left(\sum\limits_{i=1}^{n} X_i\right)^2}{n}}{n-1} \tag{3.10}$$

Variance can also be calculated as the difference between the squared value of the quadratic mean (eq. (3.5)) and the squared value of the arithmetic mean (eq. (3.4)): $s^2 = \bar{x}_Q^2 - \bar{x}^2$ (Curtis and Marshall, 2000).

Variation is often expressed as the *standard deviation* rather than variance. The standard deviation is the square root of the variance. Thus, the *population standard deviation* is

$$\sigma = \sqrt{\sigma^2} = \sqrt{\frac{\sum\limits_{i=1}^{N} (X_i - \mu)^2}{N}} \tag{3.11}$$

and the *sample standard deviation* is

$$s = \sqrt{s^2} = \sqrt{\frac{\sum\limits_{i=1}^{n} X_i^2 - \dfrac{\left(\sum\limits_{i=1}^{n} X_i\right)^2}{n}}{n-1}} \tag{3.12}$$

Both the population and sample standard deviations express dispersion in the same units as the mean, and is a measure of the average distance individuals are dispersed from the mean. Continuing with the example from Figure 3.1, the population variance is calculated to be

$$\sigma^2 = \frac{\sum\limits_{i}^{100} (X_i - \mu)^2}{100}$$

$$= \frac{(968 - 970.92)^2 + (1{,}384 - 970.92)^2 + \cdots + (908 - 970.92)^2 + (800 - 970.92)^2}{100}$$

$$= \frac{(-2.92)^2 + (413.08)^2 + \cdots + (-62.92)^2 + (-170.92)^2}{100}$$

$$= \frac{8.53 + 170635.09 + \cdots + 3958.93 + 29213.65}{100}$$

$$= \frac{4403035.36}{100} = 44030.35 = 44{,}000$$

and the population standard deviation is

$$\sigma = \sqrt{\sigma^2} = \sqrt{44030.35} = 209.83 = 210 \text{ trees / ha.}$$

For the sample of size 20, the sample variance is

$$s^2 = \frac{\displaystyle\sum_{i=1}^{20} X_i^2 - \frac{\left(\displaystyle\sum_{i=1}^{20} X_i\right)^2}{20}}{20-1}$$

$$= \frac{\left(996^2 + 1{,}304^2 + \cdots + 816^2 + 820^2\right) - \dfrac{\left(996 + 1{,}304 + \cdots + 816 + 820\right)^2}{20}}{20-1}$$

$$= \frac{21{,}986{,}096 - \dfrac{\left(20{,}420\right)^2}{20}}{19}$$

$$= \frac{21{,}986{,}096 - 20{,}848{,}820}{19} = \frac{1{,}137{,}276}{19} = 59856.63 = 59{,}900$$

and the sample standard deviation becomes

$$s = \sqrt{s^2} = \sqrt{59856.63} = 244.66 = 245 \ \text{trees}\,/\,\text{ha}.$$

(The reader should note that while we present final values rounded to the appropriate significant figures, rounding should only be conducted on the final answer, not on the intermediate calculations. This is especially important when computing statistics.)

3.4.2. Coefficient of Variation

The standard deviation can be expressed on a relative or percentage basis as the *coefficient of variation*. The coefficient of variation is the standard deviation expressed as a percentage of the arithmetic mean:

$$\text{For the population} \quad CV = 100\frac{\sigma}{\mu}$$

$$\text{For the sample} \quad CV = 100\frac{s}{\bar{x}}$$

(3.13)

The coefficient of variation for the example sample is $CV = 100 \cdot (244.66/1021) = 24\%$. The coefficient of variation is useful for expressing relative variability when comparing samples from populations where the means differ substantially. For a description of additional measures of dispersion and their associated statistical properties, see Zar (2009).

3.5. SAMPLING ERROR

If a population is not entirely uniform, single sampling units or estimates from small sample sizes are unlikely to represent exactly the parameters of the entire population. Estimates of population parameters based on samples are always subject to sampling errors resulting from the chances of selecting different individuals or samples. It is important to recognize the existence of these errors since they influence the interpretation of sample statistics.

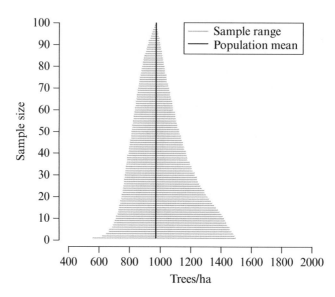

FIG. 3.4. Ranges of possible sample means versus sample size for the 25 ha forested area shown in Fig. 3.1.

By examining Figure 3.1, it is easy to see that not only are there many different samples of size 20 possible from this population, but also that each sample can result in a different estimate of the mean. Thus, for any given sample, the estimated sample mean is subject to *sampling error*, which is simply the difference between the true value of the population mean and the sample estimate.

As the sample size increases, the average sampling error becomes smaller and the reliability of the sample estimate becomes larger. This is easily illustrated in Figure 3.4 where the ranges of possible sample estimates are plotted against sample size. For each sample size, the "n" smallest and "n" largest individuals were determined. As "n" increases, larger individuals are added to the set of smaller values and smaller individuals are added to the set of larger values, and the differences between extreme estimates diminish. A useful sampling method should provide some estimate of the average sampling error. This sampling error has nothing to do with the mechanical precision with which measurements are executed or recorded. The errors referred to here are due to variations in the population and to chance selection. More consideration is given to errors in Chapter 10 on inventory sampling designs.

3.5.1. Standard Error of the Mean

The sampling error for any one particular sample is not generally of interest and is almost always unknown since the population mean is unknown. What is of interest is the distribution of this error. Figure 3.5 shows the distribution of the means based on a sample of size 20 from the population shown in Figure 3.1. From eq. (3.2), it was determined that there are 5.36×10^{20} possible combinations of samples of size 20 for this population. Using all of these sample estimates, it would be possible to calculate a variance and standard deviation for the estimated means. The standard deviation of the means is commonly called the

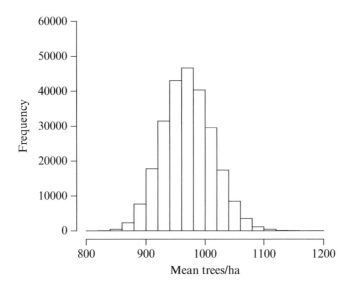

FIG. 3.5. Distribution of sample means based on a sample of size 20 from the example forest shown in Fig. 3.1. (only 250,000 of the 5.36×10^{20} possible samples are shown.)

standard error of the mean and represents the variability (or error) expected if the population was repeatedly sampled. Because the standard error represents the dispersion in estimates, it is a useful estimate of the sampling error.

By examining the example population shown in Figure 3.1, it is easy to understand how sample size influences sampling error. As shown in Figure 3.4, the range of sample mean estimates for any sample of size n can be quickly determined by calculating the mean for the smallest n values and largest n values. As more individual sampling units are included in the sample, fewer units are excluded and less variability between repeated samples is possible.

Similarly, the influence of population variability on sampling error can be understood. If all 100 plots had the exact same sample value, there would be no population variability and even a sample of size one would produce no sampling error. If only one plot had a different value, the population variance would be very low. As the number of different values increases, the population variance increases; likewise, the number of different combinations increases and the sampling error increases. If the population had a range that was twice as large as that observed in Figure 3.1, the sampling error would also increase.

The concept of sampling error, or standard error, makes sense only in repeated sampling since it is a standard deviation of different sample estimates. Strictly speaking, a standard error can only be calculated if there are two or more sample means. In most forestry situations, only one sample is taken. Fortunately, the standard error can be estimated from only one sample using the central limit theorem (CLT). The CLT is one of the most important theorems in statistics (Shiryayev, 1984). The CLT states that, for any population with a finite mean, μ, and variance, σ^2, the sample estimates of the means will follow approximately a normal distribution (as sample size increases) with mean, μ, and standard deviation of the means, $\sigma / \sqrt{n}$.

The bell shape can be observed for the distribution of means shown in Figure 3.5. It is important to note that the underlying population distribution does not have to be normal for the CLT to hold (compare Figs. 3.2, 3.3, 3.4, and 3.5). In fact, population distributions can have any shape. The population shown in Figure 3.1 is somewhat *bimodal* (i.e., has two peaks within the histogram; Fig. 3.2) and asymmetrical (skewed), yet the distribution of means shown in Figure 3.5 shows a bell-shaped distribution closely approximating a normal distribution. With a reasonably large sample size, σ^2 can be estimated using s^2 (eq. (3.9)) and standard error of the mean can be estimated using

$$s_{\bar{x}} = \frac{s}{\sqrt{n}}$$

(3.14)

where $s_{\bar{x}}$ = the standard error of the mean
$\quad\quad\quad s$ = standard deviation
$\quad\quad\quad n$ = the sample size

The standard error for the example sample of size 20 would be estimated as

$$s_{\bar{x}} = \frac{244.66}{\sqrt{20}} = \frac{244.66}{4.47} = 54.73 = 54.7 \text{ trees / ha}$$

When the entire population is measured, the sampling error is zero since the population mean can be calculated rather than estimated (eq. (3.3)). When sampling finite populations without replacement, as the sample intensity increases, eq. (3.14) tends to overestimate the sampling error. For example, if eq. (3.14) were applied to the whole example population in Figure 3.1, the standard error would be estimated as $s_{\bar{x}} = 209.83 / \sqrt{100} = 20.98$ trees/ha. However, since the entire population has been observed, the sampling error should be zero. When finite populations are sampled without replacement, the *finite population correction factor* should be applied to the estimate of standard error of the mean:

$$s_{\bar{x}} = \frac{s}{\sqrt{n}} \cdot \sqrt{\frac{N-n}{N}}$$

(3.15)

where $s_{\bar{x}}$ = standard error of the mean
$\quad\quad\quad s$ = standard deviation
$\quad\quad\quad n$ = sample size
$\quad\quad\quad N$ = population size

Using the finite population correction factor, the estimate for standard error for the example sample of size 20 becomes

$$s_{\bar{x}} = \frac{244.66}{\sqrt{20}} \cdot \sqrt{\frac{100-20}{100}} = 54.73 \cdot \sqrt{\frac{80}{100}} = 54.73 \cdot 0.89 = 48.93 = 48.9 \text{ trees / ha.}$$

3.5.2. Confidence Interval

In brief, samples selected in an unbiased manner from a population are assumed representative of the parent population. From these samples, statistics are calculated, which are estimates of the population parameters. The degree of confidence placed in these estimates depends on the standard error of the parameter in question. Thus, an estimate of a population parameter such as the arithmetic mean should properly be described, not as a single figure, but rather as a range within which we are confident that the true value of the population parameter lies. The question immediately arises, "How confident?" The answer must be in terms of probability. An estimate of a population parameter must be in terms of a range with an associated probability. This range is called a *confidence interval*. The values that define the bounds of a confidence interval are called the *confidence limits*. Confidence intervals are constructed at a given *confidence level*, which is the compliment of the significance level (α-level) used in hypothesis testing (Zar, 2009): Confidence $= 1 - \alpha$. Since the population parameter μ is a fixed value, it is the sample mean and its associated confidence interval and not μ, which vary for repeated samplings and have an associated probability distribution. The confidence interval does not predict the probability that the true population mean lies within any given interval, but rather describes the proportion of confidence intervals (resulting from repeated samples of size n) that will contain the true population mean. The confidence interval, then, is a statement used to describe an interval in which the true mean, μ, is expected to occur and to attach a probability to this interval containing the true population mean. A confidence interval can be thought of as a statement of sample precision at a stated probability, as more confidence is desired (i.e., a greater proportion of sample intervals containing the true population mean), the width of confidence interval increases, and less precision is obtained.

The limits of a confidence interval based on a sample are given by

$$\text{C.I.}\left(1-\alpha\right): \bar{x} \pm t_{\alpha/2, n-1} \cdot s_{\bar{x}} \tag{3.16}$$

The value of t depends on the probability level chosen, $1 - \alpha$, and the size of the sample, n, from which the mean was determined. The value of t can be obtained from Table A-4. Since the distribution of t depends on n, it is important to enter the table on the correct line using the appropriate *degrees of freedom*. The number of degrees of freedom to use in entering the table is the number of items in the sample less 1, that is, $(n-1)$. For a chosen probability level, such as 0.95, the t value is read under the column headed 0.05 $(1-0.95)$ in the Appendix table (the Appendix table is referred to as a two-sided t-table). The 0.05 indicates the probability that a larger value of t will occur 5 times out of 100. Conversely, this indicates that the probability of the given or smaller t value will be 0.95, the desired confidence level.

Considering the previous example of a sample of $n = 20$, the t value for a 95% confidence interval is found by locating the row corresponding to $20 - 1 = 19$ degrees of freedom and the column for 0.05 that indicates a value of $t = 2.0930$. The 95% confidence interval for the example would be

$$\text{CI}\left(0.95\right): 1021 \pm 54.73 \cdot 2.0930$$
$$1021 \pm 114 \text{ trees / ha}$$

The limits in which the population mean μ is expected to occur with a probability of 0.95 are then

$$906 \leq \mu \leq 1136 \text{ trees / ha.}$$

This confidence interval gives the range of values a sample mean will have some percentage of the time if the same population is repeatedly sampled with the same sample size. In the example above, it is expected that a sample of size 20 from the population shown in Figure 3.1 would produce a sample mean between 906 and 1136 trees/ha, 95% of the time. The calculation of confidence limits for forest inventory estimates will be presented throughout the text.

3.6. SAMPLE SIZE DETERMINATION

The precision of the estimate of a population mean, as measured by the confidence interval, depends on the variation in the population, estimated by its variance, the number of sampling units measured, and the associated probability. This fact enables the necessary number of samples to be estimated if specifications are setup regarding the precision and probability required to state the estimate of the mean. The quantity to the right of the $\pm$ sign in the confidence interval (eq. (3.16)) can be thought of as the allowable error: $E = t \cdot s_{\bar{x}}$. Since $s_{\bar{x}} = s / \sqrt{n}$, allowable error can be expressed as $E = t \cdot \left(s / \sqrt{n} \right)$. For the simplest case of random sampling from an infinite population, the sample size can be found by rearranging this formula, solving for n, yielding the following formula:

$$n = \frac{t^2 s^2}{E^2} \tag{3.17}$$

where $n =$ the minimum number of samples required to estimate μ to within $\pm E$ for a specified probability

         $t =$ the t value associated with the specified probability from Table A.4. The use of t requires the probability and degrees of freedom, $n - 1$, where n refers to the sample size. Of course, to be correct, n should really be the number of samples that is being sought.

         $s =$ the standard deviation. Some prior knowledge or estimation of the variability of the population to be sampled must be available. This can be obtained by calculating the standard deviation of a small preliminary sample or from prior experience in similar populations. It is impossible to determine sample size without having some kind of prior knowledge of the population.

         $E =$ the allowable error. Some decision must be made regarding the allowable error. This is often stated in terms of a percentage of the mean.

For a finite population the formula is

$$n = \frac{t^2 s^2 N}{\left(NE^2 + t^2 s^2 \right)} \tag{3.18}$$

where n, t, s, and E are as above and $N =$ the total population size. For large finite populations, the importance of this value decreases and for infinite populations has no meaning and the formula without N can be used (eq. (3.17)).

Instead of using the standard deviation as an expression of population variability, the coefficient of variation may be utilized and the allowable standard error of the mean expressed as a percentage of the mean. In this case, the formulas for sample size may be modified as follows:

$$\text{For an infinite population} \quad n = \frac{t^2 (CV)^2}{(E\%)^2} \tag{3.19}$$

$$\text{For an infinite population} \quad n = \frac{t^2 (CV)^2 N}{\left(N(E\%)^2 + t^2 (CV)^2\right)} \tag{3.20}$$

For example, to estimate the mean for the population shown in Figure 3.1 to within 10% of the true mean with 95% confidence, the minimum sample size required would be

1. Initially, make a reasonable guess for the value of n and determine t (we will start with 20 samples, the size of the sample we already have):

$$n_0 = 20$$
$$t = 2.0930 \, (\text{for 19 d.f. and 0.95 probability})$$

2. Calculate the first estimate of sample size using $CV = 24$ and $E = 10$:

$$n_1 = \frac{(2.0930)^2 (24)^2}{10^2} = 25.2 = 26$$

3. Compare n_0 with n_1; if equal, then n_1 is the minimum sample size; if not, calculate a new estimate of n, using n_1 to determine t:

$$n_1 = 26$$
$$t = 2.0595 \, (\text{for 25 d.f. and 0.95 probability})$$

4. and n_2 becomes

$$n_2 = \frac{(2.0595)^2 (24)^2}{10^2} = 24.4 = 25$$

5. The process is repeated until the estimates for n are the same:

$$n_2 = 25$$
$$t = 2.0639 \, (\text{for 24 d.f. and 0.95 probability})$$

6. and n_3 becomes

$$n_3 = \frac{(2.0639)^2 (24)^2}{10^2} = 24.5 = 25$$

7. Since $n_3 = n_2$, the iteration terminates.

Thus, 25 plots would be required to achieve the desired level of precision. In most situations, only a rough estimate of the minimum sample size is required. Here, a single iteration can be utilized provided the initial guess is close to the estimated sample size (e.g., using $t=2$ is often sufficient to estimate minimum sample sizes for 95% confidence). Determination of sample size for different sampling designs is covered throughout the book as different sampling schemes are presented.

3.7. INFLUENCE OF SCALAR TRANSFORMATIONS AND THE ESTIMATION OF TOTALS

Foresters often use small plots as the basis for sampling forest vegetation. In the example above, 1/4 ha plots were utilized; however, in practice, plot are often quite smaller (1/25 ha, 1/100 ha, or even smaller, are used for many applications). Traditionally, forest measurements are expressed on a per unit area basis: per acre in the imperial system and per hectare in SI. Scaling measurements from a per plot basis to a per unit area basis is straight-forward because plot size is usually known. The scalar for converting per plot measurements to per unit area measurements is $m=1/$(plot size), where plot size is expressed as a fraction of the unit area. In the example worked above, all plot measurements were scaled to per hectare measurements using $m=4$. Scalar transformations influence not only the mean, but also the variance, standard deviation, and standard error, which in turn influence the confidence interval. The scalar transformations for each parameter estimate described above are summarized in Table 3.1.

The estimation of totals for a given forest area is generally a special case of scalar transformation where per unit area estimates are multiplied by total area. In the example forest in Figure 3.1, the total area is 25 ha; therefore, totals for the forest would be obtained by multiplying the per unit area estimates by 25. The total number of trees on the forest would be estimated by multiplying the mean (1,021 trees/ha) by 25: Total trees $=25 \cdot 1,021 = 25,525$, and the confidence interval for the example sample would become $22,661 < \mu_T < 28,389$. As can be observed in Table 3.1, errors associated with mean estimates scale to their respective total estimates at the same rate as the mean. It is sometimes assumed that totals are error free since low-valued individuals and high-valued individuals, which produce the variation

TABLE 3.1. Effects of Scalar Transformations of the Form $Y=mX$ on Parameter Estimates

Parameter	Original Estimate	Transformed Estimate
Mean	$\bar{x} = \dfrac{\sum X}{n}$	$\bar{y} = m \cdot \bar{x}$
Variance	$s^2(X) = \dfrac{\sum X^2 - \dfrac{(\sum X)^2}{n}}{n-1}$	$s^2(Y) = m^2 \cdot s^2(X)$
Standard deviation	$s(X) = \sqrt{s^2(X)}$	$s(Y) = m \cdot s(X)$
Standard error	$s_{\bar{x}} = \dfrac{s(X)}{\sqrt{n}}$	$s_{\bar{y}} = m \cdot s_{\bar{x}}$
Confidence interval	$\bar{x}_L = \bar{x} - t \cdot s_{\bar{x}}$ $\bar{x}_U = \bar{x} + t \cdot s_{\bar{x}}$ $\bar{x}_L \leq \mu_X \leq \bar{x}_U$	$m \cdot \bar{x}_L \leq \mu_Y \leq m \cdot \bar{x}_U$

observed in mean estimates, cancel each other out when considering totals; however, this is an erroneous assumption since the error is associated with the estimated mean, not the individual sampling units. Additional considerations for estimating forest totals based on sample estimates are discussed throughout the text as new sample designs and approaches are introduced.

3.8. CORRELATION AND REGRESSION ESTIMATION

Often in forest mensuration, more than one aspect of a sampling unit is measured. For example, on individual trees, we might measure diameter at a specified height, total height of a tree, and perhaps height to the base of the live crown. On plots, we might calculate total stem density, basal area (the sum of stem cross-sectional areas), and total or merchantable volume. Often, we are interested in the relationships between these variables. Correlation and regression analyses are useful in evaluating association between two or more variables. The two kinds of analyses have much in common, especially in computational procedures, but there is often an important difference in their objectives. Correlation analysis measures the degree of linear association between two or more variables. The goal of correlation analysis is simply to obtain a number that expresses the degree of linear association. On the other hand, the objective of regression analysis is to quantify the relationship between a dependent variable and one or more independent variables. Regression analysis results in a mathematical equation that describes how a dependent variable changes given a unit change in one or more independent variables. While regression analysis implies a cause-and-effect relationship, it does not prove cause and effect. While both analyses have application in forest mensuration, regression analysis is widely used in developing allometric equations (Section 5.6), site index equations (Section 8.8), volume equations (Section 6.2), and many other applications.

3.8.1. Covariance and Correlation

Covariance is a measure of association between the magnitudes of two characteristics (Freese, 1974; Zar, 2009). If there is little or no association, covariance will be close to zero, while increasing negative values indicate that large values of one variable are associated with small values of the other variable. Likewise, increasing positive values of covariance indicate that large values of one variable are closely associated with large values of the other variable. As with the descriptive statistics described above, there is a population-level covariance, often denoted as σ_{XY}. Like the other population parameters, estimates of covariance can be obtained from samples. The sample covariance, denoted as covar(X,Y) or s_{XY}, is estimated using

$$
s_{XY} = \frac{\sum_{i=1}^{n}\left(X_i - \bar{X}\right)\left(Y_i - \bar{Y}\right)}{n-1}
$$

$$
= \frac{\sum_{i=1}^{n}X_iY_i - \dfrac{\sum_{i=1}^{n}X_i \cdot \sum_{i=1}^{n}Y_i}{n}}{n-1}
$$

$$(3.21)$$

Correlation is a measure of the linear association between two or more random variables. When correlation exists, the sizes of measurements of one variable are related to the sizes of measurements of another variable. The brief discussion of correlation here is limited to the relationships between measurements of two variables. The correlation between more than two variables requires multiple correlation analysis (Kutner et al., 2004).

The measure of the degree of association between two variables, X and Y, in a correlation analysis is called the *correlation coefficient*. The population correlation coefficient, ρ, can have a numerical value varying between the limits of -1 to $+1$ (Fig. 3.6). When $\rho = 0$, there is no correlation (Fig. 3.6e). When $\rho = \pm 1$, there is a perfect relationship or association. An increase in the value of one variable would result in an unvarying increase for the other variable if $\rho = +1$ (Fig. 3.6i), or an unvarying decrease if $\rho = -1$ (Fig. 3.6a). Values between 0 and ± 1 indicate something between none and perfect correlation, with increasing intensity of association as the correlation coefficient approaches 1 (Fig. 3.6). The population correlation coefficient, ρ, is estimated by the sample statistic, r, which can be obtained from sample observations. The sample correlation coefficient is estimated using $r = \mathrm{Covar}(X,Y) / \sqrt{\mathrm{Var}(X) \cdot \mathrm{Var}(Y)}$, where $\mathrm{Covar}(X,Y)$ is the covariance estimated from eq. (3.21) and $\mathrm{Var}(X)$ and $\mathrm{Var}(Y)$ are the sample variances estimated from eq. (3.10):

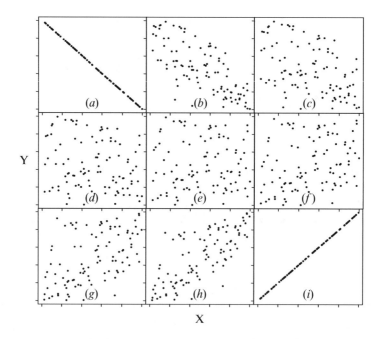

FIG. 3.6. Examples of the relationship between two variables under different levels of correlation: (a) $r=-1.00$, (b) $r=-0.75$, (c) $r=050$, (d) $r=-0.25$, (e) $r=0.00$, (f) $r=0.25$, (g) $r=0.50$, (h) $r=0.75$, and (i) $r=1.00$.

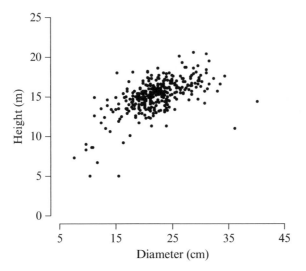

FIG. 3.7. Graph of total height versus diameter for 349 western hemlock trees in western Washington, USA.

$$r = \frac{s_{XY}}{\sqrt{s_X^2 \cdot s_Y^2}} = \frac{\dfrac{\sum XY - \dfrac{\sum X \sum Y}{n}}{n-1}}{\sqrt{\left(\dfrac{\sum X^2 - \dfrac{(\sum X)^2}{n}}{n-1}\right)\left(\dfrac{\sum Y^2 - \dfrac{(\sum X)^2}{n}}{n-1}\right)}} = \frac{\sum XY - \dfrac{\sum X \sum Y}{n}}{\sqrt{\left(\sum X^2 - \dfrac{(\sum X)^2}{n}\right)\left(\sum Y^2 - \dfrac{(\sum Y)^2}{n}\right)}}$$

$$(3.22)$$

The squared value of the correlation coefficient, r^2, is called the *coefficient of determination*. The coefficient of determination can be interpreted as indicating the percentage of variation in one variable that is associated with the other variable.

For example, Figure 3.7 shows a plot of heights versus diameters for a sample of 349 western hemlock trees in western Washington, USA. The covariance for this sample of trees would be estimated as

$$s_{XY} = \frac{\displaystyle\sum_{i=1}^{349} D_i H_i - \frac{\displaystyle\sum_{i=1}^{349} D_i \cdot \sum_{i=1}^{349} H_i}{349}}{349-1}$$

$$= \frac{118860.2 - \dfrac{(7726.2)(5278.9)}{349}}{348}$$

$$= \frac{118860.2 - 116864.9}{348} = \frac{1995.30}{348} = 5.73$$

As observed in Figure 3.7, the positive covariance value indicates that larger diameters are associated with taller trees. The magnitude of covariance is dependent upon the range and distribution of both variables. Covariance, like variance, is a measure of dispersion of pairs of points about their mean pair. As such, covariance does not give a measure of the strength of association. The correlation coefficient (eq. (3.22)), on the other hand, does provide a measure of the strength of the linear association between the two variables. For the height and diameter data shown in Figure 3.7, the correlation coefficient would be

$$r = \frac{\text{Covar}(X,Y)}{\sqrt{\text{Var}(X)\cdot\text{Var}(Y)}} = \frac{\sum XY - \dfrac{\sum X \sum Y}{n}}{\sqrt{\left(\sum X^2 - \dfrac{(\sum X)^2}{n}\right)\left(\sum Y^2 - \dfrac{(\sum Y)^2}{n}\right)}}$$

$$= \frac{118860.2 - \dfrac{(7726.2)(5278.9)}{349}}{\sqrt{\left(178314.5 - \dfrac{(7726.2)^2}{349}\right)\left(81349.81 - \dfrac{(5278.9)^2}{349}\right)}} = \frac{118860.2 - 116864.9}{\sqrt{(7271.023)(1502.288)}}$$

$$= \frac{1995.297}{3305.022} = 0.60$$

The coefficient of determination, r^2, would be 0.36, indicating about 36% of the variation in height is accounted for (or "explained") by diameter. Correlation coefficients generally greater than about 0.50 are considered indicative of the existence of a linear relationship between two variables, and formal statistical tests of significance are available (Freese, 1974; Zar, 2009). Neither high ($r \approx 1.00$) nor low ($r \approx 0.00$) values of a correlation coefficient prove the presence or absence of a relationship between two variables; it only measures the strength of a linear association between two variables. Strong curvilinear relationships may have very low linear correlation coefficients. Correlation analysis should be accompanied by visual examination of graphs of the variables of interest.

3.8.2. Simple Linear Regression

In the example shown in Figure 3.7, the correlation coefficient indicated that there might be a linear relationship between height and diameter for the sample of western hemlock trees. Relationships between two or more measurements are often of great interest in forest mensuration. The measurement of individual tree components is covered in Chapter 5. Tree diameter is often conveniently measured at a fixed height above ground and several simple tools are available to make measurement of this variable fast, accurate, and repeatable. Measurement of height, on the other hand, is often quite time consuming, prone to errors between different measurement crews, and difficult to obtain accurate measurements, even with expensive electronic equipment. If a relationship between diameter and height could be established, then height, which is an expensive and difficult measurement, could be estimated from diameter, which is an inexpensive measurement.* If height could be accurately estimated from diameter, not only could time and money be saved, but improved estimates of quantities such as individual tree volume or biomass might be obtained using the estimated heights rather than the error-prone measured heights. Regression analysis is one statistical tool used to estimate equations relating height to diameter.

The first step in regression analysis should be to plot the data for the relationship being estimated. A graph of the data will provide some visual evidence of the existence of a relationship, and, in some cases, suggest appropriate equation forms. In Figure 3.7, it can be seen that height generally increases as diameter increases. Given the trend shown in Figure 3.7, it may be appropriate, and useful, to fit an equation to predict height from diameter.

A simple linear regression equation is an equation that takes the form: $Y_i = b_0 + b_1 \cdot X_i + e_i$, where Y_i is an observed value of the dependent variable (i.e., the variable you want to predict), X_i is the independent variable (i.e., the variable that you can easily obtain), b_0 and b_1 are the regression parameters to be estimated, and e_i is the residual error. The equation is "linear" because all of the estimated parameters are linear combinations of independent variables, and the equation is "simple" because there is only one independent variable. The predicted value of Y_i is $\hat{Y}_i = b_0 + b_1 X_i$. Regression analysis estimates values for b_0 and b_1 that minimizes the sum of the errors: $e_i = Y_i - \hat{Y}_i$. The typical approach utilized in regression is often referred to the method of least squares and estimates the parameters such that the sum of the squared errors are minimized:

$$Q = \sum_{i=1}^{n} e_i^2 = \sum_{i=1}^{n} \left(Y_i - \hat{Y}_i \right)^2 = \sum_{i=1}^{n} \left(Y_i - b_0 - b_1 X_i \right)^2 \tag{3.23}$$

The solution that minimizes eq. (3.23) is obtained by taking the partial derivatives of Q with respect to each parameter, equating them to zero and solving for the parameters:

$$\partial Q / \partial b_0 = \sum_{i=1}^{n} -2\left(Y_i - b_0 - b_1 X_i \right) = 2\left(nb_0 + b_1 \sum_{i=1}^{n} X_i - \sum_{i=1}^{n} Y_i \right) = 0$$

$$\partial Q / \partial b_1 = \sum_{i=1}^{n} -2 X_i \left(Y_i - b_0 - b_1 X_i \right) = \sum_{i=1}^{n} -2\left(X_i Y_i - b_0 X_i - b_1 X_i^2 \right) = 0$$

The two partial derivatives represent a set of two equations with two unknowns that can be solved using linear algebra. Solving the first partial derivative for b_0 yields

$$b_0 = \frac{\sum_{i=1}^{n} Y_i}{n} - b_1 \frac{\sum_{i=1}^{n} X_i}{n} = \bar{Y} - b_1 \bar{X} \tag{3.24}$$

Thus, the parameter b_0, which is referred to as the y-intercept, is set such that the regression line passes through the pair of means $\{\bar{X}, \bar{Y}\}$. Substituting eq. (3.24) into the second partial derivative and solving for b_1 yields

---

*The example developed here is to introduce the concepts and processes of simple regression analysis, for a complete understanding of height-diameter relationships, their use and limitations, and the equations forms typically used to estimate them please see Section 5.6.

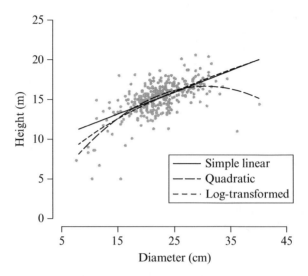

FIG. 3.8. Western hemlock height–diameter data with simple linear regression line.

$$b_1 = \frac{\sum_{i=1}^{n}(X_iY_i - X_i\bar{Y})}{\sum_{i=1}^{n}(X_i^2 - X_i\bar{X})} = \frac{\sum_{i=1}^{n}(X_i - \bar{X})(Y_i - \bar{Y})}{\sum_{i=1}^{n}(X_i - \bar{X})^2} = \frac{\text{Covar}(X,Y)}{\text{Var}(X)} \qquad (3.25)$$

Thus, the parameter b_1, which is referred to as the slope, is the ratio of the covariance between X and Y and the variance of X. For the western hemlock height–diameter data (Fig. 3.7), the slope would be estimated as

$$b_1 = \frac{\text{Covar}(X,Y)}{\text{Var}(X)} = \frac{1995.297}{7271.027} = 0.2744$$

and the y-intercept would be estimated as

$$b_o = \bar{Y} - b_1\bar{X} = 15.1258 - 0.2744 \cdot 22.1381 = 9.0511$$

Thus, the simple linear regression equation estimated for the height diameter data is $\hat{Y}_i = 9.0511 + 0.2744 \cdot X_i$. The predicted heights from this equation are shown as the solid black line superimposed on the data in Figure 3.8. The prediction from a regression equation is an estimate of a running mean. It estimates the mean of Y for any level of X, and the estimated slope, b_1, is an estimate of how the mean of Y changes given a change in X.

3.8.3. Goodness-of-Fit and Regression Diagnostics

Since one of the primary purposes of regression analysis in forest mensuration is to estimate quantities that are difficult or costly to measure, it is important to evaluate how well a regression equation fits the data. There are many goodness-of-fit criteria that measure how well a model fits (or "explains") the data. These criteria are used to not only to assess the

TABLE 3.2. Parameter estimates, standard errors, t statistics, and associated p-values for the simple linear regression, height $= b_0 + b1 \cdot$ diameter, based on the data shown in Fig. 3.7

Parameter	Estimate	Standard error	t-Value	p-Value
b_0	9.0507	0.4397	20.58	<0.0001
b_1	0.2744	0.0194	14.11	<0.0001

goodness of fit of a given model, but also to compare fits across different model formulations and to aid in final model selection. The diagnostic criteria used in evaluating regression analyses fall into three broad categories: (1) significance testing, (2) evaluation of assumptions, and (3) goodness of fit.

Significance testing is concerned with evaluating whether the regression equation is a better statistical model of the independent variable (Y) than its sample mean. Analysis of variance is often used to test the overall significance (Freese, 1974; Kutner et al., 2004). The individual parameter estimates are also tested for significance (i.e., significantly different from zero or another hypothesized value) using t-tests. Most statistical packages that perform regression analysis provide output tables for both the overall regression and the individual parameters. A typical output table for individual parameters is shown in Table 3.2 for the height–diameter equation fitted above. The t values shown in Table 3.2 were constructed under the assumption of a null hypothesis that the parameter estimates were zero. The p-values are the probability that a t value as large as was observed would occur if the null hypothesis was true. If the p-value is greater than some preset critical level (usually $p = 0.05$), then we conclude that the parameter is not significant. In Table 3.2, it can be seen that both parameters have p values <0.0001, indicating that the estimates are significantly different from zero. Generally, variables with parameter estimates not significantly different from zero are dropped from regression equations.

The method of least-squares regression provides the best, unbiased estimates of the slope and intercept terms if the following assumptions are met: (1) the independent variable (X) is known and measured without error; (2) the residual errors, e_is, are independent and normally distributed; and (3) the residual errors are homogeneously distributed across the range of independent variables. Most of the material in this text is dedicated to providing measurement methods and sampling methods that aim to insure that assumption 1 is met. Good sample designs are important for the second assumption as well. Adequate model forms and appropriate data transformations (see Section 3.8.4) both influence the last two assumptions. While formal statistical tests can be applied to testing assumptions 2 and 3, a visual assessment is generally adequate. Residuals may be graphed as a histogram (Fig. 3.9a) or as quantile plots (Fig. 3.9b) to examine normality. Like the normal distribution (Fig. 3.3), the histogram of residuals should be symmetric about zero with the tails (ends) of the distribution declining rapidly. In Figure 3.9a, it is observed that there tends to be more large negative residuals than positive ones. The slight left skewness (overprediction) in the residuals may indicate that the model does not adequately fit the data or that a transformation of the data may be necessary. Similarly, if the residuals were normally distributed, they would fall along the solid line in Figure 3.9b; however, the distribution of the residuals along the quantile plot show that there are more larger negative residuals than expected and fewer larger positive ones than expected, again indicating a left skewness to the residuals, suggesting potential problems with the current regression model. The third

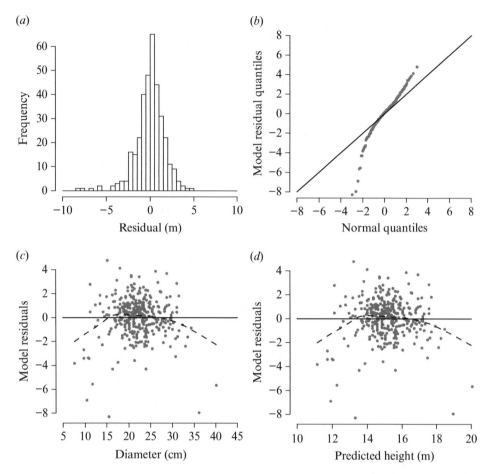

FIG. 3.9. Residual diagnostic graphs for assessing linear regression assumptions: (*a*) histogram of residuals, (*b*) normal quantile plot of residuals, (*c*) residuals versus independent variables, and (*d*) residuals versus predicted dependent variable. (The dashed lines in c and d are smoothed local regression trend lines to highlight patterns.)

assumption, homogeneity of residuals, is often examined by graphing residuals versus the independent variables (Fig. 3.9*c*) and the residuals versus the predicted values (Fig. 3.9*d*). Ideally, the residual values would be equally distributed on either side of zero (solid black line), uniformly across the ranges of all of the independent variables and the predicted values. A smoothed trend line, such as a locally weighted regression (lowess) line (Cleveland, 1994), is often used to highlight trends in residual distributions (the dashed lines in Fig. 3.9*c* and *d*). The dashed lines should be close to the solid line if homogeneity exists. In this case, there are substantial curvilinear trends indicating potential problems with the current regression model.

There are several goodness-of-fit criteria frequently used in assessing the overall fit of regression equations to data. The focus here will be on two of the simpler and commonly used criteria: coefficient of determination and root-mean-squared error (rMSE). The coefficient of determination was defined above as the square of the correlation coefficient and in simple linear regression is often denoted as r^2. For multiple regression and other forms of regression, R^2 is often used. The coefficient of determination describes the amount (percentage) of variation in the independent variable that is accounted for by the current regression equation. The coefficient of determination is calculated using

$$r^2 = \left(\frac{\Sigma(X - \bar{X})(Y - \bar{Y})}{\sqrt{\mathrm{Var}(X) \cdot \mathrm{Var}(Y)}} \right)^2 = 1 - \frac{\Sigma(Y - \hat{Y})^2}{\mathrm{Var}(Y)} \tag{3.26}$$

As noted above, the correlation coefficient, r, was 0.60 and the coefficient of determination, r^2, was 0.36. Higher values for the coefficient of determination indicate more of the variation in Y is explained and generally indicates better overall fits to the data; however, it should be noted that r^2 will increase as more independent variables are added to the regression equation even if the parameters estimates associated with the independent variable are not significant. The coefficient of determination should be viewed as a general indicator of goodness of fit.

The rMSE, also called the residual standard error, is the square root of the mean squared residual (i.e., the rMSE is the quadratic mean residual):

$$\mathrm{rMSE} = \sqrt{\frac{\sum_{i=1}^{n} e_i^2}{n - p}} = \sqrt{\frac{\sum_{i=1}^{n} (Y_i - \hat{Y})^2}{n - p}} \tag{3.27}$$

where p is the number of parameters estimated in the regression equation. In the simple linear regression example, for height–diameter $p = 2$ and the rMSE for the fitted equation is

$$\mathrm{rMSE} = \sqrt{\frac{\Sigma e^2}{n - p}} = \sqrt{\frac{954.74}{349 - 2}} = \sqrt{2.75} = 1.66$$

The rMSE is a measure of the average prediction error associated with a regression equation. It is analogous to the standard error for the mean (eq. (3.14)). rMSE is in the same units as the dependent variable. In the case of the height–diameter equation, height was measured in meters, so an rMSE of 1.66 indicates that the simple linear regression estimates height with an average error of 1.66 m.

Like r^2, rMSE is a general measure of goodness of fit, where smaller values indicate better fits. rMSE and r^2 can be useful measures for comparing goodness of fits among a set of regression equations, but must be used in conjunction with the other diagnostic criteria mentioned here. When used alone, these criteria can be misleading. Comparisons among equations are only valid when the same dependent variable is used; if transformations of the dependent variable are used (see below), then other measures of goodness of fit must be utilized.

3.8.4. Multiple Regression and Transformations

Linear regression with more than one independent variable is referred to as multiple regression. The general form of a multiple regression is $Y = b_0 + b_1 X_1 + b_2 X_2 + \cdots + b_k X_k$, and can be written in matrix notation as $[Y] = [X][B]$. The parameter estimates are the simultaneous solution to the set of linear equations. Because of the interrelatedness of many measurements

TABLE 3.3. Model formulation, parameter estimates, and goodness-of-fit criteria for various models applied to the height–diameter data shown in Fig. 3.7 (estimated lines are shown in Fig. 3.8)

| | Parameter Estimates | | | | | | |
Model	b_0	b_1	b_2	R^2	rMSE	Geometric Mean (dY)	Furnival's Index
$H = b_0 + b_1 D$	9.0507	0.2744	—	0.36	1.66	1.00	1.66
$H = b_0 + b_1 D + b_2 D^2$	0.9824	1.0271	−0.01682	0.45	1.54	1.00	1.54
$\ln(H) = b_0 + b_1 \ln(D)$	1.2518	0.4726	—	0.41	0.12	0.067	1.88

made on individual trees and plots, multiple regressions are common in forest mensuration. The quadratic equation, $Y = b_0 + b_1 X + b_2 X^2$, is a form of multiple regression frequently applied in forest mensuration. The quadratic equation fitted to the height–diameter data yields $H = 0.9824 + 1.0271 \cdot D - 0.01682 \cdot D^2$ (Table 3.3). The line associated with this equation is shown in Figure 3.8 as the long dashed line. The R^2 and rMSE both indicate that the quadratic equation fits the height–diameter data better than the simple linear equation (Table 3.3).

Linear regression analysis requires that equations be of the form: $Y = b_0 + b_1 X$. Curvilinear equations can be modeled using linear regression if an appropriate transformation can be applied. For example, power functions such as $Y = b_0 X^{b_1}$ can be linearized using a logarithmic transformation: $\ln(Y) = \ln(b_0) + b_1 \ln(X)$; similarly, exponential equations such as $Y = b_0 e^{b_1 X}$ can be linearized using logarithmic transformations: $\ln(Y) = \ln(b_0) + b_1 X$. The transformed variables are fitted using the same formulas described above. For example, consider the logarithmic transformed power function applied to the height–diameter data in Figure 3.7. The estimate for the slope (b_1) parameter would be

$$b_1 = \frac{\text{Covar}\left(\ln(X), \ln(Y)\right)}{\text{Var}\left(\ln(X)\right)} = \frac{8.1152}{17.1706} = 0.4726$$

and the intercept (b_0) would be

$$b_0 = \overline{\ln(Y)} - b_1 \cdot \overline{\ln(X)}.$$

The equation predicts $\ln(H)$, to obtain a prediction of H, the regression equation has to be back transformed: $H = \exp(b_0 + b_1 \ln(dbh))$. The resulting curve for the back-transformed equation is shown in Figure 3.9 as the short dashed line.

The r^2 for this regression was 0.41 and the rMSE was 0.12 (Table 3.3). Because $\ln(H)$ was fitted rather than H, the r^2 and rMSE are not directly comparable with the other two equations shown in Table 3.3. Both of these goodness-of-fit criteria can be used to compare different equations that have the same transformation of the dependent variable (including no transformation), but are not comparable across different transformations (Furnival, 1961). Furnival (1961) proposed a simple index of fit for comparing equations with different transformations of the dependent variable:

$$\text{FI} = \frac{\text{rMSE}}{\text{GM}\left(d(f(Y))/dY\right)} \tag{3.28}$$

where rMSE is the root mean square error from the fitted equation (eq. (3.27)); $f(Y)$ is the transformation function on the dependent variable; $d(f(Y))/dY$ is the first derivative of the transformation with respect to Y; and GM() denotes the geometric mean (eq. (3.7)). When the dependent variable is not transformed $f(Y) = Y$ and $dY/dY = 1$ and Furnival's index is simply the rMSE. For the logarithmic transformation, $f(Y) = \ln(Y)$ and $d(\ln(Y))/dY = 1/Y$ and Furnival's index is the rMSE from the logarithmic fit divided by the geometric mean of the inverse of Y. The Furnival's indices for each of the three equations we have fitted to the height–diameter data are shown in Table 3.3. Based on Furnival's index, the quadratic equation would be the best fit of the three equations we have used. Other transformations, such as inverse, square root, and power transformations of the form Y^k, are frequently encountered.

3.8.5. Advanced Topics in Regression Analysis

Regression analysis plays a pivotal role in forest mensuration. Many of the equations presented in this text contain parameters that are estimated via regression analysis. The examples presented in this chapter are relatively simple and were used to introduce the concept of linear regression. Many of the equations used in forest mensuration are complex and require more advanced regression approaches such as nonlinear regression and weighted regression. Coverage of these topics is beyond the scope of an introductory chapter on basic statistics. For a more comprehensive treatment of linear regression, the reader is referred to texts such as Freese (1974), Zar (2009), and Kutner et al. (2004), and for nonlinear regression see Motulsky and Christopoulos (2004). Recently, there has been an increase in the application of linear and nonlinear *mixed effects models* to forest mensuration. A mixed-effects model is a statistical model that contains fixed effects and random effects (Pinheiro and Bates, 2000). *Fixed effects* are factors that are controlled in the experimental or measurement process. For example, in the height–diameter example presented above, diameter is a fixed effect. *Random effects* are factors that are not directly controlled or that arise from the hierarchical nature of the sample design. For example, one of the assumptions of regression analysis discussed above was independence of the residual errors, e_i. Correlations among subsets of residual errors often arise because trees are measured in groups on plots and plots are often remeasured over time. Ignoring these hierarchical random effects results in estimates of residual errors that are too large. Residual errors that are too large can result in incorrect regression diagnostics and model misspecification. For more information on mixed effects models and their applications, see Pinheiro and Bates (2000) and Zuur et al. (2009).

3.9. USE OF COVARIATES TO IMPROVE ESTIMATION

Section 3.8 illustrated how regression equations can be developed to estimate attributes that are relatively difficult to measure. The use of covariates in estimation can be applied in a sampling context as well. Two approaches commonly used are *ratio estimation* and *regression estimation*. Details of the sample design and analysis for these techniques are presented in Section 10.8. Here, we focus on an example to illustrate how covariates can be used to improve inventory estimates using these two approaches.

TABLE 3.4. Data on root weight and crown cover of black cohosh, drawn from Chamberlain et al. (2013)

		Crown Cover (Fraction)	Dry Root Weight (g/m^2)
Full dataset ($n = 34$)	Mean	0.71	297
	Standard deviation	0.58	271
	Standard error	0.10	47
Subsample data ($m = 8$)	Plot 1	0.53	219
	Plot 2	0.68	277
	Plot 3	1.57	929
	Plot 4	0.46	76
	Plot 5	0.82	398
	Plot 6	1.07	307
	Plot 7	0.74	389
	Plot 8	0.16	71
	Mean	0.75	333
	Standard deviation	0.43	271
	Standard error	0.15	96

Black cohosh is a forest herb native to eastern North American. Its roots are harvested for medicinal uses. Increased harvest levels of black cohosh, and other forest herbs, have resulted in a need to assess long-term sustainable harvest levels of nontimber forest products (Chamberlain et al., 2002). Because the roots are the part of black cohosh utilized, estimates of belowground biomass are needed to determine sustainable levels (Chamberlain et al., 2013). Direct assessment of black cohosh root biomass requires extracting, drying, and weighing roots. Not only is the process destructive, it is time consuming and expensive. On an operational basis, only a very small number of sample plots could be assessed, leading to estimates with potentially large sampling errors. Chamberlain et al. (2013) found a close relationship between crown area and root biomass in black cohosh. Crown area is a nondestructive measure and is relatively easy to make in the field. A sampling scheme where many plots are measured for crown area and a few plots are measured for both crown area and root biomass could lead to estimates with lower errors.

At one of their study sites, Chamberlain et al. (2013) used destructive sampling to obtain the dry weight of black cohosh roots within 34 plots of 1 m^2 each. They also assessed crown area on the same plots. To illustrate the use of covariates, we suppose that crown area had been obtained on all 34 plots, but root weight was only available from a subsample of 8 plots. The details of the example data are presented in Table 3.4. If we were to estimate root weight only from those plots with data available, our estimate would be quite poor.

3.9.1. Ratio Estimation

One approach to using crown cover as a covariate would be in double sampling with a ratio estimator (see Section 10.8.2). The sample of 34 plots with crown cover data constitutes a first-phase sample, while the subsample that has both crown cover and root weight constitutes the second-phase sample. We use the second-phase sample to estimate the overall ratio between root weight and crown cover, then multiply that ratio by the better estimate

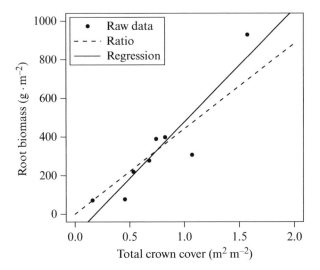

FIG. 3.10. Second-phase sample from the black cohosh data of Chamberlain et al. (2013), along with the corresponding ratio and regression relationships fitted to the data.

of mean crown cover from the larger first-phase sample. Specifically, let w_i be the value of root weight on plot i, let c_i be the corresponding value of crown cover, and let $\bar{C}$ be the average crown cover obtained from the large sample. Then, our estimate of r, the ratio of root biomass:crown cover is

$$ r = \frac{\sum\limits_{i=1}^{8} w_i}{\sum\limits_{i=1}^{8} c_i} = 442.2 $$

and our estimate of root weight per unit area, W, is

$$ \hat{W}_{\text{ratio}} = r\bar{C} = 442.2 \times 0.71 = 315\,\text{g}\,/\,\text{m}^2 $$

The raw data from the second phase sample, along with the calculated ratio relationship, are shown in Figure 3.10. By applying the equations in Section 10.8.2, we can also calculate the standard error of this estimate as $60\,\text{g/m}^2$, which is considerably smaller than the standard error of 96 we would have had if we were forced to use only the measurements of root weight without any covariates.

3.9.2. Regression Estimation

An alternative approach would be to use double sampling with a regression estimator. As before, the 34 plots with crown cover data represent a first-phase sample, while the 8 plots with both crown cover and root weight represent the second-phase sample. In regression estimation, rather than assuming a simple ratio relationship between x and y, we fit a linear regression with an intercept. The slope of that regression is then used to adjust mean $\bar{w}$ of the small second-phase sample, for the difference between the mean $\bar{c}$ of the second-phase sample and that of the first-phase sample (again denoted as $\bar{C}$, which should provide a much better estimate). For the black cohosh data, the regression line fit to the second-phase sample is

$$w = -109 + 586c$$

and the corresponding estimate of the root weight per unit area is

$$\hat{W}_{\text{regression}} = \bar{w} + b(\bar{C} - \bar{c}) = 333 + 586(0.71 - 0.75) = 309 \, \text{g} / \text{m}^2$$

By following the equations in Section 10.8.1, we can calculate the standard error of this estimate as 57 g/cm², again considerably smaller than what we obtained using the w values from the second-phase sample alone, and also slightly smaller than that of the ratio estimator.

In the original data, where all 34 plots were destructively harvested, the standard error was 47. In the smaller sample of 8 plots, the standard error increased to 97. By using canopy cover as a covariate and the strength of the 34 observations, we reduced the error from 97 to 60 for ratio estimation and to 57 for regression estimation, and substantially decreased the field work required. The use of covariates is a powerful sampling strategy with a wide range of applications. In this case, our covariate was a measured variable. This does not always need to be the case and less precise ocular estimates can be effectively used.

4

LAND AREA DETERMINATION IN FOREST MENSURATION

Generally speaking, foresters do not conduct many original property surveys; however, they often retrace old lines, locate boundaries, and run cruises lines and transects. In fact, much forest mensurational work requires the location, delineation, preparation of maps, and the determination of the magnitude of land areas and information on their characteristics and the organisms found thereon. Of prime importance are forestlands and their characteristics: topography, soils, water, forest conditions, and other vegetation and wildlife. The methods and equipment to acquire and use this information go well beyond the field of forest mensuration and involve many other disciplines that cannot be covered in depth here. Nevertheless, their existence and usefulness require brief mention.

For many years, foresters have employed simple land surveying methods in fieldwork and have utilized aerial photographs, satellite imagery, and, more recently, Light Detection and Ranging (LiDAR) technologies. Added to these techniques are the increasing use of Geographic Information Systems (GIS) and Global Positioning Systems (GPS).

GIS and GPS technologies have streamlined much of the design and implementation of forest inventories. Sample locations can be easily generated in GIS and downloaded to handheld GPS devices. Affordable handheld GPS devices are now available with 3–5 m (10–15 ft) accuracy. Permanent sample locations can be marked in the field using GPS devices and uploaded to GIS facilitating map preparation and relocation of samples for future measurements. While electronic technology has streamlined much of the field logistics for forest mensurationists, much historical data still remain in paper form. Old maps and tables of distances and bearings, showing locations of treatment units and permanent sample locations, are still common. Much of the field work of a forester requires the determination of sample plot boundaries, layout of small treatment units, and perhaps even mapping the locations of individual trees in special research studies. The accuracy required

Forest Mensuration, Fifth Edition. John A. Kershaw, Jr., Mark J. Ducey, Thomas W. Beers and Bertram Husch.
© 2017 John Wiley & Sons, Ltd. Published 2017 by John Wiley & Sons, Ltd.

for these activities is often greater than what can be attained using basic handheld GPS devices so that more expensive map quality GPS devices or specialized survey equipment would be required. Conversely, for some small projects the overhead costs of GIS and GPS are not warranted, so the ability to do work rapidly and effectively using traditional approaches is still needed. In some situations, particularly in the developing world or elsewhere when resources are limiting, the infrastructure to support GIS and GPS applications remains unaffordable. Thus, foresters often use simpler hand tools (compasses and tapes) to quickly measure distances and directions at the accuracies needed for small-scale work. Because of the continued reliance on these tools, simple distance measurement and direction determination by magnetic compass is covered since, in spite of being approximate procedures, they are still of great use in field work, especially in forest inventories. Although a text on forest mensuration cannot go into detail on aerial photographic interpretation, GIS, or GPS, mention must be made of their fundamentals and usefulness.

4.1. LAND DISTANCE AND AREA UNITS

Measurement of land distances and areas is a commonplace requirement in forest mensuration. Under the International System of Units (SI), the most important units are the meter and kilometer. Under the imperial system, lengths are expressed usually in feet, yards, chains, or miles. However, older linear units are still often used or encountered in documents and deeds. The linear unit of a *chain* (often called "Gunter's chain") is still used in much forest mensuration work and in land surveying. A chain is 66 ft long and is divided into 100 links, each 0.66 ft or 7.92 in. long. For rough work, the approximation 1 chain = 20 m is often adequate when translating between SI and imperial units. Older length units that have fallen into disuse in land surveying (but may be found in older deeds) are poles, rods, and perches: 1 pole = 1 rod = 1 perch = 25 links = 16.5 ft. Four rods equal one chain. A length of 10 chains is a furlong.

Of primary importance in forest mensuration is the expression of land area. In the imperial system, the most used expression of land area is the acre or fractions of it. An acre consists of 43,560 ft². This seemingly odd dimension is based on the linear unit of a chain as described above. One acre equals 10 square chains (10 chain × 1 chain or 660 ft × 66 ft = 43,560 ft²). A mile is 5280 ft or 80 chains (80 × 66 = 5280). One square mile equals 640 acres (5,280²/43,560 = 640 or in chains 80²/10 = 640).

In the SI, the most commonly used expression of land area is the hectare. The hectare is an area of 10,000 m² or 0.01 km². Thus, a land area of 100 × 100 m or 0.1 × 0.1 km is 1 ha. One hectare is approximately equivalent to 2.47 acres. Conversion factors for area between the imperial system and SI are given in Table A.1B.

4.2. MEASURING DISTANCES

In land surveying, the distance between two points is commonly meant to be the horizontal distance. Although slope distances are sometimes measured in the field, these distances are subsequently reduced to their horizontal equivalents since horizontal distances are the ones used in the preparation of maps and in the computation of areas. The method used in obtaining the distance between two points is largely determined by the accuracy required.

Traditional methods include pacing and chaining. Optical rangefinders and electronic distance measurement devices have been in use since the 1960s; continued improvements in accuracy and range, along with reductions in costs, have made these instruments more widely used. The accessibility and improved accuracy of global positioning systems have revolutionized measurement of distances and areas in forest inventory work. GPS is described in Section 4.8, and the other methods of distance measurement are described below.

4.2.1. Pacing

Where approximate results are sufficient, as in making tree-height measurements with certain hypsometers, in reconnaissance work, and in some types of cruising, distance can be obtained by *pacing*. Although many people think of a pace as a single step, the forester, and others who work in natural resources, often defines a pace as a double step (i.e., the distance between two steps by the same foot, be it right or left). Thus, as used here, a pace is two steps.

To graduate the pace, a recommended procedure is to establish a measured line of a convenient length such as 20 chains (1320 ft) or 400 m in terrain of the type to be encountered in fieldwork. Stakes should be set on the line at given distances: for example, in chains, at 0, 5, 10, 15, and 20 chains: in meters, at 0, 100, 200, 300, and 400 m. One should walk as naturally as possible, a minimum of four times over the line, record the number of paces between each pair of stakes, and compute the average number of paces between each pair of stakes and for each trip. From this, one can get a good picture of the consistency of the pacing and can determine the number of paces for the terrain. The pace should be graduated to meet the varying conditions encountered: wooded slopes of 0–10%, wooded slopes of 10–20%, open level woods, and so forth. On slopes over 30%, in swamps, and in logged areas with slash on the ground, it is very difficult to pace accurately.

On steep slopes, the method of *staff pacing* can be used. To staff pace, one uses a 4.125-ft or 1.25-m staff. This gives 16 staffs to a chain, or 16 staffs to 20 m. The staff is used as follows for traveling uphill: while holding the staff horizontal, the ground position of the rear end of the staff is located by plumbing by eye and the forward end of the staff by contact with the ground. For traveling downhill: while holding the staff horizontal, the ground position of the rear end of the staff is in contact with the ground and the forward end is located by plumbing by eye.

With foot or staff pacing, an experienced pacer should consistently attain an accuracy of 1/100 or better. Pacing skills need to be regularly practiced and periodically checked. While the availability of affordable electronic distance measuring devices and GPS has resulted in pacing falling into disuse, these devices are still subject to operational error and pacing can be used as an important field check to identify and correct these errors.

4.2.2. Distances with Chains and Tapes

The types of tapes commonly used by natural resource managers in the United States to measure horizontal distances are the 100-ft *steel tape* and the *steel topographic trailer tape*. For measuring distances in meters, 30- and 50-m tapes are convenient to use. Tapes are usually constructed from steel, nylon clad steel, fiberglass cloth, or plastic.

The basic procedure for measuring distance with either chains or tapes requires two people, traditionally referred to as the *head chain person* and the *rear chain person*. On level terrain, the chain (or tape) can be stretched directly on the ground. The starting point is marked with a pin and the head chain person pulls the zero end of the tape forward following the desired compass bearing. The rear chain person checks direction and warns the head chain person when the end of the tape is approaching by shouting "chain." The head chain person pulls the tape taut until the rear chain person shouts "stick." The head chain person then marks the zero position with another pin and shouts "stuck." The rear chain person picks up the pin and follows the chain to the next pin and the head chain person pulls the chain forward to the next point. The procedure is repeated until the desired distance has been measured. In rough terrain, the tape is held off the ground and plumb bobs are used to determine accurate pin placement. On very steep terrain, horizontal distances are obtained by either "breaking chain" or using a topographic trailer tape. Breaking chain is simply using shorter segments of the chain to hold a level line.

The topographic trailer tape is graduated in chains and links (Section 4.1). There are three tabs on a trailer tape: one at 0 links, one at 100 links (one chain), and one at 200 links (two chains). Beyond the two-chain tab, there is about one-half chain of tape that trails the body of the tape. The trailer is used to convert slope distance to horizontal distance. The topographic trailer tape is used with a clinometer (traditionally an Abney level that has a topographic (topo) arc). The *topographic arc* is graduated in an angular unit that represents 1 unit vertically to 66 units horizontally. For example, a topo reading of +17 indicates a vertical rise of 17 ft per 66 ft, or 17 ft per chain.

Since a slope distance is greater than its horizontal equivalent, if one measures two chains along a slope, a correction must be added to obtain two chains of horizontal distance (Fig. 4.1). The trailer on the topographic trailer tape carries corrections for converting a slope distance of two chains to two chains of horizontal distance. These corrections are on the top of the tape. Corrections for converting a slope distance of one chain to one chain of horizontal distance are on the underside of the tape beyond the one-chain tab. The corrections are applied as follows. For example, if a topo reading of 15 is obtained for a slope distance of two chains, and the trailer is let out to the 15th graduation, the slope distance becomes the hypotenuse of a right triangle whose horizontal leg is two chains. If a topo reading of 20 is obtained for a slope distance of one chain, the proper correction is applied when the tape is let out to the 20th graduation mark on the underside of the tape beyond the one-chain tab.

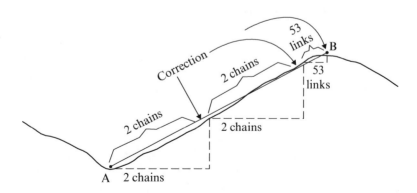

FIG. 4.1. Applying slope corrections.

When it is necessary to measure slope distances less than one chain or between one and two chains, it is easy to calculate the corrections in the field using

$$\text{Correction per chain} = \sqrt{1.0 + \left(\frac{\text{Topo reading}}{66}\right)^2} - 1.0 \qquad (4.1)$$

For example, for a topo reading of 33, the correction per chain of slope distance is $\sqrt{1 + (33/66)^2} - 1.0 = \sqrt{1.25} - 1.0 = 1.118 - 1.0 = 0.118$ chains (11.8 links). For other slope distances on the same grade, the correction equals the slope distance times 0.118. For a slope distance of 0.73 chains, the correction will be $0.118 \cdot 0.73 = 0.086$ chain (8.6 links). When $0.73 + 0.086 = 0.816$ chain (81.6 links) is measured along the slope, a horizontal distanced of 0.73 chains will be obtained. For a distance of 1.50 chains, the correction would be $0.118 \cdot 1.50 = 0.177$ chains (17.7 links). Careful chaining on horizontal distances by experienced foresters should result in accuracies of $\frac{1}{1000}$ to $\frac{1}{2500}$.

GPS and other electronic distance measurement devices have largely replaced the topographic trailer tape. The accuracy, reliability, and speed of these devices have made measurement of distances, especially in rough terrain, much easier and faster. Costs of many of these devices are now in the range of the cost of a topographic trailer tape and are substantially easier to use and lighter and more compact. Most electronic distance measurement devices have slope correction algorithms built into their onboard software; however, it is important for users to know whether slope distances or horizontal distances are being calculated when using these devices.

4.2.3. Optical Rangefinders

Optical rangefinders work on the same principle as focusing a camera. A split image or double image is created in the viewfinder using mirrors or prisms. A focusing knob is used to make the two images coincident. Distance is read from either a vernier scale on the focusing knob or calculated electronically. There are two types of optical rangefinders: fixed-based and fixed angle. In a *fixed-base rangefinder*, the distance between the mirrors is fixed and the angle is manipulated to bring the images coincident. In a *fixed-angle rangefinder*, the angle is fixed and the distance between mirrors is manipulated to bring the two images coincident. Optical rangefinders are compact, lightweight, and inexpensive. Models with a range of about 10–75 m (30–200 ft) are available for less than US$50. Models with a range of 50–1000 m (150–2000 ft) cost around US$150. Optical rangefinders must be periodically calibrated and offer an accuracy of $\frac{1}{100}$.

4.2.4. Electronic Distance Measurement Devices

A number of *laser-based rangefinders* and distance measurement devices are currently available. These devices measure distance by measuring the flight time of short pulses of infrared light. Using the speed of light, the distance to an object is estimated by measuring the time it takes a laser pulse to travel to the target and back to the receiver. Most laser rangefinders operate with or without reflectors. While many of the newer models have effective ranges of 400–800 m in open conditions, performance in forested conditions is often substantially lower. Without reflectors, the effective distance of most devices in forest conditions is about 20 m (66 ft) (Peet et al., 1997). Understory vegetation limits the effective range of most reflectorless lasers. With a reflector, most devices have an effective range of 50–100 m (150–300 ft) depending on understory vegetation and weather conditions. The use of a foliage filter can improve the range in forested conditions; however, laser-based rangerfinders require unobstructed lines of sight, which greatly limits the effective range.

Accuracies of handheld, industrial-quality, laser-based rangefinders vary from about $\frac{1}{250}$ to $\frac{1}{2000}$ depending upon laser strength, mode (reflectorless vs. reflector), weather conditions, understory vegetation, and the surface roughness of the object being sighted. Accuracies of most devices are improved using reflectors, foliage filters, and/or tripods. Laser-based rangefinders vary in cost from about US$200 to US$2000. Inexpensive recreational units are available for around $100, but the accuracy and durability of these units in forested conditions is questionable. Lasers continue to improve in quality and decrease in price, but one should check the specifications of any unit carefully to ensure it provides the accuracy needed for its intended purpose.

Ultrasonic rangefinders are also used in forestry applications. These devices emit a narrow beam of sound waves that bounce off solid targets and return to the receiver. Using the speed of sound and elapsed time, the distance to the object is estimated. Effective use of most ultrasonic rangefinders in forestry conditions requires the use of a transponder. In this case, the receiver emits a sonic pulse, and, when the transponder detects the pulse, the transponder emits a pulse back to the receiver. Most handheld sonic devices have a maximum range of about 20–30 m (60–100 ft). Sonic devices need to be frequently calibrated, especially if temperatures are fluctuating. Ultrasonic rangefinders vary in cost from about US$200 to US$2000.

Distances also can be measured using handheld GPS units. A reference point can be marked and distances from the reference point tracked in real time or distance between two marked reference points can be calculated. Unless mapping-quality GPS units are being used, most handheld GPSs have accuracies in the 3–5 m range, with 10 m (30 ft) or more likely under forest canopies. Depending upon the application and required accuracy, measurement of distances of 100 m or more can be made with handheld GPSs, but shorter distances, such as plot boundaries and inter-tree distances, cannot be made with accuracy sufficient for most forest mensuration needs.

4.2.5. Maps and Photos

Field distances can be determined using maps, aerial photographs, or other imagery of known scale. Distances along edges of cuts or other identifiable boundaries or between two points are measured on the map or photograph in inches or centimeters. These "map" distances are converted to ground distances using the map scale.

Map scales are generally given as representative fractions. The *representative fraction*, RF, specifies the number of ground units of distance represented by 1 unit of map distance. For example, an RF of 1:15,000 means that 1 in. of map distance represents 15,000 in. on the ground and 1 cm of map distance represents 15,000 cm on the ground. Generally, ground distances are specified in feet or meters rather than inches or centimeters; therefore, representative fractions are often converted to *dimensional equivalents* by dividing the right-hand side of the RF by 12 to obtain ground feet per map inch or by 100 to obtain ground meters per map centimeter. An RF of 1:15,000 would have a dimensional equivalent of 1 in. = 1250 ft and 1 cm = 150 m. Map scales may also be shown graphically in the form of a *bar scale*. The bar scale depicts the map distances for standard ground distances such as 100 ft or 100 m using alternating strips of black and white rectangular bars. Bars scales are extremely useful when maps are reduced or enlarged from their original scales since the change in bar scale will be the same as the change in map scale.

The following example illustrates the conversion of map distance to ground distance. The distance between two points on a 1:15,000 scale photo was determined to be 6.8 in. (17.3 cm). With an RF of 1:15,000, 1 in. = 15,000/12 = 1,250 ft; therefore, 6.8 in. on the photo = 6.8 · 1,250 = 8,500 ft on the ground. In SI units, with an RF of 1:15,000, 1 cm = 15,000/100 = 150 m; therefore, 17.3 cm on the photo = 17.3 · 150 = 2,595 m on the ground.

4.3. MEASURING AREA IN THE FIELD

In situations where recent maps or photographs are not available, it may be necessary to determine areas of cuts or other management blocks using simple closed traverses made with a hand or staff compass and chain or other measuring tape or electronic distance measuring device. Starting at the most reliable corner available, distances and bearings to each point, typically called a *station*, are measured and recorded. *Backsights* and frontsights are generally recorded and stakes driven into the ground at each station. The block boundary is traversed back to the origin. After the traverse is completed, the interior angles at each station are computed. If the bearings were properly read and recorded, the sum of the interior angles should be equal to $180° \cdot (n-2)$, where n is the number of sides in the traverse. All basic surveying textbooks describe methods for adjusting the interior angles for errors (e.g., McCormac et al., 2012). Once the interior angles are checked and adjusted, the traverse is plotted at a convenient scale. If the horizontal distances were correctly measured and recorded, the traverse should "close" (i.e., form a complete loop). The tract area can be determined using one of the methods described below. An introductory surveying textbook, such as Wilson (1989) or McCormac et al. (2012), should be consulted for detailed descriptions of field and analyses methods for closed traversing. GPS technology can also be used to determine areas in the field (Section 4.8.5).

4.4. MEASURING AREA USING MAPS AND PHOTOS

Areas of polygons on maps or aerial photographs can be digitized and determined using techniques provided by GIS (see Section 4.9). Alternatively, areas may be determined using manual methods such as calculation by coordinates, line transects, dot grids, or planimeters. The algorithms used in most GIS are derived from these methods. The primary advantage of using a GIS is that areas of multiple polygons can be calculated rapidly without the potential for human errors, provided the initial digitizing is accurate.

4.4.1. Area by Coordinates

Data from a closed traverse (Section 4.3) are often plotted as part of the process of checking the closure error. The points representing the stations can be converted to X–Y coordinates and the area calculated using the continuous product method. Given the direction and

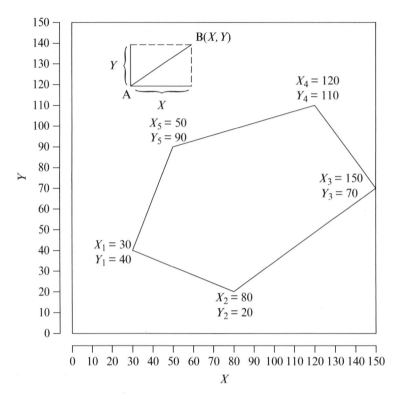

FIG. 4.2. Coordinates for vertices of a polygon.

distance between two points A and B (Fig. 4.2) of each line forming the boundary of an area, the X and Y coordinates at the ends of each line can then be calculated. The north–south vector is Y and the east–west vector is X. The coordinates of all vertices are obtained by addition or subtraction of successive vectors (Fig. 4.2), and the area of the figure, which must be closed, can be calculated by the continuous product method. If the X coordinates are designated as X_1, X_2, X_3, …, X_n and the Y coordinates as Y_1, Y_2, Y_3, …, Y_n, then the area of the polygon whose vertices are (X_1, Y_1), (X_2, Y_2), (X_3, Y_3), …, (X_n, Y_n) is

$$\text{Area} = \frac{1}{2}\left[\left(X_1 Y_2 + X_2 Y_3 + \cdots + X_n Y_1\right) - \left(X_2 Y_1 + X_3 Y_2 + \cdots + X_1 Y_n\right)\right] \qquad (4.2)$$

The vertices should be sorted counterclockwise as shown in Figure 4.2; otherwise, the negative of area will be obtained.

Using the coordinate values given in Figure 4.2, the area of the polygon is

$$\text{Area} = \frac{1}{2}[(30 \cdot 20 + 89 \cdot 70 + 150 \cdot 20 + 120 \cdot 90 + 50 \cdot 40)$$
$$- (80 \cdot 40 + 150 \cdot 20 + 120 \cdot 70 + 50 \cdot 110 + 30 \cdot 90)]$$
$$= \frac{1}{2}[35,500 - 22,800]$$
$$= 6,350 \text{ square units}$$

4.4.2. Area by Dot Grids and Line Transects

If a plane figure is drawn upon rectangular coordinate paper consisting of uniform squares of known area (e.g., 0.01 in.^{2}), the area of the figure can be estimated by counting the number of squares that fall within the boundaries of the figure. Where boundary lines include only portions of squares, one must estimate these portions and add their total to the total number of whole squares within the boundaries. If the figure represents an area drawn to a scale (e.g., a timber type), the area of the figure can be converted to the represented area by computing, for the known scale, the appropriate scale conversion (e.g., acres per square inch or hectares per square centimeter).

If a dot is placed in the center of each square on the rectangular coordinate paper and the lines are removed, a *dot grid* is formed (of course, the lines may be retained, if desired). Each dot now represents an area equal to that of the square. The area of a plane figure can then be estimated by counting the number of dots that fall within the boundaries of the figure and can be converted to the represented area by computing the appropriate scale conversion. Dot grids, generally on transparent sheets, can be prepared or purchased with varying numbers of dots per unit area.

For example, in Figure 4.3, a dot grid has been placed over a photo. There are 64 dots per square inch, so each dot represents 0.016 in.^{2}. In the area outlined, there are 375 dots counted. The area on the map is then $0.016 \cdot 375 = 5.86$ in.^{2}. The photo scale is 1:7920; therefore, 1 in. $= 7920/12 = 660$ ft and 1 in.$^2 = 660^2 = 435{,}600$ ft$^2 = 10$ acres. The ground area is then found to be $5.86 \cdot 10 = 58.6$ acres. For a more accurate estimate of area, the procedure should be repeated several times. Each time the dot grid is randomly placed over the photo and the number of dots counted. The average number of dots is then determined and the areas calculated.

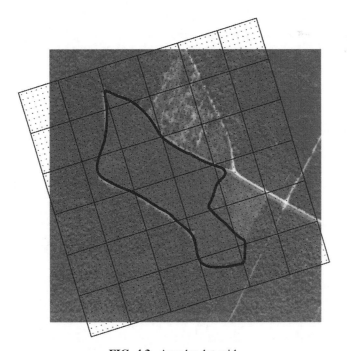

FIG. 4.3. Area by dot grid.

In addition to being used to obtain the area of individual figures, dot grids may be used to obtain area ratios. For example, a dot grid can be placed over a map or aerial photograph on which numerous forest types are outlined and the number of dots within each type counted. Then, the ratio of the number of dots in each type to the total number of dots on the map or photograph can be computed. The total area of the map or photograph can then be multiplied by the ratios to obtain the type areas.

An analogous procedure for determining the area of plane figures is the *line-transect method*, which also can be used to determine area ratios. For example, equally spaced lines are drawn on a map or an aerial photograph on which numerous forest types are outlined, and the lengths of lines within each type measured. Then, the ratios of the lengths in each type to total line length on the map or aerial photograph can be computed. These ratios can be used in the same manner as ratios computed by use of the dot grids.

4.4.3. Area by Planimeters

The *planimeter*, invented in 1854 by Jacob Amsler, is a mechanical device for measuring the area of plane figures. With this instrument, areas can be obtained on maps, photographs, drawings, and diagrams. Over the years, Amsler's polar planimeter has been greatly improved, and other forms, such as the rolling disc planimeter, have been developed. In addition, computing planimeters with built-in calculators that can be programmed for any scale ratio are available.

The traditional polar planimeter consists of a *pole arm*, a *tracing arm*, and a *carriage*. The carriage furnishes bearings for a vertical measuring wheel that revolves on a horizontal axis. When set in position, the instrument rests on three points, a fixed point at the end of the pole arm, the measuring wheel, and the tracing point. As the tracing point is moved about, the instrument pivots about the fixed point and the wheel revolves. To determine an area, one simply moves the tracing point once around the boundary of a figure and reads the resulting movement of the measuring wheel that is graduated for this purpose. In effect,

$$\text{Area} = 2\pi r (\text{TB}) n \tag{4.3}$$

where r = radius of measuring wheel W

   TB = length of planimeter tracing arm

   n = algebraic sum of rotations of measuring wheel

(Note that $2\pi r$(TB) is a constant for a given planimeter. Thus, the measuring wheel is graduated in terms of n times this constant.)

Modern digital planimeters work on a similar concept. The planimeter shown in Figure 4.4 consists of a roller arm, tracing arm, integrating wheel, and encoder. The tracing point is moved once around the boundary to obtain area. The planimeter pivots on the roller arm as the tracing point is moved around the boundary.

4.5. DETERMINATION OF PHOTO SCALE

Photographs may be taken from the air or the ground. If an aerial photograph is exposed with the camera axis vertical, or nearly vertical, it is called a *vertical photograph*. If an aerial photograph is exposed with the camera axis intentionally tilted, it is called an *oblique*

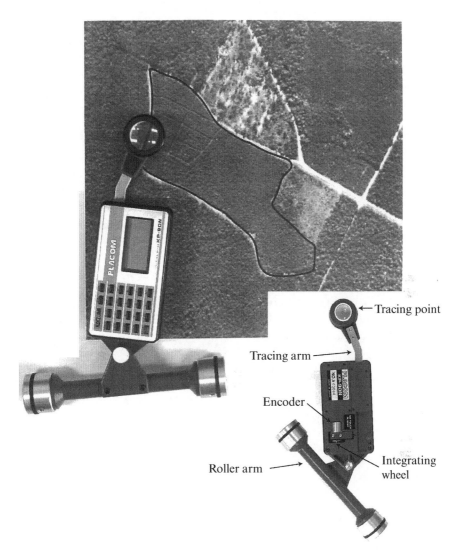

FIG. 4.4. Area measurement with planimeter.

photograph. An oblique photograph in which the apparent horizon is shown is a *high oblique*; one in which the apparent horizon is not shown is a *low oblique*. If a photograph is taken from a fixed position on the ground, it is called a *terrestrial photograph*. Foresters deal primarily with vertical aerial photographs. However, one should understand that both oblique and terrestrial photographs have important mensurational applications. The advent of affordable, high-resolution digital photography opens many applications of close-range photogrammetry (Luhmann et al., 2007, 2013) to forest mensuration and will play an increasing role in the future.

Aerial photographs play an important role in forest mensuration. Most forest stand maps are prepared based on aerial photographs. Variables used to delineate stands vary from organization to organization, but in general include species composition, crown cover, density, and height. The resulting stand maps often form the basis of many forest inventory designs (Chapter 10). In addition to mapping, many aspects of forest mensuration, such as height measurement, can be made using stereoscopic photographs (Paine and Kiser, 2012). The treatment of these subjects is beyond the scope of an introductory forest mensuration text. For a detailed treatment of the interpretation and mensuration of aerial photographs, see Paine and Kiser (2012).

Scale of an aerial photograph can be easily determined using a few simple formulas. The *scale*, or representative fraction, of a vertical photograph is the ratio of a distance on the photograph to the corresponding distance on the ground when the object planes and the image planes are parallel. From Figure 4.5, the scale S is

$$S = \frac{\text{Photo distance}}{\text{Ground distance}} = \frac{ab}{AB} \tag{4.4}$$

Scale can also be determined as the ratio of focal length (f) and altitude ($H-h$):

$$S = \frac{f}{H-h} \tag{4.5}$$

It should be noted that scale is expressed as a fraction with 1 as the numerator. Thus, a photograph that has a scale of 1:12,000 has 1 unit on the photograph equal to 12,000 units on the ground. To facilitate calculations, it is desirable to use the *photo scale reciprocal* (PSR = 1/S) instead of scale. Then,

$$\text{PSR} = \frac{1}{S} = \frac{AB}{ab} = \frac{H-h}{f} \tag{4.6}$$

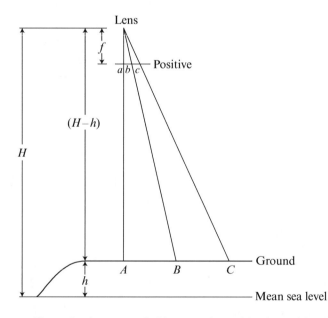

FIG. 4.5. Diagram illustrating how ground objects are imaged in the positive plane for vertical photographs. (a, b, c, are photo images of ground points A, B, C. Thus, ab is photo distance and AB is ground distance. H=height of lens above mean sea level; h=height of terrain above mean sea level; ($H-h$)=height of lens above the ground (i.e., flying height above ground).

Another useful relationship utilizes PSR and MSR (*map scale reciprocal*):

$$\frac{PSR}{MSR} = \frac{map\,distance}{photo\,distance} \qquad (4.7)$$

Note that in all of the above formulas distances must be in the same units. A few examples will illustrate the use of the above equations and some scale conversions that are best computed from PSR.

Example 1. The distance between two road intersections is 3350 ft on the ground and 4.22 in. on a photo. What is the PSR of the photo? Using eq. (4.6):

$$PSR = \frac{AB}{ab} = \frac{(12\,in./ft)(3350\,ft)}{4.22\,in.} = 9526$$

Example 2. Find the PSR of a photograph taken with a 152.36 mm focal length camera at the elevation of 1981 m above mean sea level over terrain that is 300 m above mean sea level. Using eq. (4.6):

$$PSR = \frac{H-h}{f} = \frac{(1,000\,mm/m)(1,981\,m - 300\,m)}{152.36\,mm} = 11,033$$

Example 3. A photographic crew has a camera with an 8.25-in. focal length lens. At what altitude (in feet) above the ground must they fly to produce prints with a scale of 1:20,000 (i.e., PSR = 20,000)? Using eq. (4.6) and solving for $(H-h)$:

$$(H-h) = \frac{f \cdot PSR}{12} = \frac{8.25(20,000)}{12} = 13,750\,ft$$

Example 4. Suppose that the smallest image that can be consistently distinguished on aerial photographs has a diameter of 0.01 in. If you fly some photographs with an 8.25-in. focal length camera at an altitude of 11,000 ft above ground, what would be the ground distance of the smallest tree crown you could distinguish? Using eq. (4.6) and solving for AB:

$$AB = ab\left(\frac{H-h}{f}\right) = (0.01\,in.)\left(\frac{11,000\,ft}{8.25\,in.}\right) = 13.3\,ft$$

Example 5. Assume you desire the PSR of a photograph depicting some of the same area covered by a quadrangle map. You measure the distance between two road intersections across the center of the photo and find it to be 6.35 cm. The corresponding distance on the map is 4.36 cm. If the MSR is 24,000, what is the PSR? Using eq. (4.7) and solving for PSR:

$$PSR = MSR \frac{map\,distance}{photo\,distance} = 24,000\frac{4.36}{6.25} = 16,742$$

Example 6. Compute the following scale conversions for a 23 by 23 cm vertical photo-graph with a scale of 1:20,000 (i.e., PSR = 20,000): meters per centimeter, kilometers per centimeter, hectares per square centimeter:

$$Meters\,per\,centimeter = \frac{PSR}{100} = \frac{20,000}{100} = 200$$

$$\text{Kilometers per centimeter} = \frac{\text{PSR}}{1,000 \cdot 100} = \frac{20,000}{100,000} = 0.2$$

$$\text{Hectares per square centimeter} = \left(\frac{\text{PSR}}{100}\right)^2 \left(\frac{1}{10,000}\right) = \left(\frac{20,000}{100}\right)^2 \left(\frac{1}{10,000}\right) = \frac{40,000}{10,000} = 4.0$$

When one desires the reverse of any of these equations (e.g., centimeters per kilometer instead of kilometers per centimeter), one computes the reciprocal of the appropriate value. For the example above, kilometers per centimeter=0.2, so centimeters per kilometer= 1/0.2=5.0.

4.6. DETERMINATION OF DIRECTION USING A COMPASS

The compass was one of the most important surveying instruments in the early centuries of the modern era for determining the direction of a line. It is not, however, an instrument of precision, and results of great accuracy are not to be expected in compass surveys. This is one reason the compass has largely been replaced by the GPS unit (Section 4.8). While the GPS is gaining wider use, the compass still has the advantage of speed, economy, and simplicity. It is useful in retracing lines of old surveys established by compass, in running all types of boundary lines, in maintaining the direction of cruise lines, and so forth. In dense forest canopies, a compass is often used along with a GPS unit to avoid errors associated with multipath signals (Section 4.8.3). Many cases still arise where the forester must depend on the compass for direction.

The direction of a line is generally indicated by the angle between the line and some line of reference (i.e., by an azimuth or a bearing). The line of reference is generally a true meridian or a magnetic meridian. The axis on which the earth rotates is an imaginary line cutting the earth's surface at two points: the north geographic pole and the south geographic pole. The *true meridian* at any location is the great circle drawn on the earth's surface passing through both poles and the location.

A compass consists of a magnetic needle on a pivot point, enclosed in a circular housing that is graduated in degrees. Most compasses have a sighting base attached so that it is possible to measure the angle between the line of sight and the position of the needle. An angle measured relative to the magnetic needle is referred to as a *magnetic azimuth* or *magnetic bearing*. Azimuths are horizontal angles measured clockwise from due north, while bearings are measured relative to a quadrant of the compass (i.e., NE, SE, SW, NW). Azimuths range from 0° to 360° while bearings range from 0° to 90° referenced within a quadrant. For example, a bearing of N30°E corresponds to an azimuth of 30°, while a bearing of S30°E corresponds to an azimuth of 150°.

4.6.1. Magnetic Declination

The magnetic azimuth or bearing of a line can be determined by direct reading of a compass. Converting magnetic readings to true readings by applying the magnetic declination for the location can approximate the true azimuth or bearing in question. Many compasses used by foresters have the capability of being adjusted for declination. The *magnetic declination* is the angle between the true meridian and the magnetic meridian; it is considered east if the

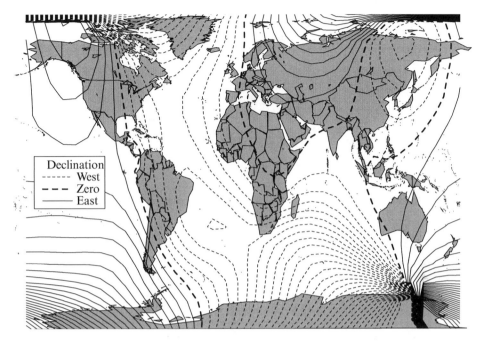

FIG. 4.6. World map showing magnetic declinations for January 1, 2010. Contours of declination are in 5° intervals. Declination data were calculated using the National Geographic Data Center's Geomagnetic Calculator (NOAA 2013).

magnetic north is east of true north and west if magnetic north is west of true north (Fig. 4.6). Declination is often called *variation of the compass* or simply *variation*. When it is desired to attach a sign to declination, east declination is considered positive, and west negative. The declination at any point can be measured with the compass when the true direction of a line can be obtained from isogonic charts such as those available from the U.S. Geological Survey. On these charts, lines called *isogonic lines* are drawn through points where the magnetic declination is the same. The *agonic line* passes through points where the declination is zero.

At any specific point, the earth's declination is continually changing. These changes are daily variation, irregular change, secular change, and annual variation. *Daily variation* is a fairly systematic and regular departure of the declination from the daily mean value. The amount of departure depends on the time of day, the season, and other factors not wholly understood. Superimposed on the regular daily variation, there are usually *irregular changes*. When they become large, we say there is a magnetic storm. These are associated with sunspots and are characterized by auroral displays and pronounced disturbances in radio-wave transmission. Since the amplitudes of daily variation and irregular variations are not predictable for any one day, these changes are not considered in most compass surveys.

In general, the average value of declination changes from one year to the next, and the change usually continues in one direction for many years. This is the *secular change*. The amount of change in one year is called the *annual variation*. Unfortunately, there is sometimes an abrupt, unpredictable change in the rate of secular change. Thus, secular change can be determined only by observations at magnetic observatories. Information from such observations are available in the United States for each state from the earliest date of valid observations to the present, at 10-year intervals up to 1900, and at 5-year intervals thereafter. This information, which is often needed for rerunning old survey lines and reestablishing corner markers in the United States, may be obtained from the National Geophysical Data Center (http://www.ngdc.noaa.gov/; last accessed August 28, 2016).

In most regions, the above changes are gradual enough so that one can use the same declination throughout an area for an entire season. But in some regions, *local disturbances* cause large differences within small areas—sometimes several degrees within a short distance. These disturbances may be artificial caused by human interference due to apparel worn by the observer, such as buckles, zippers, and glasses. Deposits of magnetite usually cause natural, local disturbances of several degrees. Other ores and geological formations cause smaller irregularities. However, even in undisturbed regions, minor irregularities are common. Almost anywhere, the declination at two points at a short distance apart, such as 100 ft, may differ by a few minutes. Such disturbances are responsible for many of the shortcomings of the compass as a surveying instrument.

A compass survey should be made with the instrument in good condition observing the instructions in the manual for the instrument. Although a compass may be in good operating condition, it may have an appreciable *index correction*; that is, there may be an angle between the real magnetic north and the direction shown by the compass. This correction for a specific compass may be determined by observation at a magnetic station (i.e., a marked point on the ground where the magnetic and true meridians have been determined accurately). In the United States, the National Geophysical Data Center can provide descriptions of nearby magnetic stations, if one specifies the location of interest.

4.7. THE U.S. PUBLIC LAND SURVEYS

The US rectangular surveying system was devised to establish legal subdivisions for describing and disposing of the public domain under the general land laws of the United States. The system uses chain units for length measurements and acres for area measurements, as described in Section 4.1. The system has been used in most of the United States with the exception of the older states along the Atlantic seaboard and in a few states where lands were in private hands before the federal government was founded. Many of the land holdings in states that do not use the US rectangular surveying system are described by *metes and bounds* (*metes* means to measure or to assign by measure; *bounds* refers to a boundary); the length and direction of each side of the boundary are determined and the corners marked with survey monuments. The description of old surveys of this type is often vague, relies on local natural or geographic features (e.g., streams, roads, and witness trees), and the corners difficult or impossible to relocate.

The Ordinance of 1785 established the rectangular surveying system. The system provides for townships 6 miles square, containing 36 sections 1 mile square. In any given region, the survey begins from an *initial point*. Through this point is run a meridian, called the *principal meridian*, and a parallel of latitude, called the *baseline* (Fig. 4.7a). With the establishment of an initial point, the latitude and longitude of the point is determined by accurate astronomical methods. Monuments are placed on the principal meridian and on the baseline at intervals of 40 chains.

Standard parallels or *correction lines* are then run for the district being surveyed. These lines, which are parallels of latitude, are established in the same manner as the baseline. They are located at intervals of 24 miles north and south of the baseline and extend to the limits of the district being surveyed. Standard parallels are numbered as First, second, third, and so on, standard parallel north, or south (Fig. 4.7a).

The survey district is next divided into tracts approximately 24 miles square by means of guide meridians. These lines are true meridians that start at points on the baseline, or standard parallels, at intervals of 24 miles east and west of the principal meridian, and extend north to their intersection with the next standard parallel. Because of the convergence of meridians, the distance between these lines is 24 miles only at the starting points. At all other points, the distance between them is less than 24 miles. Guide meridians are numbered as first, second, third, and so on, guide meridian east, or west. Note that two sets of monuments are found on the standard parallels. The monuments that were set when the parallel was first located are called *standard corners* and govern the area north of the parallel. The second set, found at the intersection of the parallel with the meridians from the south, is referred to as the *closing corners* and govern the area south of the parallel.

The townships of a survey district are numbered meridionally (east–west) into *ranges* and latitudinally (north–south) into *tiers* from the principal meridian and the baseline of the district. As illustrated in Figure 4.7a, the third township south of the baseline is in tier 3 south. Since the word *township* is frequently used instead of tier, any township in this tier is often designated as township 3 south. The fourth township west of the principal meridian is in range 4 west. By this method of numbering, any township is located if its tier, range, and principal meridian are given as Township 3 south, range 4 west, of the fourth principal meridian. This is abbreviated as T. 3S., R. 4W., 4th P.M.

In subdividing a township into 36 *sections*, the aim is to secure as many sections as possible that will be 1 mile on a side. To accomplish this, the error due to convergence of meridians is thrown as far to the west as possible by running lines parallel to the east boundary of the township, rather than running them as true meridians. Errors in linear measurements are thrown as far to the north as possible by locating monuments at intervals of 40 chains along the lines parallel to the east boundary of the township, all the accumulated error falling in the most northerly half-mile, which may be more or less than 40 chains in length.

The system used in numbering the sections of a township was established in 1796. This numbering system and the most recent order in which the lines are run to subdivide the township into sections (indicated by the numbers on the lines) are shown in Figure 4.7b. This system of subdividing a township throws the errors due to survey and losses from convergence into the extreme north and west sections. Other sections, however, may contain more or less than 640 acres due to survey errors. Nevertheless, the established boundaries are final, regardless of errors made in the original survey.

If any of the monuments of an original survey are missing, surveyors must know and observe the methods used in the original survey as well as the principles that have been adopted by the courts in order to restore the missing corners correctly. Procedures for relocating original survey lines and corners are described in the U.S. Bureau of Land Management's (2009) *Manual of Instructions for the Survey of Public Lands in the United States*.

After all the original monuments have been found or any missing ones have been replaced, the first step in the subdivision is the location of the center of the section.

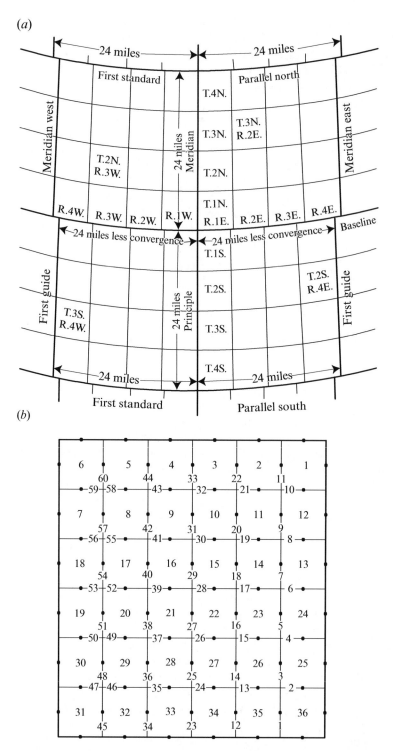

FIG. 4.7. United States Public Land Survey: (*a*) standard parallels and guide meridians, (*b*) subdivisions of a township, and (*c*) subdivisions of sections.

(c)

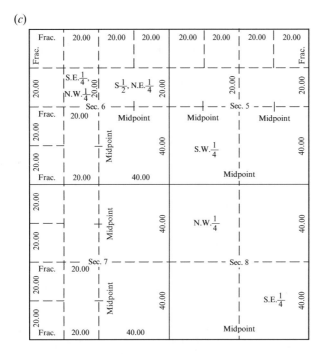

FIG. 4.7. (Continued).

Regardless of the location of the section within the township, this point is always at the intersection of the line joining the east and west quarter-section corners with the line joining the north and south quarter-section corners. By locating these lines on the ground, the section is divided into quarter sections containing approximately 160 acres each (Fig. 4.7c).

The method of dividing these quarter sections into 40-acre parcels depends on the position of the section within the township. For any section except those along the north and west sides of the township, the subdivision is accomplished by bisecting each side of the quarter section and connecting the opposite points by straight lines (e.g., Sec. 8 in Fig. 4.7c). The intersection of these lines is the center of the quarter section. The method of subdividing the sections along the north and west sides of the township is shown in Figure 4.7c. The corners on the north–south section lines are set at intervals of 20.00 chains, measured from the south, the discrepancy being thrown into the most northerly quarter mile. Similarly, the monuments on the east–west lines are set at intervals of 20.00 chains, measured from the east, the discrepancy being thrown into the most westerly quarter mile.

The rectangular system of subdivision provides a convenient method of describing a piece of land that is to be conveyed by deed from one person to another. If the description is for a 40-acre parcel, the particular quarter of the quarter section is first given, then the quarter section in which the parcel is located, then the section number, followed by the township, range, and principal meridian. Thus, the 40-acre parcel labeled in Sec. 6 of Figure 4.7c can be described as the S.E.¼, N.W.¼, Sec. 6, T.3S., R.4W., 4th P.M. The legal descriptions of other parcels appear in the figure.

4.8. GLOBAL POSITIONING SYSTEMS

GPS are used to determine locations on the earth's surface accurately. The technology has revolutionized surveying and fieldwork in almost all natural resource professions. Although the traditional methods described above are still widely used in forestry, GPS technology is being used to determine block boundaries, locate field plots, delineate special features such as wetlands, unique vegetation pockets, stream buffers and other protected areas, laying out roads and trails, and relocating survey markers. The components of the GPS system, how GPS works, and applications of GPS to forest mensuration will be briefly discussed here. For a more complete treatment of GPS theory and applications, see Leick (2003) and Oderwald and Boucher (1997).

4.8.1. Components of GPS

The GPS consists of three components: a satellite segment, a control segment, and a user segment (Fig. 4.8a). The *satellite segment* originally consisted of a network of 24 satellites in orbit around the earth at an altitude of 20,200 km. Additional satellites were added to the system to improve the precision of GPS receiver calculation. As of December 2012, there were 32 satellites in the GPS constellation. Each satellite completes an orbit around the earth every 12 hours. The network is configured so that the entire earth's surface has complete satellite coverage 24 hours a day. The satellite system is called the NAVSTAR system and is operated by the U.S. Department of Defense. Russia, the European Union, the People's Republic of China, and Japan also operate global or regional GPS satellite systems.

The *control segment* consists of a Master Control Station (MCS), six ground stations, and four dedicated ground antennas located around the world. The MCS is located at Schriever Air Force Base near Colorado Springs, Colorado. The monitoring stations are located at Colorado Springs, Hawaii, Ascencion Island, Diego Garcia, Kwajalein, and Cape Canaveral. The monitor stations send the raw data back to the MCS for processing. The dedicated ground antennas are located at Ascension Island, Diego Garcia, Kwajalein, and Cape Canaveral. The MCS receives data from the monitor stations in real time 24 hours a day and uses that information to determine whether the satellites are experiencing clock or ephemeris changes, and to detect equipment malfunctions. New navigation and ephemeris information is calculated from the monitored signals and uploaded to the satellites once or twice per day.

The *user segment* is the user and a GPS receiver. A GPS receiver is a specialized radio receiver. It is designed to listen to the radio signals being transmitted from the satellites and calculates a position based on that information. GPS receivers come in many different sizes, shapes, and price ranges. Receivers are often described by the number of receiver channels they have that determines the number of satellite signals they can lock on. GPS receivers used for forestry applications are described in Section 4.8.4.

4.8.2. How GPS Works

GPS works using triangulation (Fig. 4.8b). The GPS satellites emit a radio signal in the L-band region of the microwave spectrum. These signals travel at the speed of light (186,000 miles per second). Distances between satellites and a receiver are determined by measuring the time required for the signal to reach the receiver. Each satellite's signal

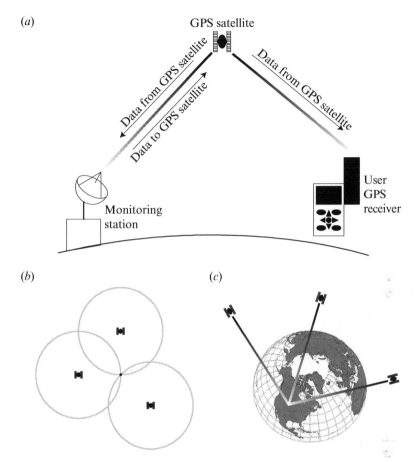

FIG. 4.8. GPS basics: (*a*) components of the system, (*b*) triangulation, and (*c*) triangulation via satellite signal.

contains an *ephemeris* (its own location), an *almanac* (the locations of other satellites), and clock information. Four satellites are required to establish the *X*, *Y*, *Z* (latitude, longitude, and elevation) coordinates for a location: one satellite to establish the time and three satellites for triangulation. The radio signals propagate from the satellite as a sphere, and the intersection of the three spheres determines the location of the receiver (Fig. 4.8*c*).

4.8.3. Accuracy of GPS

Accurate triangulation requires distances between the receiver and satellites to be determined precisely. Distance is determined in a number of ways using the various signals broadcast from the satellites. Two signals, the coarse acquisition code and the carrier phase code, are used in civilian applications. Primarily the military and other authorized personnel use the precise code, a third type of signal.

The *coarse acquisition* code (C/A) is the most widely used GPS signal. The GPS satellites have very precise clocks that are monitored and updated from the MCS. A GPS receiver generates codes that match the codes generated by the satellites at the same time.

Distance is estimated using the time between a signal being emitted from the satellite and the time the signal is received. Because the clock on the receiver never has exactly the same time as the clock on the satellite, the distances are really estimates and are referred to as *pseudoranges*. A signal takes only 1/20 of a second to travel from the satellite to the receiver, even very small differences in clock times can result in large errors in distance. Because of the error in distance estimation, satellite triangulation establishes a region in which the location is likely to be located (Fig. 4.8*b*). The *carrier phase* code is a continuous signal and distance is determined by counting the number of complete and fractional wavelengths between two locations. Carrier phase location is much more accurate than C/A location, but acquisition of carrier phase data is time consuming and requires expensive receivers.

Prior to May 2000, the U.S. government deliberately introduced error in the satellite signals. This error was called *selective availability* (SA). With SA, location accuracies in the range of ±100 m or more were common. Since May 2000, the SA has been eliminated, except on a regional basis during periods of military conflict or other crises. In 2006, the Wide Area Augmentation System (WAAS) was made available for civilian use. Inexpensive WAAS receivers now have accuracies of less than 10 m.

There are a number of factors that influence GPS accuracies:

- *Satellite clock errors.* The satellite clocks are very precise and are frequently updated from the MCS. Each satellite signal carries the clock error so that satellite time can be corrected. Uncorrected clock errors generally result in location errors of about 1 m.
- *Ephemeris data errors.* A satellite, at any given time, is never in its precisely assigned orbit. Satellite location errors generally result in location errors of about 1 m.
- *Atmospheric errors.* The ionosphere and troposphere cause signal interference. The amount of interference is dependent upon weather conditions and the location of the satellite. Satellites directly overhead have smaller amounts of interference than satellites near the horizon. Atmospheric errors generally create location errors in the range of 3–5 m.
- *Multipath errors.* Multipath errors result from delays in signal reception caused by the signal bouncing off objects between the satellite and the receiver. Large buildings, mountains, and hills cause the greatest amount of multipath error; however, trees can also result in multipath error. Location error due to multipaths generally is in the range of 2–3 m.
- *Receiver noise.* Noise errors are the combined effect of code noise (around 1 m) and noise within the receiver (around 1 m).

Other errors such as receiver malfunction and user mistakes can result in blunders of unknown size.

4.8.4. GPS Receivers

There are many different types of GPS receivers available. Handheld receivers are primarily used in field applications. Handheld GPS units consist of an antenna, the GPS engine, a microprocessor to calculate and store locations, and a battery. The main difference between most receivers is the number of receiver channels available to track satellites.

Receivers are generally categorized as either navigating receivers or mapping receivers. *Navigating receivers* are primarily intended for recreation uses. Most navigation receivers are now WAAS enabled and have accuracies that are less than 10 m. Many receivers feature colored screens and either built-in base maps or the ability to upload base maps. For most forest mensuration applications, navigation receivers are adequate. Costs, portability, and extended battery life are important considerations when selecting a navigating receiver. Navigation receivers cost between US$100 to around US$500.

Mapping receivers typically have many more options than navigating receivers including the number of channels available, multiple point readings, static and dynamic lines, increased storage capacity, and the ability to interface with multiple satellite constellations (e.g., NAVSTART and the Russian GLONASS system). Mapping quality receivers are quite expensive often costing between US$3,000 and US$12,000.

4.8.5. Using GPS Data in Forest Mensuration

GPS data have a variety of applications in forest mensuration. Plot locations can be marked and cruise lines mapped. Block boundaries can be traversed and areas determined. Property boundaries, roads, and streams can be mapped. Data can be downloaded to geographic information systems and analyzed further or used to produce maps. Plot locations and cruise lines can also be located on geographic information systems and this data uploaded to the GPS receiver. The GPS receiver can then be used as a navigation tool for traversing cruise lines and locating plots.

To effectively utilize GPS data, one must understand how the GPS data are collected and analyzed. GPS data are typically represented as either points or lines. Points are estimates of location at a particular instant in time. Points may be based on single readings or multiple readings. A *single-reading point* is an estimate of location based on all available satellites at a single point in time. A *multiple-reading point* is the average of several single-reading estimates of the same location. Multiple-reading points are generally more accurate than single-reading points; however, more field time is required to obtain multiple readings.

Lines are connected points. In most GPS receivers, lines are created by collecting single-reading points at a specified time interval. These lines are typically called *dynamic lines* because they are created as the GPS receiver is moving. Lines can be created by connecting a series of multiple-reading points. These lines are called *static lines* because the GPS receiver must be held stationary at each point during the multiple-reading phase. Static lines are more accurate than dynamic lines; however, dynamic lines are generally quite accurate because the line is based on several connecting single-reading points.

Most GPS receivers are capable of calculating areas based on line data collected in the same manner as a closed traverse. Starting at a fixed location, the boundary of the block is traversed with the GPS receiver collecting dynamic line data. The boundary is followed around to the starting point. The GPS receiver will insure closure and calculate area. Area may also be calculated by connecting a series of multiple-reading points that close.

4.9. GEOGRAPHIC INFORMATION SYSTEMS

GIS are powerful computer database programs that have the capability to input, store, manipulate and analysis, and output spatially referenced information (Longley et al., 2011). The organization and presentation of spatial information is crucial to many aspects of modern forest management (Reed and Mroz, 1997).

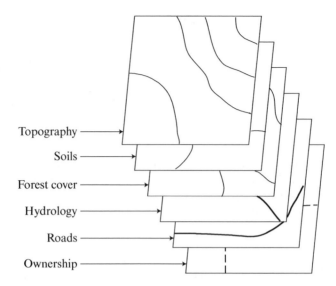

Topography
Soils
Forest cover
Hydrology
Roads
Ownership

FIG. 4.9. Data layers in GIS (adapted from Reed and Mroz (1997)).

Different types of data within a GIS are stored as *layers* (Fig. 4.9). The GIS provides the capability of overlaying different layers or combining different layers through data manipulation and analyses to produce sophisticated maps. The ability to rapidly combine and manipulate different types of data makes GIS an ideal tool for foresters. For example, a forester might combine a soils layer with a forest cover layer and a roads layer to develop a map to identify which types of machinery to use for harvesting different blocks of timber. A layer identifying the location of unique or protected areas could be added to ensure that adequate buffers are provided to protect these features.

GIS data can be stored as either vector data or raster data. *Vector-based systems* use points, lines, or polygon areas stored as series of *X–Y* coordinates. Various attributes, such as soil type, elevation, and forest cover type, are associated with the vector elements. In *raster-based systems*, the area is divided into a matrix of cells. Data attributes for a given layer are assigned on a cell-by-cell basis. Raster-based systems are generally faster than vector-based systems and perform overlays more efficiently. Vector-based systems have more compact data structures, tend to have higher spatial accuracy, and perform calculations related to the distribution of features more efficiently. Systems that can handle both types of data are available.

The applications of GIS to forestry are immense. Applications of GIS of particular interest to forest mensuration are discussed below. Descriptions of the technology involved, data sources and structures, and the manipulation and analysis of data are beyond the scope of a forest mensuration textbook. For more information, the reader is referred to one of the numerous books dedicated to GIS theory and practice (e.g., Aronoff, 1989; Johnston, 1998; Longley et al., 2011).

4.9.1. Applications of GIS to Forest Mensuration

The ability to locate points, draw lines, and measure areas makes GIS an ideal tool for inventory design. The sample designs discussed in Chapter 10 can generally be developed using a GIS. For example, if random sampling is being used, most GIS systems can generate random X–Y coordinates to use as plot centers. These coordinates can be combined with other desired layers to produce maps useful for locating the plots in the field. The data can be uploaded to GPS receivers and the receivers used to navigate to the plot locations. Some GIS packages even have the capability of determining distances and bearings between plots and can design optimal routes.

By combining desired layers, forests can be stratified in a number of ways. The areas associated with different strata can be determined and stratified sampling designs developed.

Results of inventories can be added to a GIS database as another layer. Volume per unit area or other stand parameters (Chapter 8) can be combined with other layers to produce useful maps for harvesting and other forest management activities. Data from previous inventories may be used in the design of new inventories. By comparing multiple inventories, changes such as growth, regeneration, and harvesting can be tracked.

5

INDIVIDUAL TREE PARAMETERS

5.1. AGE

Information on age is important in relation to growth and yield and as a variable in evaluating site quality. However, without specific definition, "age" can be an ambiguous term. Total age of a tree is the length of time elapsed since germination of the seed or budding of the sprout. For those trees that regenerate from sprouts, the age of the belowground portion may be considerably greater than that of the aboveground portion. Total age is difficult to determine in some situations, so age is sometimes reported as the number of years since planting, as the age measured at some defined stump height, or as measured at breast height (typically 4.5 ft in the United States and 1.3 m in other countries; see Section 5.2 for further definition). The difference between total age and other age measurements may be years or even decades, and may vary considerably from tree to tree, so clear definition of terms is important.

In certain species, branch whorls can be used to determine age. Each season's height growth starts with the bursting of the bud at the tip of the tree; this lengthens to form the leader. The circle of branchlets that grows at the base of the leader marks the height of the tree at the very start of the season's growth. This process is repeated the following year, and a new whorl appears to mark the beginning of that season's growth. A count of these branch whorls thus gives the age of the tree (Fig. 5.1a). It is only in certain coniferous species, however, that whorls are well defined, and in older trees of these species the evidence of the former whorls cannot always be distinguished.

In some hardwood species, such as maples (*Acer* sp.) and beeches (*Fagus* sp.), the terminal bud scale leaves a distinct scar in the bark around the stem. The remains of these scars can often be seen for several years. By counting the number of scars from the terminal, the age of a tree at that point may be determined (Büsgen and Munch, 1929). As with

Forest Mensuration, Fifth Edition. John A. Kershaw, Jr., Mark J. Ducey, Thomas W. Beers and Bertram Husch.
© 2017 John Wiley & Sons, Ltd. Published 2017 by John Wiley & Sons, Ltd.

(*a*)

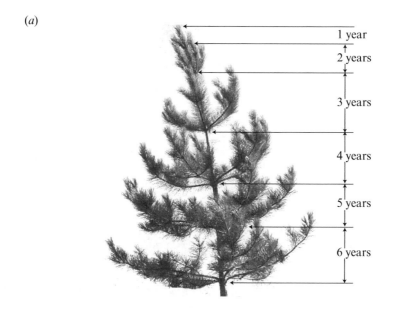

1 year

2 years

3 years

4 years

5 years

6 years

(*b*)

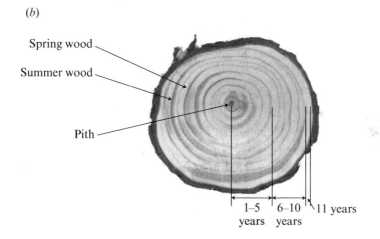

Spring wood

Summer wood

Pith

1–5 years   6–10 years   11 years

FIG. 5.1. Determining age of trees: (*a*) whorl counts and (*b*) annual ring counts.

counting whorls, this technique is limited to younger trees and branches and requires access to the upper stem to locate bud scale scars.

Annual rings afford the best method of determining tree age in temperate regions. Here, most trees grow in diameter by adding a new layer of wood each year between the old wood and bark. Formation of this layer begins at the start of the growing season and continues throughout. Woody tissue formed in the spring (springwood or earlywood) is more porous and lighter in color than woody tissue formed in the summer (summerwood or latewood). Thus, annual growths of the tree appear on a cross section of the stem as a series of concentric rings (Fig. 5.1*b*). A count of the number of rings on a given cross section gives the age of the tree above the cross section.

Consequently, if the count is made on a cross section at ground level, the count gives total tree age. If the count is made on a cross section above ground level, the number of years for the tree to grow to the height of the cross section must be added to the ring count to obtain total tree age (the number of years for a young tree to reach stump height varies from 1 year for sprouts of broadleaved species to 20 years for coniferous seedling growing on dry sites).

In tropical and subtropical regions that have alternating wet and dry seasons or other periodic fluctuations in growing conditions, growth rings similar to those that occur in temperate regions may be produced. However, these are often more variable anatomically than those of temperate trees (Rozendaal and Zuidema, 2011). Such rings may be useful for determining tree age; however, in tropical regions that do not have a regular alternations in growing conditions, any rings that may be produced have limited value for determining age. Several techniques have been proposed to estimate age of tropical trees without rings including radiocarbon dating and the use of periodic increment models (Martínez-Ramos and Alvarez-Buylla, 1998). Conventional radiocarbon dating is limited to very long-lived individuals (generally greater than 500 years). Elevated ^{14}C in the atmosphere following nuclear testing in the 1960s has allowed some authors to use isotopic techniques to estimate ages of much younger trees (Worbes and Junk, 1989). Periodic increment models are limited to species with reliable long-term growth data and results in predictions of mean ages for a cohort of trees rather than individual tree ages (Chambers and Trumbore, 1999).

Difficulties may be encountered in making ring counts. In slow-growing trees, rings may be so close together that they are hard to count. In some species, the difference in appearance between springwood and summerwood is not marked, and rings are indistinct. In addition, abnormal weather during the growing season may lead, in certain species, to the formation of false growth rings—a dry spell may interrupt annual growth, and then growth may resume when rains come. Then, too, defoliation by insects may cause a tree to produce a second set of leaves and an additional growth ring for a given year. False rings, however, are not as clear as true growth rings and often do not extend around the entire circumference of a tree cross section.

Ring counts are often made on a sawed surface such as a stump, although such surfaces are generally too rough to permit rings to be accurately counted. Consequently, rings on a smoothed strip extending from the bark to the center of a section should be cut with a sharp knife or plane and viewed with a hand lens to facilitate counting. On standing trees, an increment borer may be used (Fig. 5.2a). This consists of a hollow cutting bit that is screwed into the tree. The core of wood forced into the hollow center of the bit is removed with the extractor. The rings on this core are then counted (Fig. 5.2d). Unfortunately, total age of large trees is difficult to obtain with an increment borer because the maximum practical length of a borer is between 16 and 20 in. and because of the difficulty of finding the pith.

Average radial growth may be determined from several increment cores (cores for this purpose are normally taken at breast height), but usually only one core is taken. When a single core is taken, it should be extracted halfway between the long and short diameters of the tree. The length of the core depends on the past period for which the growth measurement is desired. If growth for the life of the tree is needed, the boring must reach the pith. If growth for a past period is needed as a basis of growth predictions, the boring should include the number of rings in the period, usually 5 or 10 years. If the width of the growth period is small (less than 2 in.), an increment hammer (Fig. 5.2b) may be used; the increment

(*a*)

(*b*)

(*c*)

(*d*)

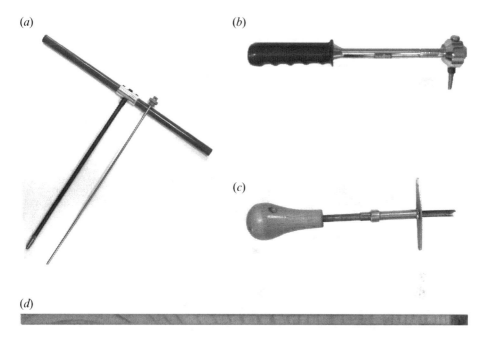

FIG. 5.2. Increment tools: (*a*) increment borer, (*b*) increment hammer, (*c*) bark gauge, and (*d*) extracted increment core.

hammer is swung against the tree and the hollow tip plunges into the wood. A small core, similar to the core from an increment borer, is extracted from the hollow tip.

Dendrochronology is the study of annual rings in living trees and aged woods to establish a time sequence in the dating of past events. Annual rings of trees have been used in dendrochronology for dating archaeological and geological events as far back as 3000 years and arranging them in order of occurrence. Tree response to varying growth conditions, such as precipitation, temperature, insect attack, or other limiting factors, is indicated by the width of annual rings; for example, narrow rings when precipitation is low, and wide rings when precipitation is high in a water-limited system. Since climatic variations tend to occur over rather large regions, patterns of characteristically narrow and wide rings in two or more trees can be matched, provided the trees grew at the same time. Recognition of this fact led to cross dating, the foundation of dendrochronology (Fritts, 1976). Dendrochronology techniques have been applied to several wide-scale environmental problems including climate change and air pollution studies. A comprehensive review of the techniques and applications of dendrochronology can be found in Cook and Kairiukstis (1990).

5.2. TREE DIAMETERS AND CROSS-SECTIONAL AREAS

A diameter is a straight line passing through the center of a circle or sphere and meeting at each end of the circumference or surface. The most common diameter measurements taken in forestry are of the main stem of standing trees, cut portions of trees, and branches. Diameter measurement is important because it is one of the directly measurable dimensions from which tree cross-sectional area, surface area, and volume can be computed.

The use of the word diameter implies that trees are circular in cross section. In many cases, however, the section is somewhat wider in one direction than another or it may be eccentric in other ways. Since for computational purposes tree cross sections are usually assumed to be circular, the objective of any tree diameter measurement is to obtain the diameter of a circle, with the same cross-sectional area as the tree.

The point at which diameters are measured will vary with circumstances. In the case of standing trees, a standard position has been established. In the United States, diameters of standing trees are measured at 4.5 ft above ground level. This is referred to as *diameter breast height* and is abbreviated to d.b.h. or dbh (Fig. 5.3*a*). In countries that use SI units, diameters of standing trees are measured at 1.3 m above ground level. This was traditionally abbreviated to the symbol *d* (IUFRO, 1959); however, it is common,

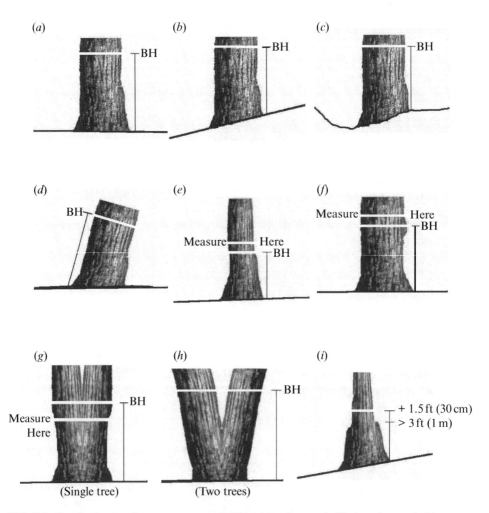

FIG. 5.3. Standard points for measurement of dbh: (*a*) level ground, (*b*) sloped ground, (*c*) uneven ground, (*d*) leaning tree, (*e*) crook at breast height, (*f*) defect at breast height, (*g*) fork at breast height—1 tree, (*h*) Fork below breast height—2 trees, and (*i*) buttressed tree. Note bh = breast height (4.5 ft in the United States, and usually 1.3 m in countries using SI units).

especially in North America, to find the symbol dbh used to refer both to diameters measured at 4.5 ft (or 1.37 m) above ground and to diameters measured at 1.3 m above ground level.* Diameters at other points along the stem of a tree are often indicated by subscripts: $d_{0.5h}$ = diameter at half total height; d_6 = diameter at 6 m above ground level.

Diameters should be qualified as d.o.b. (outside bark) or d.i.b. (inside bark); however, when this designation is omitted from breast height measurements, as it often is, measurements are assumed to be outside bark. Whether outside or inside bark diameter measurements are taken depends on the purpose for which the measurements are made. When a rule, or scale stick, is used on a cut section, it is simple to measure diameter inside or outside bark. One simply measures an appropriate line, or lines, on the section along which the bark is intact. If both d.i.b. and d.o.b. are measured, bark thickness is one-half the difference between them. If one measures d.o.b. and bark thickness, d.i.b. equals d.o.b. minus twice the bark thickness.

Bark thickness on standing trees can be determined with a bark gauge (Fig. 5.2c). The instrument consists of a steel shaft, half-cylindrical in shape, which is pushed through the bark. The cutting edge of this instrument is a half circle that is dull on one side so that the instrument can be driven through the softer bark, but not through the wood. When the instrument's sliding cross arm is pressed against the bark, its thickness can be read on a scale without removing the instrument. When the bark is thick and tough, its thickness may be obtained by boring to the wood surface with an increment borer or a brace and bit and measuring from the bark surface to the wood with a small ruler.

In the United States, tree diameters have generally been measured and recorded in inches (however, there is a growing use of SI measurements, especially in research activities). In countries where SI is used, diameters are measured in centimeters (occasionally in millimeters). In measuring dbh in the field, the following standard procedures are recommended:

1. When trees are on slopes or uneven ground, measure breast height (4.5 ft or 1.3 m) above the ground on the uphill side of the tree (Fig. 5.3b and c).
2. When a tree is leaning, breast height is measured parallel to the lean on the high side of the tree. The diameter is measured perpendicular to the longitudinal axis of the stem (Fig. 5.3d).
3. When a tree has a limb, a bulge, or some other abnormality, such as a crook, at breast height, measure diameter above the abnormality; strive to obtain the diameter the tree would have had if the abnormality had not been present (Fig. 5.3e and f).
4. When a tree consists of two or more stems forking below breast height, measure each stem separately (Fig 5.3h). When a tree forks at or above breast height, measure it as one tree. If the fork occurs at breast height, or slightly above, measure diameter below the enlargement caused by the fork (Fig. 5.3g).
5. When a tree has a buttress that extends higher than 3 ft or 1 m, it is common to measure the stem at a fixed distance above the top of the buttress, usually at 1.5 ft or 30 cm (Fig. 5.3i).
6. When a tree has a paint mark to designate the breast height point, assume that the point of measurement is at the top of the paint mark.

---

* Here, we adopt the use of dbh for both imperial and SI measurements of diameter at breast height, and use d to indicate general diameter measurements.

5.2.1. Instruments for Measuring Diameter

The most commonly used instruments for measuring dbh are calipers and diameter tapes. Less precise measurements can be made with the Biltmore stick and Bitterlich's sector fork. Multipurpose laser-based instruments can also be used to measure tree diameters.

Calipers. Calipers are often used to measure tree dbh when diameters are less than about 24 in. or 60 cm. Calipers of sufficient size to measure large trees, or those with high buttresses (commonly found in tropical regions), are awkward to carry and handle, particularly in dense undergrowth. Beam calipers (Fig. 5.4a) may be constructed of metal, plastic, or wood, and consist of a graduated beam with two perpendicular arms. One arm is fixed at the origin of the scale and the other arm slides. When the beam is pressed against the tree and the arms closed, the tree diameter can be read on the scale. For an accurate reading, the beam of the calipers must be pressed firmly against the tree with the beam perpendicular to the axis of the tree stem and the arms parallel and perpendicular to the beam.

A more advanced and precise form of calipers is the Mantax computer calipers (Fig. 5.4b), developed by the Swedish firm Haglöf. With this instrument, the diameter is measured in either imperial or SI units and stored in the instrument eliminating the need for recording the measurement on a field sheet or entering it in a field computer.

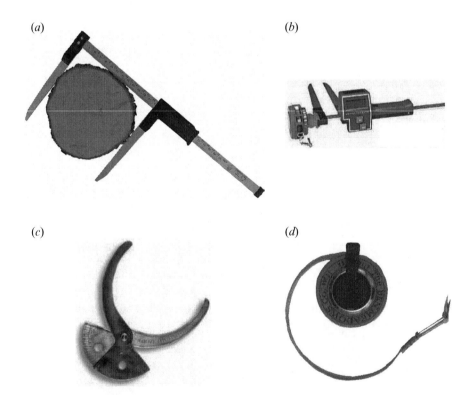

FIG. 5.4. Diameter measuring instruments: (a) beam calipers, (b) Mantax computer calipers, (c) tree calipers, and (d) diameter tape.

The stored data can be read from the instrument or downloaded to a computer or to a printer. Calipers with similar capabilities are manufactured in Finland by Masser Oy.

Other forms of calipers have been devised. The fork caliper consists of a set of fixed arms with graduations calibrated so that, when the fork is placed on the tree, the points of tangency indicate the tree diameter. Tree calipers, used for measuring small trees (<8 in.), consist of a pair of curved arms that scissor on a vertex (Fig. 5.4c). When the arms are pressed against the tree, the fan-shaped scale at the vertex gives the diameter reading. The Finnish parabolic calipers are another form. It is, of course, important that all types of calipers be held perpendicular to the axis of the tree stem at the point of measurement.

When a tree cross section is elliptical, one might measure the major and minor diameters, d_1 and d_2, and obtain the average diameter from the arithmetic mean of d_1 and d_2. This would, however, overestimate the "true" diameter of the cross section (i.e., the diameter of a circle that would have the same cross-sectional area). Measuring the perimeter and dividing by π to obtain diameter (essentially the procedure used with a diameter tape, described below) also gives an overestimate. The best practice would be to use the geometric mean (Section 3.3.3) of the two measurements: $\sqrt{d_1 \cdot d_2}$. Table 5.1 illustrates the effect of eccentricity on the cross-sectional area.

Unfortunately, tree cross sections often depart from elliptical form as well as from circular form. Consequently, for practical purposes when using calipers, the arithmetic average of the long and short "diameters," or axes, is often utilized. If it is not feasible to secure the long and short "diameters," the arithmetic average of two diameters perpendicular to each other is often used.

Diameter Tape. The diameter of a tree cross section may be obtained with a flexible tape by measuring the "circumference" of the tree and dividing by π ($D = C/\pi$). The diameter tapes used by foresters, however, are graduated at intervals of π units (inches or centimeters), thus permitting a direct reading of diameter (Fig. 5.4d). These tapes are accurate only for trees that are circular in cross section. In all other cases, the tape readings will be slightly too large because the circumference of a circle is the shortest line that can encompass any given area (Table 5.1).

The diameter tape is convenient to carry, allows measurement of large trees, and, in the case of eccentric trees, requires only one measurement. It offers high repeatability because any element of judgment in determining the "long" and "short" diameters, as with calipers, is eliminated; and the measurement obtained with a diameter tape equals the arithmetic mean of all possible caliper measurements of the same stem (Matérn, 1956, 1990). Although it is slower to use than other diameter measuring instruments, the time element is generally not important. Care must be taken that the tape is correctly positioned at the point of measurement, that it is kept in a plane perpendicular to the axis of the stem, and that it is set firmly around the tree trunk.

Biltmore Stick. The Biltmore stick (Fig. 5.5), which can only be classed as a crude measuring instrument, is an aid in estimating diameter at breast height. It consists of a straight stick, normally 24–36 in. long (60–90 cm), that is held perpendicular to the axis of the tree stem. By holding the stick so that the 0-point of the graduation at one end of the stick lies on the line EA that is tangent to the tree cross section at A, the diameter of the tree can be read at the intersection at the other end of the stick on the line EB that is tangent to

TABLE 5.1. Effects of Eccentricity on Cross-Sectional Area Estimation

Ratio of Major:Minor Axes	Relative Axis Length		Diameter from Perimeter	True Area	Major Axis	Minor Axis	Area Estimated Using			
	Major	Minor					Arithmetic Mean	Quadratic Mean	Geometric Mean	Perimeter
1:1	1.13	1.13	1.13	1.00	1.00	1.00	1.00	1.00	1.00	1.00
5:4	1.26	1.01	1.14	1.00	1.25	0.80	1.01	1.03	1.00	1.02
4:3	1.30	0.98	1.15	1.00	1.33	0.75	1.02	1.04	1.00	1.03
3:2	1.38	0.92	1.16	1.00	1.50	0.67	1.04	1.08	1.00	1.06
5:3	1.46	0.87	1.18	1.00	1.67	0.60	1.07	1.13	1.00	1.10
2:1	1.60	0.80	1.23	1.00	2.00	0.50	1.13	1.25	1.00	1.19
5:2	1.78	0.71	1.31	1.00	2.50	0.40	1.23	1.45	1.00	1.34
3:1	1.95	0.65	1.39	1.00	3.00	0.33	1.33	1.67	1.00	1.51
4:1	2.26	0.56	1.54	1.00	4.00	0.25	1.56	2.13	1.00	1.87
5:1	2.52	0.50	1.68	1.00	5.00	0.20	1.80	2.60	1.00	2.23

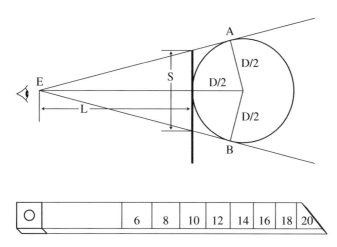

FIG. 5.5. The Biltmore stick.

the tree cross section at *B*. The distance, *L*, from the eye, *E*, to the tree is usually 25 in. (63.5 cm); however, it may be some other value. The graduations, *S*, of the stick for different values of *D* and *L* are obtained from the following formula:

$$S = \sqrt{\frac{D^2 L}{L + D}} \qquad (5.1)$$

Inaccuracies in estimating diameters with the Biltmore stick are due to (1) difficulty of holding the stick exactly the distance *L*, from the eye; (2) the failure to keep the eye at breast height level; (3) the failure to hold the stick at breast height level; and (4) the eccentricity of tree cross sections (the Biltmore stick is correct only for circular cross sections). The advantage of the Biltmore stick is its ease of use and speed. It does provide a useful check on ocular estimates of dbh in rapid work but is unsuitable for inventories where a high level of accuracy or precision is needed.

Sector Fork. Bitterlich's sector fork (Visiermesswinkel or Sektorluppe), Figure 5.6, is similar in principle to the Biltmore stick (Bitterlich, 1959). The instrument determines diameter from a sector of a circular cross section. One side of the sector is a fixed arm; one side is a line of sight. The line of sight intersects a curved scale on which diameters or cross-sectional areas are printed. It is not necessary to hold the instrument at a fixed distance from the eye because a sighting pin fixes the line of sight for any distance. The instrument is especially suited for measuring trees with diameters of less than 50 cm (about 20 in.), as in young plantations. An attachment (called the "double fork") is available that permits measurement of diameters up to 200 cm. Masser Oy manufactures a caliper with digital capabilities based on similar principles that is suitable for one-handed use.

Because many trees are eccentric in cross section, all instruments for measuring tree diameter will in the long run give results that average too large (Table 5.1). However, when an accurate determination of cross-sectional area is of prime importance, calipers will give the best results. When different people caliper the same irregular tree, there will always be some variation in the measurements. A good part of this variation results because the tree

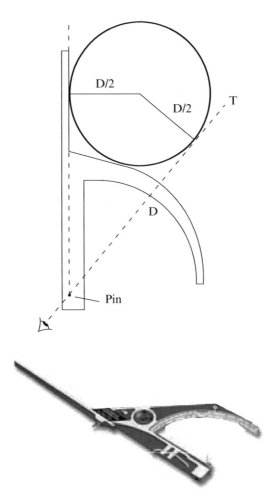

FIG. 5.6. Bitterlich's sector fork (Visiermesswinkel or Sektorluppe). *T* is the tangential line of sight, *D* is the diameter read from the scale (note: this instrument was recently renamed the Treemeter).

is not always calipered exactly in the same place and direction. Since the element of varying direction does not affect measurements with the diameter tape, tape measurements are more consistent. Such consistency is important in growth studies, when the same trees are remeasured at intervals. Then, the actual diameters of the trees are less important than the changes in diameter during the period between measurements. The diameter tape will accurately determine these changes. Errors due to eccentricity will appear in both measurements and will not significantly affect the difference between them. In any case, whenever repetitive diameter measurements are made to determine growth, it is desirable to mark the position of measurement with a scribe or paint mark. For a detailed study on the geometry of stem measurement and cross-sectional area estimation, refer to Matérn (1990). A comprehensive review of various diameter measurement tools and their associated accuracies can be found in the review by Clark et al. (2000).

When the average of a number of diameter measurements, *d*, is required, one might use the arithmetic mean. However, if the primary interest is to obtain an average for the calculation of cross-sectional area and volume, then the quadratic mean (Section 3.3.2) is more appropriate:

$$\bar{d}_Q = \sqrt{\frac{\sum\limits_{i=1}^{n} d_i^2}{n}} \tag{5.2}$$

5.2.2. Measurement of Upper-Stem Diameters

Tree stem diameters above breast height are often required to estimate form or taper and to compute the volume of sample trees from the measurement of diameters at several points along the stem. Of course, the tree can be felled and diameter measurements made at the desired points on the stem (Section 6.1.1). To avoid felling trees, measurements can be made on standing trees. Climbing a tree and using the instruments described in the previous section can obtain upper-stem diameters. In the past, diameters at the top of the first log (about 17 ft above ground) were sometimes taken by mounting calipers or a diameter tape on a pole (Ferree, 1946; Godman, 1949). These diameters were needed to determine Girard Form Class (Section 5.4).

Instruments for measuring upper-stem diameters of standing trees allow diameters to be determined from the ground at some distance from a tree. Over the years, optical forks, optical calipers, fixed-base range finders, and fixed-angle rangefinders have been used for measuring upper-stem diameters. These include the Barr and Stroud dendrometer, the Wheeler pentaprism, engineer's transits, the Spiegel relaskop, the Tele-relaskop, and the Breithaupt Todis dendrometer. A summary of these instruments can be found in Grosenbaugh (1963a), Smith (1970), Husch et al. (1983), and Clark et al. (2000). Many of these instruments and methods are mainly of historical interest since most measurements of upper-stem diameters of standing trees are made, at present, using the relaskop.

Calipers with high-intensity visible lasers are now available for measurement of upper-stem diameters. Lasers are mounted on the arms of the calipers. The beams are focused on the opposing edges of the tree by sliding the caliper arm. Diameter is read when the beams just touch the opposing edges. While these instruments represent a lower-cost alternative for obtaining upper-stem diameters, they are likely prone to user errors that increase as distance from the measurement point increases because of the reliance on visual judgment of when the beams "just touch" the opposing edges.

The relaskop (Fig. 5.7a) is an instrument of the optical fork type. An optical fork employs a fork angle on which the lines are tangent to the cross section at the level of the diameter measurement and on which the vertex is at the observer's eye. The basic geometry is shown in Figure 5.7b. Note that $\cos(\alpha/2)$ can be approximated by using $\cos(\theta/2)$. The relaskop has the advantage of adjusting the diameter measurement automatically for the angle of inclination of the line of sight. Other models and attachments are available that improve the precision of upper-stem diameters.

Several laser-based instruments that can measure upper-stem diameters as well as distances and heights are currently available (Carr, 1992; Skovsgaard et al., 1998; Clark et al., 2000). Most of these instruments utilize a reflector-less laser for distance and height measurement and a fixed-base rangefinder to measure diameters at any point along the stem.

(*a*)   (*b*)

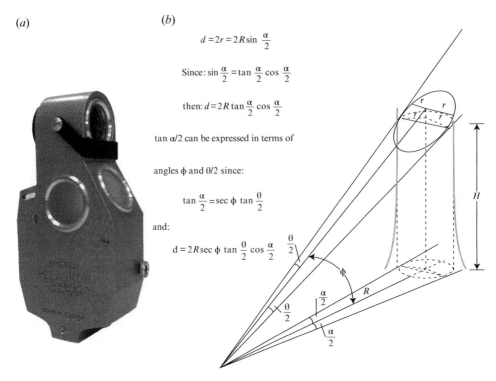

$$d = 2r = 2R\sin\frac{\alpha}{2}$$

Since: $\sin\dfrac{\alpha}{2} = \tan\dfrac{\alpha}{2}\cos\dfrac{\alpha}{2}$

then: $d = 2R\tan\dfrac{\alpha}{2}\cos\dfrac{\alpha}{2}$

tan α/2 can be expressed in terms of

angles φ and θ/2 since:

$$\tan\frac{\alpha}{2} = \sec\phi\,\tan\frac{\theta}{2}$$

and:

$$d = 2R\sec\phi\,\tan\frac{\theta}{2}\cos\frac{\alpha}{2}$$

FIG. 5.7. (*a*) The Spiegel relaskop and (*b*) the principle of the optical fork (adapted from Grosenbaugh (1963a)).

The accuracies of these devices are comparable to other optical dendrometers (Clark et al., 2000), but are not as accurate as traditional calipers and diameter tapes. A digital instrument with some of the same capabilities as the Spiegel relaskop, the Criterion RD 1000, is currently manufactured by Laser Technology Inc. (Fig. 5.8*a*). The Criterion RD 1000 currently does not have the internal capability of distance measurement and requires coupling with a laser-based range finder or manually entering the distances. The LaserAce 1000 Rangefinder, manufactured by Trimble Navigation Ltd. (Fig. 5.8*b*), is a multipurpose laser measurement device with the capability of measuring diameters using measurement offsets. A newer instrument using a holographic sight (the RC3H dendrometer) is manufactured by Masser Oy. Accuracy of most dendrometers, electronic or manual, can be improved significantly if the devices are mounted on a tripod; however, significant overestimation of diameter, especially in larger trees, often occurs (Skovsgaard et al., 1998).

5.2.3. Cross-Sectional Area

The cross-sectional areas of planes cutting the stem of a tree normal to the longitudinal axis of the stem are often desired. If the cross section of a tree is measured at breast height, it is called the *basal area*. The total basal area of all trees, or of specified classes of trees, per unit area (e.g., per acre or per hectare) is a useful characteristic of a forest stand. For example, basal area is directly related to stand volume and is a good measure of stand density (Section 8.3.2).

(*a*)   (*b*)

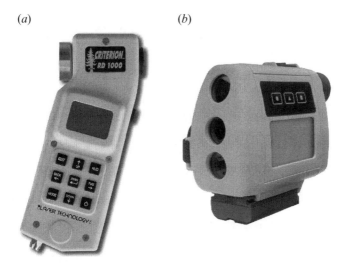

FIG 5.8. Electronic dendrometers: (*a*) the Criterion RD 1000 (photo courtesy Laser Technology Inc.) and (*b*) the LaserAce 1000 Rangefinder (photo courtesy Trimble Navigation Ltd).

When a tree cross section (for either a standing tree or a cut section) is circular, as it is often assumed to be, its area can be computed from its diameter or circumference:

$$g = \frac{\pi d^2}{4}$$

and, since $d = \frac{c}{\pi}$:

$$g = \frac{c^2}{4\pi}$$

where g = tree cross-sectional area*

d = diameter of cross section

c = circumference of cross section

In SI units diameter, d, is commonly expressed in inches, and cross-sectional area, g, in square feet (ft^2). Consequently, it is convenient to express g in square feet as a function of diameter d in inches:

$$g\left(\text{ft}^2\right) = \frac{\pi d^2}{4(144)} = 0.005454 d^2 \qquad (5.3)$$

In SI units, diameter d is commonly express in centimeters, and cross-sectional area g in square meters (m^2). In this case:

---

*g is the internationally recognized symbol (IUFRO, 1959) for tree cross-sectional area or basal area; BA and CSA also are often used, especially in North American literature. In this book, the symbol BA is used only for basal area, i.e., the cross-sectional area at breast height.

$$g\left(\mathrm{m}^2\right)=\frac{\pi d^2}{4(10,000)}=0.00007854d^2 \qquad (5.4)$$

Unfortunately, cross sections of tree stems are often not circular. Thus, when they are assumed to be circular, errors in determining cross sectional area may result (Table 5.1). For most purposes, the geometric mean of the long and short diameters, d_1 and d_2, of a section will give the most accurate results, although for practical purposes, a satisfactory practice is to take the arithmetic average of the long and short "diameters," or axes; or if it is not feasible to secure the long and short "diameters," to take the arithmetic average of the two diameters perpendicular to each other.

5.2.4. Surface Areas

Tree Bole. The exterior surface area of the stem of a tree approximates the cambial surface (i.e., the area under bark). This area represents the surface on which the wood substance accumulates and is therefore useful in the estimation of tree and stand growth. For tree stems that assume the shape of a geometric solid, the surface area can be computed by calculus or by the appropriate formula (Table A.3). For example, assume the form of a tree approximates the paraboloid generated by revolving the equation $Y^2 = 0.066X$ about the X-axis (Fig. 5.9).

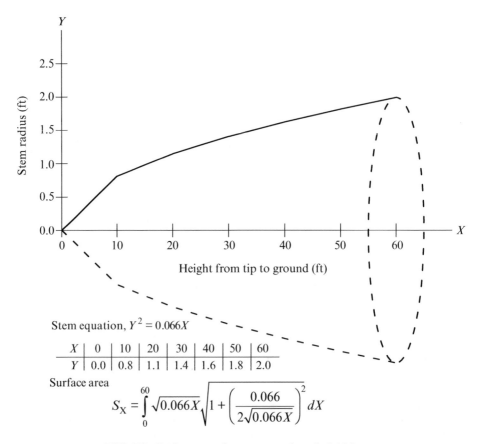

Stem equation, $Y^2 = 0.066X$

X	0	10	20	30	40	50	60
Y	0.0	0.8	1.1	1.4	1.6	1.8	2.0

Surface area

$$S_X = \int_0^{60} \sqrt{0.066X}\sqrt{1+\left(\frac{0.066}{2\sqrt{0.066X}}\right)^2}\,dX$$

FIG. 5.9. Surface area of a tree stem of paraboloid form.

In this equation, Y is the radius of the stem at a given point (e.g., at the stump) and X is the distance from the tree tip to the given point. The surface of the tree between $Y=2$ ft (i.e., $X=60$) and the tip of the tree (i.e., $X=0$) can be determined by using equation

$$S_X = 2\pi \int_a^b Y \sqrt{1 + \left(\frac{dY}{dX}\right)^2} \, dX$$

and

$$S_X = 2\pi \int_0^{60} \sqrt{0.066X} \sqrt{1 + \left(\frac{0.066}{2\sqrt{0.066X}}\right)^2} \, dX$$

$$= 2\pi \int_0^{60} \sqrt{0.066X + 0.001089} \, dX$$

$$= 2\pi \left[\frac{2(0.066X + 0.001089)^{3/2}}{(3)(0.066)} \right]_0^{60}$$

$$= 2\pi \left[10.10(0.066X + 0.00001089)^{3/2} \right]_0^{60}$$

$$= 159.25\pi \text{ ft}^2$$

$$= 500.3 \text{ ft}^2$$

In this example, the same value would have been obtained if the formula for the surface area of a paraboloid (Table A.3) had been used. For more complex stem equations, the analytical procedure used above is appropriate.

Lexen (1941) has shown that using Huber's formula or Smalian's formula and substituting circumferences for cross-sectional areas and summing the surface areas for all sections in a tree can approximate bole surface area.

Swank and Schreuder (1974) tested sampling methods to estimate foliage, branch, and stem surface areas; Gregoire et al. (1995) present some additional sampling methods. Hann and McKinney (1975) developed prediction equations for the surface area of four species in the southwestern United States by using formulas for the surface areas of a cylinder, cone, and paraboloid.

5.2.5. Applications to Understory Plants

Stem diameter, cross-sectional area, or surface area can be of interest for understory plants, especially tree regeneration and other sizeable woody vegetation. However, the small size of such material presents some advantages and disadvantages for measurement. Although it is possible to measure dbh on some understory vegetation, the notion is not meaningful for vegetation that is shorter than breast height, or for which the plane defined by breast height passes through the middle of a crown that lacks a well-defined central stem. For seedlings and other low-stature vegetation, it is customary to measure diameter near the ground surface. The *caliper* of a seedling is defined as its diameter outside bark just below the cotyledon scar (e.g., Haase, 2008). The name belies the usual instrument used for its measurement, though the large calipers used for large trees are typically too bulky and inaccurate (seedling caliper measurements are usually recorded in millimeter). Mechanical or digital machinist's calipers are fine tools for this sort of measurement. Alternatively, when small stem diameters must be recorded in a small number of diameter classes, it is simple to construct a notch-style gauge that matches the specific classes. Such gauges can be constructed out of any rigid material that will be durable under heavy use in the climatic conditions encountered. Sheet brass or aluminum are reasonable choices but a variety of plastics are much less expensive and can be worked with ordinary tools. Measurement of very small diameter stems (less than 2 in. or 5 cm) with a diameter tape tends to cause excessive wear on the tape (including kinking if the tape is made of metal) and is not recommended.

Where accurate cross-sectional areas are needed with accuracy greater than that afforded by assuming a circular shape, and destructive sampling is possible, small stems can be sectioned and the resulting surface scanned using an ordinary flatbed scanner, followed by image analysis to measure area (Kershaw and Larsen, 1992). It is important to ensure that the section is perpendicular to the axis of the stem or the cross-sectional area will be overestimated. Cutting tools must be very sharp to ensure that the stem is not crushed, shredded, or otherwise distorted in the sectioning process.

5.3. HEIGHT

Height is the linear distance of an object normal to the surface of the earth, or some other horizontal datum plane. Aside from land elevation, tree height is the principal vertical distance measured in forest mensuration. *Tree height* is an ambiguous term unless it is clearly defined. For this purpose, a logical classification of height measurements that can be applied to standing trees of either excurrent or deliquescent form is shown in Figure 5.10. Shown below are the definitions of heights and lengths.

> *Total height* is the distance along the axis of the tree stem between the ground and the tip of the tree. (In determining this height, the terminals—i.e., the base and tip of the tree—can be more objectively determined than other points on the stem. It is often difficult, however, to see the tip of a tree in closed stands and to determine the uppermost limit of a large-crowned tree.)
>
> *Bole height* is the distance along the axis of the tree stem between the ground and the crown point. (As shown in Figure 5.10, the crown point is the position of the first crown-forming branch. Therefore, bole height is the height of the clear, main stem of a tree.)
>
> *Merchantable height* is the distance along the axis of the tree stem between the ground and the terminal position of the last usable portion of the tree stem. (The position of the upper terminal is somewhat subjective. It is taken at a minimum top diameter or at a point where branching, irregular form, or defect, limit utilization. The minimum top diameter will vary with the intended use of the timber and with the market conditions. For example, it might be 4 in. (10 cm) for pulpwood and 8 in. (20 cm) for sawtimber.)
>
> *Stump height* is the distance between the ground and the basal position on the main stem where a tree is cut. (A standard stump height, generally about 1 ft or 30 cm, is established for volume table construction and timber volume estimation.)

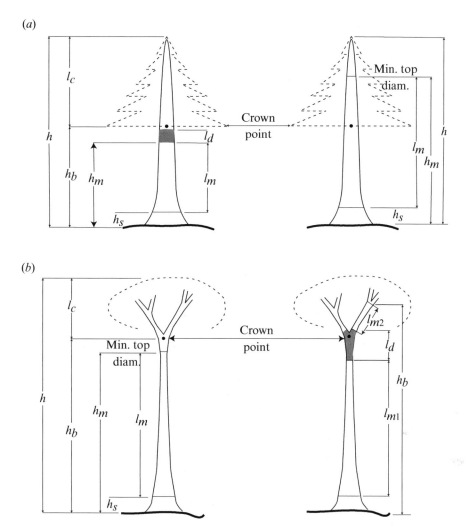

FIG. 5.10. Tree height and stem length classification: (*a*) excurrent form and (*b*) deliquescent form. h, total height; h_b, bole height; h_m, merchantable height; h_s, stump height; l_c, crown length; l_m, merchantable length; l_d, defective length. Shading denotes defective portion.

Merchantable length is the distance along the axis of the tree stem between the top of the stump and the terminal position of the last usable portion of the tree stem.

Defective length is the sum of the portions of the merchantable length that cannot be utilized because of defect.

Sound merchantable length equals the merchantable length minus the defective length.

Crown length is the distance on the axis of the tree stem between the crown point and the tip of the tree. A tree characteristic used in tree and forest health monitoring is the *live crown ratio*. It is determined by dividing the live crown length by total tree live height (Alexander and Barnard, 1994).

Generally speaking, the techniques and instruments devised for general height measurement may be applied to tree-height measurement; however, instruments must be economical, light, portable, and usable in closed stands. Total tree heights may also be estimated from measurements made on aerial photographs as described in Paine and Kiser (2012).

The heights of short trees can be measured directly with an engineer's self-reading level rod, a graduated pole, or similar devices. The heights of many taller trees can be measured directly, with the aid of sectional or sliding poles made of wood, fiberglass, plastic or light-weight metal. Although measurement with poles is slow, they are often used to measure height on continuous forest inventory plots where high accuracy is desired and merchantable length is less than 70 ft (21 m).

Most height measurements of tall trees are made indirectly with hypsometers. Hypsometers are based on the relation of the legs of similar triangles (geometric) or on the tangents of angles (trigonometric). (Note that terms such as altimeter and clinometer are also applied to instruments that are used to measure height.)

5.3.1. Hypsometers Based on Similar Triangles

A number of older instruments based on similar triangles were developed and have been used extensively over the years. These include the Christen, Merritt, Chapman, and JAL hypsometers. These instruments are less precise than the ones that utilize trigonometric relationships. The geometric-based hypsometers have the advantage of being simple and easily and cheaply constructed. The Christen and Merritt hypsometers developed in North America exemplify hypsometers based on the geometric relationships of similar triangles. In their original forms, both instruments used imperial system units; however, if desired, they can be constructed to show SI units.

The *Christen hypsometer* consists of a scale about 10 in. long (Fig. 5.11). To use the Christen hypsometer, a pole (usually 5 or 10 ft long) is held upright against the base of the tree, or a mark is placed on the tree at a height of 5 or 10 ft above ground. The hypsometer is then held vertically at a distance from the eye such that the two inside edges of the flanges are in line with the top and base of the tree. It may be necessary for the observer to move closer to or farther from the tree to accomplish this, but except for this, the distance from the tree is immaterial. The graduation on the scale that is in line with the top of the pole, or the mark, gives the height of the tree. The following proportion gives the formula for graduating the instrument:

$$\frac{A'C'}{AC} = \frac{A'B'}{AB} \tag{5.5}$$

For a given length of instrument $A'B'$ and a given pole length or mark height AC, the graduations $A'C'$ can be obtained by substituting different values of height AB in the equation.

The *Merritt hypsometer* is another simple instrument, which is often combined with the Biltmore stick. It is a convenient aid in estimating the number of logs in a tree rather than height in feet. The hypsometer consists of a graduated stick that is held vertically at a predetermined distance, usually 25 in., from the eye (Fig. 5.12). If the stick is held 25 in. from the eye along a horizontal line, then the distance to the tree should be measured on the horizontal; if the stick is held 25 in. from the eye along a line to the lower end of the stick, as shown in Figure 5.12, then the distance to the tree should be measured on the slope.

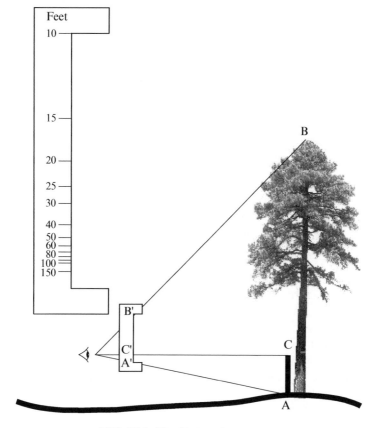

FIG. 5.11. The Christen hypsometer.

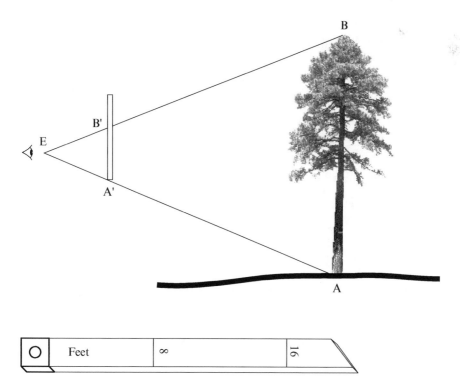

FIG. 5.12. Merritt hypsometer.

The observer must stand at a predetermined distance, such as 50 ft, from the tree. Then, the height of the tree may be read on the stick at the point where the line of sight to the top terminal on the tree intersects the scale. In Figure 5.12:

$$\frac{A'B'}{AB} = \frac{EA'}{EA} \tag{5.6}$$

If EA' is 25 in. and EA is 50 ft, for each 16 ft (the standard length of a log) of height in AB the length of $A'B'$ will be 8.0 in. The stick may thus be graduated in 8-in. intervals with the successive graduations marked 1, 2, and 3 indicating the number of logs. The stick may also be graduated using SI units, substituting centimeters and meters for inches and feet. When merchantable length is measured, as it often is with the Merritt hypsometer, one should remember that the line of sight EA must be to the top of the stump.

Although the Christen hypsometer may be used to measure any type of height, it is practical only for total height measurements. Furthermore, an examination of Figure 5.11 shows a crowding of graduations near the bottom of the scale. This makes the instruments unreliable for the determination of the height of tall trees.

In using the Merritt hypsometer, it is difficult to hold the stick in a vertical position exactly 25 in. from the eye. Small deviations in orientation result in considerable errors in height readings. Therefore, the instrument should be used only to make rough checks on ocular height estimates.

5.3.2. Hypsometers Based on Trigonometry

Although numerous hypsometers of this type have been developed, the basic principle is the same for all of them, and nearly all use the tangent method for height estimation. One first sights to the upper terminal of the height desired and takes a reading; then one sights to the base of the tree, or the top of the stump, depending on the height desired, and takes a second reading. Figure 5.13a illustrates the situation, with the distance D and the angles α_1 and α_2 known. Then, if the vertical arc of the instrument is graduated in degrees, the total height of the tree may be calculated as follows:

$$\tan(\alpha_1) = \frac{B - C}{D}$$

where $B - C$ is the elevation difference between B and C, from which we obtain

$$B - C = D \tan(\alpha_1)$$

Similarly,

$$A - C = D \tan(\alpha_2)$$

Note that because A is below C in Figure 5.13a, we are looking down from eye level when we measure α_2. So, we record α_2 as negative and $A - C$ is also negative. Since the height of the tree $B - A$ equals $(B - C) - (A - C)$,

$$B - A = D\left[\tan(\alpha_1) - \tan(\alpha_2)\right] \tag{5.7}$$

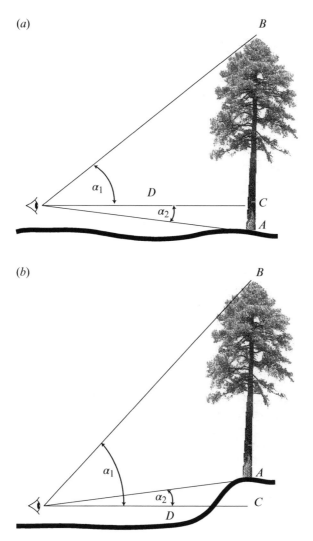

FIG. 5.13. Measuring tree height with hypsometers based on tangents of angles: (*a*) measurement on level ground and (*b*) measurement on sloped ground where the tree is above eye level.

On steep ground, a situation as shown in Figure 5.13*b* might occur. In this case, α_2 will be recorded as positive because we are looking uphill at the base of the tree and $A - C$ will also be positive. Thus, the height of the tree, $B - A$, still equals $(B - C) - (A - C)$, and eq. (5.7) still holds.

Hypsometers are often provided with arcs that are scaled rather than angular arcs. Scaled arcs, α_S, are based on angular units represented by the ratio of 1 unit vertically to D_S units horizontally. For any vertical angle α:

$$\tan(\alpha) = \frac{\alpha_S}{D_S} \tag{5.8}$$

where α = the vertical angle

$\quad\quad \alpha_S$ = the scaled arc

$\quad\quad D_S$ = the scale distance

The height of a tree is obtained using

$$B - A = \frac{D}{D_S}\left(\alpha_{S,1} - \alpha_{S,2}\right) \tag{5.9}$$

where D = the horizontal distance from the tree

$\quad\quad D_S$ = the scale distance

$\quad\quad \alpha_{S,1}$ = the upper scaled arc

$\quad\quad \alpha_{S,2}$ = the lower scaled arc (recorded as negative if point A is below eye level)

In the United States, the percentage scale ($D_S = 100\,\text{ft}$) and the topographic scale ($D_S = 66\,\text{ft}$) have been traditionally used. Hypsometers with scales of 15, 20, 25, 30, and 45 are readily available. One should note that scales of this type (and this includes percentage and topographic scales) could be used with any units: feet, yards, meters, and so on. One has only to read the scale in the same unit as one measures the horizontal baseline. However, since scale numbers usually represent convenient base distances for tree measurement with a particular unit, the unit to use is generally indicated.

Tree heights obtained using the tangent method are particularly prone to errors (see Section 5.3.3), especially when trees are leaning or it is difficult to ascertain exactly where the upper terminal is located. Bragg (2008) and Bragg et al. (2011), building on work by Grosenbaugh (1980), suggest an alternative trigonometric approach based on sines rather than tangents. Consider again Figure 5.13a, and let D_B be the distance from the observer's eye to point B. We may write $B - C$ as

$$B - C = D_B \sin\alpha_1$$

Likewise, if D_A is the distance from the observer's eye to point A, then

$$A - C = D_A \sin\alpha_2$$

Recalling that $B - A = (B - C) - (A - C)$, we may then calculate the tree height as

$$B - A = D_B \sin\alpha_1 - D_A \sin\alpha_2 \tag{5.10}$$

The sine method is much more resistant to errors than the tangent method,* but it does require measurement of two potentially challenging distances, which has limited its use until recently. At least some hypsometers that incorporate laser rangefinders are now capable of measuring D_A and D_B.

Historically, the most commonly used hypsometers in North America have been the Abney level, the Haga altimeter, the Blume–Leiss altimeter and the Suunto clinometer

---

*As pointed out in Section 5.3.3, errors associated with the tangent method will result in under- or over-estimation of tree heights, thus, as long as these errors are random, estimates of mean height should be unbiased. With the sine method, the same potential errors exists; however, these errors always produce smaller height estimates, thus estimates of means in the presence of errors using the sine method could be potentially biased.

(a)   (b)

(c)   (d)

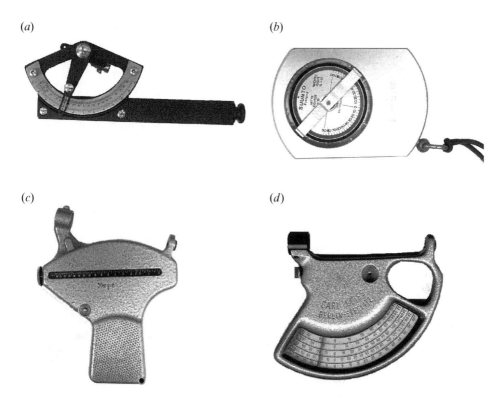

FIG. 5.14. Examples of hypsometers based on trigonometric principles: (a) Abney level, (b) Suunto clinometer, (c) Haga altimeter, and (d) Blume–Leiss altimeter.

(Fig. 5.14). The *Abney level* in its original form (Calkins and Yule, 1935) and modernized designs have been used for many years for measuring tree heights as well as land elevations. The instrument consists of a graduated arc mounted on a sighting tube about 6 in. long (Fig. 5.14a). The arc may have a degree, percentage, or topographic scale. When the level bubble, which is attached to the instrument, is rotated while a sight is taken, a small mirror inside the tube makes it possible to observe when the bubble is horizontal. Then, the angle between the bubble tube and the sighting tube may be read on the arc.

The *Suunto clinometer* is a handheld device housed in a corrosion-resistant aluminum body (Fig. 5.14b). A jewel-bearing assembly supports the scale, and all moving parts are immersed in a damping liquid inside a hermetically sealed plastic capsule. The liquid dampens undue scale vibrations. The instrument is held to one eye and raised or lowered until the hairline is seen at the point of measurement. At the same time, the position of the hairline on the scale gives the reading. Due to an optical illusion, the hairline seems to continue outside the frame and can be observed at the point of measurement. The instrument is available with a rangefinder and several scale combinations: percent and degrees, percent and topographic, degrees and topographic, and scaled distances (see eq. (5.9)).

The *Haga altimeter* (Wesley, 1956) consists of a gravity-controlled, damped, pivoted pointer, and a series of scales on a rotatable, hexagonal bar in a metal, pistol-shaped case (Fig. 5.14c). The six regular American scales are 15, 20, 25, 30, percentage, and topographic. Sights are taken through a gun-type peep sight; squeezing a trigger locks the indicator needle, and the observed reading is made on the scale. A rangefinder is available with this instrument.

The *Blume–Leiss altimeter* (Pardé, 1955) is similar in construction and operation to the Haga altimeter, although its appearance is somewhat different (Fig. 5.14*d*). The five regular SI scales are 15, 20, 30, and 40. A degree scale is also provided. All scales can be seen at the same time. The instrument is available with a rangefinder.

The mentioned hypsometers based on the tangents of angles are more accurate than those based on similar triangles. The Abney level, the Haga altimeter, the Blume–Leiss altimeter, and the Suunto clinometer are similar in accuracy. The Abney level, however, is slower to use, and large vertical angles are difficult to measure because of the effect of refraction on observations of the bubble through the tube from beneath. This makes the Abney level difficult to use in tall timber that is so dense that the tops cannot be seen from a considerable distance. A choice among the other three instruments is largely a matter of personal preference.

The *Spiegel relaskop* (Fig. 5.7*a*), originally designed for use in variable probability sampling (see Section 9.3), can be used as a clinometer with readings in degrees or percent that permit the calculation of tree height.

A variety of advanced electronic hypsometers are commercially available. Some examples of these hypsometers are shown in Figure 5.15. The Leiss BL 7 (Fig. 5.15*a*) is a digital version of the Blume–Leiss hypsometer manufactured by the German firm Höhen- und Neigungsmesser. This instrument uses an optical rangefinder to measure horizontal distances and calculates the height based on angular measurements. The TruPulse laser hypsometers (Fig. 5.15*b*), manufactured by the U.S. firm Laser Technology (Denver, CO) and the Opti-Logic Insight hypsometers (Fig 5.15*c*), manufactured by the U.S. firm Opti-Logic (Tullahoma, TN) are examples of laser-based hypsometers. Both instruments can operate in reflector-less modes or reflector modes. When operated in reflector mode, a special laser reflector has to be placed on the tree being measured. The reflector improves accuracy in heavy undergrowth. These instruments measure horizontal distances and calculates tree height based on angular measurements. The Haglöf Vertex IV hypsometer (Fig. 5.15*d*) uses ultrasonic pulses together with a transponder fixed to the tree. The instrument measures distance, angle, and horizontal distance to the transponder and displays tree height (Haglöf also manufactures a laser-based Vertex hypsometer). The LaserAce (Fig 5.8*b*) is also capable of laser-based distance and height measurement. In recent years, a large number of laser-based rangefinders with some hypsometry capability have been introduced, but not all are designed for forestry work and some lack the accuracy or precision needed for tree height measurements except of the crudest sort.

Because many of the new ultrasonic or laser-based instruments are very advanced and display results at a fine resolution (centimeters for most devices), there is a tendency to mistake precision for accuracy (Skovsgaard et al., 1998). These instruments should be initially calibrated, and the calibration checked frequently, especially after out-of-service periods and when temperatures fluctuate. When using reflector-less lasers, care must be exercised to ensure the laser beam is reflected from the target tree and not from objects between the observer and the target tree. These instruments are also prone to the same measurement errors as the mechanical hypsometers (Section 5.3.3).

The major advantage of these devices is that horizontal distances are measured very precisely and potential calculation errors are eliminated. A major disadvantage of many of these instruments is their bulkiness. The Vertex III, which is similar in size to the Suunto

(a)   (b)

(c)   (d)

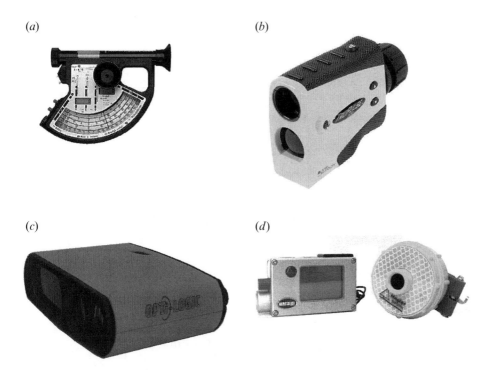

FIG. 5.15. Examples of electronic hypsometers: (*a*) Leiss BL 7, (*b*) TruPulse laser hypsometer, (*c*) Opti-Logic Insight laser hypsometer, and (*d*) Haglöf Vertex IV ultrasonic hypsometer.

clinometer, represents a major breakthrough in size for the electronic hypsometers. The costs of these instruments (US$800–US$3000) are also high relative to the mechanical instruments (US$100–US$400). However, the increased speed of measurement makes these instruments economical over the long run.

5.3.3. Special Considerations in Measuring Tree Heights

It is difficult to accurately measure the height of large, flat-crowned trees. There is a tendency to overestimate their heights (Fig. 5.16*a*) using the tangent method and underestimate heights using the sine method. Care must be exercised to focus the hypsometer on the tip of the tree, otherwise total height determination for such trees are of little value. When using the sine method, it is recommended to scan the crown profile and use the tallest measurement (Bragg et al., 2011). While this practice can be easily carried out on isolated trees, as described by Bragg et al. (2011), in dense forest conditions it might be impractical, time consuming, and result in either increased inventory costs or reduced numbers of height measurements. However, height measurement technology is changing rapidly so the utility of the sine method may be expected to improve.

In general, the optimum viewing distance for any hypsometer is the distance along the slope equal to the height to be measured. This rule of thumb, which is adapted from Beers (1974a), should be used with discretion. For example, if one were using a Suunto clinometer with a percentage scale to determine the merchantable height of a tree estimated to be

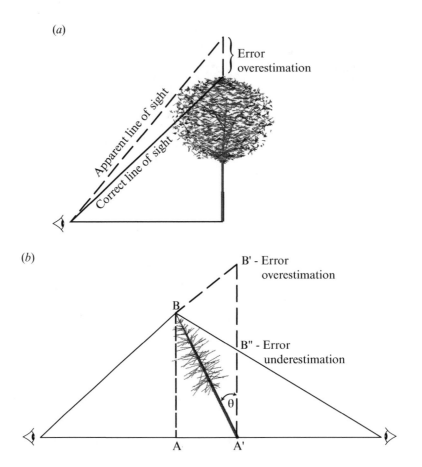

FIG. 5.16. Errors in tree height measurement: (*a*) correctly identify the top of the tree and (*b*) estimate error of leaning trees.

about 56 ft, it would be logical to use a viewing distance of about 50 ft. The best distance to use involves a balance between geometric errors and other sources of error, such as the inability to correctly identify the top of a tree. Where visibility is poor, the optimal distance may be slightly shorter than the rule of thumb would dictate.

It is more important to make sure that the angles and horizontal distances can be determined accurately. Figure 5.17 shows a contour map of errors that occur given a 2° error in angle measurement (Fig. 5.17*a*) and a 1 m error in horizontal distance measurement (Fig. 5.17*b*). Errors associated with angle measures are fairly constant across horizontal distance and more dependent upon the height above eye of the object being measured. On the other hand, errors associated with horizontal distance measurements have a much steeper slope, with errors associated with shorter distances increasing rapidly with increasing height above eye. This clearly illustrates the need for accurate horizontal distance measurements, one of the strengths of the laser- and sonic-based hypsometers. The freedom of motion afforded by laser- and sonic-based hypsometers should also allow foresters to choose a vantage point from which lines of sight are clear, and angles can be determined accurately, thus minimizing other sources of error.

(a)

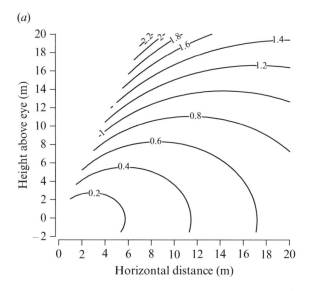

(b)

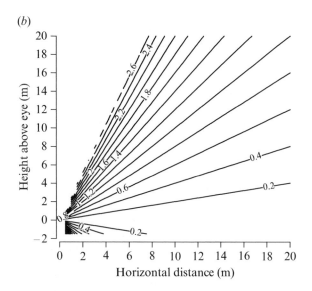

FIG. 5.17. Errors associated with height measurement: (a) effect of a 2° angle measurement and (b) effect of a 1 m horizontal distance measurement. Contour values are in meters error of tree height (adapted from Dr. David Larsen, University of Missouri; http://oak.snr.missouri.edu/advforbio/units.php; last accessed August 28, 2016).

Since all hypsometers assume trees are vertical, trees leaning away from an observer will be underestimated and trees leaning toward an observer will be overestimated (Fig. 5.16b) using the tangent method. This error will be minimized if measurements are taken such that the lean is to the left or right of the observer. If leaning trees are to be measured, one should determine the point on the ground where a plumb bob would fall if suspended from the tip of the tree (point A on Fig. 5.16b). Then, height should be measured from this point to the tip (point B on Fig. 5.16b). If stem length (distance $A'B$ on Fig. 5.16b) is desired, the measured height (AB) is multiplied by the secant of the angle of lean θ. The difference will be small except for abnormally leaning trees. Alternatively, one could use the sine method if equipment and visibility allow.

The measurement of tree height with an accurate hypsometer is slow and expensive. Consequently, it is customary to make ocular estimates whenever precision is not essential, such as for commercial timber inventories where none of the required values are precisely determined, and where the large number of trees measured makes precision of individual measurements unimportant. An experienced person can obtain reliable ocular estimates. Ocular estimates are, however, subject to serious errors because of sudden changes in the timber type or the weather. Furthermore, most people do not make reliable estimates at the start of the day or after a rest. Consequently, estimates should be checked frequently by instrumental measurements.

5.3.4. Use of LiDAR

LiDAR is an important source of remotely sensed data in modern forest inventories and will be taken up in more detail in Chapter 13. We introduce it here because it is possible, under some circumstances, to measure individual tree heights, crown attributes, and sometimes even diameters from LiDAR data. In the most common form of LiDAR data currently in use in forestry applications, a laser scanner is mounted on an aircraft (or possibly a satellite), and the scanner sweeps back and forth beneath the aircraft as it moves along its line of flight. The returns from the scanner form a *point cloud* of $[x, y, z]$ coordinates from which it is possible to extract the ground surface, the canopy surface, and (if the density of points is high enough) recognizable attributes of individual trees. A similar scanner (called a terrestrial LiDAR) can also be mounted on a tripod to yield a very high density of returns in the immediate vicinity of the scanner. General principles of LiDAR and its use in forestry applications can be found in van Leeuwen and Nieuwenhuis (2010).

Measurements of heights and crown attributes of individual trees from raw LiDAR data require, first, that the points comprising the ground be accurately identified. Then, the groups of points corresponding to individual trees must be identified in the data through a process called *segmentation*. A variety of approaches have been used for individual tree segmentation in airborne LiDAR data, including cluster analysis (Morsdorf et al., 2004) and a variety of artificial intelligence and machine vision techniques (van Leeuwen and Nieuwenhuis, 2010). Once the points corresponding to an individual tree have been isolated from the overall cloud, it is a seemingly straightforward operation to find total tree height as the difference between the highest z-coordinate and that of the ground, and to characterize the area, shape, and spread of the crown. Success depends somewhat on the density of returns in the point cloud, and the shape of the tree crown. Individual tree heights measured with airborne LiDAR tend to show a negative bias because there is no guarantee that any returns in the point cloud will correspond precisely to the apex of the crown. The problem is worse with conifers than with broad-crowned hardwoods, with low-density point clouds, and when scanning is conducted during leaf-off periods (which is often done to maximize information about the ground surface). The challenge of reconstructing tree heights and crowns from airborne lidar is also more difficult for understory trees.

5.4. FORM

The form of the main stem of a tree has been the subject of many studies and methods of expression in forest mensuration. The stimulus for studying form is its relation to the cubic volume of the tree (Chapter 6). The form of the main stem of trees varies due to differences in rates of diminution in diameter from the base to the tip. The diminution in diameter, known as *taper*, which varies with species, dbh, and age of trees and with sites, is the fundamental reason for variation in volume. In a definitive study, Larson (1963) discussed the biological concept of stem form by a comprehensive review of the literature. Prodan (1965) presented a detailed summary of European approaches to the description of stem form, which was updated in Prodan et al. (1997). Form factors, form quotients, form point, and taper tables, curves, and formulas can express stem form. In all cases, the ultimate objective is to utilize the expressions in the estimation of tree cubic volume.

5.4.1. Form Factors

A *form factor* is the ratio of tree volume to the volume of a geometrical solid, such as a cylinder, a cone, or a cone frustum, that has the same diameter and height as the tree (the diameter of the geometrical solid is taken at its base; the diameter of the tree is taken at breast height). A form factor is different from other measures of form in that it can be calculated only after the volume of the tree is known. In formula form, the form factor f is

$$f = \frac{V}{V_{gs}} \tag{5.11}$$

where V = volume of tree

V_{gs} = volume of geometrical solid of same diameter and height

Early in the nineteenth century, it was recognized in Europe that the form of tree stems approached that of the solids discussed in Chapter 6. It was also recognized that there were many variations in form, and that a tree rarely was of the exact form of one of these solids. Thus, the form factor was conceived as a method of coordinating form and volume. The main objective of this early work was to derive factors that would be independent of diameter and height, and by which the volume of standard geometrical solids could be multiplied to obtain the tree volume. For example, the ratio of the volumes of a paraboloid to a cylinder is 0.5 when the base diameter and the height of the two solids are equal: the volume of the paraboloid is obtained by multiplying the volume of the cylinder by 0.5.

The *cylindrical form factor*, f_c, which has been the most commonly used, may be expressed by the equation

$$f_c = \frac{V}{gh} \tag{5.12}$$

where V = volume of tree in cubic units

g = cross-sectional area of cylinder whose diameter equals tree dbh (eqs. (5.3) or (5.4))

h = height of cylinder whose height equals tree height

This form factor has also been called the "false" form factor. The form factor based on cross-sectional area determined from the tree diameter at one tenth of total height ($H/10$) has been termed the "real" or "true" form factor (Prodan et al., 1997). The usefulness of form factors to estimate the volume of trees of variable form is limited. However, there are some uses for the rapid approximation of volume as well as the volume of trees of little form variation.

5.4.2. Form Quotients

A *form quotient* is the ratio of a diameter measured at some height above breast height, such as one-half tree height, to diameter at breast height. In formula form the form quotient q is

$$q = \frac{d_h}{d} \tag{5.13}$$

where d_h = diameter at height h above breast height
         d = dbh

Next to diameter at breast height and total height, form quotient is the most important variable that can be used to predict volume of a tree stem. Thus, it may be used as the third independent variable in the construction of volume tables (Chapter 6).

The original form quotient (Schiffel, 1899) took diameter at one-half total tree height $d_{0.5h}$ as the numerator, and dbh as the denominator. This was termed the *normal form quotient $q_{0.5}$*:

$$q_{0.5} = \frac{d_{0.5h}}{\text{dbh}} \tag{5.14}$$

For this form quotient, as tree height decreases, the position of the upper diameter comes closer to the breast height point until, for a tree whose height is double breast height, they coincide. To eliminate this anomaly, Jonson (1912) changed the position of the upper diameter to a point halfway between the tip of the tree and breast height, $d_{0.5(h-bh)}$ and called the ratio the *absolute form quotient q_a*:

$$q_a = \frac{d_{0.5(h-bh)}}{d} \tag{5.15}$$

The absolute form quotient is a better measure of stem form than the normal form quotient; however, it is not a pure expression of stem form. It is not independent of diameter and height and varies within a diameter–height class for a given species. For most species, absolute form quotients diminish with increasing diameters and heights, varying between 0.60 and 0.80. Absolute form quotient is 0.707 for a paraboloid, 0.500 for a cone, and 0.354 for a neiloid when diameter is taken at the base. These values hold irrespective of the diameter and height of the solid.

In determining the normal or absolute form quotient, the two diameters may be taken either outside or inside bark. Although it is difficult to obtain an accurate upper-stem diameter inside bark, the ratio of the inside bark measurements is a better index of form than the outside bark measurements because variable bark thicknesses potentially distort the ratio.

Originally, the term *form class* was applied to classes of absolute form quotients, as in a frequency table. For example, form classes with intervals of 0.05 have been laid out as follows: 0.575–0.625, 0.625–0.675, 0.675–0.725, and 0.725–0.775. The midpoints of these classes, 0.60, 0.65, 0.70, and 0.75 were used to name the classes, and a tree falling into a particular class was said to have the form quotient of the midpoint of the class. Tree–volume computations were classified by absolute form class as well as by diameter and height. Form class may be related to the density of the stand (Jonson, 1912) or to form point.*

In North America, the term form class has been used in a different sense. Girard (1933), in the course of work with the U.S. Forest Service, developed a form quotient for use as an independent variable in volume table construction. This measure, termed *Girard form class*, q_g, is the percentage ratio of diameter inside bark at the top of the first standard log, d_u, to dbh outside bark. When 16-ft logs are taken as the standard, the upper diameter, $d_{u17.3}$, is taken at the height of a standard 1-ft stump plus 16.3 ft (allowing an extra 0.3 ft for trim on the log). Thus,

$$q_g = 100\left(\frac{d_{u17.3}}{\text{dbh}}\right)$$ (5.16)

Girard form class is a useful form quotient that has been widely employed in US forestry practice.

Efforts to develop form quotients that can be computed from more accessible diameters have led to a number of form quotients of the type advocated by Maass (1939). Maass' form quotient is the ratio of the diameter at 2.3 m above the ground to dbh. Unpublished studies by Miller (1959) of similar form quotients indicated that this quotient, whether measurements are inside or outside bark, is too variable for trees of the same species, diameter, and height to be of practical use.

5.4.3. Taper Tables, Curves, and Formulas

If sufficient measurements of diameters are made at successive points along the stems of a sample of trees, one can prepare average taper tables that give a good picture of stem form. In its simplest form, taper can be mathematically expressed as (Ormerod, 1973)

$$\frac{d}{D_r} = \left(\frac{H-h}{H-H_r}\right)^{b_1}$$ (5.17)

where d = diameter measured at height h

   D_r = diameter measured at a reference height (usually breast height)

   h = height of diameter d

   H_r = reference height (usually breast height, 1.3 m or 4.5 ft)

   H = total tree height

   b_1 = taper coefficient

---

* Form point is the percentage ratio of the height to the center of wind resistance on the tree, approximately at the center of gravity of the crown, to the total tree height (Jonson, 1912). The greater the form point, the more nearly cylindrical will be the form of the tree. Fogelberg (1953) reported a study where form points where estimated for the various diameter classes of a stand. At present the form point is rarely used in forest mensuration.

The reference height is usually breast height (1.3 m or 4.5 ft) and D_r = dbh. The structure of eq. (5.17) insures the $d = D_r$ when $h = H_r$ and $d = 0$ when $h = H$; however, the equation also assumes that tree taper is constant over the entire length of the stem. Tree stems can be viewed as a composite of geometric solids (Section 6.1.3), thus simple taper functions, such as eq. (5.17), are often modified or expanded to include additional diameter and/or height measurements along the stem.

For example, Flewelling et al. (2000) describe a method for predicting upper-stem diameters based on three or more measurements: dbh, total height, and one or more upper-stem diameter measurements. Their multipoint profile system allows for upper-stem measurements at any height and constructs a smooth profile prediction that passes through all measurement points. Examples of other more complex taper functions can be found in Max and Burkhart (1976), McClure and Czaplewski (1986), Kozak (1988), and Rustagi and Loveless (1991). Kozak (2004) provides a brief history of taper studies and offers suggestions for fitting and evaluating taper functions.

While the use of taper equations have become more common with increased computing capabilities, historically, taper was often summarized tabularly. The ultimate purpose of all taper equations and tables is to show upper-stem diameters, which can then be used to calculate the volume of the sections of a tree, and the entire tree as described in Section 6.1.1. Taper tables can assume several forms. Table 5.2 is an example of a taper table showing upper-log taper rates inside bark for standard log lengths according to dbh and merchantable height of standard logs. Table 5.3 is an example of another form of taper table. It shows the diameters of the stem at increasing heights as percentages of dbh. Separate tables are prepared for species and total height classes. Within each height class, table percentages are shown for different dbh classes.

5.4.4. Slenderness

The slenderness coefficient S is defined as the ratio of tree height to dbh when both are expressed in the same units. For example, if a tree has $d = 40$ cm and $h = 20$ m, then S is $20/0.4 = 50$. The slenderness coefficient has been used in a variety of studies as an indicator

TABLE 5.2. Average Upper-Log Taper Inside Bark (inches) in 16-ft Logs

dbh (in.)	Two-Log Tree Second Log	Three-Log Tree Second Log	Three-Log Tree Third Log	Four-Log Tree Second Log	Four-Log Tree Third Log	Four-Log Tree Fourth Log
10	1.4	1.2	1.4			
12	1.6	1.3	1.5	1.1	1.4	1.9
14	1.7	1.4	1.6	1.2	1.5	2.0
16	1.9	1.5	1.7	1.2	1.6	2.1
18	2.0	1.6	1.8	1.3	1.7	2.2
20	2.1	1.7	1.9	1.4	1.8	2.4
22	2.2	1.8	2.0	1.4	2.0	2.5
24	2.3	1.8	2.2	1.5	2.2	2.6
26	2.4	1.9	2.3	1.5	2.3	2.7
28	2.5	1.9	2.5	1.6	2.4	2.8
30	2.6	2.0	2.6	1.7	2.5	3.0

Source: Mesavage and Girard (1946).

TABLE 5.3. Taper Percentages (Percent of dbh at Height on Tree) According to dbh Class, Height, Location of Tree and Species

Height Location on Tree (m)	dbh Class (cm)			
	20–25	25–30	...	45–50
1	102.9	103.0		103.5
3	92.6	92.0		88.4
5	86.8	86.5		81.4
⋮	⋮	⋮		⋮
27	18.8	17.8		14.7
30	0.0	0.0		0.0

Source: Adapted from Prodan et al. (1997).

of vulnerability to wind or snow damage. Its earliest use seems to have been in Europe (Brünig, 1974) but its simplicity and intuitive use has lead to its rapid spread to other regions. Although the slenderness coefficient does not account for all the factors that predispose a tree or stand to windthrow, several studies have found it to be an apt predictor (e.g., Cremer et al., 1982; Wonn and O'Hara, 2001). Diameter growth tends to be more affected by competition than height growth, so dominant trees typically have a lower S than trees in lower crown classes (Oliver and Larson, 1996). Because slenderness is strongly influenced by stand density, it can be used as a guide to thinnings. Castedo-Dorado et al. (2009) included S in a density management diagram for *Pinus radiata* plantations in Spain.

A similar ratio, calculated from height and the caliper of a seedling, can be considered an index of vigor or sturdiness (Haase, 2008). Seedlings with a high ratio are thin and spindly, and are more vulnerable to damage from poor handling and adverse weather conditions.

5.5. CROWN PARAMETERS

Research over the past 30 years has begun to focus on understanding the mechanisms involved in forest growth and dynamics. Since foliage is the primary source of photsynthates in most forest vegetation, the use of crown dimensions, as a surrogate for foliage or resource use, has become an integral part of many forest growth and yield studies. Crown dimensions are also useful for predicting wildlife habitat value and monitoring forest health. Moreover, remotely sensed data, including aerial photographs and satellite imagery, often contain information directly linked to tree crowns. Thus establishing a link between tree crowns and other variables of interest can be a key step in using such information.

5.5.1. Crown Length

Crown length is a useful measure of tree vigor (e.g., Briegleb, 1943; Keen, 1943) and has been suggested to be a good predictor of foliage and branch biomass (Mäkelä and Valentine, 2006). In principle, crown length should be straightforward to measure as the difference between total height of the tree and the bole height, or between total height and the base of the crown (Section 5.3). However, there is some variation in the exact definition of crown length. For example, Curtis and Reukema (1970) identified crown base as the height of the first whorl of branches having live foliage in at least 3 of 4 quadrants around the stem. Consistent measurements of crown length can be especially challenging in trees where lower branches may persist for extended periods with very small amounts of foliage or in trees with decurrent form.

5.5.2. Crown Diameter and Area

International definitions of forest area depend critically on the amount of tree crown area per unit land area (e.g., FAO, 2006). Crown diameter measurements on aerial photographs or other high-resolution imagery can be used to estimate diameter at breast height, tree volume, and other tree parameters (Francis, 1986). Estimates of crown diameter are also used to determine the ratio of crown diameter to dbh and for estimating the crown competition factor (Section 8.9.4). Measurements can be made on vertical aerial photographs, if available, or by field measurements.

Field determination of crown diameter or area is difficult because of the irregularity of a tree crown's outline. The field technique is to project the perimeter of the crown vertically to the ground and to make either "diameter" or "radius" measurements on this projection (Fig. 5.18a). The usual procedure is to take the average of the diameter of the

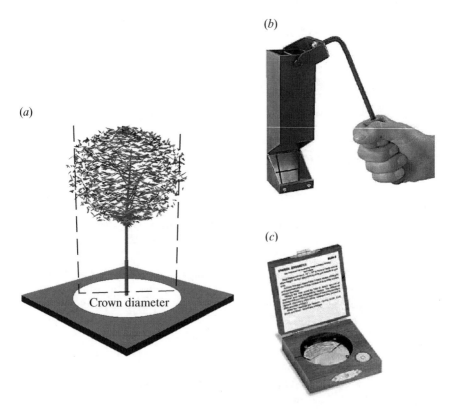

FIG. 5.18. Crown projection measurement: (*a*) crown projection area, (*b*) right-angle prism densitometer, and (*c*) spherical crown densitometer.

crown's widest point and a second measurement at right angles; this is similar to the procedure used for measuring tree diameter with calipers and assuming an elliptical form. Alternatively, one can begin at the tree stem (or other suitable location in the interior of the crown) and take multiple measurements of radius. If the measurements are taken at random azimuths or systematically from a point in the interior, then using the quadratic mean of those measurements as the radius of an equivalent circle gives an unbiased estimate of crown area without any assumptions about crown shape (Gregoire and Valentine, 1995). Most instruments used to achieve the vertical projection incorporate a mirror, a right-angle prism, or a pentaprism (Fig. 5.18b); they may be handheld or staff-mounted. Instruments that employ the pentaprism are recommended because they do not invert or revert images, and because slight movement of the prism will not affect the true right angle of reflection. Instruments and methods for determining vertical crown projection have been described by Husch (1947), Nash (1948), Raspopov (1955), Shepperd (1973), Cailliez (1980), and Tallent-Halsell (1994).

Crown diameters measured on traditional vertical aerial photographs are more clearly defined than those measured on the ground, although crown diameters measured on aerial photographs are smaller than those measured on the ground, because parts of the crown are not resolved on photographs; however, the photo measurement is probably a better measure of the functional growing space of a tree and is better correlated with tree and stand volume.

Crown cross-sectional area is most often calculated from the formula for the area of a circle using one of the following: the average of the maximum and minimum crown diameters; two diameters at right angles; or from the maximum diameter and another at right angles to it. The accuracy of these approaches depends on the shape of the crown, which is rarely either circular or elliptical. The sampling strategy presented by Gregoire and Valentine (1995) is unbiased for any crown shape.

5.5.3. Crown Surface Area and Volume

Crown surface area and volume can be estimated from crown diameter and length. Hamilton (1975) gives formulas for several assumed geometrical shapes. Schreuder et al. (1993) indicate that the crown surface area and volume for conifers and young hardwoods can be estimated as cones from the following:

$$S = \pi d_b \left(\frac{L}{2} \right) \tag{5.18}$$

and

$$V = \pi d_b^2 \left(\frac{l}{12} \right) \tag{5.19}$$

where S = crown surface area in m^2 or ft^2

  V = crown volume in m^3 or ft^3

  d_b = diameter at crown base (in m or ft)

  L = sloping length (in m or ft) from the apex of the crown to its base

  l = crown length (in ms or ft)

Maguire and Hann (1989) suggest that crown surface area should be a particularly good predictor of foliage area and biomass.

5.5.4. Foliage Area and Biomass

Leaf Area. The surface area of the foliage of forest trees is a useful measure for the study of precipitation interception, light transmission through forest canopies, forest litter accumulation, soil moisture loss, and transpiration rates (e.g., Wang and Jarvis, 1990). Leaf surface area also has become an important component to many growth and yield studies and the models resulting from these studies (Vose and Allen, 1988). Leaf area may be expressed as a quantity (e.g., square meter or square feet) or as a ratio of leaf area to crown projection area (Section 5.5.2) or stand area, commonly referred to as leaf area index.

Measurement of the surface area of detached leaves, even on a small sample of foliage is extremely time consuming. A variety of methods have been employed to measure leaf surface area; a description of the more commonly utilized methods can be found in the review by Larsen and Kershaw (1990). Because weighing foliage is generally easier than measuring surface areas, the typical approach to determining leaf surface area is to directly measure the surface area on a small sample of fresh foliage. This sample is then dried in an oven and the dried foliage weighed. The ratio of fresh surface area to dried weight is then used to estimate surface area on larger samples of dried foliage.

Because of the difficulty in directly measuring leaf surface areas, a number of techniques for estimating individual tree and stand-level leaf areas have been developed. These techniques fall into two broad classes: (1) subsampling methods and (2) canopy estimation methods (Larsen and Kershaw, 1990).

Subsampling techniques require destructive sampling of all or a proportion of the canopy of interest. The simplest method is the percentage sample where a fixed proportion of foliage is sampled for direct measurement. The sample units may be individual leaves or branches. Sample units may be selected by random, systematic or stratified selection (see Chapter 10 for a description of these sample designs). Total foliage is estimated by scaling the sample values by the percentage sampled (Larsen and Kershaw, 1990). Another subsampling technique is the stratified clip method. The stratified clip method uses a small plot projected vertically as a column through the canopy. All of the foliage within this space is carefully clipped, usually divided into preset height strata. The surface area of the foliage is directly measured or obtained by applying ratios of surface area to dry weight to the oven-dried samples. Total leaf surface area is obtained by scaling these measures to the whole tree or to the per unit area level (MacArthur and Horn, 1969). Valentine et al. (1984) developed a subsampling scheme for hardwood trees based on ratios of stem cross-sectional areas. This approach assumes that stems of a given size support similar amounts of foliage and the ratio of stem cross-sectional area to branch cross-sectional area is used to determine a sampling probability.

Litter traps are an alternative, nondestructive subsampling technique. Fine-meshed nets or screens are placed under the canopy and the surface area of foliage litter falling into the traps is used to estimate total foliage. Usually, multiple traps are needed to capture the spatial variability in litterfall even for a single relatively homogenous plot. Several years of sampling may be required to obtain accurate estimates of foliage in species that retain their leaves for more than one growing season unless foliar demography can be assessed through marked foliage or destructive measurement of a subsample of branches (Trofymow et al., 1991; Innes et al., 2005). Sorting the material that accumulates in litter traps, either to obtain separate estimates by species or to eliminate the contributions of twigs, branches, cones, and other material, can be extremely labor-intensive. Thus, weighing all traps in bulk, and sorting only a subsample, can be an important tool to control costs (Dellenbaugh et al., 2007).

Several indirect methods have been developed to estimate the amount of foliage in crowns and canopies. The most widely developed technique is the use of allometric relationships, to which we will return in Section 5.6. Light interception models (Ross, 1981; Campbell, 1986; Pierce and Running, 1988) have often been used to estimate canopy properties. Many other indirect methods are based on closely related theory originally developed around the idea of inserting probes into the canopy and counting contacts (Wilson, 1959a, 1959b, 1965). This theory was subsequently broadened to deal with situations in which only the distance to the first contact, or the proportion of probes making a contact, would be available. Hemispherical photography (Ondok, 1984; Wang and Miller, 1987; Jonckheere et al., 2004) and methods based on rangefinding with a camera (MacArthur and Horn, 1969; Aber, 1979) have been widely used. Similar rangefinding methods have also employed using handheld lasers (Radtke and Bolstad, 2001; Maynard et al., 2013), while terrestrial and airborne LiDAR both offer possibilities for estimating LAI and its vertical structure (Lovell et al., 2003; van Leeuwen and Nieuwenhuis, 2010). It is also possible to predict foliage area from reflectance data derived from satellite or airborne optical remote sensing data (Running et al., 1986; Peterson et al., 1987; Chen et al., 1997).

5.5.5. Other Crown Characteristics

Other crown characteristics used in tree and forest health monitoring are crown density; live crown ratio, crown dieback, and foliage transparency. Examples of techniques for their measurement are presented in USDA Forest Service Forest Inventory and Analysis Field Methods (FIA, 2011). Measurements for each variable are made on individual trees. The measurements are estimated into 5% rating classes ranging from 0 to 100%. The ratings are based on the estimated percentage of a tree's crown that meets the definition of the variable.

Crown density is the amount of crown branches, foliage, and reproductive structures that block light visibility through the crown. A crown density-transparency card is used for making this estimate. The higher the rating, the denser and, presumably, healthier the tree. A spherical crown densiometer (Fig. 5.18c) is also used for estimating crown density and canopy coverage.

Live crown ratio is the percentage of total height supporting live foliage that is effectively contributing to tree growth. It is the ratio of live crown length to total tree height.

Crown dieback is recent mortality of branches with fine twigs, which begins at the terminal portion of a branch and proceeds toward the trunk and/or base of the live crown. The lower the rating, the lower the mortality and presumably the healthier the tree.

Foliage transparency is the amount of background (skylight, foliage of other trees) visible through the live normally foliated portion of the crown or branch. The lower the rating, the thicker the foliage and presumably the healthier the tree.

5.6. REGRESSION AND ALLOMETRIC APPROACHES

5.6.1. Allometry of Standing Trees

As the preceding sections make clear, there are a variety of ways of measuring several attributes of standing trees. Many of these methods are time-consuming and potentially error-prone, so it rapidly becomes impractical to measure several attributes on all standing trees in a sample unless the sample is very small. As a result, allometric or regression approaches are often employed. The subject of allometry for estimating the weight and biomass of entire trees will be taken up in much more detail in Chapter 6; here, we consider a few examples and salient issues for using allometry to estimate the variables considered in this chapter, including height, crown dimensions, and foliage area or mass. Although allometric relationships can be of interest in their own right, our primary purpose in considering them here is to replace an expensive or time-consuming measurement with a predictive relationship based on a much less expensive one, at least for a fraction of the trees in a sample.

Tree height and diameter are fundamentally linked through mechanical constraints on tree form (Greenhill, 1881). It is extremely common in inventories to measure dbh for all trees in a sample, but to measure height only for a subsample, thus speeding the overall work. By doing so, a double-entry volume table can be converted into a local single-entry volume table, allowing volume to be estimated from dbh only (see Section 6.2.2). Alternatively, a regression equation developed elsewhere may be used to predict the heights of all trees; however, the use of off-site height–diameter equations is always subject to some error because site quality and the history of competition in a stand exert a strong influence over relationships between these two variables.

There are many different equation forms used to model height–diameter relationships. One common equation form for modeling height–diameter relationships is the Schumacher equation (Schumacher and Hall, 1933):

$$H = \exp\left(b_0 + \frac{b_1}{\text{dbh}} \right) \tag{5.20}$$

where the b_i are regression parameters. Although the Schumacher equation appears non-linear, it can be fit using linear regression (Section 3.8) as $Y = b_0 + b_1 X$, where $Y = \ln(H)$, and $X = (1/\text{dbh})$. The logarithmic transformation of H often helps equalize the variability of the height residuals, thus better satisfying the assumptions of least squares regression, while the reciprocal transformation of dbh creates a sigmoid curve that reproduces the intuitive shape of a height–diameter curve. In addition to eq. (5.20), other equation forms include

$$H = \text{bh} + b_0 \text{dbh}^{b_1} \tag{5.21}$$

$$H = \text{bh} + \frac{b_1 \text{dbh}}{b_0 + \text{dbh}} \tag{5.22}$$

$$H = \text{bh} + b_0 \left(\frac{\text{dbh}}{1 + \text{dbh}} \right)^{b_1} \tag{5.23}$$

$$H = \text{bh} + \frac{b_0}{1 + b_1 e^{-b_2 \text{dbh}}} \tag{5.24}$$

$$H = \text{bh} + b_0 \left(1 - e^{-b_1 \text{dbh}}\right) \tag{5.25}$$

$$H = \text{bh} + b_0 \left(1 - e^{-b_1 \text{dbh}}\right)^{b_2} \tag{5.26}$$

where bh = breast height. Equations (5.21)–(5.26) ensure the H = bh when dbh = 0. No single equation form has proven best in empirical studies. Figure 5.19 shows eq. (5.20) through (5.26) fitted to the western hemlock height–diameter data used in Section 3.8. All of the equations make similar predictions near the middle of the range of data and differ most noticeably near the tails of the data. Choice of equation form often depends upon the

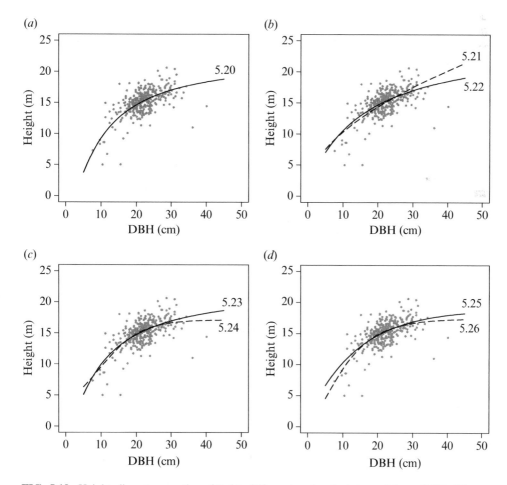

FIG. 5.19. Height–diameter equations fitted to 349 western hemlock trees: (*a*) eq. (5.20), (*b*) eqs. (5.21) and (5.22), (*c*) eqs. (5.23) and (5.24), and (*d*) eqs. (5.25) and (5.26).

range of data sampled and the need to extrapolate beyond that range. For a more comprehensive list of height–diameter equations, see Huang et al. (2000).

Because height–diameter relationships are subject to the influence of stand age and competition, stand-level variables often improve predictions (Staudhammer and LeMay, 2000; Trincado and Leal, 2006; Kershaw et al., 2008; Rijal et al., 2012). Various stand and site parameters are incorporated into developing a more mechanistic approach to height–diameter relationships remains an active area of research (Valentine et al., 2012a).

Crown dimensions also can be time-consuming and error-prone to measure, so allometric approaches often are used with these variables (Bechtold, 2003). Crown radius or crown width is often linear with diameter, and hence crown area is often approximately quadratic with diameter, though a power function is also frequently used. For crown length, care must be taken to ensure that the base of the live crown is not predicted to fall below the base of the tree, or above its tip, so alternative equation forms are often used (Temesgen et al., 2005). As with height–diameter allometry, including some stand characteristics often improves predictions because of the influence of competition (e.g., Bonnor, 1964a; Condés and Sterba, 2005).

The amount of foliage (measured as either area or biomass) is of prime interest for understanding and predicting growth, but is exceptionally difficult to measure directly. Diameter at breast height can be used in allometric relationships to predict leaf area or biomass, but such relationships are often unstable outside the specific site and stand conditions where they were developed. Many allometric models utilize sapwood area as a predictor of leaf area (Turner et al., 2000). The sapwood is the living portion of the tree bole and is the zone where active transport of water and nutrients occurs. A functional relationship between sapwood area and foliage area, called the pipe model theory, was formulated by Shinozaki et al. (1964). The basic premise of this theory is that a unit of foliage requires a corresponding unit of sapwood for support. For many species, a linear relationship between sapwood area at breast height and total foliage area has been observed, and a number of equations predicting leaf area from sapwood area at breast height have been developed (e.g., Waring et al., 1982; Marshall and Waring, 1986; O'Hara and Valappil, 1995); however, several researchers have observed significant biases in the predictions of leaf area from sapwood area at breast height (e.g., Dean et al., 1988). To address these biases, the basic relationship between sapwood area and foliage area has been modified to incorporate measures of stand density, height to crown base, total tree height, crown width, and other tree dimensions (e.g., Dean and Long, 1986; Long and Smith, 1988; Maguire and Batista, 1996; Turner et al., 2000). Mäkelä and Valentine (2006) suggest that crown length can be used directly as a predictor of foliage area and biomass and offer theory as well as empirical results to bolster that claim.

5.6.2. Applications to Seedlings, Saplings, and Understory Plants

The application of allometric approaches to predicting height, crown dimensions, and foliage to seedlings and other plants in the understory is straightforward, except that the measurements used are often different. For example, seedling caliper rather than dbh often is used. For shrubs and other plants without a strong central stem, height may be the preferred measurement to use as a predictor of other variables because height is much more simply measured in the field for low-stature vegetation. Log-log or power function regressions often are used to relate variables allometrically (e.g., Paton et al., 2002).

TABLE 5.4. Diameter at Breast Height and Dry Foliage Biomass for *Poecilanthe effusa* Trees in the Eastern Amazon

dbh (cm)	Foliage Biomass (g)
1.1	66
1.1	131
1.2	274
1.4	28
1.7	91
3.5	762
3.7	1085
3.9	696
5.8	582
6.9	3340

Source: Adapted from Ducey et al. (2009).

As an example, Table 5.4 gives the dbh and foliage biomass for 10 stems of *Poecilanthe effusa*, a leguminous understory tree, in a secondary forest in the eastern Brazilian Amazon (Ducey et al., 2009). Using regression to develop a predictive allometric equation, one could estimate foliage biomass of other trees of this species in the same stand; use of the equation outside the population within which it was developed would require verification that the equation is applicable to the new population.

By taking a logarithmic transformation of both dbh and foliage biomass (*f*), one can fit a power–function allometric equation using linear regression. The resulting equation is

$$\ln(f) = 4.1596 + 1.7588 \ln(\text{dbh})$$

or, after reversing the log transformation,

$$f = 64.0 \cdot \text{dbh}^{1.759}$$

Although log transformation provides a convenient way to fit allometric equations, the predictions are biased when transformed back to the original scale (Baskerville, 1972); the bias depends on root mean square error (rMSE). A correction multiplier can be calculated as

$$c = e^{\left(\text{rMSE}^2/2\right)}$$

In this case, the errors around the regression line are quite large, so $c = 1.4012$. Multiplying through by the correction factor yields a bias-corrected allometric equation,

$$f = 89.7 \cdot \text{dbh}^{1.759}$$

The least-squares regression, the corrected regression, and the original allometric equation of Ducey et al. (2009), fit using a maximum-likelihood technique that accounts for bias and also unequal variance, are shown in Figure 5.20.

Examination of Figure 5.20 reveals several features that are common in fitting allometric equations. First, the log-transformation does help the data better satisfy the ordinary least squares assumptions of equal variance and normally distributed errors, and creates a

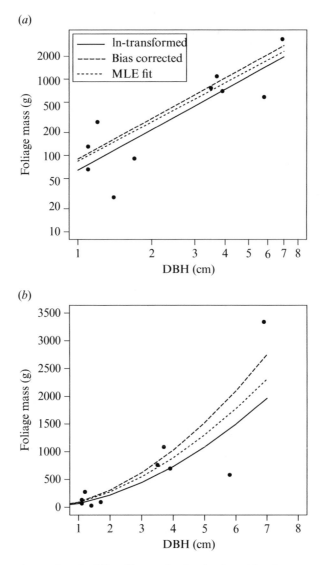

FIG. 5.20. Allometric equations for foliage biomass fitted to data from 10 understory trees in a secondary Amazonian rainforest: (*a*) natural log-transformed space and (*b*) normal measurement space.

more linear relationship. Second, however, log-transformation can also conceal errors between the regression and individual data values that are quite large. These data are noisy, and large prediction errors are to be expected for trees at the large end of the dbh range. Finally, note that the bias-corrected ordinary least-squares regression is very close to that fitted by a far more complex maximum likelihood technique ($f = 84.4 \cdot \text{dbh}^{1.700}$). In this case, a relatively simple statistical approach is more than adequate to reveal a useful predictive relationship.

6

DETERMINATION OF TREE VOLUME, WEIGHT, AND BIOMASS

Throughout most of the nineteenth and twentieth centuries, the most important portion of a tree, in terms of usable wood, has been the stem. The opening decades of the twenty-first century have seen an interest in estimating biomass and carbon content, and the determination of values for the whole tree rather than just the main stem. In forest mensuration, a tree has been traditionally considered as consisting of four primary components: *roots, stump, stem,* and *crown.* Much of the current work on biomass estimation divides the tree into roots (often further divided into coarse and fine roots), stemwood, branch wood, foliage, and bark components. The *roots* are the underground part of the tree that supplies it with water, nourishment, and mechanical support. The *stump* is the lower end of the tree that is left above ground after the tree has been felled. In many biomass studies, the stump is often considered as part of the root component (Smith et al., 2003); however, methods vary (Snowdon et al., 2000). The *stem* is the main ascending axis of the tree above the stump (trees that have the axis prolonged to form an undivided main stem, as exemplified by many conifers, are termed *excurrent*; trees that have the axis interrupted in the upper portion due to branching, as exemplified by many broadleaved species, are termed *deliquescent*). The *crown* consists of the primary and secondary branches growing out of the main stem, together with twigs and foliage. In most biomass studies, the crown is separated into branch wood and foliage components (Jenkins et al., 2003, 2004; Smith et al., 2003). In the past, since the roots and stumps of trees were less often utilized, little attention was given to the determination of their volume. Now, with increased interest in complete tree utilization and quantification of total biomass and carbon content, estimation of the volume, weight, or biomass of the roots and stump, as well as the crown of trees is often as important as the estimates for the stem (Jenkins et al., 2003).

Traditionally, volume has been the most widely used measure of wood quantity in forest mensuration. Volume, as defined in Chapter 2, is the three-dimensional space occupied by an object. Tree volume may be expressed in terms of *total cubic volume* (ft^3 or m^3) or in

Forest Mensuration, Fifth Edition. John A. Kershaw, Jr., Mark J. Ducey, Thomas W. Beers and Bertram Husch.
© 2017 John Wiley & Sons, Ltd. Published 2017 by John Wiley & Sons, Ltd.

terms of *merchantable volume*, which only measures the portion of the tree utilizable for timber products. Merchantable volume may be expressed in terms of cubic volume or in *board-foot* measures (see Chapter 7). Total volume may include bark volume or just the volume of wood, while merchantable volume almost always is a measure of wood volume excluding bark. Recent interests in forests as a source of renewable energy and as a sink for CO_2 emissions has motivated interest in measuring and estimating tree weight, biomass, and carbon content (Brown, 1999; Jenkins et al., 2004; Smith et al., 2006).

Weight is defined in this text as the green or fresh weight of trees or other forest vegetation (i.e., the weight of the wood, bark, and water in freshly cut trees). While estimation of volume has been a primary focus of forest mensuration, interest in estimating weight is not new (Guttenberg et al., 1960; Husch, 1962). Weights of individual trees are not easily measured (see Section 6.1.7); however, weights of truckloads of logs are quickly and accurately determined using scales. *Weight scaling* is widely used and volume:weight ratios (Section 7.7) or ratio sampling procedures (Section 10.8) often are used to convert weight measures to volume measures.

Biomass is defined as the oven dry weight of trees and other forest vegetation. (In some contexts, especially forest-based bioenergy markets, bulk material scaled by weight is also called "biomass"; in this text, when biomass is used without qualification the term refers to dry weight.) Biomass has long been of interest to ecologists as a measure of site productivity (Satoo and Madgwick, 1982). The last 20 years has seen growing interest in biomass estimation in forest management as markets have emerged for biomass energy and carbon mitigation (Parresol, 1999; Jenkins et al., 2004; Lambert et al., 2005). The growing interest in using forests as carbon sinks to offset carbon emissions from fossil fuels and the potential to trade carbon offsets as a commodity has necessitated the need for methods to measure carbon quantity in forest vegetation and to estimate totals across forest management units (MacDicken, 1997), and evaluate long-term forest management plans based on maximizing carbon storage potential (Brown, 2002). As a result, estimation of carbon and other nutrient contents have become an emerging aspect of forest mensuration.

The distribution of volume, weight, and biomass varies depending upon the type of tree (conifer vs. hardwood) and tree size (Fig. 6.1). The cubic volume of a merchantable stem

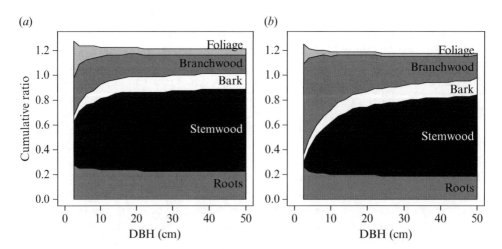

FIG. 6.1. Cumulative component ratios (% above-ground biomass) by tree component for: (*a*) conifer species and (*b*) hardwood species (adapted from Chojnacky et al. (2009)).

(to a 4-in. or 10-cm top diameter) of an average tree is about 55–65% of the cubic volume of the complete tree, excluding foliage, and the cubic volume of the crown of an average tree, excluding foliage, is about 15–20% of the cubic volume of the complete tree, excluding foliage, for softwoods and 20–25% for hardwoods. Most mensurational studies of roots have focused on determination of root biomass rather than volume (Snowdon et al., 2000). These studies indicate a range of 10–65% of the total biomass of a tree is found in roots (Barton and Montagu, 2004), and an average of 25% is frequently used when estimating below-ground biomass from above-ground (Snowdon et al., 2000).

6.1. MEASUREMENT OF INDIVIDUAL TREES

Direct determination of volume, weight, biomass, or carbon and nutrient content is usually made on sample trees to obtain basic data for the development of relationships between the various tree dimensions and the component of interest (Sections 6.2 and 6.3). Relationships of this type are then used to estimate these components for other standing trees. In the past, sample-tree measurements were often made on trees cut during harvesting operations. Volume relationships developed from such measurements may lead to bias because they may not be representative of all trees in a stand. Preferably, measurements should be taken on a representative sample of all standing trees across the full range of sizes of interest.

Direct determination of volume, weight, or biomass of any part of a tree involves clearly defining the part of the tree of interest, and carefully making measurements in accordance with constraints imposed by the definition. For example, for purposes of measurement, we might include the portion of the stem above a fixed-height stump to a minimum upper-stem diameter outside bark, or on stems that do not have a central tendency, to the point where the last merchantable cut can be made. For roots, we might include roots larger than some minimum diameter, and, for tops, we might include the branches and the tip of the stem to some minimum diameter outside the bark.

Generally speaking, a tree must be felled and the limbs cut into sections before one can obtain measurements to directly determine crown values. To obtain measurements to directly determine root values, roots must be lifted from the ground and the soil removed. For stem and stump volumes, the necessary measurements may be obtained from either standing or felled trees; however, weight and biomass require felling.

6.1.1. Stem Dissection

Once the components of interest are defined and a sample of representative trees selected, the measurements required to calculate volume, weight, and/or biomass must be collected. Most measurements require felling sample trees and cutting them into sections small enough to measure. Despite a large body of literature on volume, weight, and biomass estimation, very little consensus on methods exists. Most guidelines for estimating biomass and/or carbon content focus on per unit area estimates rather than individual tree measurements (e.g., Brown, 1997; Ravindranath and Ostwald, 2008), and individual tree methods are scattered throughout the literature.

The following general guidelines are recommended for the dissection and measurement of individual stems:

1. Select a representative sample across the full range of tree sizes of interest.
2. Establish clear, quantitative definitions for each tree component. If trees are felled, separate each tree into the different components prior to measurement or further dissection.
3. If subsampling is used for some tree components (such as branches and foliage), make sure the selection strategy is clear and easily implemented in the field. Ensure that adequate information is gathered on the nonsample portions so that the sub-sample can be scaled to the full tree.
4. Where possible, measurements should be made at equal intervals along the stem, branches, and roots. If trees are being dissected into smaller sections for weighing or volume displacement, these segments should be as large as feasible to minimize the number of cuts made. Each cut results in a loss of volume, weight, and biomass. For conifer species, measurements are generally made at equal intervals along the stem below the live crown, and at the midpoint between whorls within the live crown. For hardwood species, measurements are generally made at equal intervals along the stem up to the point of dividing into major branches.
5. Stem, branch, and root diameters can be measured using calipers (Section 5.2.1); however, care must be taken so that stem eccentricity does not influence caliper measures, especially on large logs lying on the ground.
6. If subsamples of stems (such as thin disks, increment cores, or cross-sectional slices) are being taken for weight, biomass, or carbon and nutrient content assessment, make sure that subsamples are clearly labeled using a logical system that enables the tree and location within the tree to be easily determined.
7. If biomass or carbon and nutrient contents are going to be determined, then the sub-samples will need to be returned to a laboratory, oven-dried, weighed, and chemical analyses conducted.

For volume estimation, measurements are typically made at intervals corresponding to normal log lengths (Bruce, 1982); for weight or biomass determination trees are dissected into pieces that are generally manageable by hand. Satoo and Madgwick (1982) give additional factors to consider when measuring biomass.

6.1.2. Volume Determination by Displacement

The most accurate method of measuring volume of an irregularly shaped solid is by measuring the volume of water that it will displace (Young et al., 1967; Martin, 1984) using principles developed by Archimedes in approximately 250 BCE. For example, the cubic volume of any part of a tree may be found by submerging it in a tank in which the water displacement can be accurately read. Water displacement may be read directly on the tank, calculated from a depth stick, or collected in a smaller container and read there. To use such a tank, which is termed a *xylometer*, to measure the volume of a tree, it is necessary to fell the tree and cut it into sections that are small enough to fit into the tank.

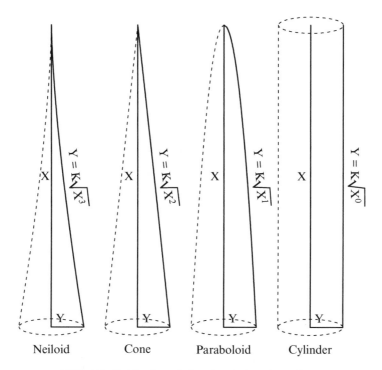

Neiloid   Cone   Paraboloid   Cylinder

FIG. 6.2. Solids of revolution descriptive of tree form.

Because of the logistics of submerging parts of trees and accurately measuring displacement, xylometric studies are not very common. Generally xylometric studies are used to confirm choice of volume calculation methods such as volume determination by formula (Section 6.1.3). Examples of the use of a xylometer to determine tree volume are found in Young et al. (1967), Martin (1984), Filho et al. (2000), Özçelik et al. (2008), and Akossou et al. (2013).

6.1.3. Volume Determination by Formulas

Solid objects may assume the form of polyhedrons, solids of revolution, and solids of irregular shape. The standard formulas shown in Table A.3 can be used to compute volumes of polyhedrons, such as cubes, prisms, and pyramids, and volumes of solids of revolution, such as cones, cylinders, spheres, paraboloids, and neiloids. Stems of excurrent trees are often assumed to resemble neiloids, cones, or paraboloids—solids that are obtained by rotating a curve of the general form $Y = K\sqrt{X^r}$ around the X-axis (Fig. 6.2). As the form exponent, r, in this equation changes, different solids are produced. When $r=1$, a paraboloid is obtained; when $r=2$ a cone; when $r=3$, a neiloid; and when $r=0$, a cylinder. Stems of excurrent trees are seldom exactly cones, paraboloids, or neiloids. In general, the form of most excurrent stems falls between a cone and a paraboloid. Merchantable portions of stems of deliquescent trees also resemble frustums of neiloids, cones, or paraboloids (occasionally cylinders). In general, they fall between the frustum of a cone and the frustum of a paraboloid.

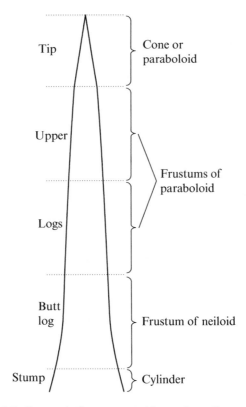

FIG. 6.3. Geometric forms assumed by portions of a tree stem.

It is more realistic to consider the stem of any tree to be a composite of geometrical solids (Fig. 6.3). For example, when the stem is cut into logs or bolts, the tip approaches a cone or a paraboloid in form, the central sections approach frustums of paraboloids, or, in a few cases, frustums of cones or cylinders, and the butt log approaches the frustum of a neiloid. Although the stump approaches the frustum of a neiloid in form, for practical purposes it is often considered to be a cylinder.

Formulas to compute cubic volume of the solids that have been of particular interest to mensurationists are given in Table 6.1. A number of other formulas are frequently found in older literature and may still be found in use outside of North America. Prodan (1965) and Prodan et al. (1997) give examples of these formulas used to compute cubic content of logs or other stem segments. All of these alternative formulas, as with the formulas shown in Table 6.1, calculate a weighted average cross-sectional area and determine volume by multiplying this average cross-sectional area by length. The formulas only vary by the number of cross sections used, the location of these cross sections, and the weight each cross section is given. For example, Smalian's formula (eq. (6.5), Table 6.1) uses the cross-sectional areas at the base and upper end, each cross section has equal weight, and the arithmetic average is computed, while Newton's formula (eq. (6.9), Table 6.1) uses cross-sectional area at the base, middle, and upper ends, and places four times the weight on the middle cross sections as on the ends. The variations in formulas are the result of differing assumptions about tree form, local conditions, and measurement and calculation technologies available (Fonseca, 2005).

TABLE 6.1. Equations to Compute Cubic Volume of Important Solids

Geometric Solid	Equation for Volume V (Cubic Units)[a]	Equation Number
Cylinder	$V = A_b H$	(6.1)
Paraboloid	$V = \dfrac{1}{2}(A_b H)$	(6.2)
Cone	$V = \dfrac{1}{3}(A_b H)$	(6.3)
Neiloid	$V = \dfrac{1}{4}(A_b H)$	(6.4)
Paraboloid frustum	$V = \dfrac{L}{2}(A_b + A_u)$ (Smalian's formula)	(6.5)
	$V = A_m L$ (Huber's formula)	(6.6)
Cone frustum	$V = \dfrac{L}{3}\left(A_b + \sqrt{A_b \cdot A_u} + A_u\right)$	(6.7)
Neiloid frustum	$V = \dfrac{L}{4}\left(A_b + \sqrt[3]{A_b^2 \cdot A_u} + \sqrt[3]{A_b \cdot A_u^2} + A_u\right)$	(6.8)
Neiloid, cone or paraboloid frustum	$V = \dfrac{L}{6}(A_b + 4A_m + A_u)$ (Newton's formula)	(6.9)

[a] A_b=cross-sectional area at base of segment, A_m=cross-sectional area at middle, A_u=cross-sectional area at upper end, H=height, and L=length.

The quarter-girth or Hoppus formula, which was widely used throughout the British Empire (Fonseca, 2005), is another approach frequently encountered in older literature. In imperial units, the quarter-girth formula is $V = (C/4)^2 (L/144)$ and in SI units the formula is $V = (C/4)^2 (L/10,000)$ (sometimes referred to as the Francon formula), where C is the midpoint girth (circumference) in inches or centimeters, and L is length in feet or meters. Many have attributed the girth divisor of 4 as a factor accounting for volume loss when sawing logs into boards (Section 7.3); however, the quarter-girth formula was most likely intended as a quick, easily calculated approximation rather than an exact calculation (Fonseca, 2005). Use of the quarter girth formula can still be found; however, use has declined with increased globalized wood markets and the need for standardization of estimates of volume and carbon content.

Newton's formula is exact for all the frustums we have considered. Smalian's and Huber's formulas are exact only when the solid is a frustum of a paraboloid. For example, if the surface lines of a tree section are more convex than a paraboloid frustum, Huber's formula will overestimate volume while Smalian's formula will underestimate volume; however, if the surface lines of a tree section are less convex than a paraboloid frustum, as they often are, Smalian's formula will overestimate volume and Huber's formula will underestimate volume. Assuming Newton's formula gives correct volume values, it can be shown by subtracting Newton's formula first from Smalian's formula and then from Huber's formula that the error incurred by Smalian's formula is twice that incurred by Huber's formula and opposite in sign ($[V_N - V_S] = -2 \cdot [V_N - V_H]$).

In a study by Young et al. (1967) on 8- and 16-foot softwood logs of 4- to 12-in. diameters, volumes calculated by Newton's, Smalian's, and Huber's formulas were compared with volumes determined by displacement (Section 6.1.2). Average percent errors of about

0% were obtained for Newton's formula, +9% for Smalian's formula, and −3.5% for Huber's formula. They also found that there were no significant errors for any of the three formulas when using 4-foot bolts. In a study by Miller (1959) on 16-foot hardwood logs of 8- to 22-in. diameters, volumes calculated by the three formulas were compared with volumes determined by graphical techniques (Section 6.1.4). Average percent errors of about +2% were obtained for Newton's formula, +12% for Smalian's formula, and −5% for Huber's formula. Other studies have found that Huber's formula performs superior to Newton's. Filho et al. (2000) found that Huber's formula performed best in a study of logs from 1 to 6 m in length. Newton's formula provided similar results, but Smalian's formula had the poorest performance. For large diameter logs, Huber's was consistently superior (Filho et al., 2000). In a recent study of Teak logs (Akossou et al., 2013), Huber's formula was found to be most effective for short bolts (0.5 m) while Newton and Smalian's formulas were better on longer bolts (1–3 m in length).

It should now be apparent that in calculating cubic volume of trees and logs, mensurationists should select their methods carefully. Unless one is willing to accept a rather large error, Smalian's formula should not be used unless it is possible to measure sections of the tree in short lengths (maximum lengths of 4 ft or about 1 m). For longer lengths, such as logs of 8 or 16 ft (about 3 to 6 m), Newton's or Huber's formulas will give more accurate results.

Newton's formula will give accurate results for all sections of a tree except for butt logs with excessive butt swell. For such butt logs, Huber's formula will generally give better results. Either the paraboloidal formula (eq. (6.2), Table 6.1) or the conic formula (eq. (6.3), Table 6.1) is appropriate to determine the volume of the tip. The cylindrical formula (eq. (6.1), Table 6.1) is normally used to compute the volume of the stump, although the stump actually approaches the neiloid frustum in form. Newton's and Huber's formulas cannot, of course, be applied to stacked logs because it is not possible to measure middle diameters.

Newton's formula may be used to compute volume of the merchantable stem or of the total stem. If the sections are of the same length, the procedure can be summarized in a single formula. To illustrate, consider a stem with diameters measured from the top of the stump to a point where the last merchantable cut will be made, d_0, d_1, d_2, d_3, d_4, d_5, and d_6, located at intervals of h units of length (Fig. 6.4a). To give each section three diameters, the volume is computed by sections of $2h$ length (Fig. 6.4b). With Newton's formula the cubic volume, V, is then

$$V = \frac{2h}{6}c\left(d_0^2 + 4d_1^2 + d_2^2\right) + \frac{2h}{6}c\left(d_2^2 + 4d_3^2 + d_4^2\right) + \frac{2h}{6}c\left(d_4^2 + 4d_5^2 + d_6^2\right)$$

$$= \frac{2h}{6}c\left(d_0^2 + 4d_1^2 + 2d_2^2 + 4d_3^2 + 2d_4^2 + 4d_5^2 + d_6^2\right)$$

$$= \frac{2h}{3}c\left(\frac{d_0^2}{2} + 2d_1^2 + d_2^2 + 2d_3^2 + d_4^2 + 2d_5^2 + \frac{d_6^2}{2}\right)$$

where d_i = diameter (in. or cm) at ith measurement point

      h = distance (ft or m) between measurement points

      c = diameter to cross-sectional area conversion factor (0.005454 for imperial units and 0.00007854 for SI units)

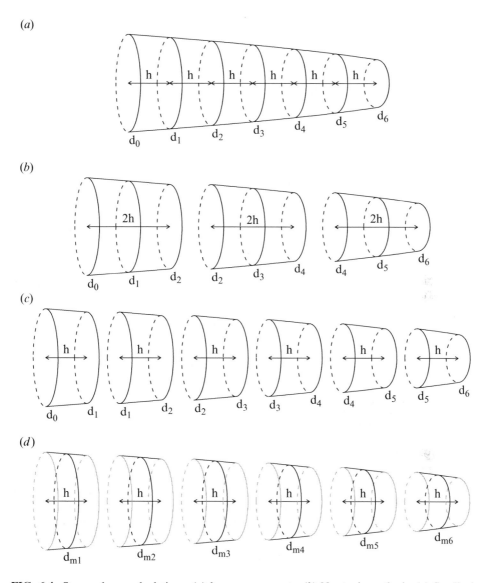

FIG. 6.4. Stem volume calculations: (*a*) log measurements, (*b*) Newton's methods, (*c*) Smalian's method, and (*d*) Huber's method.

This formula may be extended for as many sections as desired, provided there are an odd number of diameters (i.e., an even number of sections of h length); however, if the number of diameters measured is even, the last interval of h cannot be computed by Newton's formula because it will have only two end diameters. Thus, its volume must be found by Smalian's formula and added to the previous formula. For eight diameters, or seven intervals of h length, this yields

$$V = \frac{2h}{3}c\left(\frac{d_0^2}{2} + 2d_1^2 + d_2^2 + 2d_3^2 + d_4^2 + 2d_5^2 + \frac{5d_6^2}{4} + \frac{3d_7^2}{4}\right).$$

Grosenbaugh (1948) described a systematic procedure using this method.

The total or merchantable volume of the stem also can be calculated with good accuracy using Smalian's formula if the stem is divided into short sections. To illustrate, consider a stem with diameters measured from the top of the stump to a point where the last merchantable cut will be made, $d_0, d_1, d_2, \ldots, d_n$, located at intervals of h units of length along the stem (Fig. 6.4c). Then according to Smalian's formula:

$$
\begin{aligned}
V &= \frac{h}{2}c\left(d_0^2 + d_1^2\right) + \frac{h}{2}c\left(d_1^2 + d_2^2\right) + \cdots + \frac{h}{2}c\left(d_{n-1}^2 + d_n^2\right) \\
&= \frac{h}{2}c\left(d_0^2 + 2d_1^2 + 2d_2^2 + \cdots + 2d_{n-1}^2 + d_n^2\right) \\
&= hc\left(\frac{d_0^2}{2} + d_1^2 + d_2^2 + \cdots + d_{n-1}^2 + \frac{d_n^2}{2}\right)
\end{aligned}
$$

To adapt Huber's formula to the computation of merchantable stem volumes, the diameter measurements are made at the midpoints of the sections (Fig. 6.4d). When diameter measurements $d_{m1}, d_{m2}, \ldots, d_{mn}$ are taken at the midpoints of sections of h length, Huber's formula yields

$$
\begin{aligned}
V &= hcd_{m1}^2 + hcd_{m2}^2 + \cdots + hcd_{mn}^2 \\
&= hc\left(d_{m1}^2 + d_{m2}^2 + \cdots + d_{mn}^2\right)
\end{aligned}
$$

An example illustrating the calculations for each method is shown in Table 6.2 for a sample stem section.

6.1.4. Determination of Cubic Volume by Graphical Methods and Integration

A graphical method can be used for measuring the volume of solids that have circular cross section but that vary in diameter along an axis normal to the cross sections. To obtain the volume of a tree or section, one must have diameter measurements along the stem or section, preferably both inside and outside bark. With suitable measurements for a given tree or section, the cross-sectional area or diameter squared, inside and outside bark, should be plotted over height or length on graph paper for each cross section measured. Then, the points should be connected by smooth lines to give a profile that is analogous to that of one side of a longitudinal section taken through the center of the tree or section. A separate graph is prepared for each tree stem or section.

For the graph of a tree stem, it is useful to label diameters at important points, such as top of stump, breast height, log ends, and merchantable limit of the stem, and to record pertinent information on species, locality, observers, date, and so forth. Measuring the required area on the graph and applying the appropriate conversion factor can obtain volume for the entire stem, or for any section, either inside or outside bark. For example, in Figure 6.5, diameter squared is plotted over length for a section of a tree. The area under the curve can be obtained by using a planimeter or a dot grid (Section 4.4.3), and converted to cubic volume by multiplying by cubic feet per square inch (or cubic meters per square centimeter) represented by the graph. Specifically,

$$
C_f = A \cdot L
$$

TABLE 6.2. Sample Stem Measurements for Determination of Volume Using Newton's, Smalian's, and Huber's Methods

Measurement Number	Diameter (cm)	Length (m)	d^2 (cm²)	Volume (m³)		Huber's		
				Newton's	Smalian's	Midpoint d (cm)	d_m^2 (cm²)	Volume (m³)
0	24.6		605.16					
1	23.4	1.0	545.36	0.0855	0.0452	23.9	569.62	0.0447
2	21.9	1.0	479.14		0.0402	22.5	505.75	0.0397
3	20.1	1.0	403.95	0.0641	0.0347	21.2	447.89	0.0352
4	18.8	1.0	355.13		0.0298	19.5	378.76	0.0297
5	17.0	1.0	287.70	0.0459	0.0252	17.9	319.32	0.0251
6	15.8	1.0	248.10		0.0210	16.3	266.05	0.0209
Total				0.1956	0.1962			0.1954

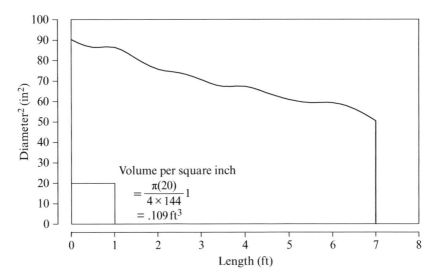

FIG. 6.5. Graphical estimation of the volume of a section of a tree.

where C_f = the conversion factor for ft³ per in² of graph (or m³ per cm² of graph)

  A = cross-sectional area per in. or cm of Y axis

  L = length per in. or cm of X axis

In the example, the conversion factor for cubic feet per square inch of graph is

$$C_f = \frac{\pi d^2}{4(144)} L$$

$$= 0.005454(20)(1)$$

$$= 0.109 \ \text{ft}^3.$$

Since the area under the curve of the section is 24.75 in.², the volume of the section is 24.75(0.109) = 2.70 ft³.

The volume of a solid with known cross sections can also be obtained by summation of the volumes of cross-sectional slices of the solid. As shown in Section 6.1.3, a sum of such volumes approximates the volume of the solid. If we take the integral that is suggested by this sum to be the volume, we get

$$V = \int_a^b A(X) dX$$

in which the volume of a slice is expressed as a function, $A(X)$ of X, times thickness, dX, where thickness approaches zero. This procedure can be used for solids of any shape for which cross-sectional area can be expressed as a function. The scanning technology employed in many mills today for measuring log volumes and optimizing sawing strategies use integration approaches in their software systems (Skog and Oja, 2007).

Whenever the profile of a tree or section can be expressed by an equation (Section 5.4.3), the formula for the stem volume obtained by rotating the graph of the equation $Y=f(X)$ about the X axis (between $X=a$, and $X=b$) may be written as

$$V = \pi \int_a^b Y^2 dX$$

In the integral, the radius, Y, of the solid of revolution must be expressed as a function of X. A cylinder is formed by rotating a rectangle around one side; a cone by rotating a right triangle around the vertical leg; a sphere by rotating a semicircle around the diameter; and a paraboloid by rotating a second-degree polynomial around its axis. The generalized formulas for the volumes of the solids shown in Table A.3 can be derived from the integral shown above by substituting the appropriate $f(X)$ for Y. For solids of revolution formed by curves of other shapes, simple formulas such as these are not available; integration is required to obtain volume.

To illustrate the procedure by a simple example, we will use Figure 5.9 and will compute the volume of the stem shown in the figure between a 1-ft stump and an upper limit that is 10 ft from the top (in Fig. 5.9, length is measured from the tip of the tree, therefore, the integral is over the range 10–59). The volume would then be

$$V = \pi \int_{10}^{59} 0.066 X dX$$

Integrating and evaluating yields

$$V = \left(\frac{\pi 0.066 X^2}{2} \right)_{10}^{59} = \left(0.104 X^2 \right)_{10}^{59}$$
$$= \left(0.104 \right)\left(59^2 \right) - \left(0.104 \right)\left(10^2 \right)$$
$$= 351.6 \text{ ft}^3$$

Since this solid is the frustum of a paraboloid, eq. (6.5) in Table 6.1 will yield the same result.

6.1.5. Determination of Crown and Root Volumes

The techniques described above for stem volume determination can be applied to determining volume of branches and roots. Because of the number of branches and roots and their complexity of measurement, generally subsamples are made. Frequency- or size-based subsampling schemes can be used and appropriate expansion factors developed to scale from branch or root measures to whole tree estimates. In conifer trees, branches are generally stratified by position in the crown, by, for example, thirds, quartiles, or annual whorls (e.g., Weiskittel et al., 2009b) or selected systematically throughout the entire crown (e.g., Kershaw and Maguire, 1995). In hardwoods, branches are usually stratified by branch size (e.g., Clark and Schroeder, 1977) or branch order (e.g., Van Pelt et al., 2004). Alternatively, randomized branch sampling can be used in either hardwood or conifer crowns (Valentine et al., 1984).

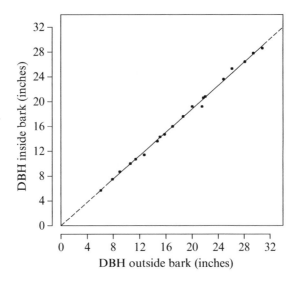

FIG. 6.6. Relationship between corresponding diameters inside and outside bark of white oak (see Table 6.3).

Direct determination of root volume requires excavating roots, removing all of the soil and either measuring root diameters at periodic intervals as described for main stems, or sectioning and using a xylometer. Barton and Montagu (2004) describe the use of ground penetrating radar for measuring root diameters. These diameter measures could potentially be used to estimate volumes without needing to extract the roots from the soil. Snowdon et al. (2000) describe a number of additional sampling strategies for estimating roots.

6.1.6. Determination of Bark Volume

Bark volume will average between 10 and 20% of unpeeled volume for most species. Often, one needs to know bark volume more accurately than this to determine peeled stem or log volume from unpeeled stem or log volume, the quantity of bark residue that will be left after the manufacturing process has been completed, and, in some cases, where the bark has value, the quantity of bark available. Of course, unpeeled stem or log volume can be computed from outside bark diameter and peeled stem or log volume from inside bark diameter. The difference between the two volumes is a good estimate of bark volume. Miles and Smith (2009) provide tables of bark volumes, weights, and biomass for several North American tree species.

The bark-factor method, which deserves more consideration than most foresters give it, is easier to apply and gives sufficiently accurate results for most purposes. Bark thickness, which must be accurately determined to obtain reliable bark factors, may be determined as described in Section 5.2. The accuracy of bark measurements is increased if single-bark thickness is measured at two or more different points on a given cross section of the stem and the average single-bark thickness b computed. Then, diameter inside bark d_{ib} may be computed from diameter outside bark d using $d_{ib} = d - 2b$. When d_{ib} is plotted as a function of d, the relationship will be linear, or close to linear, with a Y intercept approximately 0. Figure 6.6 shows this relationship for white oak for the cross section at breast height.

TABLE 6.3. Diameter and Bark Measurements of 20 White Oak Trees

d	$2b$	d_{ib}
Dbh	Double Bark	Dbh
Outside Bark (in.)	Thickness (in.)	Inside Bark (in.)
30.8	2.2	28.6
14.7	1.1	13.6
24.8	1.2	23.6
20.0	0.8	19.2
21.7	1.1	20.6
7.9	0.4	7.5
15.8	1.1	14.7
12.7	1.3	11.4
18.6	1.0	17.6
17.0	1.0	16.0
26.1	0.8	25.3
28.1	1.7	26.4
10.6	0.6	10.0
6.1	0.4	5.7
29.4	1.6	27.8
15.1	0.8	14.3
21.5	2.3	19.2
22.0	1.2	20.8
11.4	0.7	10.7
9.0	0.3	8.7
Sum 363.3	21.6	341.7

$\sum(d \cdot d_{ib}) = 7179.62; \sum d^2 = 7628.93.$

It is reasonable to assume that the prediction equation for this relationship may be written in the general form $d_{ib} = kd$. Because the regression coefficient k is normally determined at stump or breast height, it can be called the *lower-stem bark factor*. Such bark factors range from 0.87 to 0.93, varying with species, age, and site. Since the major portion of the variation can be accounted for by species, it is reasonable and convenient to assume that this bark factor will remain the same, for a given species, at all heights on the stem. For many species, upper-stem bark factors are often not the same as lower-stem bark factors. To account for this, one might develop multiple regression equations to predict upper-stem bark factors from such variables as tree age, tree dbh, height above ground, and diameter outside bark at the cross section for which bark factor is desired. Examples of these equations can be found in Cao and Pepper (1986), Maguire and Hann (1990), and Muhairwe (2000). The development of such equations requires much time spent on obtaining measurements in the field. Consequently, the common practice is to assume that the bark factor is the same at all heights on the stem.

Whether we assume that the bark factor is the same at all heights on the stem or that the bark factor will be different at different heights on the stem, once the bark factor has been obtained, the method of using it to obtain bark volume is the same. In the following explanation, it is assumed that k is the same at all heights on the stem.

An average value of k, to be reliable, should be based on 20–50 bark-thickness measurements and corresponding diameter-outside bark measurements. By the method of least squares, k is determined so that the sum of the squared deviation of the individual d_{ib}s (Ys) about the fitted regression line is a minimum (Section 3.8). In this case, where we assume that when $d=0$ then $d_{ib}=0$, it is appropriate to use the following equation to determine k:

$$k = \frac{\Sigma(d \cdot d_{ib})}{\Sigma d^2} \tag{6.10}$$

When the variation of the dependent variable is proportional to the independent variable, as it generally is when d_{ib} is plotted over d, Meyer (1953) has shown that the following formula will give similar results:

$$k = \frac{\Sigma d_{ib}}{\Sigma d} \tag{6.11}$$

For the trees listed in Table 6.3 and plotted in Figure 6.6, eq. (6.10) gives

$$k = \frac{7179.62}{7628.93} = 0.941.$$

Equation (6.11) gives

$$k = \frac{341.7}{363.3} = 0.941.$$

Agreement between the two formulas will not always be this good. In general, eq. (6.11) will, for practical uses, always give satisfactory results.

The bark thickness b, corresponding to an average value of k, can be determined as follows for any diameter d:

$$b = \frac{1}{2}(d - d_{ib})$$

and since $d_{ib} = kd$, then

$$b = \frac{1}{2}(d - kd) = \frac{d}{2}(1 - k) \tag{6.12}$$

Thus, for a white oak cross section of 14.0 in. in diameter, the bark thickness is estimated to be

$$b = \frac{14.0}{2}(1 - 0.941)$$

$$= 0.413 = 0.4 \text{ in.}$$

The average value of k can also be used to obtain cubic bark volume V_B from cubic volume outside bark V for a given stem section. If diameter outside bark at the middle of the section, d_m, diameter inside bark at the middle of the section, $d_{m,ib}$, and section length, L, are all in the same units, we have

$$V = \frac{\pi d_m^2}{4}(L) \quad \text{and} \quad V_{ib} = \frac{\pi d_{m,ib}^2}{4}(L).$$

Because $d_{m,ib} = kd_m$, V_{ib} is then

$$V_{ib} = \frac{\pi}{4}(kd_m)^2 (L) = k^2 V \tag{6.13}$$

Finally,

$$V_B = V - V_{ib} = V(1 - k^2) \tag{6.14}$$

and percent bark volume can be expressed as

$$V_B(\%) = (1 - k^2)100 \tag{6.15}$$

When V_{ib} is determined by eq. (6.13), it will be theoretically correct. However, V_B determined by eq. (6.14) will be greater than the actual value. This is because V_B includes air spaces between the ridges of the bark. A study by Chamberlain and Meyer (1950) shows that the difference in volume between stacked-peeled and stacked-unpeeled cordwood is, on the average, 80% of the volume given by eq. (6.13). One might not expect this result. However, it comes about because in a stack of wood the ridges of the bark of one log will mesh with the ridges of another log and because the weight of the logs will compress the bark. Thus, for practical purposes, eq. (6.14) can be rewritten to give the bark volume of cordwood in stacks V_{Bs},

$$V_{Bs} = 0.8V(1 - k^2)$$

6.1.7. Weight Determination

The weight of objects (precisely speaking, the force on the objects due to gravity) can be measured by any one of five general methods (Doebelin, 1966):

1. By balancing the object against the gravitational force on a standard mass. This method employs the well-known weighing machines: equal-arm balances, unequal-arm balances, and pendulum scales.
2. By measuring the acceleration of a body of known mass to which the unknown force is applied. This method, which uses an accelerometer for force measurement, is of restricted application since the force determined is the resultant of several inseparable forces acting on the mass.
3. By balancing against a magnetic force developed by the interaction of a current-carrying coil and magnet.
4. By transducing the force to a fluid pressure and then measuring the pressure. This method is exemplified by a container filled with a liquid, such as oil, under a fixed pressure. Application of a load increases the liquid pressure that can then be read on a gauge. Instruments of this type can be constructed for measuring large weights.
5. By applying the force to some elastic members and measuring the resulting deflection. This method permits the measurement of both static and dynamic loads while the previously described methods are restricted to static or slowly moving loads. By using a deflection transducer, the force applied to some elastic member will cause it

to move, and this motion can then be transformed to an electrical signal. The various devices differ principally in the form of the elastic element and in the displacement transducer, which generates the electrical signal. The movement of the elastic element may be a gross, ocularly perceptible motion, or it may be a very small motion that requires the use of strain gauges to sense the force.

In the field, weight can be determined using a spring scale or strain gauge. Weights of small sections of stems or branches can be obtained using a handheld scale. Weights of larger pieces can be obtained using a scale attached to a block and tackle or other device to lift larger, heavier objects off the ground. Very large sections of stems, or even whole trees, can be weighed using a strain gauge attached to a front-end loader or harvester arm.

Generally, smaller pieces can be weighed with more precision (g or ounces) than larger pieces (kg or lb). If precise values are required, small sections, such as thin disks cut from stem sections, may be weighed and the volume determined precisely using a xylometer. The weight:volume ratios of these small sections can then be multiplied by the volume (obtained using one of the methods described above) to estimate weights. While this technique is better than using published averages, the sampling error associated with weighing small pieces may negate improvements associated with more precise weighing.

For most conifer species, whole branches can be weighed using handheld scales or scales attached to a block and tackle. For hardwood species, larger branches may need to be sectioned into smaller pieces before weighing. If fresh foliage weights are required, the foliage must be removed from the branches and twigs and weighed. Whole root systems can be weighed using a strain gauge attached to a front-end loader or harvesting boom, but like root volume determination requires the root system to be excavated and the soil removed.

Green weights of trees and their component parts can be estimated using published tables of species specific gravities and moisture contents (Haygreen and Bowyer, 1996; Miles and Smith, 2009). *Specific gravity (SG)* is the ratio of the density of the solid content (lb/ft³ or kg/m³) divided by the density of water (62.4 lb/ft³ or 1000 kg/m³). Density is the oven dry weight of wood or other tree components divided by the associated volume. Because wood shrinks when dried, volume differs with differing moisture contents, thus, while the density is always based on oven-dried values, the volume may be based on green volume (SG_g), volume at some specified moisture content ($SG_{MC\%}$), or oven dry volume (SG_d) (Haygreen and Bowyer, 1996). Estimates of weight are then calculated as

$$\mathrm{WT} = SG_g \cdot \mathrm{Den}_{water} \cdot V \cdot \left(1 + \frac{MC_g\%}{100}\right) \qquad (6.16)$$

where SG_g = specific gravity, green volume

Den_{water} = density of water (62.4 lb/ft³ or 1000 kg/m³)

V = volume of the section (green basis)

$MC\%$ = green moisture content

Because wood shrinks when drying while the dry weight of the solid component is constant, $SG_d \geq SG_{MC\%} \geq SG_g$. If SG_g values are not available, the SG_d or $SG_{MC\%}$ will need to be adjusted to a green volume basis using wood shrinkage (Haygreen and Bowyer, 1996):
$SG_g = SG_X \cdot (1 - \%\text{Shrinkage } X/100)$.

While this technique is often used, the resulting weight values can only be considered approximate because wood weight depends on a number of factors including site, season, past management, age, and many others (Haygreen and Bowyer, 1996) and varies within trees (Schniewind, 1962; Heger, 1974; Antony et al., 2010). The best measures of weight will be obtained by direct weighing trees and their components; however, this process is time consuming and expensive. Therefore, weight estimates obtained from volume conversions are frequently used (Miles and Smith, 2009).

Clutter et al. (1983) describe a sampling procedure for estimating weight that is derived from the procedures described above for estimating volume of logs or bolts. Using an adaptation of Newton's formula (eq. (6.9)), the green weight of wood and bark W_g for a section is

$$\text{WT} = \frac{L}{6}\left(A_l D_{gu} + 2A_l D_{gl} + 2A_u D_{gu} + A_u D_{gl}\right) \tag{6.17}$$

where A_l = cross-sectional area of wood and bark at the lower end of the section
  A_u = cross-sectional area of wood and bark at the upper end of the section
  D_{gl} = green-weight density at the lower end of the section
  D_{gu} = green-weight density at the upper end of the section
  L = length of section

If only wood weight is desired, the cross-sectional area of the wood alone and its associated green-weight density would be substituted into eq. (6.17).

6.1.8. Biomass Determination

Direct determination of biomass requires collecting field samples, drying the samples, and weighing them. Because oven-dry weight is required, generally only small subsamples are used to determine SG or green weight:dry weight ratios. Then, the volume or green weight of the whole tree or its components are used to scale up the biomass estimate from the subsamples. Small disks, cross-sectional slices, and even increment cores can be used as subsamples for determining biomass (Wiemann and Williamson, 2013). To determine SG, volume of the subsample must be determined. For easiest application, volume of the subsample should be determined in green condition (Miles and Smith, 2009); otherwise, moisture content and wood shrinkage will need to be determined (Haygreen and Bowyer, 1996). Density of the subsample is calculated using

$$\text{Den}_g = \frac{\text{WT}_d}{V_g} \tag{6.18}$$

where Den_g = density of the sample on a green volume basis (lb/ft^3 or kg/m^3)
  WT_d = dry weight (lb or kg)
  V_g = green volume (ft^3 or m^3)

Oven-dried volume or volume at some specified moisture content also can be used in the denominator of eq. (6.18); however, estimates of biomass based on these wood densities need to be adjusted for wood shrinkage. SG is calculated by

$$\text{SG}_g = \frac{\text{Den}_g}{\text{Den}_{water}} \tag{6.19}$$

where SG_g = specific gravity, green volume basis

Den$_g$ = density of the sample on a green volume basis (lb/ft^3 or kg/m^3)

Den$_{water}$ = density of water (62.4 lb/ft^3 or 1000 kg/m^3)

SG is a unitless value and has the same value in imperial or SI units. Procedures for determining SG are described in Besley (1967), Wenger (1984), and Haygreen and Bowyer (1996); also see Section 6.1.7. Biomass for whole trees or tree components are then calculated using

$$BM = SG_g \cdot Dens_{water} \cdot V \tag{6.20}$$

Biomass also can be calculated by multiplying eq. (6.18) by V. Because of vertical and horizontal variation in SG, samples must be collected across the range of tree components. For branch biomass estimates, foliage is generally separated from branchwood and weights of each component obtained.

Like green weight, biomass can be determined by multiplying published species averages by the volume estimates obtained using one of the methods described above. The specific gravities of commercial woods in North America range between 0.3 (19 lb/ft^3, 300 kg/m^3) and 0.8 (50 lb/ft^3, 800 kg/m^3). For specific gravities of commercial woods in the United States, see Wenger (1984), Haygreen and Bowyer (1996), Alden (1997), and Miles and Smith (2009). The range of specific gravities of woods from other parts of the world is much wider, ranging from 0.1 (6 lb/ft^3, 100 kg/m^3) to 1.1 (69 lbs/ft^3, 1100 kg/m^3). Specific gravities for most tropical timbers of the world are given in Chudnoff (1984). A global wood density database is described by Chave et al. (2009). Using published averages of SG can result in estimates with substantial error. SG varies both vertically from tree tip to stump and horizontally from pith to bark as growth rates and wood densities vary (Schniewind, 1962; Heger, 1974; Antony et al., 2010). Within a species, SG often varies regionally with genotype and growth conditions.

6.1.9. Carbon and Nutrient Content Determination

Direct determination of carbon and other nutrient contents requires oven-dried samples, ground to a fine powder, and analyzed using elemental analysis or mass spectral analysis. Direct determination of carbon and nutrient contents is not typically carried out in forest mensurational studies. Usually, published average content values are used to convert biomass to carbon and nutrient contents:

$$\begin{aligned} Mass_X &= [X\%] \cdot BM \\ &= [X\%] \cdot SG_g \cdot Den_{water} \cdot V \end{aligned} \tag{6.21}$$

where   Mass$_X$ = mass of the element of interest (nutrient or carbon)

BM   = biomass

SG$_g$ = specific gravity on a green volume basis

Den$_{water}$ = density of water

V_g = volume

TABLE 6.4. Nutrient Concentrations (%) by Species and Tree Component

Species	Nutrient[a]	Tree Component				
		Foliage	Stem Wood	Bark	Branch Wood	Roots
Gmelina arborea[b]	N	1.970	0.334		0.337	0.323
	P	0.107	0.025		0.025	0.030
	K	0.622	0.337		0.342	0.289
Pinus taeda (41-year-old)[c]	N	1.065	0.064	0.176	0.188	
	P	0.106	0.005	0.011	0.018	
	K	0.233	0.043	0.049	0.085	
Pinus taeda (22-year-old)[d]	N	1.075	0.139	0.218	0.280	
	P	0.139	0.013	0.023	0.030	
	K	0.538	0.073	0.073	0.114	

[a] K, potassium; N, nitrogen; and P, phosphorous.
[b] Swamy et al. (2004).
[c] Van Lear et al. (1984).
[d] Rubilar et al. (2005), first rotation results.

As with using published values of average SG, the use of average carbon and nutrient values is prone to errors.

There are many nutrients found in trees and their components. Nitrogen, phosphorous, and potassium are three important macronutrients because of their importance in photosynthesis and tree growth and development (Kozlowski et al., 1991). These nutrients are mobile nutrients, meaning they can be transported throughout the tree, thus their concentrations can vary throughout the growing season and from year to year (Rytter, 2002; Swamy et al., 2004; Rubilar et al., 2005). Concentrations of these nutrients are higher in the foliage than in other components of the tree (Table 6.4). Concentrations often decrease from the top of the tree to the base (Rytter, 2002), but generally are stable over time (Swamy et al., 2003, 2004); however, concentrations may differ significantly between rotations (Rubilar et al., 2005). Understanding the spatial and temporal dynamics of nutrient concentrations is important when designing sampling strategies to estimate concentrations in the different tree components. Understanding distributions of nutrient concentrations is required when trying to assess impacts of biomass harvesting on long-term site productivity (Rytter, 2002; Swamy et al., 2004; Rubilar et al., 2005).

Carbon content of woody stems has typically been assumed to equal about 50% of the biomass (IPCC, 1996, 2006; Birdsey, 2004; Thomas and Martin, 2012); however, this assumption has been criticized as an oversimplification (Lamlom and Savidge, 2003). Lamlom and Savidge (2003) report average stem carbon content values of 44–50% for 22 North American hardwood species and 47–55% for 19 conifer species; however, their values were obtained from multiple samples extracted from single blocks of wood for each species tested. Elias and Potvin (2003) report stem carbon content values of 44–49% for 34 neotropical species. In their study, species differences accounted for about 39% of the variation in carbon content while site differences only account for about 3% of the differences (Elias and Potvin, 2003). The Intergovernmental Panel on Climate Change revised their carbon content values in 2006 (IPCC, 2006) based on biome and species type (Table 6.5). Thomas and Martin (2012) summarized published values using the

TABLE 6.5. Carbon Content of Selected Hardwood and Conifer Species by Biome (adapted from Thomas and Martin, 2012): Comparison of IPCC (2006) and Review of Published Values

Biome	Species Type	Species Number	IPCC (2006) Carbon (%)	Published Carbon (%) Stemwood	Published Carbon (%) Volatile
Tropical	Hardwood	134	49	47.1	2.5
	Conifer	1	49	49.3	
Subtropical/Mediterranean	Hardwood	18	49	48.1	
	Conifer	10	49	50.5	
Temperate/boreal	Hardwood	54	48	48.8	1.3
	Conifer	36	51	50.8	2.1
All biomes	Hardwood	206		47.7	2.3
	Conifer	47		50.8	2.1
All biomes/all species		253	47	48.2	2.3

same biome–species-type classification (Table 6.5). For most biomes, average carbon content for hardwoods was about 48% and for conifers was about 51% (Thomas and Martin, 2012).

While nutrient values differ significantly between tree components (Table 6.4), carbon values do not differ as much. Ritson and Sochacki (2003) report carbon content values of 51% for branches and foliage, 50% for stem wood, and 49% for roots of maritime pine (*Pinus pinaster*) growing in Australia. Similarly, Tolunay (2009) reports values of 53% for foliage, 55% for branches, 54% for bark, and 51% for stem wood in Scot's pine. Given that the bulk of the volume and biomass is contained within the main stem (Fig. 6.1) using carbon content values based on stem wood to estimate carbon mass of other tree components will not introduce significant errors. Use of a constant value of 50% carbon content will result in overestimates of carbon storage in many tree species. Using either species-specific carbon content values (e.g., Elias and Potvin, 2003; Lamlom and Savidge, 2003) or averages within species types and biomes (e.g., IPCC, 2006; Thomas and Martin, 2012) will reduce biases in carbon estimates by about 2.5% (Thomas and Martin, 2012).

6.2. ALLOMETRIC EQUATIONS FOR VOLUME, WEIGHT, AND BIOMASS

The measurements described in Section 6.1 can only be carried out during focused mensurational studies aimed at developing new relationships between tree content (volume, weight, biomass, or carbon and nutrient content) and the individual tree measurements described in Chapter 5. The intensive, and often destructive, measurements required to obtain these measures of content restrict the application to studies aimed at developing or testing new equations or tables to predict these values based on common individual tree measurements. The approaches to developing equations to predict various tree contents are similar (in fact, the approaches used to predict weight, biomass, or carbon and nutrient content are extensions of the approaches used to predict volume).

This makes sense given the long history and experience with developing equations to predict volume.

Heinrich Cotta, an early nineteenth-century German Forester, remarked that

Tree volume is dependent upon diameter, height, and form. When the correct volume of a tree has been determined, it is valid for all other trees of the same diameter, height, and form.

Since the time of Cotta, hundreds of volume functions have been constructed and used. In mathematical models, the volume of the stem of trees is considered a function of the independent variables, diameter, height, and form in the expression

$$V = f(D,H,F)$$

where $V =$ volume in cubic units or board feet

   $D =$ dbh

   $H =$ total, merchantable or height to some specific limit

   $f =$ a measure of form such as Girard form class or absolute form quotient

All equations and tables developed for predicting tree content are basically variants of the above equation and fall into three classes: standard functions; local functions; and form class functions.

6.2.1. Standard and Form Class Functions

Standard functions estimate volume or content according to diameter and height. *Form class functions* include an additional measure of form. Behre (1927, 1935) and Smith et al. (1961) concluded that no practical advantage is gained from the use of a measure of form in addition to dbh and height. Clutter et al. (1983) have given the following reasons that models using only diameter and height are preferred:

1. Measurement of upper-stem diameters is time consuming and expensive.
2. Variation in tree form has a much smaller impact on tree volume or weight than height or dbh variation.
3. With some species, form is relatively constant regardless of tree size.
4. With other species, tree form is often correlated with tree size, so that the dbh and height variables often explain much of the volume (or weight) variation actually caused by form differences.

Some commonly used standard functions are

$$\text{Constant form factor}\quad V = b_1 D^2 H \qquad (6.22)$$

$$\text{Combined variable}\quad V = b_0 + b_1 D^2 H \qquad (6.23)$$

$$V = b_1 D^{b_2} H^{b_3} \qquad (6.24)$$

Logarithmic transformed*

$$V = \ln(b_1) + b_2 \ln(D) + b_3 \ln(H)$$ (6.25)

Honer's[†] transformed variable

$$V = \frac{D^2}{\left(b_0 + b_1 H^{-1}\right)}$$ (6.26)

Models that include form as a variable are

$$V = b_0 + b_1 FD^2 H$$ (6.27)

$$V = b_0 F^{b_1} D^{b_2} H^{b_3}$$ (6.28)

The constants, b_i, are obtained using regression analysis techniques and the data collected from stem dissection studies. While the above equations depict volume as the dependent variable, weight, biomass, or nutrient content can all be estimated using equations of similar form (e.g., Yerkes, 1966; Rytter, 2002; Lambert et al., 2005). These equations, fitted to volume, also can be converted to weight, biomass, or carbon and nutrient content as described in Sections 6.1.7 through 6.1.9 (Smith et al., 2003).

6.2.2. Local Functions

Local volume functions utilize the single independent variable D, generally dbh, or some transformation of this variable. The simplest model for a local volume function is

$$V = b_0 D^{b_1}$$ (6.29)

where V and D are as above, and b_i = regression constants. This can be linearized using the logarithmic transformation:

$$\ln(V) = \ln(b_0) + b_1 \ln(D)$$

Other local functions, which have been used principally in Europe, as reported by Prodan (1965) and Prodan et al. (1997), include

$$V = b_0 + b_1 D^2$$

$$V = b_0 + b_1 D + b_2 D^2$$

$$V = b_0 + b_1 CSA$$

In the last model, CSA is cross-sectional area (or tree basal area). This model has been called the volume line since the use of basal area has linearized the volume–diameter relationship

---

*Logarithms to base 10 may also be used
[†] Honer (1967)

(Hummel, 1953). Height or crown area also can be used as variables in local functions. Functions based on these variables are often used in conjunction with measurements from aerial photographs (e.g., Husch, 1947; Gingrich and Meyer, 1955; Paine and Kiser, 2012) or with LiDAR data (e.g., Lefsky et al., 2002; Kwak et al., 2010; Lindberg et al., 2010; Dassot et al., 2011).

Kittredge (1944) was one of the first to apply local functions to predict foliage weight in trees. Since his work, many biomass studies have utilized local functions (e.g., Baskerville, 1965; Satoo and Madgwick, 1982; Jenkins et al., 2004) of the form shown in eq. (6.29) (or its logarithmic form). dbh-based local functions dominate much of the national biomass and carbon accounting work in the United States (Jenkins et al., 2003, 2004; Chojnacky et al., 2009; Westfall, 2012), and in Europe (Zianis et al., 2005). Lambert et al. (2005) incorporated height into the Canadian nation biomass equations; however for most species, the parameter estimates associated with height were not significant. Many studies have justified the use of local functions on the basis of lack of consistently available height data (e.g., Lambert et al., 2005). For others, the influence of sample size is important in understanding the lack of statistical significance of height. Tree biomass data are variable, height is correlated with diameter, and the effort required for destructive harvesting of trees often precludes sampling a sufficient number of trees across a sufficient number of stand conditions to separate the influence of height and diameter in a regression model. Failure to detect a significant influence of height on biomass with a given dataset does not mean height is unrelated to biomass or that height would not improve predictions if sufficient data were available. For some components, such as crown biomass, some studies have indicated that including crown measures in the regression equations can significantly improve biomass estimation (e.g., Monserud and Marshall, 1999).

6.2.3. Volume Functions to Upper-Stem Diameter Limits

The prediction of stem volumes (or weights) to different upper stem diameter limits (such as merchantability limit) has until recently required the preparation of separate functions and volume tables for each limit. Models have now been developed that permit the adjustment of volume to specified upper stem diameter limit as follows:

$$V = f(D,H)$$

$$V_u = r \cdot V$$

where D = dbh

H = total height

V = total stem volume

V_u = stem volume to a specified upper stem diameter

r = proportion of volume up to the upper stem diameter

The total volume V is multiplied by the adjustment factor r to obtain the volume to the specified upper diameter limit. Several functions have been developed for r, some using only diameter as the independent variable and others with both height and diameter as independent variables. An example of the first approach is the model of Burkhart (1977):

$$r = 1 + b_0 \left(\frac{d_u^{b_1}}{d^{b_2}} \right) \qquad (6.30)$$

where d =dbh

   d_u=upper stem diameter

An example of a function using both diameter and height is that of Matney and Sullivan (1982):

$$r = b_0 \left(\frac{d_u}{d} \right) + b_1 \left(\frac{d_u}{d} \right)^2 h_u \tag{6.31}$$

where d and d_u are as above

   h_u=height at the upper stem diameter

Models including height as an independent variable can be transformed to determine the height h_u at which a specified upper stem diameter occurs. For the Matney and Sullivan function, this has been done with the restriction that d be at 4.5 ft and that $r=1$ when $d_u=0$, yielding the equation

$$h_u = h + \left(\frac{d_u}{d} \right)(4.5-h) + \left(\frac{576}{\pi} \right) \left(\frac{V}{d^2} \right) \left(\frac{d_u}{d} \right) \left[b_1 \left(\left(\frac{d_u}{d} \right) - 1 \right) \right.$$
$$\left. + b_2 \left(\left(\frac{d_u}{d} \right)^2 - 1 \right) + b_3 \left(\left(\frac{d_u}{d} \right)^3 - 1 \right) + b_4 h \left(\left(\frac{d_u}{d} \right)^2 - 1 \right) \right]$$

For a number of other models to determine r and h_u, see Prodan et al. (1997). Using these principles, similar ratios could be developed for weight, biomass, or nutrient and carbon content distributions.

6.3. TABULAR ESTIMATION

Prior to the advent of readily available computers and handheld calculators, volume equations were often presented in tables for easy look up of volume given species, dbh, and height. While volume tables are discussed here, tabulated values for weights, biomass, or nutrient and carbon contents could also be developed using equations and the techniques described here. Volume tables can also be converted to other measures of content using the processes described in Sections 6.1.7–6.1.9.

Numerous methods have been used to construct volume tables. However, since the middle of the twentieth century, there has been a trend, particularly for broadleaved species, to reduce the number of volume tables used by adopting composite volume tables, tables applicable to average timber, regardless of species. Where the same standards of utilization are employed, differences in tree volumes among species are often of no practical consequence. Excellent examples of composite volume tables are Beers' (1973) for hardwoods in Indiana, and Gevorkiantz and Olsen's (1955) for timber in the Lake States. These tables have been extensively tested and have been found to replace individual species tables, especially for the estimation of volume on large tracts. Adjustment factors can be used for individual species that vary from the average.

Why have so many volume tables been constructed? Why has so much research gone into the development of volume tables? The answer is that foresters have been looking for methods that are simple, objective, and accurate; however, because trees are highly variable geometric solids, no single table, or set of tables, could possibly satisfy all of these conditions, regardless of method of construction. Consequently, one by one the older methods of volume table construction have been abandoned. For example, the once popular harmonized-curve method (Chapman and Meyer, 1949), which requires large amounts of data to establish the relationships and considerable judgment to fit curves, is rarely used today. The alignment-chart method, another subjective method, has been generally discarded. Other discarded methods have been described by Spurr (1952). Today, interest has focused on the use of mathematical functions, or models, to prepare volume tables. There is no advantage for the majority of foresters in using any other method.

In the United States, a majority of volume tables give volumes in board feet of lumber (Section 7.3.6), although tables that give volumes in cubic feet are available. In nations where the SI is used, expression of volume in board feet is still common, although many volume tables give volume in cubic meters.

Standard Volume Tables. Standard volume tables give volume in terms of diameter at breast height and merchantable or total height. Tables of this type may be prepared for individual species, or groups of species, and specific localities. The applicability of a standard volume table, however, depends on the form of the trees to which it is applied rather than on species or locality; for each diameter–height class, the form of the trees to which the table is applied should agree with the form of the trees from which the table was prepared. Table 6.6 shows a typical standard volume table.

Local Volume Tables. Local volume tables give tree volumes in terms of diameter at breast height only. The term *local* is used because such tables are generally restricted to the local area for which the height–diameter relationship hidden in the table is relevant. Although local volume tables may be prepared from raw field data—that is, from volume and diameter measurements for a sample of trees—they are normally derived from standard volume tables. A local volume table may be derived from a standard volume table by "localizing" the heights by dbh classes using the following procedure.

1. Measure the heights and dbhs of a sample of trees representative of those to which the local volume table will be applied. Record dbh to the nearest 0.1 in. (or 0.1 cm), and height to the nearest foot (0.1 m).
2. Prepare a height–diameter equation (Section 5.6.1).
3. Calculate average heights from the height–diameter equation for each dbh class.
4. Interpolate from the standard volume table the volume of the tree of average height for each dbh class, or if a standard volume equation is available, substitute the appropriate values in the equation and compute volume for each dbh class.

Table 6.7 shows a local volume table derived from Honer's SI standard volume equations (Honer et al., 1983).

TABLE 6.6. Gross Volume[a] **in Cubic Feet (Excluding Bark) in 8-ft Height Intervals for Hardwood Species in the Central States**

dbh	Merchantable Height in Feet (Number of 8-ft Bolts)									
Class (in.)	8 (1)	16 (2)	24 (3)	32 (4)	40 (5)	48 (6)	56 (7)	64 (8)	72 (9)	80 (10)
5	0.9	1.5								
6	1.3	2.2	3.1							
7	1.8	3.1	4.2	5.3						
8	2.4	4.0	5.5	6.9	8.1					
9	3.1	5.1	7.0	8.8	10.3	11.5				
10	3.8	6.3	8.7	10.9	12.8	14.3				
11	4.6	7.7	10.6	13.3	15.5	17.4				
12	5.5	9.2	12.7	15.9	18.6	20.8	22.7			
13	6.5	10.8	15.0	18.7	21.9	24.6	26.7	28.3		
14	7.6	12.6	17.5	21.8	25.5	28.6	31.1	33.0	34.3	
15	8.7	14.6	20.2	25.1	29.4	33.0	35.9	38.1	39.5	40.2
16	10.0	16.7	23.1	28.7	33.7	37.8	41.1	43.5	45.2	46.0
17	11.3	18.9	26.2	32.6	38.2	42.8	46.6	49.4	51.3	52.2
18	12.8	21.3	29.5	36.7	43.0	48.3	52.5	55.6	57.7	58.8
19	14.3	23.8	33.0	41.1	48.2	54.0	58.8	62.3	64.6	65.8
20	15.9	26.5	36.7	45.8	53.6	60.2	65.4	69.3	72.0	73.3
21	17.6	29.4	40.7	50.7	59.4	66.6	72.5	76.8	79.7	81.2
22	19.5	32.4	44.9	55.9	65.5	73.5	79.9	84.7	87.9	89.5
23	21.4	35.6	49.3	61.4	71.9	80.7	87.7	93.0	96.5	98.3
24	23.4	39.0	53.9	67.2	78.7	88.3	96.0	101.8	105.6	107.5
25	25.5	42.5	58.7	73.2	85.8	96.2	104.6	110.9	115.1	117.2
26	27.7	46.1	63.8	79.6	93.2	104.6	113.7	120.5	125.1	127.3
27	30.0	50.0	69.2	86.2	101.0	113.3	123.2	130.6	135.5	138.0
28	32.4	54.0	74.7	93.2	109.1	122.4	133.1	141.1	146.4	149.1
29	34.9	58.2	80.5	100.4	117.6	131.9	143.4	152.0	157.8	160.6
30	37.5	62.6	86.6	107.9	126.4	141.8	154.2	163.5	169.6	172.7
31	40.3	67.1	92.8	115.8	135.6	152.1	165.4	175.3	181.9	185.2
32	43.1	71.8	99.4	123.9	145.1	162.8	177.0	187.7	194.8	198.3
33	46.0	76.7	106.2	132.4	155.0	173.9	189.1	200.5	208.0	211.8
34	49.1	81.8	113.2	141.2	165.3	185.5	201.6	213.8	221.8	225.8
35	52.3	87.1	120.5	150.2	175.9	197.4	214.6	227.5	236.1	240.4
36	55.5	92.6	128.0	159.7	187.0	209.8	228.1	241.8	250.9	255.5
37	58.9	98.2	135.8	169.4	198.4	222.6	242.0	256.6	266.2	271.0
38	62.4	104.0	143.9	179.5	210.2	235.8	256.4	271.8	282.1	287.1
39	66.0	110.1	152.3	189.9	222.3	249.5	271.2	287.5	298.4	303.8
40	69.8	116.3	160.9	200.6	234.9	263.6	286.6	303.8	315.3	321.0

Source: Adapted from Beers (1964a).

[a] Gross volume in cubic feet (excluding bark) for a 1-ft stump height to a minimum top diameter of 4 in.

Volume calculated using $V = 79 \left[\dfrac{dbh^2 (dbh + 190)}{10^5} \right] \cdot \left(\dfrac{1}{10^2} \right) \cdot \left[\dfrac{H(168 - H)}{64} + \dfrac{32}{H} \right].$

TABLE 6.7. Local Volume Table (Total Volume, m³) for Noonan Forest in Central New Brunswick, Canada

dbh Class (cm)	Spruce		Balsam Fir		White Pine		Maple		White Birch		Yellow Birch	
	HT (m)	Vol (m³)	HT (m)	Vol (m³)	HT (m)	Vol (m³)	HT (m)	Vol (m³)	HT (m)	Vol (m³)	HT (m)	Vol (m³)
2	2.2	0.000	2.0	0.000	1.8	0.000	1.3	0.000	1.4	0.000	1.4	0.000
4	4.2	0.003	3.9	0.003	3.4	0.002	2.6	0.001	2.8	0.002	2.7	0.002
6	6.0	0.009	5.6	0.008	5.0	0.007	3.7	0.005	4.0	0.006	3.9	0.005
8	7.6	0.020	7.2	0.019	6.5	0.016	4.8	0.011	5.2	0.013	5.0	0.012
10	9.1	0.037	8.6	0.034	7.9	0.029	5.8	0.021	6.3	0.024	6.1	0.023
12	10.5	0.060	10.0	0.055	9.2	0.049	6.8	0.034	7.3	0.040	7.0	0.038
14	11.8	0.090	11.2	0.082	10.5	0.075	7.7	0.052	8.3	0.060	8.0	0.058
16	12.9	0.128	12.4	0.116	11.7	0.109	8.5	0.075	9.2	0.087	8.8	0.083
18	14.0	0.173	13.4	0.156	12.8	0.150	9.3	0.103	10.1	0.119	9.6	0.113
20	14.9	0.226	14.4	0.203	13.9	0.199	10.1	0.137	10.8	0.156	10.3	0.149
22	15.8	0.286	15.3	0.257	14.9	0.257	10.8	0.176	11.6	0.200	11.0	0.190
24	16.6	0.355	16.1	0.318	15.8	0.324	11.4	0.221	12.3	0.251	11.6	0.238
26	17.4	0.432	16.9	0.386	16.7	0.399	12.0	0.272	12.9	0.308	12.2	0.291
28	18.0	0.517	17.5	0.460	17.6	0.484	12.6	0.329	13.6	0.371	12.8	0.350
30	18.7	0.610	18.2	0.542	18.4	0.579	13.1	0.392	14.1	0.441	13.3	0.416
32	19.2	0.710	18.8	0.631	19.2	0.683	13.6	0.461	14.7	0.518	13.8	0.487
34	19.7	0.819	19.3	0.726	19.9	0.797	14.1	0.536	15.2	0.601	14.2	0.565
36	20.2	0.935	19.8	0.828	20.6	0.921	14.5	0.617	15.7	0.691	14.6	0.649
38	20.6	1.060	20.3	0.938	21.2	1.054	15.0	0.705	16.1	0.788	15.0	0.739
40	21.0	1.192	20.7	1.054	21.8	1.198	15.3	0.799	16.5	0.892	15.4	0.835
42	21.4	1.332	21.1	1.176	22.4	1.351	15.7	0.899	16.9	1.002	15.7	0.937
44	21.7	1.479	21.5	1.306	22.9	1.514	16.0	1.006	17.3	1.119	16.0	1.045
46	22.0	1.634	21.8	1.442	23.5	1.688	16.4	1.118	17.6	1.242	16.3	1.159
48	22.3	1.796	22.1	1.585	24.0	1.871	16.7	1.237	17.9	1.372	16.6	1.279
50	22.5	1.966	22.4	1.734	24.4	2.064	16.9	1.362	18.2	1.509	16.8	1.405

Species

(*Continued*)

TABLE 6.7. (Continued)

dbh Class (cm)	Spruce HT (m)	Vol (m³)	Balsam Fir HT (m)	Vol (m³)	White Pine HT (m)	Vol (m³)	Maple HT (m)	Vol (m³)	White Birch HT (m)	Vol (m³)	Yellow Birch HT (m)	Vol (m³)
										Species		
52	22.8	2.143	22.7	1.890	24.9	2.267	17.2	1.493			17.1	1.537
54	23.0	2.328	22.9	2.052	25.3	2.480	17.4	1.630			17.3	1.675
56	23.2	2.519	23.1	2.221	25.7	2.703	17.7	1.773			17.5	1.818
58	23.3	2.718	23.3	2.396	26.0	2.935	17.9	1.922			17.7	1.967
60	23.5	2.924	23.5	2.577	26.4	3.177	18.1	2.077			17.9	2.122
62	23.6	3.137			26.7	3.429	18.3	2.238			18.0	2.282
64	23.8	3.357			27.0	3.690	18.5	2.404			18.2	2.449
66	23.9	3.584			27.3	3.961	18.6	2.577			18.3	2.620
68	24.0	3.818			27.6	4.242	18.8	2.755			18.5	2.798
70	24.1	4.059			27.9	4.532	18.9	2.939			18.6	2.980
72					28.1	4.831						
74					28.3	5.139						
76					28.6	5.457						
78					28.8	5.784						
80					29.0	6.121						

Source: Unpublished inventory data collected during the summer, 2001.

Average heights predicted from a species-specific height–diameter equation using dbh class. Volumes predicted from Honer et al.'s (1983) SI total volume equations using dbh class and average height.

Form Class Volume Tables. Form class volume tables give volumes in terms of diameter at breast height, merchantable or total height, and some measure of form, such as Girard form class or absolute form quotient. Such tables come in sets, with one table for each form class. The format of each table is similar to that of a standard volume table. Note that if a single form class table is chosen as representative of a stand, volume determinations may be in error because it is unlikely that all trees will be of the same form class. Furthermore, since form class varies with each tree size, species, and site, it is unlikely that variation in form class will be random. Thus, it is difficult to obtain an accurate average form class for a stand and it is therefore undesirable to use a single form class table for any extensive area. Ideally, form class volume tables should be used in conjunction with field measures of form; however, they are often used with published form values for different species that are assumed to be representative for a particular locality or region.

6.3.1. Descriptive Information to Accompany Tables

A volume table should include descriptive information that will enable one to apply it correctly. This information includes

1. Species, or species group, to which the table is applicable, or the locality in which the table is applicable.
2. Definition of dependent variable, that is, volume, including units in which volume is expressed.
3. Definitions of independent variables, including stump height and top diameter limit, if merchantable height is used.
4. Author.
5. Date of preparation.
6. Number of trees on which table is based.
7. Extent of basic data.
8. Method of determining volumes of individual trees (in basic data).
9. Method of construction.
10. Appropriate measures of accuracy.

Tables 6.6 and 6.7 include these items.

The first three items in the above list should always be given. The remaining items are of less interest and are sometimes omitted, but they can be helpful for users in evaluating the adequacy of a table for a particular task. When measures of accuracy are given, they should be understood to be measures of accuracy of the table when it is applied to the data used in its construction. Such measures give no assurance that a volume table will apply to other trees. Thus, when an accurate estimate is required, a table should be checked against the measured volumes of a representative sample of trees obtained from the stands to be estimated.

6.3.2. Checking Applicability of Tables

In an applicability check, one should compare the volume of sample trees with the estimated volume from the volume table to be checked. Three conditions should be observed in selecting sample trees:

1. Sample trees for a given species, or species group, should be distributed through the timber to which the volume table will be applied.
2. No sizes, types, or growing conditions should be unduly represented in the sample.
3. If a sample of cut trees is used, this sample, if not representative of the timber, should be supplemented by a sample of standing trees.

Definite rules for measuring sample trees should be established. As an example, the following rules are satisfactory for the eastern United States:

1. Diameters along the tree stem, inside and outside bark, should be measured at 8-ft intervals above a 1-ft stump, and at stump height, breast height, and merchantable height.
2. Diameter should be measured to nearest 0.1 in. and bark thickness to nearest 0.05 in.
3. Knots, swellings, and other abnormalities should be avoided at points of measurement by taking measurements above or below them.
4. Total or merchantable heights should be measured to nearest foot. (Utilization standards for the timber in question should be considered in determining the upper limit of merchantable height.)

For practical purposes, the aggregate difference of a test sample should not exceed $2 \cdot CV/\sqrt{n}$, where CV is the coefficient of variation of the volume table being tested, and n is the number of trees used in the test. If desired, checks may be made by diameter classes. Of course, more complicated statistical tests, such as the chi-square goodness-of-fit or the Kolmogorov–Smirnov test (Zar, 2009), might be used. The above procedure, however, is generally satisfactory. When a table is judged to be inapplicable, one should adjust the table or obtain a better table. Practical methods of making adjustments are described by Gevorkiantz and Olsen (1955).

6.3.3. Conversion of Volume Tables to Weight, Biomass, or Carbon and Nutrient Tables

Weight, biomass, or content tables can be constructed using predictions from allometric equations as discussed above. Table 6.8 is an example of a weight table constructed from a tree weight equation. Volume tables or equations also can be converted to other measures of content if reliable volume–content conversions can be established. For example, a cubic volume table can be converted to green-weight using eq. (6.16) and converted to biomass using eq. (6.20). The weight or biomass obtained would correspond to the part of the tree for which the volume table was prepared for, usually either total main stem volume or merchantable main stem volume. The *biomass expansion factor* approach described by Heath et al. (2009) might be used to convert main stem volume to total biomass.

TABLE 6.8. Weight in Pounds of Green Wood in Merchantable Stem of Red Pine to a 4-in. Top

dbh (in.)	Total Tree Height (ft)					
	16	24	32	40	48	56
6	122	168	215	261	307	
8	194	276	358	441	523	605
10		415	543	672	800	929
12			769	954	1139	1324
14			1037	1288	1540	1792
16			1345	1674	2002	2331

Source: Adapted from Cody (1976).
Weight $= 29.5872 + 0.16055$ dbh^2H; $R^2 = 0.946$; SE $= 49.416$.

TABLE 6.9. Average Distribution of Tree Volumes by Logs According to Log Position

Useable Length (16-ft logs)	Percent of Total Volume in Each Log, by Position					
	1st	2nd	3rd	4th	5th	6th
1	100					
2	59	42				
3	42	33	25			
4	34	29	22	15		
5	29	25	21	15	10	
6	24	23	20	16	11	6

Source: Mesavage and Girard (1946).

Biomass expansion factors can simply convert stem volume to stem biomass, as described above, or convert stem volume to stem biomass, then to total tree biomass using an estimate of the ratio of stem biomass to total tree biomass (Somogyi et al., 2009). The applicability of a table, whether derived from allometric equations or converted from existing volume tables, should be verified using the methods outlined in Section 6.2.3 for verifying volume tables.

6.4. VOLUME AND BIOMASS DISTRIBUTION IN TREES

6.4.1. Methods for Estimating Stem Volume Distribution

Knowledge of volume distribution over a tree stem can be used to improve volume estimates and to aid in estimating volume losses from defects. (Volume deduction for defect is treated in Section 7.4.1.) Table 6.9 illustrates a method of expressing volume distribution. This method shows the percentage of volume according to each 16-ft log in trees with heights of 1–6 logs.

Although the percentages vary slightly with tree diameter and unit of volume, they may be used without serious error for merchantable trees of all sizes that are measured in cubic or board-foot volume. Note that Table 6.9, which is for 16-ft logs, provides a satisfactory guide when heights are measured in 8- or 12-ft lengths. Table 6.10 shows another generalized volume structure of tree stems according to cylindrical form factors and section lengths of 0.2 height.

In another approach, Honer et al. (1983) developed volume percentages based on both height ratios and diameter ratios. In their model, ratios are predicted using a quadratic equation:

$$r = b_0 + b_1 X + b_2 X^2$$

where r = the volume ratio to a specified upper limit
   X = either the height ratio (h_u/H) or the diameter squared ratio $\left(d_u^2 / D^2\right)$
   h_u = upper height limit
   H = total height
   d_u = upper diameter limit inside bark
   D = dbh outside bark
   b_i = constants

Table 6.11 shows the volume percentage by diameter ratio and height ratio for the combined species volume predictions based on Honer's transformed variable equation (Eq. (6.26)).

The main advantage of this system is that products may be specified by both a minimum upper diameter and length. The proportion of total volume in each product can be determined by using the two volume percentages iteratively. For example, sawlogs might have a minimum upper diameter of 12 in. and are sold in 16-ft lengths. The proportion of sawlog volume contained in a tree with dbh=30 in. and total height of 90 ft would be found using the following steps:

1. Determine the volume percentage to the minimum top diameter:

$$\frac{d_u}{D} = \frac{12}{30} = 0.4$$

$$\%V = 96.22$$

2. Using the volume percentage, determine the height ratio to the minimum top diameter:

$$\%V = 96.22$$

$$\frac{h_u}{H} \approx 0.80$$

$$h_u = 0.80 \cdot 90 = 72$$

3. Subtracting 1 ft for the stump and allowing 0.3 ft for trim, then number of 16-ft logs are determined:

$$L = \left(\frac{72 - 17.3}{16.3}\right) + 1$$

$$= 3.3 + 1 = 4 \ \text{logs}$$

TABLE 6.10. Tree Stem Volume Structure According to Cylindrical Form Factors

	Percentages of Volume for Cylindrical Form Factors											
	0.407		0.442		0.500		0.550		0.600	
Sections of $0.2h$	Vol%	Σ	Vol%	Σ	Vol%	Σ	Vol%	Σ	Vol%	Σ
I	49.1	49.1	45.3	45.3	40.0	40.0	36.4	36.4	33.3	33.3
II	29.7	78.8	29.3	74.6	28.2	68.2	29.0	65.4	27.1	60.4
III	15.1	93.9	16.3	90.9	19.6	87.8	20.8	86.2	22.0	82.4
IV	5.5	99.4	7.4	98.3	10.5	98.3	11.6	97.8	14.4	96.8
V	0.6	100.0	1.7	100.0	1.7	100.0	2.2	100.0	3.2	100.0

Source: Prodan et al. (1997).

TABLE 6.11. Proportions of Total Stem Volume for Specified Ratios of Upper Height:Total Height (h_u/H) and Upper Diameter:dbh (d_u/D)

Height Ratio (h_u/H)	Volume (%)	Diameter Ratio (d_u/D)	Volume (%)
0.10	21.76	0.10	100.00
0.20	39.60	0.20	99.96
0.30	55.12	0.30	98.88
0.40	68.30	0.40	96.22
0.50	79.12	0.50	91.59
0.60	87.70	0.60	84.07
0.70	93.92	0.70	72.52
0.80	97.80	0.80	55.63
0.90	99.36	0.90	31.87

Source: Honer et al. (1983).

4. The height at the top of the last log is determined and the final volume percentage interpolated from the table:

$$h_u = 1.0 + 16.3 \cdot 4 = 66.2$$

$$\frac{h_u}{H} = \frac{66.2}{90} = 0.73$$

$$\%V = 93.92$$

Thus, the percent volume in sawlogs for this tree is 93.92% of total volume. Volume percentages for smaller products such as pulp could then be determined for this tree. The method is easily automated and applicable to any dataset where both diameter and heights are available. Kershaw et al. (2007) have developed software that implements this procedures for multiple species and product definitions.

6.4.2. Distribution of Weight and Biomass in Trees

In volume studies, the focus is generally on the distribution of volume within the main stem. Different products, which have potentially very different values, are often determined by minimum size considerations. Estimating the numbers of logs in the different product classes can be useful for determining total value, log sorts, and appropriate harvesting equipment. In weight or biomass studies, the focus is often on whole trees. When harvesting for biomass utilization, much more of the tree is potentially useful. When assessing carbon storage, both above-ground and below-ground tree components are important. Because the different components of a tree have different specific gravities and carbon and nutrient contents, biomass and carbon estimation is often summarized by tree component. Generally, trees are divided into roots, stem, bark, branch wood, and foliage (Young et al., 1964; Chojnacky et al., 2009). Table 6.12 shows the distribution of weight divided into eight components for a red spruce and red maple tree.

As described in Section 6.2, allometric equations are frequently applied for estimating weight, biomass, and carbon content in trees. When distributing the content among tree components, two approaches are generally used: (1) separate estimation equations for each component or (2) an equation to estimate total tree content and ratio equations to divide the total content into its component parts. Choice of approach depends on many factors including the data available for fitting equations and the purpose for making predictions.

TABLE 6.12. Example of Proportional Weights of a Tree in Eight Components

	Weights[a]			
	Red Spruce		Red Maple	
	12-in. dbh, 70 ft		12-in. dbh, 70 ft	
	Total Height		Total Height	
Tree Component	Pounds	Percent	Pounds	Percent
Roots less than 1 in. in diameter	55	3	62	3
Roots from 1 to 4 in. diameter	115	6	96	5
Roots larger than 4 in. to base of stump	115	6	115	6
Stump—from 6 in. above ground to large roots	109	5	159	8
Merchantable stem from stump to 4 in. upper diameter	1218	60	1224	63
Branches larger than 1 in. diameter	76	3	109	6
Branches smaller than 1 in. diameter including leaves	320	16	118	6
Stem above merchantable portion	20	1	58	3
Total tree	2028	100	1941	100

Source: Young et al. (1964).
[a] Weights are based on moisture conditions when freshly cut and include bark.

Ker (1984) presents an example of fitting separate equations to predict total biomass and the biomass of boles (with and without bark), foliage, and branches for several species in northeastern North America. In this work, total biomass and each component had the same basic allometric equation form: $BM_C = b_0 dbh^{b_1}$. Because the equations were fitted independently, sums of the components did not equal the estimates obtained using the total equations. Parresol (1999) presented an analysis of several linear regression models and their statistics for evaluating goodness-of-fit and for use in comparing alternative biomass models. In this study, model form was allowed to vary between components and the following "best" models were derived for willow oak (*Quercus phellos* L):

$$\widehat{BM}_{wood} = b_{wood0} + b_{wood1} dbh^2 H$$

$$\widehat{BM}_{bark} = b_{bark0} + b_{bark1} dbh^2 H$$

$$\widehat{BM}_{crown} = b_{crown0} + b_{crown1} \frac{dbh^2 H \cdot LCL}{1000} + b_{crown2} H$$

For total tree biomass, the "best" individual equation was

$$\widehat{BM}_{total} = b_{total0} + b_{total1} dbh^2 H$$

where $\widehat{BM}_i$ = biomass of component i (kg)
      dbh = diameter at breast height (cm)
      H = total height (m)
      LCL = live crown length (m)

Parresol (1999) addressed additivity using two methods. The first was simply summing the components and computing a joint estimation error. Comparing the joint estimation error to the error obtained from independently fitting the best equation for total biomass showed that this approach was inefficient and produced unacceptably large estimation errors. The second approach used seemingly unrelated regression estimation (Srivastava and Giles, 1987). In this case, each component was fitted using the best equation shown above; the total was fitted as the sum of the individual components:

$$\widehat{BM}_{wood} = b_{0wood} + b_{1wood}dbh^2 H$$

$$\widehat{BM}_{bark} = b_{0bark} + b_{1bark}dbh^2 H$$

$$\widehat{BM}_{crown} = b_{0crown} + b_{1crown}\frac{dbh^2 H \cdot LCL}{1000} + b_{2crown}H^2$$

$$\widehat{BM}_{total} = \left(b_{0wood} + b_{0bark} + b_{0crown}\right) + \left(b_{1wood} + b_{1bark}\right)dbh^2 H + b_{1crown}\frac{dbh^2 H \cdot LCL}{1000} + b_{2crown}H^2$$

and the joint system of equations fitted simultaneously. This approach resulted in estimation errors for total biomass that were smaller than the errors resulting from fitting the best equation to total biomass directly. Parresol (2001) found similar results for nonlinear biomass equations applied to slash pine biomass estimation.

Similarly, Lambert et al. (2005) proposed two sets of seemingly unrelated regression equations for the Canadian national biomass estimators, one based solely on dbh:

$$\widehat{BM}_{wood} = b_{1wood}dbh^{b_{2wood}}$$

$$\widehat{BM}_{bark} = b_{1bark}dbh^{b_{2bark}}$$

$$\widehat{BM}_{stem} = \widehat{BM}_{wood} + \widehat{BM}_{bark}$$

$$\widehat{BM}_{foliage} = b_{1foliage}dbh^{b_{2foliage}}$$

$$\widehat{BM}_{branches} = b_{1branches}dbh^{b_{2branches}}$$

$$\widehat{BM}_{crown} = \widehat{BM}_{foliage} + \widehat{BM}_{branches}$$

$$\widehat{BM}_{total} = \widehat{BM}_{wood} + \widehat{BM}_{bark} + \widehat{BM}_{foliage} + \widehat{BM}_{branches}$$

And one based on dbh and height:

$$\widehat{BM}_{wood} = b_{1wood}dbh^{b_{2wood}} H^{b_{3wood}}$$

$$\widehat{BM}_{bark} = b_{1bark}dbh^{b_{2bark}} H^{b_{3bark}}$$

$$\widehat{BM}_{stem} = \widehat{BM}_{wood} + \widehat{BM}_{bark}$$

$$\widehat{BM}_{foliage} = b_{1foliage}dbh^{b_{2foliage}} H^{b_{3foliage}}$$

$$\widehat{BM}_{branches} = b_{1branches}dbh^{b_{2branches}} H^{b_{3branches}}$$

$$\widehat{BM}_{crown} = \widehat{BM}_{foliage} + \widehat{BM}_{branches}$$

$$\widehat{BM}_{total} = \widehat{BM}_{wood} + \widehat{BM}_{bark} + \widehat{BM}_{foliage} + \widehat{BM}_{branches}$$

The equation set including height generally fitted the biomass data better; however, for some species, the parameters associated with height were either not significantly different from zero or were the wrong sign (Lambert et al., 2005).

In the United States, the national biomass estimation equations utilize a component ratio approach for predicting biomass in the various tree components (Jenkins et al., 2003, 2004; Westfall, 2012; Chojnacky et al., 2014). In this approach, above-ground biomass is predicted using a local function (eq. (6.29)) or a standard function (eq. (6.22)), or alternatively using biomass expansion factors (Heath et al., 2009), and the ratio of biomass for each component is prediction using (Jenkins et al., 2003)

$$r_C = e^{\left(b_0 + \frac{b_1}{\text{dbh}}\right)}$$

(6.32)

where r_C = component ratio of tree component C
dbh = diameter at breast height
b_i = regression parameters

Because of the lack of consistent data across species, component ratios used in the United States were only estimated for softwood and hardwood species groups (Jenkins et al., 2003; Chojnacky et al., 2009). Figure 6.1 shows the component ratio predictions by species group. Component ratios were predicted for foliage, stem bark, stem wood, and coarse roots, while branch biomass (wood and bark) was obtained by subtraction (Jenkins et al., 2003).

The equation form shown in eq. (6.32) was chosen because it provides asymptotic ratios as tree size (dbh) increases. Thus, a constant proportion of the above-ground biomass will be estimated for each component as tree size increases. While this may be appropriate across a large range of tree sizes and appropriate for certain tree components (such as stem wood and bark), it will result in overestimates of components such as foliage and branch wood as tree crowns senesce (Monserud and Marshall, 1999).

6.5. OTHER METHODS OF ESTIMATING TREE CONTENT

6.5.1. Determination of Volume by Height Accumulation

The height accumulation concept was conceived and developed by Grosenbaugh (1948, 1954), who stated that the system can be applied by selecting tree diameters above breast height in diminishing arithmetic progressions, say 1- or 2-in. taper intervals, and estimating, recording, and accumulating tree height to each successive diameter. The system uses diameter as the independent variable instead of height, is well adapted to use with computers, and permits segregation of volume by classes of material, log size, or grade; however, since optimum log lengths for top log grades depend on factors other than diameter, the best grades may not be secured.

To apply the system, one must know the following information:

L = the number of unit height sections between taper steps
H = the total number of L values below a given diameter
H' = the cumulative total of H values below a given diameter

If volume inside bark is desired, the mean bark factor k (see Section 6.1.6) is also required. Figure 6.7 illustrates the method for estimating cubic feet. In Figure 6.7, diameter has been measured at stump height (0.5 ft) and at 4-ft intervals up the stem. In this example, 2-in. taper steps and 4-ft unit height lengths are utilized. The stem measurements are used to determine the heights at which each taper step occurs. The first taper step ($d=8$ in.) occurs at the top of the first unit length, thus $L_1 = 1$, $H_1 = 1$, and $H_1' = 1$. The second taper step ($d=6$ in.) occurs at the top of the fourth height segment and $L_2 = 3$, $H_2 = 4$, and $H_2' = 5$. The complete height accumulation data for this tree would be

Diameter	L	H	H'
8	1	1	1
6	3	4	5
4	2	6	11
2	1	7	18
Total	7	18	35

Volume for this tree is then obtained from the equation

$$V = A\left(\Sigma H'\right) + B\left(\Sigma H\right) + C\left(\Sigma L\right)$$ (6.33)

where, A, B, and C are the height accumulation coefficients
         L, H, and H' are as above

The height accumulation coefficients for 2-in. taper steps and 4-ft unit heights are shown in Table 6.13 for various dib/dob ratios. Using the dib/dob ratio of 1 (i.e., volume outside bark), the example tree's volume would be estimated to be

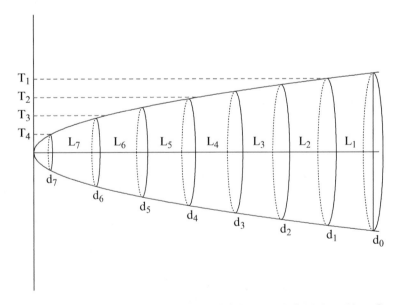

FIG. 6.7. Measurements for determination of volume by height accumulation (adapted from Grosenbaugh (1954)).

TABLE 6.13. Height Accumulation Coefficients, A, B, and C, to Compute Cubic-Foot Volume by 2-in. Taper Steps, 4-ft Unit Heights and Various Mean dib/dob Ratios

| Mean Ratio (dib/dob) | Volume Coefficients for Cubic Feet | | |
	A	B	C
1.00[a]	0.175	0	0.0291
0.95	0.158	0	0.0263
0.90	0.141	0	0.0236
0.85	0.126	0	0.0210

Source: Grosenbaugh (1954).
[a] When computing volume outside bark, use coefficients for ratio of 1.00.

TABLE 6.14. Sample Tree Data for Computation of Volume by Height Accumulation by 2-in. Taper Steps and 4-ft Unit Heights

| Dbh | Dob Taper Steps | | | | | Sum |
	10	8	6	4	2	
9.5	1	3	4	2	0	10
7.7		1	5	2	0	8
10.5	1	5	3	2	0	11
$L =$	2	9	12	6	0	29
$H =$	2	11	23	29	29	94
$H' =$	2	13	36	65	94	210

$$V = 0.175(35) + 0(18) + 0.0291(7)$$
$$= 6.125 + 0 + 0.204$$
$$= 6.329 \text{ ft}^3.$$

The height accumulation method has no advantage for the simple volume estimation shown in Figure 6.7. The technique has real utility in more complicated timber cruising applications. For example, Table 6.14 gives sample tree data needed to compute the volume V in cubic feet for a number of trees.

Volume, V, can be calculated for individual trees or for all trees combined using eq. (6.33) and the coefficients in Table 6.13. For the trees given in Table 6.16, the total cubic foot volume, inside bark, for a mean dib/dob ratio of 0.90 is

$$V = 0.141(210) + 0(94) + 0.0236(29) = 30.3 \text{ ft}^3$$

Similarly, individual-tree cubic-foot volume is

$$V_{9.5} = 0.141(75) + 0(33) + 0.0236(10) = 10.8 \text{ ft}^3$$
$$V_{7.7} = 0.141(46) + 0(23) + 0.0236(8) = 6.7 \text{ ft}^3$$
$$V_{10.5} = 0.141(89) + 0(38) + 0.0236(11) = 12.8 \text{ ft}^3$$
$$\overline{ 30.3 \text{ ft}^3}$$

Grosenbaugh (1954) also gives coefficients for 1-in. taper steps and 1-ft unit heights, coefficients for determination of board-foot volume and surface area, formulas to calculate coefficients for other cases, and the theory of height accumulation. The potential utility of this unique system has been generally overlooked.

6.5.2. Importance Sampling and Centroid Methods

In Section 6.1.3, it was shown that volume can be obtained by summing volumes of cross-sectional slices. Cross-sectional area, $A(X)$, multiplied by section length, h, is an estimate of section volume, and the sum of these volumes equals the volume of a log, merchantable stem, or total stem, depending upon the length of stem over which the cross-sectional slices are summed:

$$V = \sum_{i=1}^{n} A(X_i) h_i \qquad (6.34)$$

In Section 6.1.4, this concept was extended to the idea that if cross-sectional area, $A(X)$, could be described by a continuous function of height or distance from tree tip, then volume could be obtained by integrating this continuous function over the section of interest:

$$V = \int_a^b A(X) dX \qquad (6.35)$$

where $A(X)$ is the cross-sectional area at X, and X is either height above ground or distance from tree tip. From eq. (6.34), it can be shown that the average cross-sectional area $(\overline{A(X)} = \sum A(X)/n)$ multiplied by the section length $(L = n \cdot h)$ is also an estimate of volume for that section. All of the formulas shown in eqs. (6.1) through (6.9) attempt to estimate volume by estimating average cross-sectional area and multiplying by section length. Theoretically, volume of a section or whole stem could be estimated by measuring average cross-sectional area and multiplying by length or height; however, determining the average cross-sectional area is not a trivial task (hence, one of the reasons there are so many formulas for estimating section volumes). Importance sampling (Gregoire et al., 1986) generalizes this concept, and rather than trying to estimate average cross-sectional area, samples cross-sectional areas with probability proportional to their size (see Chapter 11 for a complete discussion on sampling with probability proportional to size). An estimate of volume is then calculated as

$$\hat{v} = \frac{A(X)}{f(X)} \qquad (6.36)$$

where $A(X)$ is cross-sectional area and $f(X)$ is the probability density associated with $A(X)$:

$$\int_a^b f(X) dX = 1$$

An unbiased estimate of volume would then be obtained from

$$\overline{v} = \frac{1}{n} \sum_{i=1}^{n} \frac{A(X_i)}{f(X_i)} \qquad (6.37)$$

The probability density function, $f(X)$, referred to as a *proxy function* (Gregoire et al., 1986), can be any function, but is often derived from a taper function or one of the formulas for a regular solids (e.g., a paraboloid); however, the more closely $f(X)$ mimics $A(X)$, the more precise the estimate of $\bar{v}$ will be. Gregoire et al. (1986) developed their probability density function based on a simple taper function:

$$S(h) = G\left(\frac{H-h}{H-b}\right)$$

where $S(h)$ = cross-sectional area at height h (m² or ft²)
     H = total height (m or ft)
     h = height (m or ft)
     G = cross-sectional area at breast height (i.e., basal area; m² or ft²)
     b = breast height (1.3 m or 4.5 ft)

Integrating the taper function over the height range (H_l, H_u) yields $V(H_l, H_u) = \left[G/(2(H-b))\right]K$ where $K = 2H(H_u - H_l) + (H_l^2 - H_u^2)$ and the proxy function becomes

$$f(h) = \frac{2(H-h)}{K}$$

Sampling is then carried out by selecting a random number, U, from a uniform [0,1] distribution and solving $V(H_l, \theta_U) = U \cdot V(H_l, H_u)$ for θ_U:

$$\theta_U = H - \sqrt{(H-H_l) - U \cdot K}$$

θ_U is height at which cross-sectional area needs to be determined, $A(\theta_U) = c(d_\theta^2)$ where c is the cross-sectional area conversion factor ($c = 0.005454$ in the imperial system and $c = 0.00007854$ in SI) and d_θ is the diameter at height θ_U. The volume estimate, based on this measurement, is then

$$\hat{v} = \left(\frac{K}{2(H-\theta_U)}\right) A(\theta_U)$$

Average volume estimated from multiple samples from a tree will produce more precise estimates:

$$\bar{v} = \frac{1}{n}\sum_{i=1}^{n} \hat{v}_i = \frac{1}{n}\sum_{i=1}^{n}\left(\frac{K}{2(H-\theta_i)}\right) A(\theta_i) = \frac{K}{2n}\sum_{i=1}^{n}\frac{A(\theta_i)}{(H-\theta_i)}$$

Importance sampling is easy to implement in the field. Most handheld calculators and smartphones have the capability of generating random numbers and solving the required equations for the heights at which measurements must be made. Field data loggers can be easily programmed to calculate and store volume estimates. The volume estimate potentially requires a measurement of diameter at heights other than breast height. Section 5.2.2 describe a number of instruments that can be used to measure upper stem diameters, and

Gregoire et al. (1986) describe a few additional tools. Measurement of diameter at breast height is a tradition of convenience, importance sampling focuses diameter measurement at points on trees more related to volume. While implementing importance sampling is more costly than traditional breast height measurements and applying allometric relationships, it is less costly than detailed dendrometry studies (Section 6.1) required to develop such relationships (Gregoire et al., 1986).

The *centroid method*, originally derived from importance sampling, has been suggested as an alternative method to estimating tree volume without the use of species-specific volume functions. Wood et al. (1990) noted that variability in importance sample estimates was lowest when single-diameter measures were made at the height on the stem where half of the predicted volume was below and half above the measurement height. Wood et al. (1990) called this point the *centroid point* and developed the centroid method as a technique for estimating volume based on measures made at the centroid point. So, rather than randomly selecting the measurement point, the measurement point is fixed at the midpoint of the volume distribution. When the upper diameter of the tree is zero (i.e., when estimating total volume), the height (or distance from large end) of the centroid position is

$$H_C = \left(1 - \frac{1}{\sqrt{2}}\right) H \approx 0.293H$$

and volume is estimated as

$$V = A_C \left(\frac{H}{\sqrt{2}}\right) \approx 0.707 A_C H$$

The centroid method can be implemented on logs where the upper diameter is not zero; however, the algebra associated with deriving total volume is more complex (Wiant et al., 1992a). Wiant et al. (1992b) provide practical guidelines for implementing the centroid method.

Forslund (1982) developed a similar approach based on a stem model shape that was between a cone and a paraboloid. Forslund (1982) termed this shape a *paracone* and was derived based on empirical observation that 0.3H was the height of the center of gravity of stems without branches. The volume estimate associated with the paracone method is $V = 0.693 A_{0.3H} H$. Wiant et al. (1991) compared the paracone method with the centroid method and found that both methods gave similar results, with the paracone method performing slightly better. Given the similarity of the total volume formulas, this was not unexpected (Wiant et al., 1991; Ducey and Williams, 2011). Ducey and Williams (2011) pointed out that the centroid method and the paracone method were both similar to the formula proposed by Hossfeld, a German forester, in the 1800s.

Importance sampling, the centroid method, and the paracone method all provide approaches for estimating stem volume without the need for volume functions or detailed dendrometric measurements. While measurement of stem diameters at heights other than breast height are more time consuming, they are not impossible or difficult to obtain. The approach is easily adaptable to estimating weight, biomass or nutrient and carbon contents, especially if a biomass expansion factor approach is used (Heath et al., 2009; Somogyi et al., 2009). As technology continues to improve, these methods will have more direct applications in many forest inventories.

6.6. APPLICATIONS TO SEEDLINGS AND UNDERSTORY VEGETATION

Volume is not typically measured or estimated for seedlings and understory vegetation (Whittaker and Woodwell, 1968). Biomass is more often of interest because it better reflects productivity (Whittaker and Woodwell, 1968) and is often more important in terms of management since understory vegetation, in addition to their contribution toward total biodiversity, are often of interest because of their potential for browse and forage material for wildlife and domestic animals (Paton et al., 2002) or for other nontimber forest products (Chamberlain et al., 2002, 2013). Because of their size, direct determination of weight and/or biomass is generally much easier. Above-ground portions of plants can generally be easily clipped, bagged, returned to the lab and dried. Below-ground portions are also easier to extract than for trees. While logistically easier, the process is still time consuming, costly, and destructive to the individual plants harvested. As result, allometric relationships are often developed (e.g., Whittaker and Woodwell, 1968; Dickinson and Zenner, 2010). The equations are similar to those used for estimating trees; however, the independent variables are often different. Except for the taller shrub species, most understory plants do not develop diameters at breast height, therefore, diameters at root collar are sometimes used (Brown, 1976) or at some additional height, such as 18 cm (e.g., Smith and Brand, 1983). Others use total height or some measure of canopy area or size (e.g., Paton et al., 2002; Chamberlain et al., 2013).

Paton et al. (2002) compared maximum plant height, relative plant height (plant height/ maximum plot height), minimum crown diameter, maximum crown diameter, canopy area ($\pi \cdot D_{min} \cdot D_{max}$), and canopy volume ($CA \cdot H$) as predictors for 22 Mediterranean shrub species. Using a local equation similar to eq. (6.30), and substituting each of these variables in place of dbh, canopy volume was the best predictor for 12 of the 22 species, canopy area was best for 3 species, while plant height was best for only 1 species. Olson and Martin (1981) also found that canopy volume was a superior predictor of above-ground biomass for understory plants in western Washington, USA. Chamberlain et al. (2013) used canopy area to predict the root biomass for the medicinal herb *Actaea racemosa*.

6.7. APPLICATIONS TO SNAGS AND DOWN WOODY MATERIAL

The measurement of volume, biomass, and carbon content of standing dead trees (snags) and downed woody material presents some additional challenges, even though both types of material are derived primarily from tree stems. Following mortality, the structural and density changes associated with decay tend to create irregular shapes that gradually diverge in volume and content from those that would be found in live tree stems with similar basic dimensions. Biases that would be relatively unimportant in most applications with live trees, such as assuming a solid circular cross section of the stem, become more important as decay progresses. Thus, special care is needed if accurate estimates of standing and downed dead wood attributes are needed.

6.7.1. Standing Dead Trees

In principle, for freshly killed standing dead trees, any of the usual methods for estimating volume of standing live trees should be applicable. Likewise, whole-tree biomass equations also should provide sufficient accuracy, though if a component ratio method is used, deduction of the biomass of live foliage would be warranted. For trees that have lost most of their branches, or where volume and biomass estimates are only of interest for the main stem, ordinary dendrometry (including measurement of dbh, height of the remaining stem, and diameter at the point of breakage) will often suffice to estimate gross volume. However, even for trees that have recently died, there is an elevated probability of rotten or hollow cull. As a standing tree decays, the following factors become increasingly important:

1. Loss of volume due to breakage of the top, sloughing of bark and branches, and internal decay.
2. Loss of wood density due to decay.
3. Changes in relative abundance of nutrients due to preferential breakdown of cellulose, hemicellulose, and lignin, as well as leaching.

Unfortunately, these losses are not always well characterized for a given species and region. As a result, Smith et al. (2003) recommend not only the elimination of biomass or carbon from leaves, but also 10% of stem wood and bark, 33% of branches, and 20% of roots, as average corrections for standing dead trees in North America.

The prediction of volume, biomass, and carbon content, as well as attributes important for wildlife habitat, can be enhanced for standing dead trees by recording decay class. A variety of decay class systems for snags are in use, reflecting differences in the decay trajectories of individual species as well as the influence of the original investigators who developed them. Table 6.15 shows the decay class system used by the USDA Forest

TABLE 6.15. Decay Class System for Standing Dead Trees

Decay Class	Branches	Top	Bark Remaining	Sapwood	Heartwood
I	Present	Pointed	100%	Intact; sound or incipient decay; hard, original color	Sound, hard, original color
II	Few limbs, no fine branches	May be broken	Variable	Sloughing; advanced decay; fibrous, firm to soft, light brown	Sound at base, incipient decay in outer edge of upper bole, hard, light to reddish brown
III	Stubs only	Broken	Variable	Sloughing; fibrous, soft, light to reddish-brown	Incipient decay at base, advanced decay throughout upper bole, fibrous to cubical, soft, dark reddish brown
IV	Few or no stubs	Broken	Variable	Sloughing; cubical, soft, reddish to dark brown	Advanced decay at base, sloughing from upper bole, fibrous to cubical, soft, dark reddish-brown
V	None	Broken	<20%	Absent	Sloughing, cubical, soft, dark brown, or fibrous, very soft, dark reddish brown, encased in hardened shell

Service, Forest Inventory and Analysis program (FIA, 2012), which is ultimately derived from that presented by Thomas (1979).

Note that the sapwood and heartwood characteristics in Table 6.15 are specifically targeted to Douglas-fir; individual species vary considerably in the progression of wood decay. Gross structural descriptions (such as those for branches, top, and bark in Table 6.15) are much easier to use and also tend to provide classifications that are more consistent across species. Although the characteristics used are qualitative, and ratings by different field personnel are certain to vary, relatively few inventories have attempted to develop strict guidelines based solely on quantitative measurements, in part because of the inherent variability of dead woody material. However, some studies (e.g., Eaton and Sanchez, 2009; Angers et al., 2012) have highlighted the variability between species and suggested that species-specific approaches that reflected different decay pathways could provide better predictions of key variables such as wood density.

If the species of a standing dead tree can be identified, then its wood density and biomass can often be estimated by applying a *density reduction factor* that accounts for the effects of species and decay stage. Specifically,

$$BM = DRF \cdot SG_g \cdot Dens_{water} \cdot V \qquad (6.38)$$

where DRF is the density reduction factor, and all other terms are as defined for live trees in eq. (6.20). The quantity $\left(DRF \cdot S_g \cdot Dens_{water}\right)$ gives the density of the dead, decayed wood. Typically, DRF starts near 1 for freshly killed trees and declines with advancing decay class, though the rate of decline may depend on species. Harmon et al. (2011) present some species- and decay-class density reduction factors for standing dead trees in North America.

6.7.2. Downed Woody Material

Fallen logs, logging slash, and other downed woody material presents some similar challenges to those offered by standing dead trees. A primary advantage of downed woody material is that it is usually accessible for measurement, so that (at least in theory) it should be possible to obtain accurate volume estimates. However, because all downed woody material is subject to breakage and decay, and some downed woody material includes branches, the variety of solid forms presented by downed woody material is highly variable. Many studies of downed woody material nonetheless employ the classic volume formulae for portions of the main stem of a live tree, including Huber's, Smalian's, and Newton's formulae, as well as that for the frustum of a cone. Fraver et al. (2007) also explored an average-of-ends formula and a novel conic-paraboloid formula. Their conic-paraboloid required no midpoint measurement, but showed high accuracy in the material they encountered. However, given the variety of solids presented by downed woody material, even Newton's formula occasionally shows substantial bias. Moreover, all conventional volume formulae required a measurement of piece length, which is generally time consuming and can be a source of error in extensive inventories, especially since fallen wood is often curved, and retained branches can make measurement along the stem challenging (Westfall and Woodall, 2007). Thus, as we shall see in Chapter 12, there can be considerable advantages in the use of inventory techniques for downed woody material that do not require a volume equation.

TABLE 6.16. Decay Class System For Downed Dead Wood

Decay Class	Structural Integrity	Texture of Rotten Portions	Color of Wood	Invading Roots	Branches and Twigs
I	Sound, freshly fallen	Intact, no rot	Original color	Absent	If branches are present, fine twigs are still attached
II	Sound	Mostly intact; sapwood may be partly soft	Original color	Absent	Most fine twigs absent
III	Heartwood sound; supports its own weight	Hard, large pieces; sapwood, if present, may be pulled apart	Reddish-brown or original color	Sapwood only	Branch stubs will not pull out
IV	Heartwood rotten; will not support its own weight but maintains shape	Small, soft blocky pieces	Reddish or light brown	Throughout	Branch stubs pull out
V	No longer maintains shape, spreads out on ground	Soft; powdery when dry	Red-brown to dark brown	Throughout	Branch stubs have usually rotted out

Even for those inventory methods that require no volume equation, estimates of one or more cross-sectional areas are required. Although such measurements are often referred to as "diameter," consistent with the measurement techniques for standing trees, it is not generally safe to assume circular cross sections in downed wood, especially as decay becomes advanced and pieces flatten. For example, Fraver et al. (2007) found that the cross-sectional area of freshly fallen material was 91% of what it would be under a circular assumption (treating the thickness of the log, calipered horizontally as it is most accessible, as the diameter), but declined to only 38% by the most advanced stages of decay. Harmon and Sexton (1996) recommend the use of calipers and an elliptical approximation when trees have flattened due to decay, but obtaining the vertical "diameter" measurement can be challenging when heavy pieces of fallen wood are in contact with the ground. For material in the most advanced stages of decay, a sharpened metal rod engraved with graduated marks can be used to probe the depth, testing for changes in resistance in a fashion akin to that of a bark gauge. In any case, the use of a diameter tape is nearly impossible with most downed woody material.

Just as with standing dead trees, decay class is often recorded for downed dead wood both to facilitate estimation of biomass and carbon content, and to describe the attributes of the material for habitat, fuels, and seedbed characteristics. A wide range of decay class systems have been employed, but most of those used in North America are broadly similar to that used by the USDA. Forest Service, Forest Inventory and Analysis (FIA, 2011). The five-class system is simplified and summarized in Table 6.16.

The qualitative nature of the decay indicators, possible differences in rates and pathways of decay between species, and the heterogeneity of decay within an individual piece have all been cited as challenges for simple decay class systems (Pyle and Brown, 1999). The accessibility of downed woody material has prompted a number of attempts to use more objective, quantitative measures to predict density, carbon content, and other attributes. These have included measurements using a variety of penetrometers and drill resistance (Creed et al., 2004; Kahl et al., 2009; Larjavaara and Muller-Landau, 2010; Mäkipää and Linkosalo, 2011) or using ultrasonic waves (Bütler et al., 2007). Although these approaches have shown some success, most inventory work is still done using decay class systems. Examples of density tabulations by species and decay class, or density reduction factors for downed logs, can be found in Næsset (1999), Sandström et al. (2007), Harmon et al. (2008), and Paletto and Tosi (2010). Nonetheless, the use of local and species-specific information can have a substantial impact on estimates of downed wood mass and carbon content (Weggler et al., 2012). As an alternative to the use of tabulated density and content data, it is possible to integrate such measurements directly in a sample-based inventory (Chapter 12).

Once a volume estimate of suitable accuracy has been obtained, direct use of tabulated wood densities by decay class or density reduction factors can be employed to estimate biomass and carbon content. The computations follow the procedures used for standing dead trees exactly.

7

MEASUREMENT OF PRIMARY FOREST PRODUCTS

7.1. UNITS OF MEASUREMENT OF FOREST PRODUCTS

7.1.1. Board Foot

The board foot, which is still widely used to express the volume of trees, logs, and lumber, is defined as a piece of rough green lumber 1 in. thick, 12 in. wide, and 1 ft long, or its equivalent in dried and surfaced lumber. Thus, a board foot is a nominal volume unit. For example, in the purchase of lumber, the standard finished size of a kiln-dried 1 in. by 12 in. board would be ¾ in. by 11¼ in.: a kiln-dried 2 in. by 4 in. board would be 1½ in. by 3½ in.; and so on; however, the board foot volume of a piece of lumber would be computed from the nominal thickness and width, not the exact dimensions, by the following formula:

$$\text{Board feet} = \frac{W \cdot T \cdot L}{12} \tag{7.1}$$

where W = nominal width (in.)
  T = nominal thickness (in.)
  L = actual length (ft)

When log and lumber volumes in commercial transactions, inventory reports, and so forth involve large quantities of material, the volumes are normally given in thousands of board feet. In this case, the abbreviation M.B.F. (1000 board feet) or M.B.M. (1000 ft, board measure) is used. For example, 11,234,000 board feet = 11,234 M.B.F. Occasionally, especially in Canada, the abbreviation F.B.M (foot board, thousands) is used.

Forest Mensuration, Fifth Edition. John A. Kershaw, Jr., Mark J. Ducey, Thomas W. Beers and Bertram Husch.
© 2017 John Wiley & Sons, Ltd. Published 2017 by John Wiley & Sons, Ltd.

The board foot is the standard unit of lumber measurement in the United States (and is used in many other countries). Lumber that is thinner than 1 in. in thickness is usually sold on the basis of surface measure. Finished lumber, such as trim and molding, is sold by the linear foot.

Although there is an inexactness in the measurement of the board foot volume of lumber, board foot log rules showing the estimated number of board feet of lumber that can be sawed from logs of given lengths and diameters are frequently used to estimate the contents of logs and trees. This is rather unique. Few raw materials are measured in terms of the finished product. It is somewhat like trying to measure the yield of a field of corn in terms of the boxes of corn flakes.

The number of board feet that can be produced from a log depends primarily on its length and diameter. In addition, there are other factors that affect the yield of a log. Among these are

1. Efficiency of the sawmill machinery and, in particular, the thickness of the kerfs cut by the various saws.
2. Efficiency of the workers, particularly the sawyers, edgers, and trimmers.
3. Market conditions. (When markets are good, efforts will be made to utilize small pieces, and the output will be raised. When the ratio of thick lumber cut to 1-in. lumber cut increases, output will be raised.)
4. Amount of defect and taper in logs.

Other units, which have largely disappeared, are the *Hoppus foot* and the *Blodgett foot*. The Hoppus foot is a cubic volume unit that was used in the United Kingdom, Australia, New Zealand, and many of the Caribbean Islands. The formula to determine Hoppus feet, often called the quarter-girth formula, is obtained by squaring one quarter of the girth (circumference) and multiplying by length divided by 144. The Hoppus formula gives 78.5% of the actual cubic volume of a log, and a Hoppus foot is considered equivalent to 10 board feet. To obtain cubic volume, a divisor of 113 is used in place of 144 in the quarter-girth formula (Fonseca, 2005). The Blodgett foot, formerly employed in the northeastern United States, is defined as a cylindrical block of wood 16 in. in diameter and 1 ft in length. It is equivalent to 1.4 ft^3.

7.1.2. Volume Units for Stacked Wood

The *cord* is a unit of measure widely utilized in North America to express the volume of stacked wood. The need to measure relatively small pieces of low value, rough wood, stacked at scattered locations in the woods gave rise to this unit.

A *standard cord* is a pile of stacked wood 8 ft long and 4 ft high that is made up of 4 ft pieces (The standard cord occupies 128 ft^3.) The actual solid wood content is generally less than 100 ft^3 and varies by species, method of stacking, form of wood, length and diameter of wood, bark thickness, and other factors. A *long cord* measures 8 ft long and 4 ft high and is made up of pieces longer than 4 ft. Long cords commonly consist of pieces 5 ft long, or 5 ft 3 in. long, and occupy 160 or 168 ft^3, respectively. A *short cord* or *face cord* is a unit smaller than the standard cord and is usually used to measure fuelwood that is cut less than 4 ft long. A *rick* of fuelwood is often considered to be an 8 ft by 4 ft pile made up of 12-in. pieces (four ricks per standard cord) or an 8 ft by 4 ft pile made up of 16-in. pieces (three ricks per standard cord).

A *pen*, which was used in the southern United States, consists of sticks stacked to a height of 6 ft in the form of a square enclosure with two sticks to a layer. Five pens of 4-ft sticks are assumed to equal a standard cord; five pens of 5-ft sticks are assumed to equal a long cord of 160 ft³. The *stere* or metric cord, which is used in some countries that employ the SI, is a pile of stacked wood 1 m long and 1 m high that is made up of 1 m pieces (the stere occupies 1 m³ of space).*

7.2. LOG RULES

A *log rule* is a table or formula that gives the estimated volume of logs of specified diameters and lengths. In North America, most log rules give volumes in board feet of lumber, although tables that give volumes in cubic units are available. In nations where the SI is used, expression of volume in board feet is still common, although many log rules give volume in cubic meters.

The cubic volume of a log may be determined by any of the methods given in Section 6.1. However, for log scaling, gross cubic volumes are usually computed by Huber's or Smalian's formula. Bell and Dilworth (1993) mention other formulas used on the west coast of the United States (two-end conic rule, subneiloid rule, and the Bruce butt log formulas). Traditionally, logs rules have been presented in tabular form; however, the formulas can be used directly. Formulas are generally used in computer processing of inventory data rather than lookup tables that occupy large amounts of computer memory and are generally slower than direct computation.

In the preparation of cubic meter tables, volumes are most often computed for 2-cm diameter classes and 0.2-m length classes. In the preparation of cubic foot tables, volumes are most often computed for 1-in. diameter classes and 1-ft length classes. Diameter is often given as the midpoint diameter inside bark. (When the midpoint diameter cannot be measured, the average of the diameters inside bark at the two ends of the log is generally used for midpoint diameter.) In some cases, volume is given in terms of small-end diameter inside bark.

7.3. BOARD FOOT LOG RULES

It is not easy to prepare a board foot log rule that is universally applicable because of variations in the dimensions of lumber produced from logs, the equipment used to saw logs, the skill of operators, computer programs used to saw logs, and the logs themselves. In the early years of the lumber industry in the United States and Canada, there were a number of independent marketing areas, and no industrial organization or governmental agency had control over the measurement of lumber and logs. As a result, different areas devised rules to fit specific operating conditions. Freese (1973) pointed out in his excellent report on log rules, "In the United States and Canada there are over 95 recognized rules bearing about 185 names." Most of these rules have long since been forgotten. Only five or six rules are important in present-day use. There are perhaps half a dozen others that may be encountered in certain localities.

---

* Other units to measure stacked wood are employed in different countries. For example, the unit of stacked wood in Chile is the "*metro ruma*," which consist of a 1 m by 1 m pile of 2.44 m sticks occupying 2.44 m³ of space. The solid wood content is approximately 1.66 m³.

Board foot log rules are used to estimate the contents of logs. This constitutes an attempt to estimate, before processing, the amount of lumber in logs. Thus, we must distinguish between the measurements of the board foot contents of sawn lumber, the *mill tally*, and the estimation of the board foot contents of logs, the *log scale*. The board foot mill tally, though not exact, is a well-defined unit; the board foot log scale is an ambiguous unit. Consequently, the amount of lumber sawed from any run of logs rarely agrees with the scale of the logs. This variation, O_v, may be expressed in board feet: O_v (in board feet)=mill tally − log scale. When O_v is positive, a mill has produced an *overrun*; when O_v is negative, a mill has produced and *underrun*.

O_v is generally expressed as a percentage of log scale:

$$O_v = 100\left(\frac{MT-LS}{LS}\right) = 100\left(\frac{MT}{LS}-1\right) \tag{7.2}$$

where MT=mill tally

LS=Log scale

The quantity MT/LS is commonly referred to as the *overrun ratio*, that is, the number of board feet mill tally per board foot scale.

In the construction of all known board foot log rules, three basic methods have been used: (1) mill study, (2) diagram, and (3) mathematical. For the board foot log rules in present-day use, generally the yield of logs V in board feet is estimated in terms of lumber 1-in. thick, from average small-end diameter inside bark D in inches, and log length L in feet.

7.3.1. Mill-Study Log Rules

In this method of constructing log rules, a sample of logs is first measured prior to sawing. Then, as each log is sawn, the boards are measured to determine the board foot volume yield of each log. The log rule is prepared by relating board foot yields, the dependent variable, with log diameters and lengths, the independent variables. The problem may be solved graphically or by regression analysis (Section 3.8).

A rule of this type should give good estimates for mills that cut timber with certain characteristics or for those that use specific milling methods. The method, however, has never been widely used. The Massachusetts log rule, one of the few constructed by this method that is still in use, was based on 1200 white pine logs. This rule was constructed for round- and square-edged boards sawed from small logs (¼ in. saw kerf). Some boards over 1-in. thick were included, so the values are slightly high for 1-in. boards.

Large modern sawmills, which use laser scanning of individual pieces and computer optimization of the sawing pattern, are capable of providing the data for a mill-study log rule and its daily and hourly variation. Often, such data are typically treated as proprietary and secret, and are not used openly in the buying and selling of logs.

7.3.2. Diagram Log Rules

The procedure for the construction of a diagram log rule is simple:

1. Draw circles to scale to represent the small ends of logs of different diameters inside bark. Assume logs are cylinders of a specific length, such as 8 ft.

2. Use definite assumptions on saw kerf and shrinkage, and board width, and draw boards (rectangles) 1-in. thick within the circles.

3. Compute the total board foot content for each log diameter.

4. Determine the board foot contents of other log lengths by proportion.

When a diagram log rule is prepared for any given log length, it will be found that increases in volume, from one diameter to the next, will be slightly irregular. Preparing a freehand curve, or a regression equation, to predict volume from diameter for each length, may eliminate these irregularities.

The Scribner log rule, the most widely used diagram log rule, was first published in 1846 by J.M. Scribner, a country clergyman. The rule was prepared for 1-in. lumber with a 1/4-in. allowance for saw kerf and shrinkage. The minimum board width is unknown. The original table gave board foot contents for logs with scaling diameters (diameter inside bark at small end) from 12 to 44 in., and with lengths from 10 to 24 ft. Log taper was not considered. A few years after the original rule was published, Scribner modified the rule by increasing the slab allowance on larger logs. This is the rule in use today. The Scribner rule gives a relatively high overrun (up to 30%) for logs under 14 in. Above 14 in., the overrun gradually decreases and flattens out around 28 in. to about 3–5%.

A regression equation was prepared from the original table by Bruce and Schumacher (1950). The equation, which gives volume in board feet V, in terms of scaling diameters in inches D and log length in feet L is

$$V = \left(0.79D^2 - 2D - 4\right)\frac{L}{16} \tag{7.3}$$

Some so-called Scribner tables contain values based on this equation. Values in this table differ slightly from the original Scribner values because Scribner did not smooth the values he obtained from his diagrams. Note that the values in Table 7.1 are from the original Scribner table.

TABLE 7.1. Comparison of Volumes of 16-Foot Logs from Different Log Rules

Log Diameter (in.)	Mill Tally[a] (Board Feet)	Log Scale				
		Scribner	Maine	International 1/4-in. Rule	Doyle	Spaulding
6		18	20	20	4	
10	75	50	68	65	36	
14	145	114	142	135	100	114
18	229	213	232	230	196	216
22	382	334	363	355	324	341
26	578	500	507	500	484	488
30	665	657	706	675	676	656
34	862	800	900	870	900	845
38	1037	1068	1135	1095	1156	1064

[a] Average yield of logs sawed in an Indiana bandmill.

Since calculating machines were not generally available in the nineteenth century, scalers found the adding of long columns of figures laborious. Consequently, the Scribner rule was often converted into a *decimal rule* by dropping the units and rounding the values to the nearest 10 board feet. Thus, the value for 114 board feet was written as 11 and for 159 board feet was written as 16.

Because the original Scribner rule did not give values for logs less than 12 in. in diameter, a number of lumber companies extrapolated to derive volumes for small logs. Finally, the Lufkin Rule Company prepared three tables using different assumptions to extend the rule to cover small logs. They published these as decimal rules and called them Scribner decimal A rule, Scribner decimal B rule, and Scribner decimal C rule. The decimal C rule is the only one of these rules still widely used.

There are other log rules based on diagrams. The best known are the Spaulding rule, which was devised by N.W. Spaulding of San Francisco in 1868, and the Maine rule, which was devised by C.T. Holland in 1856. The Spaulding rule is used on the Pacific Coast of the United States; the Maine rule is used in northeastern United States. The Spaulding rule, which closely approximates the values of the Scribner rule, may be expressed by the following regression equation, where V is volume in board feet, D is scaling diameter in inches, and L is log length in feet:

$$V = \left(0.778D^2 - 1.125D - 13.482\right)\frac{L}{16} \tag{7.4}$$

7.3.3. Mathematical Log Rules

In this method of constructing log rules, one makes definite assumptions on saw kerf, taper, and milling procedures and prepares a formula that gives board foot yield of logs in terms of their diameters and lengths.

Doyle Log Rule. This is one of the most widely used, one of the oldest, and one of the most cursed log rules. It was first published in 1825 by Edward Doyle. The rule states: "Deduct 4 in. from the scaling diameter of the log, D, in inches, for slabbing, square one-quarter of the remainder, and multiply by the length of the log, L, in feet." As Herrick (1940) pointed out, when Doyle deducted 4 in. from the diameter of the log for slabbing, he was squaring the log. Then, he calculated the board foot contents of the squared log or cant as follows:

$$(D-4)(D-4)\frac{L}{12}$$

To allow for saw kerf and shrinkage, he reduced the volume of the cant by 25% to obtain the final rule:

$$V = \frac{(D-4)^2 L}{12}(1.00 - 0.25) = \left(\frac{D-4}{4}\right)^2 L \tag{7.5}$$

For 16-ft logs, the *Doyle rule of thumb* is

$$V = (D-4)^2$$

The simplicity and ease of application are the reasons for the wide acceptance of this rule.

When the Doyle rule is applied to logs between 26 and 36 in. in diameter, it gives good results. When the rule is applied to large logs, it gives an underrun; when the rule is applied to small logs, it gives a high overrun. This comes about because the 4-in. slabbing allowance is inadequate for large logs and excessive for small logs.

International Log Rule. This is one of the most accurate mathematical log rules. It was developed by J.F. Clark and published in 1906 (Clark, 1906). In the derivation of the original rule, one first computes the board foot contents of a 4-ft cylinder in terms of cylinder diameter in inches, D, assuming the cylinder will produce lumber at the rate of 12 board feet per cubic foot. The solid board foot content of a cylinder is

$$V = \frac{\pi D^2}{4(144)}(4)(12) = 0.262 D^2$$

To allow for saw kerf and shrinkage, one assumes that for each 1-in. board cut 1/8 in. will be lost in saw kerf and 1/16 in. in shrinkage. Thus, the proportion lost from saw kerf and shrinkage is

$$\left(\frac{3/16}{1+3/16}\right) = 0.158$$

When reduced by 0.158, the volume of the 4-ft cylinder becomes

$$V = 0.262 D^2 (1.000 - 0.158) = 0.22 D^2$$

Thus, the losses from saw kerf and shrinkage are proportional to the end area of the log (i.e., to D^2).

Clark determined that losses from slabs and edgings constitute a ring-shaped collar around the outside of the log and that they are proportional to the surface area, or the diameter, D, of the log. From a careful analysis of the losses occurring during the conversion of sawlogs to lumber, Clark found that a plank 2.12 in. thick and D in. wide would give the correct deduction. (The thickness of the collar T is about 0.7 in. for all values of D.) This can be determined from the following equation:

$$2.12D = \pi \frac{D^2}{4} - \frac{\pi(D-2T)^2}{4}$$

which leads to $T^2 - DT + 0.6748D = 0$. Therefore, in terms of cylinder diameter in inches D, the board foot deduction is computed using eq. (7.1):

$$\frac{2.12(D)(4)}{12} = 0.71D$$

Thus, the net board foot volume V of a 4-ft cylinder is

$$V = 0.22 D^2 - 0.71D \tag{7.6}$$

After studying a number of tree species, Clark decided to allow a taper of 1/2 in. for each 4-ft section. With this assumption, the basic formula was expanded to cover other log lengths:

$$V(8-\text{ft logs}) = 0.44D^2 - 1.20D - 0.30 \qquad (7.7)$$

$$V(12-\text{ft logs}) = 0.66D^2 - 1.47D - 0.79 \qquad (7.8)$$

$$V(16-\text{ft logs}) = 0.88D^2 - 1.52D - 1.36 \qquad (7.9)$$

$$V(20-\text{ft logs}) = 1.10D^2 - 1.35D - 1.90 \qquad (7.10)$$

Lengths over 20 ft are scaled as two or more logs.

The following equation, developed by Grosenbaugh (1952a) using the method of least squares, can be used for the determination of the board foot volume (international rule, 1/4 in. kerf) for logs of any length:

$$V = 0.0497621LD^2 + 0.006220L^2D - 0.185476LD + 0.000259L^3 - 0.011592L^2 + 0.042222L$$

The original international log rule may be modified to give estimates for saw kerfs other than 1/8 in. For example, for a kerf of 1/4 in. (shrinkage of 1/16 in.), the proportion lost is

$$\left(\frac{5/16}{1+5/16} \right) = 0.238$$

So, the original rule may be converted to a 1/4-in. rule by multiplying the values by the following factor:

$$\frac{1.000 - 0.238}{1.000 - 0.158} = 0.905$$

The factor to convert to a 7/64-in. rule is 1.013; the factor to convert to a 3/16-in. rule is 0.950.

When Clark (1906) published his log rule in table form, he rounded all values to the nearest multiple of 5 board feet. This is the form that most international log rule tables appear today.

7.3.4. Combination Log Rules

This type of rule combines values from different log rules. Such rules take advantage of the best, or the worst, features of the rules used. For example, the *Doyle–Scribner rule*, a combination of the Doyle and Scribner rules, was prepared for use in defective and over mature timber. Since the Doyle rule gives an overrun for small logs, its values were used for diameters up through 28 in. Since the Scribner rule gives an overrun for large logs, its values were used for diameters greater than 28 in. Thus, the Doyle–Scribner rule gives a consistently high overrun that is supposed to compensate for hidden defects. The *Scribner–Doyle rule*, exactly opposite to the Doyle–Scribner rule, gives a consistently low overrun.

7.3.5. Comparison of Log Rules

Because different methods and assumptions are used in the construction of log rules, different rules give different results, none of which will necessarily agree with the mill tally for any given log (Table 7.1). This points out that the board foot log scale, by any rule, is a unit of estimate, and not a unit of measure.

7.3.6. Tabular Presentation of Log Rules

Section 6.3 describes the use and construction of cubic volume and content tables. Board foot log rules also are often presented in tables. Because of the various approaches and assumptions associated with log rules, it is very important to ensure that the information list in Section 6.3.1 is included. Like cubic volume tables, board foot volume tables may be presented as standard tables (Table 7.2) or local tables (Table 7.3).

TABLE 7.2. Example of a Standard Volume Table, Using Board Foot Volume, International 1/4-in. Rule, for Red Oak (*Quercus rubra*) in Pennsylvania

dbh (in.)	Merchantable height—number of 16-ft Logs											
	½	1	1½	2	2½	3	3½	4	4½	5	5½	6
8	8	18	28	37	47	57						
9	11	23	35	48	60	73						
10	13	29	44	59	75	90	105	121				
11	17	35	54	72	91	109	128	146				
12	20	42	64	86	108	130	153	175	197			
13	24	50	76	102	128	153	179	205	231			
14	28	58	88	118	148	178	208	238	268	298	328	
15	33	67	102	136	170	205	239	274	308	343	377	
16	37	77	116	155	194	233	273	312	351	390	429	469
17		87	131	175	219	264	308	352	396	441	485	529
18		97	147	197	246	296	345	395	445	494	544	593
19		109	164	219	275	330	385	440	496	551	606	661
20		121	182	243	304	366	427	488	549	611	672	733
21			201	268	336	403	471	538	606	673	741	808
22			220	295	369	443	517	591	665	739	813	887
23			241	322	403	484	565	646	727	808	889	970
24			263	351	439	527	616	704	792	880	968	1057
25			285	381	477	572	668	764	859	955	1051	1147
26			309	412	516	619	723	826	930	1033	1137	1240
27				445	556	668	780	891	1003	1114	1226	1338
28				478	598	718	838	959	1079	1199	1319	1439
29				513	642	771	900	1028	1157	1286	1415	1543
30				549	687	825	963	1101	1238	1376	1514	1652
31					734	881	1028	1175	1322	1470	1617	1764
32					782	939	1096	1253	1409	1566	1723	1880
33					832	999	1165	1332	1499	1666	1832	1999
34					883	1060	1237	1414	1591	1768	1945	2122
35					936	1124	1311	1499	1686	1874	2061	2249
36					990	1189	1387	1586	1784	1983	2181	2379

Source: Adapted from Bartoo and Hutnik (1962).
Stump height, 1 ft. Top diameter 8.0 in., inside bark.
Block indicates extent of basic data. Basis, 210 trees.
Sample trees scaled as 16-ft logs; top section measured to nearest foot.
Standard error of regression coefficient=0.00261.
Proportion of variation accounted for by the regression=0.974.
Tabular values derived from regression $V = -1.84 + 0.01914D^2H$.

TABLE 7.3. Example of a Local Volume Table for Yellow Poplar (*Liriodendron tulipifera*) in Stark County, Ohio, Using International Rule (1/4-in. Kerf)—Merchantable Stem to a Variable Top Diameter

dbh Outside Bark (in.)	Volume Per Tree (Board Feet)	Merchantable Length (ft.)	Basis in Trees (Number)
10	30	19.5	4
11	50	23	5
12	70	26.5	13
13	95	30	9
14	125	33	9
15	155	36.5	1
16	190	40	5
17	235	43	7
18	285	45.5	6
19	345	48	4
20	405	51	2
21	480	53.5	1
22	555	56	2
23	635	58	3
24	720	60	1
25	800	62	–
26	885	64	1
27	975	65.5	2
28	1065	67	1
29	1155	69	5
30	1245	70	9
31	1340	71.5	7
32	1435	72.5	7
33	1535	73.5	1
34	1630	74.5	1
35	1725	75	–
36	1825	76	–

Source: Adapted from Diller and Kellogg (1940).

Trees climbed and measured by personnel of Work Projects Administration Official Project 65-1-42-166 – the Ohio Woodland Survey. Measurements taken at 16-ft log lengths above a 2.0-ft stump height. Scaled as 16-ft logs, and additional shorter top logs; top sections less than 8 ft in length scaled as fractions of an 8-ft log. Basis, 107 trees.

Table prepared in 1939 by curving volume of merchantable length over dbh.

Aggregate difference: Table is 0.8% low. Average percentage deviation of basic data from table, 19.4%.

Tarif tables represent another way of presenting volume tabulated from log rules. The term *tarif*, which is Arabic in origin, means tabulated information. In continental Europe, the term has been applied for years to volume table systems that provide, directly or indirectly, a convenient means for obtaining a local volume table for a given stand (Garay, 1961).

British tarifs (Hummel, 1955), which have been quite successful, stimulated the preparation of "comprehensive tree-volume tarif tables" by Turnbull et al. (1963). This clever system, which is summarized in Table 7.4, merits wider consideration in all types of

TABLE 7.4. Specimen of a Comprehensive Tree Volume Tarif Table

Height—dbh Access Table for Douglas-fir					Instructions

dbh	Total Height (ft)				
	60	62	64	66	68
12.2	22.2	23.1	23.9	24.8	25.7
12.4	22.0	22.9	23.7	24.6	25.5
12.6	21.8	22.7	23.6	24.4	25.3
12.8	21.6	22.5	23.4	24.2	25.1
13.0	21.5	22.3	23.2	24.0	24.9
13.2	21.3	22.1	23.0	23.8	24.7
13.4	21.1	22.0	22.8	23.7	24.5
13.6	20.9	21.8	22.6	23.5	24.3
13.8	20.8	21.6	22.5	23.3	24.2
14.0	20.6	21.4	22.3	23.1	24.0
14.2	20.4	21.3	22.1	23.0	23.8
14.4	20.3	21.1	22.0	22.8	23.6
14.6	20.1	21.0	21.8	22.6	23.5

Instructions:

1. Measure height and dbh of sample trees representative of stand
2. Lookup tarif numbers of sample trees in appropriate height—dbh access table and average them. For example:

Height	dbh	Tree Tarif No.
60	12.2	22.2
68	14.3	23.7
...	...	...

Mean = 24.5

3. In tarif book, find table with mean tarif number

Tarif Table No. 24.5

	Total Tree Volume							Volume to a 6-in. Top			
dbh (in.)	Including Top and Stump (ft³)		Including Top Only (ft³)		Volume to a 4-in. Top (ft³)		(ft³)		Scribner (Board Feet)		International ¼-in. Rule (Board Feet)
	Vol A	GM A	Vol B	GM B	Vol C	GM C	Vol D	GM D	Vol E	GM E	Vol F	GM F
2	0.3	0.2	0.2	0.2	—	—	—	—	—	—	—	—
3	0.7	0.7	0.6	0.6	—	—	—	—	—	—	—	—
4	1.5	1.0	1.4	1.0	—	—	—	—	—	—	—	—
5	2.6	1.4	2.5	1.3	1.4	1.5	—	—	—	—	—	—
6	4.1	1.7	4.0	1.6	3.0	1.8	—	—	—	—	—	—
7	5.9	2.0	5.7	2.0	4.9	2.1	1.9	2.2	6	7.2	9	10.7
8	8.1	2.4	7.8	2.3	7.1	2.4	4.3	2.8	14	9.8	21	13.9
9	10.5	2.7	10.2	2.6	9.6	2.7	7.2	3.1	25	11.9	36	16.2
10	13.3	3.0	12.9	2.9	12.3	3.0	10.5	3.4	38	13.7	53	18.0
dbh	V/BA Ratio		V/BA Ratio	% of Vol A	V/BA Ratio	% of Vol B	V/BA Ratio	% of Vol C	V/BA Ratio	B/CU Ratio	V/BA Ratio	B/CU Ratio
2	9.3		8.3	89.0	—	—	—	—	—	—	—	—
3	12.5		11.6	92.5	—	—	—	—	—	—	—	—
4	16.3		15.4	94.4	—	—	—	—	—	—	—	—
5	19.0		18.1	95.5	9.7	51.0	—	—	—	—	—	—
6	20.7		19.9	96.1	14.9	71.9	—	—	—	—	—	—
7	22.0		21.2	96.4	18.1	82.1	6.8	37.7	21.1	3.1	32.0	4.7
8	23.0		22.2	96.6	20.1	87.6	12.3	61.1	40.6	3.3	59.9	4.9
9	23.7		23.0	96.7	21.5	90.7	16.3	75.6	56.8	3.5	81.5	5.0
10	24.3		23.6	96.7	22.5	92.6	19.1	84.6	69.6	3.6	97.5	5.1

Source: Adapted from Turnbull and Hoyer (1965).

This tarif table gives volume in cubic feet for entire tree and volume in cubic and board feet to various merchantable limits. Volume/basal area ratios for horizontal point sampling and GMs to determine growth are also given. The letters A, B, C, ... that follow Vol and GM are used for convenient identification of columns.

inventories.* It requires no curve fitting to obtain a local volume table; it provides a convenient method of converting from one unit of measure to another or from one merchantable limit to another. To determine average annual volume increment per tree in any desired unit of volume and merchantable limit, one simply multiplies the average annual diameter increment in inches by the growth multiplier (GM). These tarif tables, or their updated versions, are still used in western Washington and Oregon (Bowers et al., 2013), though their use has largely been supplanted by taper equations.

7.4. LOG SCALING

Scaling is the determination of the gross and net volumes of logs in board feet, cubic feet, cubic meters, or other units. Determination of gross scale consists of measuring log length and diameter, and then determining the volume using a log rule. The gross volume may be read from a table or from a *scale stick*: a flat stick that has volumes for different log diameters and lengths printed on its face. Deductions are then made for defects to obtain the net volume.

7.4.1. Board Foot Scaling

In board foot scaling, a maximum length of 40 ft is standard for the western regions of the United States; 16 ft is standard for the eastern regions. When logs exceed the maximum scaling length, they are scaled as two or more logs. If a log does not divide evenly, the butt section is assigned the longer length. The scaling diameter for the assumed point of separation can be estimated from the taper of the log. Although logs are most commonly cut and measured in even lengths (i.e., 8, 10, 12, 14, and 16 ft), they may be cut and measured, particularly with hardwood logs, in both odd and even lengths (i.e., 8, 9, 10, 11, and 12 ft). Logs must be cut longer than standard lumber lengths because it is impossible to buck logs squarely and because there is logging and transportation damage to log ends. This extra length, which will range from 3 to 6 in., depending on the size of timber, products sawed, and logging methods, is called *trim allowance*.

Most board foot log rules call for diameter measurements inside bark, to the nearest inch, at the small end of the log. When a log is round, one measurement is enough. When a log is eccentric, as most logs are, the usual practice is to take a pair of measurements at right angles across the long and short axes of the log end and to average the results to obtain the scaling diameter.

To determine net scale, one must deduct from gross scale the quantity of lumber, according to the log rule used that will be lost due to defects. These deductions do not include material lost during manufacturing or defects that affect the quality of the lumber. Instead, they include those defects that reduce the volume of lumber.

---

*The tarif system is based on the concept that a tree with a dbh of 4 in. (equivalent to a basal area of 0.087 ft²) has zero volume. The tarif number is the predicted volume for a tree of 1.0 ft² of basal area. The volume function is $V = b_1(g-0.087)$ and the tarif number T is

$T = b_1(1.0-0.087)$; so that $b_1 = T/0.913$.

For a given tarif number T and tree basal area g, the volume of a tree is $V = \dfrac{T}{0.913}(g-0.087)$.

Detailed procedures have been worked out to estimate the volume of deductions in defective logs, and many of these procedures are prescribed by legislation in some jurisdictions. Bell and Dilworth (1993) provide a comprehensive summary of procedures often employed. Grosenbaugh (1952a) proposed a logical and generally applicable system. In this method, the amount of material lost in defect is estimated by multiplying the gross scale by the proportion of the log affected. The system works, with minor modifications, regardless of the units in which the log is measured: board feet by any log rule, cubic feet, cubic meters, cords, pounds, kilograms, and so forth. The procedure for common defects can be summarized in the following five rules, where scaling diameter is defined as the average inside bark diameter at the small end, and measurements are in imperial units.

1. When defect affects entire section, the proportion P lost is

$$P = \frac{\text{Length of defective section}}{\text{Log length}}$$

Example:

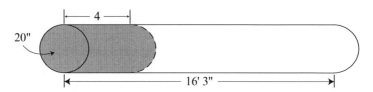

Gross scale, Scribner decimal C rule =   28

$$Cull = \left(\frac{4}{16}\right)(28) = 7 \qquad -7$$

Net scale   21

2. When defect affects wedge-shaped sector:

$$P = \left(\frac{\text{Length of defective section}}{\text{Log length}}\right)\left(\frac{\text{Central angle of defect}}{360°}\right)$$

Example:

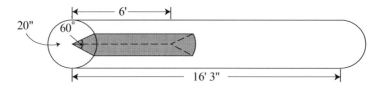

Gross scale, Scribner decimal C rule =   28

$$Cull = \left(\frac{6}{16}\right)\left(\frac{60}{360}\right)(28) = 2 \qquad -2$$

Net scale   26

3. When log sweeps (ignore sweep less than 2 in.).

$$P = \frac{\text{Maximum departure} - 2}{\text{Scaling diameter}}$$

Example:

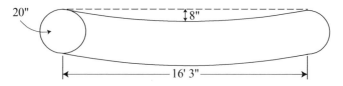

Gross scale, Scribner decimal C rule = 28

$$Cull = \left(\frac{8-2}{20}\right)(28) = 8 \qquad -8$$

Net scale   20

4. When log crooks.

$$P = \left(\frac{\text{Maximum deflection}}{\text{Scaling diameter}}\right)\left(\frac{\text{Length of deflecting section}}{\text{Log length}}\right)$$

Example:

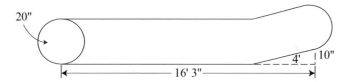

Gross scale, Scribner decimal C rule = 28

$$Cull = \left(\frac{10}{20}\right)\left(\frac{4}{16}\right)(28) = 3 \qquad -3$$

Net scale   25

5. When the average cross section of interior defect is enclosable in an ellipse or circle.

$$P = \left[\frac{(\text{Major diameter} + 1)(\text{Minor diameter} + 1)}{(\text{Scaling diameter} - 1)^2}\right]\left(\frac{\text{Defect length}}{\text{Log length}}\right)$$

(Defect in peripheral inch of log (slab collar) can be ignored.)

Example:

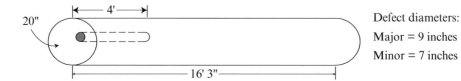

Defect diameters:

Major = 9 inches

Minor = 7 inches

Gross scale, Scribner decimal C rule =   28

$$Cull = \left[\frac{(9+1)(7+1)}{(20-1)^2} \right]\left(\frac{4}{16}\right)(28) = 2 \qquad \frac{-2}{}$$

Net scale   26

7.4.2. Cubic Volume Scaling

Cubic volume scaling estimates the total contents of a log and not the products that can be obtained, as is the case with board foot scaling. The total cubic contents of a log can then be converted to the unit of measure appropriate to each manufacturing plant with less uncertainty than in converting board feet log scale to board feet of lumber, or from board feet log scale to square feet of veneer.

In cubic volume scaling, log diameters and log lengths are measured as explained in Section 6.1. In making deductions for defects, the five rules given in Section 7.4.1 are applicable. If cubic feet are used, diameter is average small-end diameter in inches and length is in feet and the rules are used without modifications. If cubic meters are used, diameter is average small-end diameter in centimeters and length is in meters, and the third and fifth rules are changed as follows.

3. When log sweeps (ignore sweep less than 5 cm)

$$P = \frac{\text{Maximum departure} - 5}{\text{Scaling diameter}}$$

5. When average cross section in interior defect is enclosable in an ellipse or circle.

$$P = \left[\frac{(\text{Major diameter} + 2)(\text{Minor diameter} + 2)}{(\text{Scaling diameter} - 2)^2} \right]\left(\frac{\text{Defect length}}{\text{Log length}} \right)$$

where P equals the proportion lost due to defect.

For cubic foot scaling rules prepared by the United States Forest Service, see USDA (1991).

7.4.3. Unmerchantable Logs

The definition of a cull, or unmerchantable, log is largely a local matter. Merchantability varies with species, economic conditions, and other factors. No matter what units are employed, specifications for a merchantable log should give the minimum length and minimum diameter allowed, and the minimum percent of sound material left after deductions are made for cull. For example, a cull log might be defined as any log less than 8 ft long, less than 6 in. in diameter, or less than 50% sound.

7.4.4. Sample scaling

Under conditions where the scaling operation interferes with the movement of the logs, or where scaling costs are high, sample scaling should be considered. Sample scaling is generally feasible when (1) logs are fairly homogeneous in species, volume, and value; (2) logs are concentrated in one place so they can be scaled efficiently; and (3) total number of logs is large. Sampling can be applied to individual logs or to truckloads.

Once one has decided to use sample scaling, there are two basic questions:

1. How many logs or truckloads must be scaled to determine the total scale within limits of accuracy acceptable to both buyer and seller?
2. How should the sample logs or truck loads be selected?

The number of logs or truckloads to measure can be calculated from the eq. (3.20), the minimum sample size formula applicable to a finite population:

$$n = \frac{t^2 CV^2 N}{\left(N(E\%)^2 + t^2 CV^2 \right)}$$

where   CV=coefficient of variation of the volume of logs or truckloads, expressed as a percent

t=t-value corresponding to chosen probability

N=total number of logs or truck loads in the population

$E\%$=desired error of mean expressed as a percent of the mean (acceptable sampling error)

As an example, assume CV is 50% for a population where N=10,000 logs, and the acceptable sampling error is 3% (both CV and N are estimates), then, if we let t=2, giving approximately 1 in 20 odds that a chance discrepancy between the estimated and true value will not exceed 3%, we obtain

$$n = \frac{\left(50^2\right)\left(2^2\right)(10,000)}{10,000\left(3^2\right)+\left(50^2\right)\left(2^2\right)} = 1,000$$

A practical procedure to obtain the 1000-log sample would be to scale every tenth log; that is, take a systematic sample. Of course, to obtain total volume, every log must be counted since the total number of logs used to calculate the sample size n is an estimate. The same procedure can be applied to truckload sampling, where N is the total number of truckloads, CV is the variation in volume among truckloads, and n is the number of truckloads to be scaled.

Although random sampling is required if one desires to calculate valid sampling errors, it is not essential if the sole purpose of sampling is to obtain an unbiased estimate of the average volume per log or per truckload and the total volume. Johnson et al. (1971) described how a 3P sample selection procedure (see Section 10.10) could be applied to sample log scaling.

7.5. SCALING STACKED VOLUME

Stacked volume (Section 7.1.2) has traditionally been obtained for firewood, pulpwood, excelsior wood, charcoal wood, and other relatively low-value products that are assembled in stacks. In scaling a stack of wood in cords, one first records the length—the average of measurements taken on both sides of the stack, to the nearest 0.1 ft. Then, stack height is obtained by averaging measurements taken at intervals of about 4 ft. The height, which is reduced about

1 in./ft by some scalers to compensate for settling and shrinkage, is recorded to the nearest 0.1 ft. Finally, piece lengths are checked to see if they vary from the lengths specified in the sale or purchase contract (standard lengths for pulpwood cut in the United States are 4 ft, 5 ft, 5 ft 3 in., and 8 ft 4 in.). If they do, the procedure given in the contract should be followed.

The volume in standard cords V_c of a stack of wood is calculated as follows:

$$V_c = \frac{L_s H_s W}{128} \tag{7.11}$$

where L_s = stack length (ft)
        H_s = stack height (ft)
        W = stick length (ft)

If stacks are piled on slopes, the length and height measurements should be taken at right angles to one another.

If the stacked volume is measured in cubic meters ($1\,m^3 = 1$ stere), stack length and height are usually measured in 2-cm classes, and piece lengths are check as described above. Then, gross volume of a stack in cubic meters V_m is*

$$V_m = L_S \cdot H_s \cdot W \tag{7.12}$$

where L_s, H_s, and W are as above, measured in meters rather than feet.

Since the above procedure gives gross stacked volume, to obtain net volume, deductions must be made for defective wood and poor stacking. The definitions of defects and procedures for allowing for defects will vary from one organization to another. In general, deductions are made for *defective sticks* and *loose piling*. Defective sticks include rotted, burned, undersized, and peeled sticks with excessive bark adhering. Loose piling may occur when knots have been improperly trimmed, when excessively crooked wood is present, and when sticks have been carelessly piled.

When making deductions for defective sticks, the scaler examines each stick in a pile and notes which sticks do not meet specifications. These sticks are then culled by deducting the cubic space they occupy from the gross cubic space occupied by the pile—either a stick is acceptable or it is not acceptable. Estimating the cubic space that would be occupied by sticks that could be included in the loose pile and subtracting this volume from the gross cubic space occupied by the pile makes deductions for loose piling.

The term *rough wood* is used to designate wood with bark in contrast to the term *peeled wood*, which refers to wood with bark removed. It should be made clear in a sales contract whether wood is to be measured *rough* or *peeled*. If the sale price is based on rough wood volume, then if peeled wood must be measured, volume must be increased 10–20%, depending on bark thickness (Section 6.1.6).

7.6. VOLUME UNIT CONVERSION

Board foot–cubic foot conversions for logs vary by log rule, log and tree size, and form. Table 7.5 illustrates the nature of these variations for 16 ft logs. In general, the ratios increase rapidly from small to large diameters and level off once the larger diameters have

---

* Other units of stacked volume may be used in different countries.

TABLE 7.5. Board Foot–Cubic Foot Ratios for 16-ft Logs by Taper Rate and Scaling Diameter

Scaling	2 in. per Log					4 in. per Log			
Diameter (in.)	Volume[a] (ft³)	International 1/4-in.	Scribner	Doyle		Volume[a] (ft³)	International 1/4-in.	Scribner	Doyle
10	10.6	6.710	5.166	3.381		12.9	5.532	4.259	2.787
12	14.8	7.221	5.781	4.314		17.5	6.138	4.914	3.667
14	19.7	7.598	6.229	5.071		22.7	6.604	5.414	4.407
16	25.3	7.887	6.569	5.690		28.6	6.974	5.808	5.031
18	31.6	8.117	6.836	6.205		35.3	7.273	6.126	5.560
20	38.6	8.303	7.052	6.637		42.6	7.520	6.387	6.012
22	46.2	8.457	7.229	7.005		50.6	7.728	6.606	6.402
24	54.6	8.586	7.378	7.322		59.3	7.904	6.792	6.741
26	63.7	8.697	7.504	7.598		68.8	8.057	6.952	7.039
28	73.5	8.792	7.613	7.839		78.9	8.189	7.091	7.302
30	83.9	8.875	7.707	8.053		89.7	8.305	7.212	7.536
32	95.1	8.948	7.790	8.242		101.2	8.408	7.320	7.745
34	107.0	9.013	7.863	8.412		113.4	8.500	7.416	7.933
36	119.6	9.071	7.928	8.565		126.4	8.582	7.501	8.104
38	132.8	9.122	7.987	8.704		140.0	8.656	7.578	8.259
40	146.8	9.169	8.039	8.830		154.3	8.723	7.648	8.400

[a] Solid wood volume determined by Smalian's formula.

been reached. Although any fixed conversion factor should be used with care, a "rule-of-thumb" factor generally used is 6 board feet (Scribner log rule) per ft^3. Tarif tables, such as those proposed by Turnbull and Hoyer (1965), Table 7.4 provide a systematic approach for conversion between cubic feet and various log rules.

Cubic foot–cubic meter conversions can be easily accomplished by using an exact mathematical relationship:

$$1\,\text{ft}^3 = 0.028317\,\text{m}^3$$
$$1\,\text{m}^3 = 35.3145\,\text{ft}^3$$

See Table A.1 for a complete list of conversion factors.

Cubic foot–cord conversions vary greatly. Approximations of between 60 and 94 ft^3 per standard cord are generally used for rough wood, depending on species, method of stacking, size of wood, and bark thickness. The average bark volume will vary from 10 to 20% of unpeeled volume for most species. Good working averages are between 75 and 85 solid ft^3 per standard cord for green, rough southern pine, between 80 and 90 solid ft^3 per standard cord for green, rough Douglas-fir or western hemlock, and between 73 and 85 solid ft^3 per standard cord for green, rough hardwoods.

Board foot–cord conversions are unreliable and are used infrequently. One rule-of-thumb conversion factor used at times is 1000 board feet (Scribner log rule) equals 2 standard cords of rough stacked wood. For more accurate conversion, however, consideration must be given to diameter and length of logs, bark thickness, and other factors. For example, one might obtain a variation in yield, per 1000 board feet, of 1.8 cords with logs of 30 in. diameter and larger, to 3.3 cords with logs of 6 in. diameter.

Cubic meter–stere conversions are analogous to cubic foot–cord conversions. On a percentage basis, the solid cubic contents of a stere can be expected to vary, as does the cord. For example, if averages are between 73 and 85 solid cubic feet per standard cord for green, rough hardwoods—that is, between 57 and 66% of the 128 ft^3 occupied by the cord is solid wood—averages would be between 0.57 and 0.66 solid m^3 per stere for green, rough hardwoods.

Conversion of volume per unit area. The following conversion factors can be used for cubic conversions between the SI and imperial systems:

$$1\,\text{ft}^3/\text{acre} = 0.06997\,\text{m}^3/\text{ha}$$
$$1\,\text{m}^3/\text{ha} = 14.29\,\text{ft}^3/\text{acre}$$

An approximate conversion from m^3/ha to board feet/acre (assuming 5 board feet/ft^3) is $1\,\text{m}^3/\text{ha} = 14.3\,\text{ft}^3/\text{acre} = 75$ board feet/acre.

7.6.1. Determination of Solid Cubic Contents of Stacked Wood

It is often necessary to know the solid cubic contents of wood that is stacked (standard cord, stere, etc.) on the ground or on trucks. Although average conversion factors, such as those given above, are often used, better factors are generally required. These can be determined by the following methods.

1. *Direct measurement.* The cubic volume of individual sticks, or of groups of sticks, can be determined by displacement (Section 6.1.2). Using Huber's, Smalian's, or Newton's formula can also compute the cubic volume of individual sticks. In any case, when the cubic space occupied by a pile is known, the ratio of solid cubic volume to total cubic volume can be calculated from the equation:

$$f = \frac{\text{Solid cubic volume of pile}}{\text{Total cubic volume of pile}}$$

The factor f multiplied by the space occupied by a cord, or by an entire pile, will give the solid cubic volume of wood in the stack.

2. *Photographic methods.* The factor f can be estimated from photographs of the ends of the sticks in a pile. The camera is located a fixed distance from the pile with the optical axis of the lens perpendicular to the side of the pile. Normally, only a portion of a stack or truckload is included in a single photograph. A transparent grid with systematically spaced dots is superimposed on the photograph, either manually using a print or electronically using an image analysis system, and the number of dots falling on air space can be counted (Fig. 7.1). The factor f (percent solid wood) is computed as follows:

$$f = 1 - \left(\frac{\text{Total dots in air spaces}}{\text{Total dots in photograph}} \right)$$

In Figure 7.1, there are 67 dots falling in air spaces and a total of 182 dots on the photograph. The percent solid wood $f = 1 - (67/182) = 0.63$ and the solid wood volume per stack cord $= (0.63)(128) = 81 \, \text{ft}^3$.

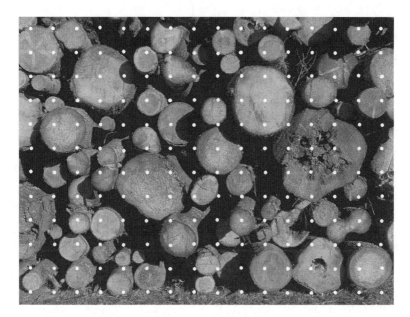

FIG. 7.1. Photographic method of determine solid wood content in stacked wood.

Although one photograph of a truckload will usually give an adequate sample, several photographs of large stacks may be required. For example, Garland (1968) determined that a 20% photo sample was required to estimate the solid wood content to within ±2.4% accuracy at 95% confidence. With the advent of high-quality, affordable digital cameras and image analysis software, the photographic method is an affordable and easily implemented method of determining solid wood content. More advanced image analysis techniques such as image segmentation can be applied as well (Burger and Burge, 2008).

3. *Angle-gauge method.* By "projecting" an angle of about 23° parallel to the face of a stack from randomly selected points, the conversion factor f may be quickly and efficiently obtained. This method, which is a modification of horizontal point sampling, is discussed by Loetsch et al. (1973).

7.7. SCALING BY WEIGHT

When a tree is felled and cut into logs, the wood immediately begins to lose moisture. If permitted to air-dry, its moisture content will reach about 12%. The rates of drying for logs vary with air temperature, humidity, species, log size, knottiness, method of stacking, presence of bark, and other factors. Besley (1967), Nylinder (1967), and Adams (1971) reported varying weight losses over time for several species in North America and Scandinavia. The results of Adam's study to determine the weight loss of red-oak logs are shown in Figure 7.2.

If only the weight of wood is desired, bark and foreign material such as ice, snow, mud, and rocks must be removed before measurement or their weight must be deducted from the gross weight to obtain the wood weight. Green bark was found to be between 9 and 19% of the weight of green, rough (i.e., unpeeled) hardwood pulpwood in Maine (Hardy and Weiland, 1964). Oven-dry bark was found to be between 3.8 and 9.1% of the weight of green, rough logs for several species in Minnesota (Marden et al., 1975).

7.7.1. Weight Measurement of Pulpwood

The use of stacked measure (e.g., cord or stere) has long been used to measure bulk forest products, such as fuelwood and pulpwood. Although stacked measure is not an accurate measure of the solid cubic contents, when measurements must be made in the woods, particularly at scattered locations, it is often a necessary compromise; however, if measurement can be done on trucks, and weighing facilities are available, weighing provides a cheaper, more accurate, and more objective method of scaling. Where bioenergy, chemical feedstocks, and other uses of bulk wood have developed, they have often done so using the infrastructure of the pulpwood-based industries, and weight scaling procedures are essentially identical. Some industries place standards on the cleanliness or bark fraction of roundwood or chips, but these do not affect the basic methods and procedures of weight scaling.

The green weight and moisture content of wood provide good measures of the wood available for pulp. In spite of some problems in obtaining moisture content, weight as a measure of pulpwood quantity has several advantages:

1. It permits an accurate determination of the yield of pulp, which is measured in weight.
2. It encourages wood suppliers to promptly deliver wood to the mill to reduce weight loss (fresh, green pulpwood is preferred at the mill).
3. It encourages better loading of trucks since there is no advantage in loose piling.
4. It is faster and more economical than volume scaling.
5. It eliminates personal judgment.

When weight scaling is employed, a dependable procedure for obtaining moisture content is essential. If rough wood is being handled, estimates of bark weight must also be made. There are two ways to deal with these problems.

1. *Develop average moisture and bark percentages for green wood at the time of cutting and at intervals since cutting.* To use the percentages for reducing fresh, green-wood weight, it is best that weighing be done immediately following cutting. In practice, the conversion from green weight with bark to dry weight without bark often is not made. Instead, the wood is weighed immediately after cutting, and prices are based on this weight. Although the quantity of dry wood is not explicitly

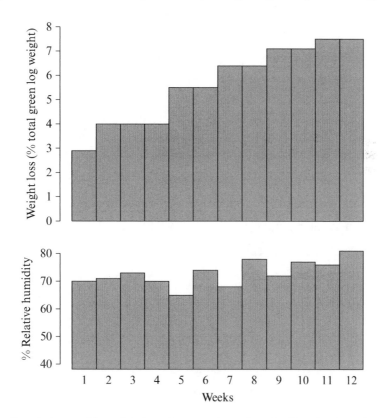

FIG. 7.2. Cumulative weekly weight loss due to moisture loss for 21 red oak sawlogs, and the associated relative humidity record (adapted from Adams (1971)).

determined, it is implicit in its effect on the price per unit weight. Some moisture percentages for several species used for pulpwood in the southern and northeastern parts of the United States are shown in Table 7.6. The percentages shown are based on freshly cut green wood without bark. When weighing is done at varying times following cutting, moisture and bark percentages are required at intervals of time since cutting. Percentages for the moisture component are more variable and less reliable than those for fresh green wood since season, weather, method of stacking, and so on will affect moisture content.

2. *Determine current moisture and bark percentages at the time of weighing.* This approach requires that an estimate be made at the time of weighing, of the contribution of moisture and bark to the total weight. Sampling procedures must be utilized to provide these estimates. When moisture percentages have been determined, the dry weight (biomass) of wood can be calculated using: $BM = WT \cdot (1 - MC\%_g/100)$. Nylinder (1958) found that sample disks cut 10 cm from the ends of logs gave satisfactory estimates of log moisture contents. Braathe and Okstad (1967) found that a thin triangular segment of the log cross-sectional cut with a chain saw, or a sample taken from the log radius with a drill, gave satisfactory estimates of log moisture content of logs. Electrical moisture meters are available but they have not been too satisfactory for moisture determination of pulpwood or sawlogs.

Large platform scales capable of determining the gross weight of a truck and its load are commonly used for weight scaling. The gross weight of the truck and load minus the weight of the empty truck, the *tare weight*, equals the weight of wood, including bark and moisture. The dry weight of wood can be estimated by applying corrections for moisture and bark. Since pulpwood quantity in terms of stacked volume may be desired after the wood has been weighed, estimates of weight per cord are available such as shown in Table 7.6.

TABLE 7.6. Sample Weights and Moisture Contents of Some Species Used for Pulpwood (All Weights Are per Cord)

	Weight per Cord Green (lb)			Oven-Dry Weight	Percent Moisture
Species	Unbarked	Bark	Barked	Barked Wood (lb)	Content Barked Wood
Longleaf pine	6374	660	5714	2920	95.8
Shortleaf pine	5669	675	4994	2037	145.1
White ash	5031	795	4236	2992	41.5
Beech	5584	446	5138	3157	62.7
Grey birch	5690	677	5013	2700	85.6
White birch	5731	705	5026	2788	80.3
Yellow birch	6090	786	5304	3013	76.0
Elm	5857	763	5094	2627	93.9
Red maple	5482	720	4762	2877	65.5
Sugar maple	5977	801	5176	3178	62.0

Source: Adapted from Taras (1956) and Swan (1959).

7.7.2. Weight Measurement of Sawlogs

Weight scaling of sawlogs can be used to (1) estimate the amount of lumber, veneer, and so on, in logs, and (2) to estimate the total quantity of wood in logs. In both cases, the determination of weight is an intermediate step in the estimation of volume because the wood-using industries, particularly in North America, customarily measure sawlogs in volume units. Weight scaling of sawlogs, therefore, normally consists of weighing and converting the weight to volume. The same problems with moisture content, bark, and foreign material that pertain to weight scaling of pulpwood pertain to weight scaling of sawlogs.

Generally, weights of entire truckloads are determined and converted to volume equivalents, although weights of individual logs can be determined. Weight scaling in truckload lots is most suitable for logs of a single species, with uniform diameter, uniform length, and uniform quality. This is the reason weight scaling has been most widely used for southern pine in the southeastern United States. Weight scaling of truckloads eliminates the need for measurement of individual logs, speeds up scaling, and reduces errors of judgment. It also encourages the delivery of freshly cut logs since price is almost always based on green weight. When logs are variable in size and quality, however, prices must be adjusted for these factors. Consequently, objections have been raised to weight scaling of hardwood logs because they vary more in size, shape, and quality than softwood logs.

In North America, weight scaling of sawlogs has generally been viewed as a more rapid method of obtaining volume estimates than conventional log scaling. Thus, the main efforts have been to develop volume–weight relationships. Relationships of this type can be developed from studies that relate volume of a sample of sawlogs to their weights (volumes may be estimated from log rules or obtained from mill studies). Guttenberg et al. (1960) developed a regression equation showing the relationship of the board feet mill tally to weight in pounds for individual shortleaf and loblolly pine sawlogs in Arkansas and Louisiana (Table 7.7). Yerkes (1966) developed a regression equation showing the relationship of cubic feet to weight in pounds for Black Hills ponderosa pine sawlogs: cubic foot volume$=-2.09+0.020W$, where W is total log weight in pounds.

TABLE 7.7. Board Foot Lumber Yields from Loblolly and Shortleaf Pine

Log Weight (lb)	Predicted Green Lumber Yield (Board Feet)
200	9
600	91
1000	94
1400	134
1800	184
2000	207
2200	231
2600	278
3000	328

Source: Adapted from Guttenberg et al. (1960).

Based on the equation: Board feet $= \dfrac{\text{Weight}_{\text{lbs}}}{9.88} + \dfrac{\text{Weight}^2_{\text{lbs}}}{254,362} - 10.96$.

For an industrial application of sawlog weight scaling, it is desirable to develop relationships of volume to weight for truckloads of logs (Row and Fasick, 1966; Row and Guttenberg, 1966; Timson, 1974; Adams, 1976; Donnelly and Barger, 1977). Donnelly and Barger (1977) found that the accuracy of conversion from weight to volume was improved by including number of logs per load as an additional independent variable.

In applying truckload weighing, sampling procedures are generally used to determine, or adjust, the conversion factors. The number of truckloads to measure is first determined for an acceptable sampling error. Each load sampled is chosen randomly or systematically, weighed, and scaled log by log for volume. Average volume–weight ratios are then calculated for converting weight to volume for all the truckloads in the operation (Chehock and Walker, 1975; Adams, 1976; Donnelly and Barger, 1977).

7.7.3. Weight Measurement of Pulp

In the United States, pulp is usually measured in air-dry tons (imperial system). The air-dry ton consists of 90%, 1800 lb, of dry pulp, and 10%, or 200 lb, of water. The oven-dry ton consists of 2000 lb of oven-dry pulp. The air-dry metric tonne consists of 90%, or 900 kg, of dry pulp and 10%, or 100 kg, of water. The oven-dry metric tonne, also known as bone-dry metric tonne, consists of 1000 kg of oven-dry pulp (dried to 0% humidity at 103°C for 24 hours).

To convert the oven-dry weight of pulp to cubic volume of wood equivalent, the oven-dry weight of pulp is divided by the yield factor to obtain the oven-dry weight of wood. This figure is then divided by the species' density (oven-dry weight per green cubic unit) to give the green volume equivalent:

$$W_r = \frac{W_p}{Y}$$

$$V_g = \frac{W_r}{SG_d}$$

where W_r = oven-dry weight of raw material
        W_p = oven-dry weight of pulp
          Y = yield factor
        V_g = green volume equivalent
      SG_d = species density (oven-dry weight per green cubic unit)

For example, assume that we desire to determine the number of cubic feet of solid green wood of a given species that is required to produce 1 air-dry ton (1800 lb) of unbleached kraft pulp. Assuming the yield of oven-dry unbleached kraft pulp is 54% of oven-dry wood, and that the oven-dry weight is 25 lb per green cubic foot, then

$$W_r = \frac{1800}{0.54} = 3333\,\text{lb}$$

$$V_g = \frac{3333}{25.0} = 133\,\text{ft}^3$$

To express the wood requirement in cords, one simply divides the cubic-foot volume by the appropriate number of solid cubic feet per cord. For example, assume that a cord of rough pulpwood contains 80 solid ft^3 of wood. The cord equivalent for the above example would be $133/80 = 1.66$ cords per air-dry ton of pulp.

7.7.4. Weight Measurement of Other Forest Products

The weights of sawn lumber, plywood, veneer, wood chips, and sawmill residues may be desired. Such weights may be determined by weighing, or indirectly by estimating volume and converting to weight.

The process of indirectly determining weight can be illustrated by an example. Assume that we desire to know the weight of a 1000 board feet of rough 2 by 4 in. lumber with moisture content of 18%. Since the cubic foot equivalent of 1000 board feet of this lumber is 64.8, and the density of this lumber at 18% moisture content is 35.7 lb/ft^3, the estimated weight per 1000 board feet is $(64.8)(35.7) = 2313$ lb. The weights of other products can be estimated in a similar manner.

8

STAND PARAMETERS

A *stand* is a group of trees that occupy a given area and that has some common characteristic or combination of characteristics, such as origin, species composition, size, or age that set it apart from other groups of trees. A number of stands taken together form a forest.

Stand structure is the distribution of species and tree sizes within a stand or forest area. A stand's structure is the result of the species' growth habits and of the environmental conditions and management practices under which the stand originated and developed. Traditionally, stand structures have been broadly classified on the basis of tree ages: *even-aged* versus *uneven-aged*. An even-aged stand, as the name implies, is a stand where all trees have roughly the same age. Even-aged stands are composed of trees that originated within a short period of time, generally following a major disturbance such as wildfire or clear-cut harvesting. Uneven-aged stands are composed of trees having many different ages. These stands are generally the result of smaller-scale disturbances that influence single trees or small groups of trees, such as windthrow, selective harvesting, natural senescence, and inter-tree competition.

Stand structure also can be described using species composition, diameter distribution, height distribution, and crown classes (Oliver and Larson, 1996). For example, stands might be composed of a single species or a mixture of species. These stands might consist of a single predominant canopy layer or be vertically stratified into two or more distinct canopy layers.

The most important stand parameters characterizing structure include age, species composition, diameter and basal area, height and crown closure, density and stocking, volume, weight, and site quality. In this chapter, the most common methods of expressing these parameters are presented. For details of the measurement of these parameters, refer to Chapters 5 and 6, and for aspects of sampling and estimation, refer to Chapters 9–11.

Forest Mensuration, Fifth Edition. John A. Kershaw, Jr., Mark J. Ducey, Thomas W. Beers and Bertram Husch.
© 2017 John Wiley & Sons, Ltd. Published 2017 by John Wiley & Sons, Ltd.

8.1. AGE

8.1.1. Even-Aged and Single-Cohort Stands

In the conventional definition, an even-aged stand is a group of trees that originated within a short period of time. The trees in an even-aged stand thus belong to a single age class. The limits of the age class may vary, depending on the length of time during which the stand formed. A natural stand may seed-in over a period of several years. Rarely will an age class be only 1 year, except in plantations. More commonly, the age class for even-aged stands growing in moderate conditions will extend to 10 or 20 years, and it can be much longer in cases where site conditions or disturbances prolong the period of initial regeneration (stand initiation stage, Oliver and Larson, 1996). In some cases, a stand may appear even-aged because the trees show size uniformity. For example, stands growing slowly on poor sites may consist of trees of widely diverse ages; yet have little variation in size. Even-aged stands may be composed of shade-tolerant or shade-intolerant species, or a mixture. When species of widely different shade tolerance and growth rate occur together, there may also be a wide range in tree sizes even though the range in age is comparatively small. Thus, one cannot reliably distinguish even-aged and uneven-aged stands based on diameter distribution alone. Even-aged stands arise out of, or are perpetuated by, environmental conditions that allow trees to become established within a comparatively short, definable period. An even-aged forest may consist of several even-aged stands belonging to different age classes.

If a stand, such as a plantation, is absolutely even-aged, a count of the annual rings of a single tree will give the age. Stands originating through natural reproduction, which generally takes 5–15 years or more, contain trees of various ages. There are several conceptions of the age of such a stand. The average age of all the trees in the stand, as determined by sampling, is sometimes used. The objection to this concept is that younger trees should not be given the same weight as the larger, older trees that make up the major part of the basal area or volume of the stand. One method suggested is to take the age of several sample trees whose volume is the average for the stand. However, a simpler concept, with many advantages, is to consider the age of the stand to be the average age of the dominant and codominant trees (i.e., the largest individuals). A straightforward sampling alternative would be to use the average age of trees selected with horizontal point sampling, thus giving an average weighted by basal area.

A challenge even with stands that have arisen following a single major disturbance is that some trees in the stand may have been present before the disturbance in the form of advance regeneration. The aboveground age of such trees will exceed the age of those trees that seeded in following the disturbance, often by decades. The group of trees that established or was released following a single disturbance is called a *cohort* (Oliver and Larson, 1996). Because a cohort may include trees of a very wide range of ages, the terms cohort and age class are not truly identical. If the disturbance event that released trees in a cohort is identifiable from the width of their rings, it may be possible to speak of the age of a cohort not only in terms of its biological age, but its age since the disturbance that defines it.

8.1.2. Uneven-Aged and Multicohort Stands

A stand consisting of trees of many ages and corresponding sizes is said to be uneven-aged (sometimes called *all-aged*, but many uneven-aged stands have cohorts of trees that regenerated or were released during pulses following disturbance events, rather than as a continuous uniform process). The trees in an uneven-aged stand originate at different times,

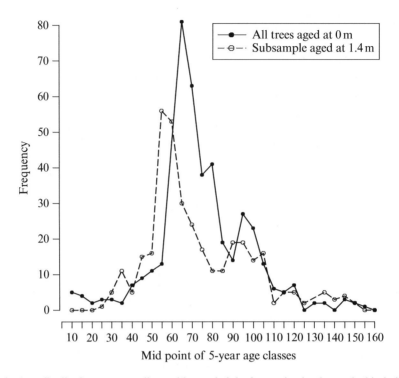

FIG. 8.1. Age distribution at root collar and breast height for a mixed oak–maple–birch–hemlock stand in New England (adapted from Oliver and Stephens (1977)).

in contrast to the single or short reproductive period characterizing a classic even-aged stand. This continuing or periodic source of new trees produces trees of ages varying from germinating seedlings to large veterans that may be centuries old. Consequently, it is challenging to speak meaningfully of the *age* of an uneven-aged stand as if it were a single value. Stands that consist of two or more age groups are often called *multiaged*. Where the trees in an uneven-aged stand became established or were released following discrete disturbance events, one can speak of the stand as a multicohort stand. Just as in an even-aged stand, the age range within an individual cohort of a multicohort stand can be broad.

Because the ages of individual trees vary within uneven-aged stands, age is often not considered an important stand parameter. If age is summarized at all for these stands, it is generally expressed as an average age by diameter class, height or canopy position, or by species or species group. Occasionally, the age distribution might be shown in the form of a frequency polygon or histogram (Fig. 8.1). As shown in Figure 8.1, the frequency polygons, while similarly shaped, are different for age determined at breast height (4.5 ft or 1.3 m) versus age determined at ground level.

8.2. SPECIES COMPOSITION

The species present in a stand has always been an important parameter in describing forest stands. Different species represent not only different forest products and values, but are also important indicators of wildlife habitat, site quality, and disturbance history.

Environmental issues, such as global climate change and the apparent declining biodiversity, have required foresters to be more explicit in documenting the species present. Forest certification procedures and monitoring protocols require foresters to document what species are present, and how forest management activities are modifying their abundance and distribution. Three distinct challenges present themselves: how to describe the overall composition of a stand, how to identify the number and diversity of species in a stand, and how to assign a type label that characterizes its composition.

8.2.1. Describing Species Composition

Typically, foresters express species composition as the distribution of individuals among the different species present in a stand. Species composition may be expressed using number of individuals, basal area, volume or biomass, and can be either the sum of these parameters or a percentage of the total.

The resulting species composition values can vary greatly depending upon which stand parameter is used. For example, Table 8.1 shows species composition for a mixed species stand in New Brunswick, Canada. Beech (*Fagus grandifolia*) makes up 43% of the species, in terms of number of stems, but only 22% of the basal area and 16% of the volume. Red spruce (*Picea rubrum*), on the other hand, makes up only 2% of the species, in terms of number of stems, but 12% of the basal area and 14% of the volume.

The best parameter to use to express species composition depends upon data availability, and on how composition values will be used and interpreted. For example, if species composition is going to be used as a parameter to stratify stands into classes for harvesting or other silvicultural activities, volume might be a more meaningful measure of composition than number of stems. In some ecological contexts, biomass might well replace volume in that role.

From the perspective of traditional community ecology, species composition can be viewed as having three components: frequency, abundance, and dominance. *Frequency* is the number of sampling units in which a species is found. *Abundance* is the number of individuals in a population, and *dominance* is an expression of the size of individuals in a population. Each of these parameters is often expressed on a relative basis as a percentage of the totals for a stand or forest area. Comparable values of abundance and dominance can be obtained using many of the types of sampling units identified in Chapter 9, but values

TABLE 8.1. Species Composition for a Mixed Species Stand in New Brunswick, Canada

Species	Trees (number/ha)		Basal Area (m²/ha)		Volume (m³/ha)	
	Number	%Total	Sum	%Total	Sum	%Total
Fagus grandifolia	269	43.0	6.1	22.4	43.9	15.8
Betula papyrifera	31	5.0	2.7	9.9	26.7	9.6
Acer saccharum	56	9.0	4.8	17.6	44.7	16.0
Acer rubrum	238	38.0	9.9	36.3	120.3	43.2
Acer pennsylvanicum	13	2.0	0.5	1.7	3.0	1.1
Picea rubra	13	2.0	3.2	11.7	39.0	14.0
Abies balsamea	6	1.0	0.1	0.5	1.0	0.3
Total	625		27.2		278.5	

for frequency will depend critically not only on the type of sampling unit employed, but also its size (e.g., plot size in the case of fixed-area plots). Importance value has been widely used as a measure of species composition that combines frequency, abundance, and dominance (Greig-Smith, 1957):

$$I_j = 100 \left(\frac{n_j}{N} + \frac{d_j}{D} + \frac{x_j}{X} \right) \tag{8.1}$$

where I_j = importance value of the jth species

n_j = the number of sampling units where the jth species is present

N = total number of sampling units

d_j = the number of individuals of the jth species present in the sample population

D = total number of individuals in the sample population ($D = \Sigma d_j$)

x_j = the sum of the size parameter (generally basal area or volume) for the jth species

X = the total of the size parameter across all species ($X = \Sigma x_j$)

Importance has a range of 0–300. The value 300 would occur in stands composed of a single species. In many situations, importance is calculated using relative frequency (n_j/N) and either relative density (d_j/D) or relative dominance (x_j/X). In this case, importance has a range of 0–200. Although the label "importance" is traditional, importance value may or may not be relevant to any particular ecological or management question. Such synthetic measures should be used carefully, and their uncritical use should be looked on with suspicion.

8.2.2. Number and Diversity of Species

With the increased interest in biodiversity and monitoring impacts of forest management on species diversity, there is growing interest in assessing the number and diversity of species in stands or forests. The simplest of all measures, at least at first glance, is total species richness or number of species, S.

A problem arises, however, when the information available about stands or forests comes from samples and not from exhaustive surveys. Let S_n be the species richness calculated from a sample of n sample units. All that can be said with complete certainty is that $S_n \leq S$. If the area included in the plots is a small fraction of total area, then the difference between S_n and S may be relatively large; on the other hand, as the area included in the sample approaches a complete survey, S_n will approach S quite closely. This problem was first addressed by Arrhenius (1921), who examined the influence of sample size on S_n, and extrapolated curves fit to the species–area relationship to estimate S. Likewise, MacArthur and Wilson (1967) suggested that the number of species could be expressed as a power–law relationship with the area sampled, that is, $S_n \propto A^b$, where A is the total area sampled and b is an empirical coefficient typically between 0.1 and 0.4. This approach is still widely used, though it has definite pitfalls. Extrapolating species–sample size or species–area curves can severely over- or underestimate total species richness, and the results are sensitive both to the data used and to the form of the regression equation used. Gotelli and Colwell (2001) review some of the challenges for this approach, both for estimating total species richness and for comparing two ecological communities when both are represented by sample data.

An alternative approach, which turns out to give very simple estimators for S despite the complexity of the underlying theory, is to use a jackknife estimator. Jackknife estimators are derived from theoretical work in mark-recapture sampling in wildlife (Burnham and Overton, 1978, 1979; Brose et al., 2003). The first-order jackknife estimator for S is

$$\hat{S}_{JACK,1} = S_n + \frac{(n-1)}{n}Q_1 \tag{8.2}$$

where Q_1 is the number of species that occur on only one plot. An estimate of its sampling variance is

$$s^2_{S_{JACK,1}} = \left(\frac{2n-1}{n}\right)^2 Q_1 + (S_n - Q_1) - S_{JACK,1}$$

The second-order jackknife estimator is similar, but depends also on Q_2, the number of species found on exactly two plots:

$$\hat{S}_{JACK,2} = S_n + \frac{(2n-3)}{n}Q_1 - \frac{(n-2)^2}{n(n-1)}Q_2 \tag{8.3}$$

and its sampling variance can be estimated as

$$s^2_{S_{JACK,2}} = \left(\frac{3n-3}{n}\right)^2 Q_1 + \left[\frac{(n-2)^4}{n^2(n-1)^2}\right]Q_2 + (S_n - Q_1 - Q_2) - S_{JACK,2}$$

The second-order jackknife is more complex than the first-order but may provide superior estimates in some circumstances.

For example, suppose we have conducted an inventory in a diverse forest using 30 fixed-area plots. We found $S_{30}=20$ species. Of those, $Q_1=2$ were found on only one plot, and $Q_2=3$ were found on exactly two plots. The first-order jackknife estimator gives

$$\hat{S}_{JACK,1} = 20 + \frac{(30-1)}{30}2 = 21.93$$

and we would calculate the sampling variance as

$$s^2_{S_{JACK,1}} = \left(\frac{2 \cdot 30 - 1}{30}\right)^2 2 + (20-2) - 21.93 = 3.80$$

Taking the square root of the sampling variance yields the standard error, 1.95. Using the second-order jackknife, we would estimate the total species richness as

$$\hat{S}_{JACK,2} = 20 + \frac{(2 \cdot 30 - 3)}{30}2 - \frac{(30-2)^2}{30(30-1)}3 = 23.61$$

and we would obtain the sampling variance of this estimate as

$$s^2_{S_{\text{JACK},2}} = \left(\frac{3 \cdot 30 - 3}{30}\right)^2 2 + \left[\frac{(30-2)^4}{30^2 (30-1)^2}\right] 3 + 15 - 23.61 = 10.65$$

Again, the square root of the sampling variance yields the standard error, 3.26. Obviously, the actual species richness S must be a whole number; based on the jackknife estimators, we would safely conclude that the true number of species in the stand is at least 20, quite likely around 22 or 23, and, with 95% confidence, less than 26 (using the first-order jackknife) or less than 30 (using the second-order jackknife).

It is possible to estimate species richness using higher-order jackknife estimators, but these have been shown to perform poorly in simulated sampling (Lam and Kleinn, 2008). Lam and Kleinn (2008) also show that the first- and second-order jackknife estimators can be badly biased for small sample sizes (n less than about 20). At present, there is no estimator for S that can be shown to be unbiased under all possible conditions. Estimation of species richness, however, does remain a very active area of research. For example, Magnussen et al. (2010) provide an estimator that is somewhat more complicated computationally than the jackknife estimators, but does seem to show good performance in a range of tests.

By itself, the total number of species S gives an incomplete accounting of the diversity of a forest. For example, consider two different forests with $S = 10$. In the first forest, each of the species comprises 10% of the total abundance; in the second, one species represents 91% of the total abundance with the remaining nine each contributing 1%. The latter forest is very nearly a monoculture with only incidental contributions from other species, while the former seems much more diverse. To capture these sorts of differences, a variety of *diversity indices* have been proposed (additional discussions can be found in Pielou (1966), Peet (1974), and Krebs (1989); for a review, see Magurran (2003)). As with the basic description of species composition (Section 8.2.1), species diversity can be expressed based on the proportional contribution of each species to number of individuals, basal area, or biomass of a stand or forest, and the results may depend quite strongly on which attribute is used. The choice of attribute should depend on the particular ecological or management questions at hand.

The Berger–Parker index (Berger and Parker, 1970) is perhaps the simplest of all diversity indices, though it is not very widely used in ecology. The Berger–Parker index B is the proportion of the population in the most common species. For example, consider the mixed-species stand described in Table 8.1, and let us consider diversity based on the proportion of basal area in each species. The most common species is *Acer rubrum* with 36.3% of the total basal area. Therefore, the Berger–Parker index is $B = 0.363$. Note that the higher the Berger–Parker index, the less diverse the stand. When a stand is almost completely dominated by a single species, the Berger–Parker index approaches 1; if the stand is comprised of S equally abundant species, the Berger–Parker index approaches its minimum value of $1/S$. Thus, the higher the value of the Berger–Parker index, the more concentrated and less diverse is the species composition.

By contrast, Shannon's index (also called Shannon entropy, Shannon–Weiner entropy, and Shannon–Weaver diversity) has a much more complicated mathematical formula. The index was originally derived by Claude Shannon (1948) as a profound result in the theory of information, which was subsequently recognized as having deep mathematical connections to thermodynamics and statistics. Let p_i be the proportion (of individuals, of basal area, or whatever characteristic is being used) in species i. Then, Shannon's index is defined as

$$H = -\sum_{i=1}^{S} p_i \ln p_i \qquad (8.4)$$

If a species is absent ($p_i=0$), the quantity $p_i \ln p_i$ is taken to equal 0 as well. Note that as written here (using the natural logarithm), Shannon's index has the rather unfamiliar units of *nats*. Sometimes, the base 10 logarithm is used, in which case Shannon's index is in *digits*. In computing applications, one often sees base 2 logarithms, in which case Shannon's index is in *bits*. One must be careful in comparing values of Shannon's index across publications and studies to ensure that the logarithms (and hence units) are consistent. Considering again the mixed-species stand from Table 8.1, and letting p be the proportion of basal area for each species (calculated to slightly higher precision than in the table, to avoid rounding errors),

$$H = -\left(0.2236\ln 0.2236 + 0.0988\ln 0.0988 + 0.1761\ln 0.1761 + \cdots + 0.0052\ln 0.0052\right)$$
$$= 1.584$$

If a stand is almost completely dominated by a single species, then H approaches its minimum value of 0. Conversely, if a stand is composed of S equally abundant species, then H approaches its maximum value of $\ln S$. Consequently, the value $H/\ln S$ is sometimes called Shannon evenness (Pielou, 1966). Similarly, $\exp(H)$ can be considered as an "effective" number of species. (Note that if base 10 or base 2 logarithms are used to calculate H, the calculation of evenness and effective number of species must use the same base.) For our example stand, $S=7$, so the maximum value of H that could have occurred is 1.946, and the Shannon evenness is $1.584/1.946=0.814$. Its effective number of species is $\exp(1.584)=4.875$.

Another commonly used diversity index is Simpson's (1949) index. Simpson's index is calculated as

$$q = \sum_{i=1}^{S} p_i^2 \qquad (8.5)$$

For the example stand in Table 8.1, Simpson's index is

$$q = 0.2236^2 + 0.0988^2 + 0.1761^2 + \cdots + 0.0052^2$$
$$= 0.236$$

When the proportions are defined in terms of trees per hectare or trees per acre, Simpson's index can be interpreted as the probability that two randomly drawn trees will be of the same species. Simpson's index approaches a maximum value of $q=1$ when nearly all of the community is comprised of a single species, and equals $q=1/S$ when all S species have an equal proportion. Thus, as with the Berger–Parker index, the higher the value of Simpson's index, the more concentrated and less diverse is the community. To remedy this, some authors suggest using $1-q$ (Whittaker, 1965; Pielou, 1969); indeed, one can sometimes find the quantity $1-q$ called "Simpson's index" in the ecological literature (occasionally, q is referred to as Simpson's similarity index). One should use caution in comparing or interpreting values of Simpson's index without confirming which form has been used.

At first glance, these indices appear to have little in common, and, in fact, often rank different stands or communities differently, creating a sense of confusion or a feeling that the quantitative descriptions of diversity are entirely arbitrary (e.g., Hurlbert, 1971). However, these indices can be viewed within a unifying framework. Rényi (1961) entropy, a generalization of Shannon's entropy, defines entropy of order a as

$$H_a = \frac{1}{1-a} \ln\left(\sum_{i=1}^{S} p_i^a\right) \qquad (8.6)$$

All of the diversity measures described here can be shown to be special or limiting cases of Rényi entropy:

$$\lim_{a \to 0} H_a = -\ln\frac{1}{S} = \ln S$$

$$\lim_{a \to 1} H_a = H$$

$$H_2 = -\ln q$$

$$\lim_{a \to \infty} H_a = -\ln B$$

Moreover, for any order a, we can define an effective number of species as

$$S_{\text{effective},a} = \exp\left(H_a\right)$$

For any given stand or plant community with more than one species, Rényi entropy will be a monotonically declining function of the order a, and, as a increases, Rényi entropy becomes more and more sensitive to the proportion of the most abundant species.

8.2.3. Assigning Stand Types

Stand types based on species composition have many applications in forest management and ecology. Stand types may be used in advance as strata in a stratified inventory design (Section 10.5), may be used after-the-fact for reporting and description of inventory results, and are often used in tactical and strategic forest management planning. The distribution of stand types by area is, in itself, often used as an indicator of wildlife habitat, biodiversity, and the degree of "naturalness" of forested landscapes (Barbati et al., 2014). Although a great deal of debate among early community ecologists focused on the question of whether ecological communities were really discrete types or varied along a continuum, it is now broadly recognized that classification can be useful for description and conservation purposes even when the underlying community structure is continuous (Whittaker, 1978; Gauch, 1982). Any classification of a forest stand or plant community into a type discards information about the specific composition and structure of that particular stand or community; the adequacy of a stand-typing system depends on how well it trades off needed detail against practical complexity and reliability for a given set of objectives.

Assigning stand types is straightforward in plantations where species composition is relatively pure or composed of a mixture that is tightly controlled by management. Likewise, it is also reasonably simple in natural monocultures or near-monocultures. In such stands, it is common to define the stand type using the most common species. However, in mixed-species stands the situation is more complex. Indeed, there is, at present, no clear and uniform definition of what "mixed-species" is. Toumey and Korstian (1947) defined stands as "pure" whenever at least 80% of the overstory was dominated by a single species, but this definition is not universally accepted or applied. Olsthoorn et al. (1999), in a European context, required additional structural criteria and indicated that interspecific competition should be a dominant factor driving stand dynamics and silvicultural decisions. Once a stand is defined as "mixed," there is even less agreement on how many species are needed to label a particular stand type or whether basal area, canopy cover, or other attributes should be used to determine the order of importance of different species. Bravo-Oviedo et al. (2014) provide an extended review and propose additional definitions for mixed-species forests.

With these factors in mind, it should come as no surprise that a wide range of stand-typing approaches exist, ranging from very formal and complex to informal and simplified. An example of the former is the stand-typing algorithm used by the USDA Forest Service, Forest Inventory and Analysis (FIA) program (Arner et al., 2003), which uses a complex decision tree to sort individual plots into a nationally consistent, hierarchically structured set of forest-type groups and forest types. Likewise, a harmonized approach for broad-scale reporting of forest indicators in Europe has been developed (Barbati et al., 2007, 2014), though individual countries continue to use their own, typically more detailed, forest typing systems for internal reporting. At the other extreme, some-stand typing systems on large ownerships in North America only include three main forest types: "hardwood," "softwood," and "mixedwood," typically subjectively assigned by inspection of aerial photographs. The naming of these types clearly reflects a utilitarian, rather than ecological, focus.

Regardless of the purpose or sophistication of a stand-typing system, when the type assignment is based on sample data from the field, one should bear in mind that the assigned labels will include errors that propagate from the sampling variability (and other sources of error) in the field data. Inaccurate stand-type assignments can be problematic even in simple classifications. For example, MacLean et al. (2013) found that a minimum of six horizontal point samples per stand were needed to obtain acceptable accuracy in a very simple hardwood–softwood–mixedwood stand classification system based on proportion of basal area in hardwoods and softwoods. The more refined the typing system, the more intense the sampling may need to be to provide accurate stand-type assignments.

8.3. DIAMETER

Diameter is the most widely used descriptor of stand structure. Diameter may be summarized into a single parameter, generally the average diameter, or used to compute cross-sectional area and summed to yield an estimate of basal area. The distribution of diameters might be summarized into a stand table (Sections 9.2.5 and 9.3.2) depicting number of trees per unit area by diameter class or used to obtain parameter estimates for mathematical distribution functions, which are subsequently used to describe stand structure.

8.3.1. Expressions of Mean Diameter

The average diameter (dbh) of a stand may be expressed by the arithmetic mean or by the quadratic mean (Curtis and Marshall, 2000). If the primary interest is to obtain an average for the calculation of stand basal area and volume, then the quadratic mean is more appropriate. If trees are selected with equal probability (i.e., using fixed-area plots; see Section 9.2), then the arithmetic mean stand diameter is

$$\bar{d} = \frac{\sum\limits_{i=1}^{n} d_i}{n}$$

and the quadratic mean stand diameter is

$$\bar{d}_Q = \sqrt{\frac{\sum\limits_{i=1}^{n} d_i^2}{n}}$$

where d_i = dbh of the ith tree
        n = number of trees measured

The quadratic mean diameter can also be calculated using

$$\bar{d}_Q = \sqrt{\frac{\text{BA}}{c(n)}}$$

where   BA = basal area per unit area (m²/ha or ft²/acre)
        n = number of trees per unit area
        c = conversion factor for diameter in cm or inches to cross-sectional area in m² or ft² (c = 0.00007854 in the SI system and c = 0.005454 in the imperial system)

This second approach can also be used when trees have been sampled with unequal probability, as with horizontal point sampling, or using nested fixed-area plots.

Alternative descriptions of stand diameter include the basal area-weighted mean and the basal area-weighted median. The arithmetic mean diameter of trees selected using horizontal point sampling (Section 9.3.1) is actually an unbiased estimate of the basal area weighted mean because trees are selected with probability proportional to their basal area. A weighted mean using tree factor as the weight should be used to obtain an estimate of the mean unweighted diameter when estimating average diameter from horizontal point sampling data. Similarly, the median diameter of trees selected using horizontal point sampling provides a natural estimate of the basal area-weighted median.

8.3.2. Basal Area

The total basal area of all trees, or of specified classes of trees, per unit area is a useful characteristic of a forest stand. For example, basal area is directly related to stand volume, is correlated with biomass, and is a commonly used measure of stand density and competition. The parameter incorporates the number of trees in a stand and their diameters. Although basal area is perhaps second only to tree density as a frequently reported parameter, this is partly due to its ease of measurement: in nearly all the roles it serves, it stands as a proxy for or correlate of other stand characteristics that are of primary interest. Stand basal area can be calculated from measurements of the dbh of all trees or of those of particular interest in a known area such as a fixed-area plot. Alternatively, it can be calculated using horizontal point sampling where tree dbh need not be measured (Section 9.3.1).

8.3.3. Diameter Distributions

A stand table shows the number of trees per unit area (or for a total area) according to designated characteristics. The most common form is a table that shows the number of trees per unit area by species and dbh class (Table 8.2). Knowledge of stand structure is useful for deciding on silvicultural measures and for estimating the yield of different products that may be obtained from a stand.

TABLE 8.2. Number of Trees per Hectare by Species and 2-cm dbh Classes for a Mixed Species Stand in New Brunswick, Canada

dbh Class (cm)	Species[a]							
	Fgrd	Bpap	Asac	Arub	Apen	Prub	Abal	Total
10	81	0	0	25	0	0	0	106
12	25	6	0	19	0	0	0	50
14	31	0	6	31	0	0	0	69
16	44	0	0	38	0	0	6	88
18	25	0	0	50	6	0	0	81
20	19	6	0	6	0	0	0	31
22	19	0	13	13	0	0	0	44
24	13	0	6	6	6	0	0	31
26	0	0	6	19	0	0	0	25
28	0	0	6	0	0	0	0	6
30	6	0	0	13	0	0	0	19
32	0	6	0	0	0	0	0	6
34	0	0	0	0	0	0	0	0
36	6	0	0	0	0	0	0	6
38	0	0	0	0	0	0	0	0
40	0	6	0	6	0	0	0	13
42+	0	6	19	13	0	13	0	50
Total	269	31	56	238	13	13	6	625

[a] Abal, *Abies balsamea*; Apen, *Acer pensylvanicum*; Arub, *Acer rubrum*; Asac, *Acer saccharum*; Bpap, *Betula papyrifera*; Fgrd, *Fagus grandifolia*; and Prub, *Picea rubra*.

Trees of the same species in an even-aged stand are fairly consistent in height with variations depending on their crown position (Fig. 8.2); more variability can be expected in trees of greater shade tolerance. Diameters, however, show wider variation than do heights for most species. As even-aged stands grow older, the diameter class distribution changes. The total number of trees in the stand decreases, with trees appearing in larger diameter classes not previously represented. Figure 8.2 shows the diameter distribution of an even-aged mixed oak stand at several ages. The diameter distribution for a typical pure even-aged stand is *unimodal*, as shown in the figure, with a tendency to asymmetry or skewness toward the right side of the curve. The diameter distribution in an even-aged stand may be *bimodal* or *multimodal* in stands of mixed species especially if one is tolerant and the other intolerant or if the species differ in growth rates. However, when many species are present, and especially if shade-tolerant species are abundant, the modes in a multimodal stand may run together, leading to a reverse-J shape.

In an uneven-aged forest, the trees are of many heights, resulting in an irregular stand profile as viewed from a vertical cross section (Fig. 8.3a). The more shade-tolerant species tend to form uneven-aged stands, as do intolerant species on very poor sites where the ability of dominant trees to exclude new regeneration through competition is limited. Cutting methods that remove only scattered individuals at short intervals sometimes maintain forest conditions favorable to shade-tolerant species and an uneven-aged stand. The typical diameter distribution for an uneven-aged stand is a large number of trees in the smaller diameter classes with decreasing frequency as diameters increase, as shown in Figure 8.3b.

Diameter distributions for small areas of uneven-aged forests may show considerably greater irregularity. As the area of the uneven-aged stand or forest increases, the irregularities tend to even out and the inverse J-shaped diameter distribution of an uneven-aged forest becomes apparent. For the forest shown in Figure 8.3, the inverse J-shaped distribution was apparent on a scale of about 0.24 ha (Loewenstein, 1996).

Early investigations of stand diameter distributions for uneven-aged forests were carried out by de Liocourt, a French forester, in 1898 (de Liocourt, 1898). He found that the ratios of number of trees in successive diameter classes varied and described the ratio between the numbers of trees in successive diameter classes using a ratio *q*. Kerr (2014) provides an excellent summary of de Liocourt's original work and subsequent interpretations. Meyer (1953), basing his work on the investigations of de Liocourt, studied the structure of what he termed a *balanced uneven-aged forest*. His definition was "one in which current growth can be removed periodically while maintaining the diameter distribution and initial volume of the forest." Meyer stated that a balanced uneven-aged forest tends to have a diameter distribution whose form can be expressed by the negative exponential equation:

$$Y = ke^{-aX} \tag{8.7}$$

By taking the natural logarithm of both sides, a linear form of the equation is obtained:

$$\ln Y = \ln k - aX$$

where Y = number of trees per diameter class

   X = dbh class

   a, k = constants for a characteristic diameter distribution

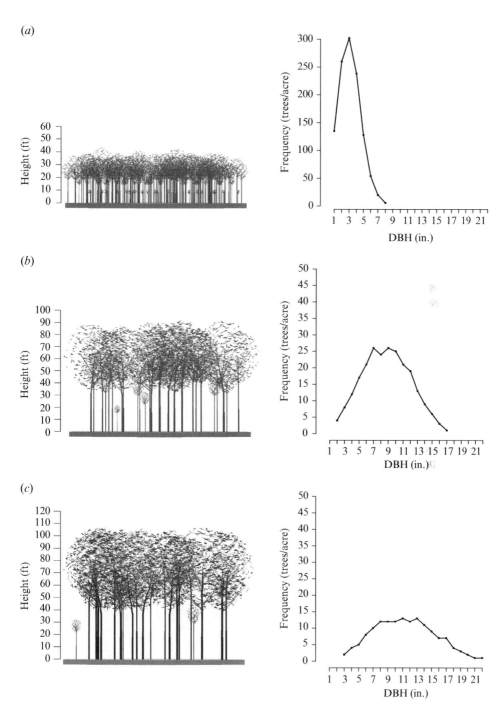

FIG. 8.2. Examples of stand profiles and diameter distributions for even-aged oak stands at three different ages: (*a*) 20-year-old stand, (*b*) 60-year-old stand, and (*c*) 100-year-old stand (stand data for site index 80, Schnur, 1937).

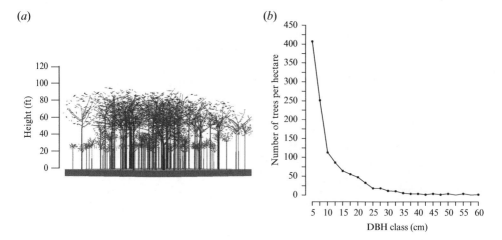

FIG. 8.3. Diameter distribution for an uneven-aged stand of oak–mixed hardwoods in Missouri: (*a*) stand profile and (*b*) diameter distribution (adapted from Loewenstein et al. (2000)).

A balanced distribution implies that the number of trees in successive diameter classes follows a geometric series of the form $m, mq, mq^2, mq^3, \ldots$, where q is the ratio of the series (as originally described by de Liocourt) and m is the number of trees in the largest diameter class considered. The q of de Liocourt is related to the constant a in eq. (8.7) by $a = \log(q)/W$, where W is the diameter class width.

If the logarithm of number of trees is plotted over diameter class, a balanced distribution will show a linear trend line (Fig. 8.4). Thus, the parameters, a and k, are sometimes obtained by fitting to the log-transformed data using regression analysis (though maximum likelihood estimation is also used). Excluding the trees greater than 42.5 cm dbh, the resulting equation for the data shown in Figure 8.4 is $\ln N = 6.4786 - 0.1399 (\text{dbh})$. The current value of the constant q can be obtained using the estimated value of a and the diameter class width:

$$q = e^{aW} \tag{8.8}$$

where a = the estimated slope coefficient

W = the diameter class width

In Figure 8.4, the diameter class width is 2.5 cm, so $q = e^{(0.1399 \cdot 2.5)} = 1.42$.

For many silvicultural applications, foresters are interested in determining the diameter distribution for a specified q value, maximum diameter class, and basal area. Brender (1973) gives a numeric example for calculating m for the series $m, mq, mq^2, mq^3, \ldots$:

$$m = \frac{g}{c \sum_{i=1}^{n} D_i^2 q^{(i-1)}} \tag{8.9}$$

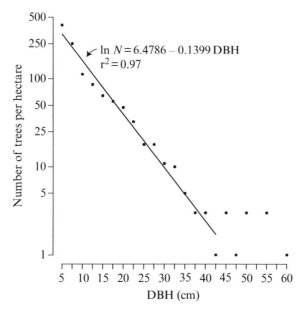

FIG. 8.4. Natural log of number of trees per hectare by diameter class for the forest depicted in Figure 8.3. (Note: diameters greater than 42.5 cm were excluded only for illustration purposes.)

where m = the number of trees in the largest diameter class

   g = specified basal area (ft²/acre or m²/ha)

   c = the cross-sectional area conversion factor (0.005454 in the imperial system, 0.00007854 in SI)

   D_i = the diameter class midpoint (D_1 = the class midpoint of the largest class and D_n = the class midpoint of the smallest class)

   q = the series ratio (eq. (8.3))

Moser (1976) presented generalized methodology for calculating the parameters a and k of eq. (8.5) given a specified q value, maximum diameter class, and stand density specified as either basal area, tree area ratio (Section 8.9.3), or crown competition factor (CCF) (Section 8.9.4).

Stand diameter distributions may be represented mathematically by *probability density functions*. A number of functions have been used or tried including the normal, exponential, binomial, Poisson, Charlier, Fournier series, normal logarithmic, Johnson's S_B, Pearl, Reed, Schiffel, Gamma, Beta, Burr, and Weibull. For detailed treatments of probability density functions in forest mensuration, see Loetsch et al. (1973), Schreuder et al. (1993), Prodan et al. (1997), and Johnson (2001). Of the numerous functions used to describe diameter distributions, the Weibull function has received the greatest attention.

The probability density function for the Weibull distribution is

$$f(D) = \frac{c}{b}\left(\frac{D-a}{b}\right)^{\frac{1}{c}} e^{-\left(\frac{D-a}{b}\right)^{c}}$$ (8.10)

where, $f(D)$ = probability density

   a = location parameter (theoretical minimum population value)

   b = scale parameter

   c = shape parameter

   D = diameter

The cumulative probability distribution function is given by

$$F\left(d \leq D\right) = 1 - e^{-\left(\frac{D-a}{b}\right)^c}$$ (8.11)

where $F(d \leq D)$ = the probability of observing a value d less than or equal to D

Equations (8.5) and (8.6) are often referred to as the three-parameter Weibull distribution because three parameters (a, b, and c) are required to specify the distribution. The two-parameter Weibull model is a special case of the three-parameter model where a, the location parameter, is assumed to equal 0.

The Weibull distribution has been widely applied in forest mensuration because of its flexibility. The Weibull distribution can exhibit a variety of shapes depending on the value of c.

Value of c	Shape of the curve
$c < 1$	Inverse J-shape
$c = 1$	Exponential decreasing
$1 < c < 3.6$	Unimodal with positive asymmetry (right skewed)
$c = 3.6$	Symmetric
$c > 3.6$	Negative asymmetry (left skewed)

By fitting the Weibull distribution to the data shown in Figures 8.2 and 8.3, this flexibility can be observed (Fig. 8.5).

The parameters of the Weibull distribution can be estimated directly from a list of diameter measurements or from stand table summaries (number of trees by diameter class). In many cases, the location parameter is assumed known. Since the location parameter is the minimum value of the distribution, it is logical to set this parameter to equal the smallest value observed or the lower limit of diameter measurement. When diameter measurements are grouped into classes, the lower bound of the smallest diameter class is often used as the location parameter. For example, if diameters were measured in 2 cm classes with 10 cm as the smallest midpoint, then the location parameter would equal 9.0 cm. Alternatively, one can set the location parameter to zero and attempt to estimate the number of unmeasured trees in smaller classes from the shape of the distribution itself (McGarrigle et al., 2011).

The other two parameters can be estimated using parameter prediction, parameter recovery, or percentile estimation techniques. The parameter prediction technique involves directly predicting the parameters based on stand characteristics such age, site index, mean dbh, quadratic mean dbh, density, or total basal area (Bailey and Dell, 1973). An efficient algorithm for moment-based parameter recovery is given by Burk and

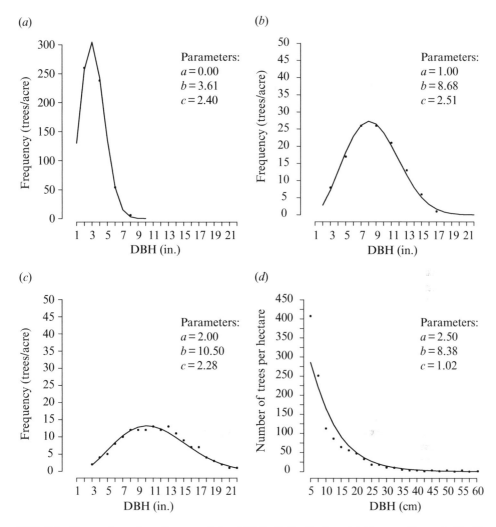

FIG. 8.5. Observed number of trees and fitted Weibull distribution: (*a*) 20-year-old stand, (*b*) 60-year-old stand, (*c*) 100-year-old stand, and (*d*) uneven-aged stand.

Newberry (1984). Zarnoch and Dell (1985) discuss the use of percentile estimators for obtaining Weibull parameters. Using the estimators specified by Zanakis (1979), they compared percentile estimators to moment-based estimators. While the moment-based estimators were superior, in terms of accuracy, the percentile estimators are simpler to obtain and, when $c < 2$, are as accurate as moment-based estimators, especially if sample size is small. With the advent of faster computers and more efficient numerical methods, however, the maximum-likelihood technique has gained in prevalence (e.g., Gove and Fairweather, 1989). When diameters are drawn from a horizontal point sample rather than a fixed-area sample, maximum-likelihood estimation of the so-called "size-biased distribution" (which reflects the unequal weighting of large trees in the sample) is much more efficient (Gove, 2003).

8.4. HEIGHT

Height is another widely used stand structure parameter. Height is an important factor in determining individual tree and total stand volumes. Height is widely used as a measure of site quality and stand productivity. Vertical structure (i.e., the height distribution) is an important factor in many silvicultural prescriptions and in assessing wildlife habitat.

For single-species, even-aged stands, variation in heights is typically less than variation in diameters (Fig. 8.2). For mixed-species or uneven-aged stands, height distributions are often similar to the diameter distributions, though generally not as wide (Fig. 8.3).

Vertical structure can be described using actual heights or using a classification based on crown position. Mean height may be calculated or the height distribution may be specified using many of the same models as used for diameter. In many applications, trees are classified based on their competitive position within the canopy. Kraft (1884) proposed the following definitions that are still widely used today:

> *Predominant.* Trees with exceptionally well-developed crowns generally above the main canopy and with few restrictions on growth.
>
> *Dominant.* Trees that have generally well-developed crowns and form the main part of the canopy.
>
> *Lower codominant.* Trees that have crown form similar to the predominant and dominant trees, but potentially showing some loss of form or density due to competition, and representing the lower boundary of the main canopy.
>
> *Dominated.* Trees with weak crowns, typically hemmed in from two or more sides, asymmetrical, or partially overtopped, and falling behind the main canopy. Kraft further divides these into *intermediate* trees (those that are free from overtopping but suffer severe lateral restriction) and *partially overtopped* trees.
>
> *Suppressed.* Trees with crowns completely overtopped by the main canopy and able to sustain growth only in shade-tolerant species.

In much common use in North America, Kraft's original five-class system is condensed and rearranged into four classes: *dominant*, *codominant*, *intermediate*, and *suppressed* (e.g., Oliver and Larson, 1996). Often, the predominant and dominant layers are merged, and the entire category that Kraft terms *dominated* is merged into a single *intermediate* class. The definitions of classes may vary from protocol to protocol; however, the intent is to give a qualitative, but repeatable, description of relative competitive position. In even-aged stands composed of a single canopy layer, the crown classes generally reflect relative height with the predominants and dominants being the tallest and the suppressed being the shortest (Fig. 8.6a); differentiation into crown classes reflects the long-term outcome of competition. In uneven-aged stands and in many mixed-species even-aged stands, the trees often form rough canopy layers referred to as *strata* (Oliver and Larson, 1996). In stratified even-aged mixtures, the strata reflect differences in the growth rate of individual species. Strata may be distinct or may blend together somewhat. Within each of these strata, dominant, codominant, intermediate, and suppressed trees may be found (Fig. 8.6b).

(a)

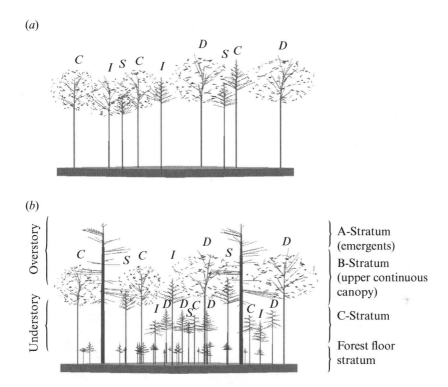

(b)

FIG. 8.6. Crown class and canopy strata: (*a*) single stratum canopy and (*b*) multistrata canopy (*D*, dominant; *C*, codominant; *I*, intermediate; and *S*, suppressed) (adapted from Oliver and Larson (1996)).

8.4.1. Expressions of Mean Height

A single value to characterize the height of a stand is useful in estimating stand volume or biomass from a stand volume or biomass function and to determine site index from a curve or equation. The following are methods that have been used to obtain the average height of a stand.

1. Measure and average the heights of all trees, or a sample of trees, regardless of their size or relative position in the stand. This corresponds to an unweighted mean stand height.

2. Measure and average the heights of dominant trees, or of dominant and codominant trees. (In Australia and New Zealand, these are called the *predominant height* and *dominant height*, respectively.) Since the selection of a dominant or codominant tree is subjective, different observers often obtain different results. The average height of dominant and codominant trees is the expression commonly used in determining the site index in North America.

3. Measure and average heights of a fixed number of the largest trees (usually, largest in diameter, although it could be largest in height) per unit area. (In much of European practice, as well as in Australia and New Zealand, the mean total height of the 100 tallest trees per hectare is called the *top height*.)

4. Determine the average height weighted by basal area called *Lorey's mean height* h_L

$$h_L = \frac{n_1 g_1 h_1 + n_2 g_2 h_2 + \cdots + n_z g_z h_z}{G} = \frac{\sum_{i=1}^{z} n_i g_i h_i}{\sum_{i=1}^{z} n_i g_i} \qquad (8.12)$$

where G = total basal area per unit area

        n_i = number of trees in the ith diameter class

        g_i = average basal area of the ith diameter class

        h_i = average height of the trees in the ith diameter class

The arithmetic average height of the trees selected in horizontal point sampling (Section 9.3.1) yields Lorey's mean height as shown by Kendall and Sayn-Wittgenstein (1959), and later by Beers and Miller (1973).

In addition to the above, several expressions of average stand height have been developed that utilize a height–diameter curve constructed from sample tree data representative of the stand. After preparing the curve, one decides what average diameter is representative of the stand and reads the height of the tree of this diameter from the curve. For example, the diameter of the tree of average basal area might be used (symbolized h_g); the arithmetic mean diameter might be used (symbolized h_d); the median diameter might be used (symbolized h_{dM}); or the diameter of the tree of median basal area might be used (symbolized h_{gM}). In New Zealand, the quadratic mean diameter of the largest 100 trees is determined. The corresponding height for this diameter tree is then estimated from a height–diameter curve.

8.4.2. Height–Diameter Curves

Height–diameter curves (previously introduced in Section 5.6.1) are often needed for preparing local volume tables and can be of interest in their own right in evaluating stand structure. Historically, they were plotted free hand; the relationships are now expressed by mathematical functions and fitted using regression analysis. Since the curve form may vary from one forest stand to another, numerous functions have been developed. Some examples are

$$h = 4.5 + b_1 D + b_2 D^2 \text{ (Trorey, 1932)}$$
$$\log h = b_0 + b_1 \log D \text{ (Stoffels and Van Soest, 1953)}$$
$$h = b_0 + b_1 \log D \text{ (Hendricksen, 1950)}$$
$$\log h = b_0 + b_1 D^{-1} \text{ (Schumacher and Hall, 1933)}$$

Prodan et al. (1997) describe these additional models:

$$h - 1.3 = \frac{D^2}{\left(b_0 + b_1 D\right)^2}$$

$$h = b_o D^{b_1} \quad \text{or} \quad \log h = \log b_0 + b_1 \log D$$

$$h = b_0 \left(1 - e^{-b_1 D}\right)$$

$$h - 1.3 = b_0 \left(\frac{D}{1+D}\right)^{b_1}$$

$$h - 1.3 = b_0 e^{-b_1/D}$$

$$\ln(h - 1.3) = \ln b_0 - b_1 \frac{1}{D}$$

$$h = b_0^{\left(b_1 \ln D - b_2 \ln(D)^2\right)}$$

$$\frac{1}{(h-1.4)^{0.4}} = b_0 + \frac{b_1}{D} \quad \text{or} \quad h = 1.4 + \left(b_0 + \frac{b_1}{D}\right)^{-2.5} \quad (\text{Petterson}, 1955)$$

where b_i = constants

h = total height

D = dbh

In any of these models, height may be expressed in feet or meters and diameter in inches or centimeters; the choice of units will affect the value of any constants in the equations. Often, $h - bh$, where bh = breast height, is modeled to insure h = bh when $D = 0$.

In ecological work, equations of the form $h = b_0 D^{b_1}$ are often used, and a great deal of theory has been advanced to justify particular choices of the "scaling exponent" b_1, which is commonly expected to be near 2/3 (e.g., Greenhill, 1881; West et al., 1999; Enquist and Niklas, 2002). However, empirical results usually point to considerable variation in b_1 based on species, site, and stand characteristics (e.g., Muller-Landau et al., 2006; Ducey, 2012). Thus, if one desires to use a mathematical function to describe the height–diameter relationship for a particular stand, one should test to see which function is most applicable and use appropriate regression techniques to estimate any needed parameters.

8.4.3. Height–Diameter Ratio

The ratio of height to diameter, when both measurements are taken in the same units, is termed the "slenderness coefficient" when applied to an individual tree and is often used to assess resistance to wind or snow damage (Section 5.4.4). Similarly, in single-cohort monocultures, the ratio of some measure of stand height to a comparable measure of stand average diameter can be calculated and is often used to evaluate stability and resilience to wind. For example, Cremer et al. (1982) used the ratio between the mean height and mean diameter of the 200 largest trees per hectare to diagnose wind-firmness in *Pinus radiata* plantations in New Zealand. The calculation of a stand-level height–diameter ratio can be strongly influenced by the choice of "average height" (top height, Lorey's height, mean height of dominant trees, mean height of all trees) as well as "average diameter" (typically arithmetic or quadratic mean diameter, sometimes of dominant trees only and sometimes of the entire stand), so care must be taken in interpreting such ratios or in comparing measured values for a stand to a reference or guideline. Castedo-Dorado et al. (2009) summarize some of the primary considerations in choosing which height and diameter values to use. In particular, a focus on dominant trees tends to emphasize those with the greatest economic value. However, the height–diameter ratio of dominant trees may be least sensitive to stand density and competition. Moreover, wind-damaged trees in a stand may come primarily from lower crown classes (Hinze and Wessels, 2002). Thus, the choice should be made by carefully weighing tradeoffs between sensitivity of the height–diameter ratio, on the one hand, and the particular ecological and economic values at stake, on the other hand.

8.5. VOLUME, WEIGHT, AND BIOMASS

8.5.1. Volume

An estimate of the total volume of trees in a stand is an important parameter for providing information necessary for making forest management decisions. The stand components included in the volume must be specified (e.g., species, minimum dbh and top diameter of the main stem, branches, etc.). Likewise, so must the units, typically either those of cubic volume (cubic meter or cubic feet) or of primary forest products (e.g., board feet). For even-aged stands, age should also be specified. Note that in even-aged forests, stand volume at a given age is indicative of site quality, though the volume–age relationship can be modified by a wide variety of factors including management and disturbance (Skovsgaard and Vanclay, 2008).

Stand volume can be estimated by various procedures:

Ocular estimates. Rough estimates of stand volume can be made by experienced foresters or timber cruisers who have worked in similar stands with known volumes. Such estimates can be rapid and quite accurate depending on observer skill; however, it is impossible to substantiate the accuracy of ocular estimates objectively, so they are usually unsuitable for work with scientific, financial, or legal implications.

Average tree procedures. Determining the tree of average volume and multiplying by the number of trees can estimate stand volume. In principle, the method is simple and direct. The practical difficulty in applying this procedure is to determine what the volume of the mean tree might be, and also to know the total number of trees in the stand. Of course, it would be possible to measure a number of sample trees and calculate their mean volume but the benefit of a quick estimate is lost as the number of sample trees measured increases and the procedure becomes the same as measuring forest inventory plots. In structurally simple forests, it may be possible to develop a sample frame of individual trees from segmentation of remotely sensed data and to select trees at random from this frame. A variation of the mean tree volume procedure is to divide the volume of what is chosen as the average tree by its basal area and then multiply this ratio by the basal area of the stand: $V = (v_t/g_t)G$, where V = stand volume, v_t = mean tree volume, g_t=mean tree basal area, and G=stand basal area. Estimation of stand basal area and the average basal area per tree is readily obtainable by point sampling (see Section 9.3 and Chapter 11). The volume of the tree of average basal area can then be estimated from a basal area–volume relationship.* The use of average tree methods was developed and used primarily in European forestry. Several procedures for the application of this technique using dbh or forest classes are described by Prodan (1965).

Inventory procedures. Of course, stand volume may be estimated by sample plot procedures discussed in Chapter 9 using one of the inventory designs discussed in Chapter 10. Because of the close correlation between volume and basal area of individual trees, horizontal point sampling (Chapter 11) is especially effective for volume estimation, provided field personnel are adequately trained and nonsampling errors can be minimized.

Stand volume equations. The cubic volume of a stand can be estimated as the product of the total stand basal area, the average stand height, and some expression of the stand form factor. Then,

$$V = G \cdot H \cdot F,$$

where V=total stand cubic volume

G=total basal area of the stand

H=average stand height

F=form factor for the stand, for example, cylindrical form factor[†]

The objective is to obtain a stand volume estimate rapidly without the necessity of measuring all trees or obtaining their volumes from tables. The accuracy of this method is dependent upon how representative the values of basal area, and average height and form factor are for a specific stand. Stand volume functions can also be developed using several of the models for tree volume equations substituting stand basal area instead of dbh^2 and stand height instead of tree height (e.g., see eq. (6.24)).

Unless individual trees are destructively sampled to obtain their volumes by direct measurement (Section 6.1.1), the mean-tree and inventory approaches depend on the availability of tables or equations to predict individual tree volume from field measurements (Sections 6.2 and 6.3). Depending on the quality of such tables or equations and their ability to capture site- or stand-specific differences in allometry, the volume tables or equations can quickly become limiting factors for the accuracy of stand-level volume estimates as the number of trees or sample plots measured increases. For example, widely used volume equations often have errors of 10–30% when applied to individual forest stands (Wiant et al., 1992b; Flewelling et al., 2000). These potential errors have motivated a range of approaches for obtaining more refined volume estimates (e.g., Sections 6.5 and 11.2). The same concerns apply to a stand-level volume equation if the basis for

---

*The linear regression of the volumes of individual trees on their respective basal areas is called the *volume line* (Hummel, 1953, 1955). From a volume–line relationship, the volume of the tree of mean basal area can be estimated.

[†] In Australian forest practice, stand form factors F have been determined empirically for *Pinus radiata* plantations from the relationship $F = V/(HG)$, where H is stand height and G is stand basal area (Brack, 1988).

developing the equation is plot-level estimates that are derived in turn from a volume table or equation: a distinction must be made between how well such an equation performs on average and how accurately it can provide volume estimates on a site-specific basis.

8.5.2. Weight, Biomass, and Carbon Content

Changes in markets for raw materials from forests, a shift to weight scaling for many bulk materials, and a growing interest in forest biomass and carbon content have stimulated a broader interest in estimating the weight, biomass, and carbon or nutrient content of forested stands and landscapes. Just as with stand volume, the components included in an expression of weight, biomass, or other contents of a stand must be specified. For simplicity here, we follow the same terminology used in Chapter 6: *weight* will be used for the green or fresh weight of trees, while *biomass* will be used for oven dry weight. For the moment, we restrict our attention to the weight, biomass, or other content of live trees; dead trees will be considered in Section 8.5.3, while other vegetation will be taken up in Section 8.7.

Just as for stand volume, there are a variety of possible approaches for estimating the weight, biomass, carbon, or nutrient content of a stand. Indeed, when dealing with a single-species stand, and if the primary concern is with the weight or biomass of stemwood, one logical starting point would be a cubic volume estimate for the stand obtained with an appropriate approach from Section 8.5.1. Multiplication of estimated stand volume by wood density would yield a consistent estimate for stemwood biomass, and further multiplication by $(1+(MC\%/100))$, where MC% is the green moisture content, gives an estimate of stemwood weight (see eq. (6.16)). Likewise, multiplication of the stemwood biomass for the stand by a representative carbon or nutrient concentration (on a dry weight basis) would give a consistent estimate of carbon or nutrient content.

However, in mixed-species situations, or where stemwood is not the sole focus of interest, inventory approaches remain by far the most common for estimating stand-level weight, biomass, carbon, or nutrient content. In principle, any of the equal or unequal probability approaches discussed in Chapter 9 are applicable, provided an appropriate set of weight or biomass equations is available for individual trees. For forest carbon, some authors have emphasized the use of fixed-area plots (e.g., Pearson et al., 2007). This emphasis appears to reflect two concerns. First is the potential for fixed-area plots to be more robust to field errors than horizontal point sampling or other methods when implemented by personnel who lack specific training. Second is a presumed lack of familiarity with basic forest mensuration and inventory among those responsible for some emerging carbon markets, such that simplified field procedures are preferred to maintain transparency. Notwithstanding such concerns, the close correlation between individual tree basal area and weight or biomass makes horizontal point sampling extremely efficient when correctly implemented (e.g., Ducey, 2009a).

It is possible to construct stand-level weight, biomass, or carbon equations similar to those that have been developed for estimating stand volume. Given the importance of stemwood to stand weight, especially when merchantable weight is of concern, it should not be surprising that many of the same equation forms used to predict stand cubic volume are also used to predict stand green weight. For example, Burkhart et al. (1972) use regression to predict green weight of loblolly pine stands as a function of stand age, average height of dominant and codominant trees, total basal area of the stand, and proportion of basal area actually comprised of loblolly pine. Such equations allow prediction of yield from a simplified set of inventory variables. Such an approach can easily be extended to the estimation of stand-level biomass or carbon content.

When stand-level estimates of weight, biomass, or related quantities depend on individual tree estimates taken not by direct sampling but from an allometric equation, the quality of that allometry represents a major source of uncertainty in the results. For example, MacLean et al. (2014) compared the carbon content of live trees in forests in the northeastern United States as calculated using three different approaches, using the same basic inventory data from the US Forest Service, FIA program: first, using the simplified allometric equations of Jenkins et al. (2003), the approach used by FIA prior to 2009; second, using a component ratio method based on merchantable volumes as implemented by FIA after 2009 (Woodall et al., 2011); and finally, using the carbon calculation tools in the Forest Vegetation Simulator (Crookston and Dixon, 2005; Rebain, 2010), which are widely used in reporting for voluntary carbon registries. Results for the region as a whole differed by as much as 28% between methods. At the plot level, discrepancies were variable and potentially even greater, and no simple relationship emerged to facilitate conversion between one set of values and another. Reducing such uncertainties is made especially hard by the challenges of direct measurement of tree weight in the field (Section 6.1.7), which are much more severe than those for volume. While a suitable dendrometer can be used to obtain upper-stem diameters and reduce errors in volume estimates in the field, in situ measurement of wood density, carbon, or nutrient content is not usually possible, and taking field samples back for laboratory analysis can be costly. Consequently, while some advanced sampling methods that can be used to reduce uncertainty in stand volume estimation are also theoretically useful for weight, biomass, and carbon content, in practice they are fraught with difficulty.

8.5.3. Volume, Weight, and Biomass of Dead Wood

Estimating the volume, weight, biomass, or other contents of standing and downed dead wood is similar in some ways to the process for live trees, but complicated by the irregular form that dead material can take and by the variability in material properties created by decay (Section 6.7). Unless there is a particular emphasis on estimating the economic losses associated with recently killed trees, such as those from fresh insect kill or logging waste, the volume of standing and downed dead wood is almost invariably expressed in terms of cubic volume rather than product-oriented units such as board feet. An exception may be where there is interest in utilizing such materials for bioenergy, in which case, estimates may be desired in terms of weight or cords.

In principle, estimating the volume of standing dead trees can follow exactly the same inventory procedures as used for standing live trees, though additional measurements (including assessment of decay and breakage) may be needed to estimate individual tree volume, weight, or biomass (see Section 6.7.1). Any of the standard sampling approaches outlined in Chapter 10 can be used, and commonly standing dead trees are tallied on the same plots or sample points as standing live trees (though estimates of tree density, basal area, and volume from dead trees are reported separately from those of live trees, and are not included in conventional stand totals). In practice, however, since standing dead trees are usually much less abundant than standing live trees and are often patchily distributed, the sampling errors associated with standing dead material are greater than those for live trees. Reducing the sampling variability for standing dead trees may require a greater number of sample points, a larger plot area (for fixed-area plot sampling), or a smaller basal area factor (for horizontal point sampling). Alternatively, modified sampling approaches (such as modifications of horizontal line sampling) may be useful to increase the sampling intensity for dead trees (Ducey et al., 2002; Kenning et al., 2005). Any such changes increase the cost and complexity of an inventory and should be weighed against the values inherent in standing dead trees to determine whether they are truly warranted.

Reporting of volume, weight, biomass, and nutrient content for downed dead wood closely parallels that for standing dead wood. However, the inventory of downed dead wood is complicated—and in some ways facilitated—by the fact that it is lying on or near the ground. Historically, inventory of downed dead wood has been viewed as tedious and time-consuming, but recent decades have seen an explosion of methodological research driven by the interest in downed wood as habitat, biomass, and fuel for wildfires. The inventory of downed dead wood is treated in detail in Chapter 12.

8.6. CROWN AND CANOPY MEASUREMENTS

Examinations of production ecology, wildlife habitat, water resource use and protection, and forest health often depend on aspects of the crown density and structure of forest stands. International definitions of forest area (FAO, 2006) depend on canopy attributes. The canopy represents the biophysical link between the stand and many remote sensing measurements, so characterizing the canopy can be a useful step in mapping forest attributes using data from airborne or spaceborne platforms. Some crown measurements for individual trees were reviewed in Section 5.5. Certain stand-level canopy measurements can be viewed as aggregations of individual tree crown attributes, but even so it is sometimes useful to consider alternatives to an individual tree approach to measurement. For other canopy attributes, the position and spatial relationship of crowns affects the results so that a simple summation is not appropriate.

8.6.1. Crown Closure and Canopy Cover

Foresters and ecologists have been interested in the gross dimensional relationships of crowns within the forest canopy for decades, if not centuries, but a lack of clarity in terminology has often contributed to confusing and occasionally inappropriate measurements. Some aspects of this history have been reviewed by Jennings et al. (1999), and we conform closely to their terminology, along with that of Stenberg et al. (2008), here.

Canopy closure is the fraction of the sky obscured by vegetation when measured from a single point. The fraction of the sky that is not obscured by vegetation is *canopy openness*, that is, one – canopy closure. Traditionally, canopy closure has been measured with a spherical densiometer (Lemmon, 1956, 1957), essentially a curved mirror with a grid of lines etched on the surface. The densiometer is held with the center of the mirror oriented vertically (often with the aid of a bubble level), and the etched lines are used to estimate area in a fashion similar to that of using a dot grid to estimate area on a map or photograph (Section 4.4.2). The widespread availability of high-quality digital cameras has made them an attractive alternative; image analysis can, in principle, provide a much more accurate and repeatable measure than counts on a grid, and the photograph itself preserves a record of canopy conditions. Photographic techniques for canopy closure measurement are sensitive to the techniques used for thresholding, or the separation of the image into sky and nonsky

components, and can also be sensitive to illumination conditions. Such issues are less important for terrestrial laser scanning, which uses an active rather than passive approach. All
measurements of canopy closure depend on the angle over which the measurement is taken
(which may range from a relatively narrow cone about the vertical, to a full hemisphere).

Canopy cover is the proportion of a horizontal surface that is covered by the projection
of the forest canopy onto that surface. Strictly speaking, canopy cover must be measured
vertically and the measurements must be spread over many points in the stand, most commonly on a systematic grid (though Jennings et al., 1999 recommend a random array of
points). A variety of instruments have been designed to facilitate canopy cover measurement,
often integrating a level with a right-angle mirror or pentaprism to ensure a vertical line of
sight while avoiding neck strain on the observer. The "moosehorn" sighting tube is one
familiar example (Bonnor, 1967). Korhonen and Heikkinen (2009) describe an approach
for estimating canopy cover from digital photographs using image processing techniques
to eliminate small gaps within individual tree crowns. Although a camera-based
measurement is, in the strict sense, a canopy closure measurement, Korhonen and Heikkinen
(2009) reported small errors when a relatively narrow field of view was used (implying
measurements that are nearly vertical).

Crown cover is the total projected area of individual tree crowns onto a horizontal surface (sometimes presented as m^2/ha or $ft^2/acre$; but if the areas of crowns are converted to
the same units as those of ground area, then crown cover may be expressed as a unitless
value, i.e., ha/ha or acre/acre). In a truly single-layer canopy, crown cover is identical to
canopy cover. However, where individual tree crowns overlap, crown cover will be greater
than canopy cover, and in many forests with complex canopies, crown cover can take
values greater than 1. Crown cover is best estimated by measuring the crown areas of
individual trees that have been selected using an appropriate plot or point sampling technique (Chapter 9), and expanding the individual tree measurements to a per-unit area basis.
Stenberg et al. (2008) describe a technique for estimating crown cover using an angle
gauge or relascope; that technique seems most useful in relatively open stand conditions.

Korhonen et al. (2006) provide a recent comparison of several alternative methods for
evaluating canopy closure and canopy cover. In general, they found strong tradeoffs between time requirements, on the one hand, and accuracy and objectivity, on the other. When
measurements of canopy closure or cover are needed in an inventory, careful attention to
the variables needed, requirements for objectivity and repeatability, and sampling design
are all warranted.

8.6.2. Leaf Area Index

Leaf area index (LAI), the total area of foliage in a stand on a per unit area basis (e.g., m^2/m^2),
is a fundamental variable in production ecology and a key to understanding the exchange
of CO_2, water, and energy with the atmosphere. In many coniferous forests, LAI has been
shown to be a strong linear predictor of stemwood and total biomass production, and can
also help diagnose nutrient limitations and responsiveness to fertilization (e.g., Waring
et al., 1981; Oren et al., 1987; Vose and Allen, 1988; Long and Smith, 1992; Jokela and
Martin, 2000; Innes et al., 2005). LAI can be defined in terms of projected (i.e., one-sided)
or all-sided leaf area (i.e., twice projected leaf area in hardwood stands, and slightly more
in stands of conifers having needles that are not entirely flattened). Most forestry applications assume projected LAI.

Direct measurement of LAI requires destructive sampling: individual trees must be selected and felled, the foliage removed and weighed, and a subsample of foliage taken to determine specific leaf area (area per unit weight). The labor costs involved, as well as the effect on the remaining stand, usually make direct measurement impossible except in a handful of research applications. Indirect methods for estimating LAI have been reviewed by Larsen and Kershaw (1990) and Jonckheere et al. (2004). Unfortunately, all known indirect methods also present theoretical, practical, or methodological challenges.

Allometric methods use a regression relationship (Section 5.6) to predict the leaf area (or biomass) of an individual tree from easily obtained measurements. The individual tree values are then expanded to obtain per-unit area estimates based on a sound sampling unit design (Chapter 9). Although many allometric equations use dbh as the predictor, dbh-foliage allometries are notoriously unstable and stand-specific; therefore, many allometric equations for foliage use sapwood area at breast height, which the pipe model theory of Shinozaki et al. (1964) suggests should provide more stable predictions. However, using sapwood area implies (at a minimum) extensive coring of individual trees, which is labor-intensive and may compromise values of the residual stand. Moreover, due to a lack of published sapwood area-foliage area equations for many species, and lingering concerns over the stability of predictive equations, it may still be necessary to obtain destructive direct leaf area measurements for a number of trees in order to build purpose-specific equations. The use of component ratio equations (e.g., Woodall et al., 2011) to estimate foliage biomass from total tree biomass, while warranted for obtaining rough estimates over large areas, is not likely to prove adequate for stand- or site-specific work.

Litterfall techniques can be quite attractive, especially in stands that hold exactly one year's foliage and then drop that foliage in a short period of time at the end of the growing season. Traps of known dimension (often $1\,m^2$ or smaller) are laid out in an appropriate sampling design within the stand of interest, and material is collected, dried, weighed, and subsampled to obtain the area-to-weight ratio. In principle, litter traps function as very small fixed-area plots, so the basic sampling considerations outlined in Chapter 10 apply. However, local climate must be considered in the design of traps and in the frequency of collection (to avoid loss due to wind, decay, and decomposition that might bias estimates). Without detailed study of foliar demography (e.g., Innes et al., 2005), litter traps cannot be used to estimate the LAI of stands that retain more than one year's foliage or to estimate intra-annual variation in LAI. Finally, the need to visit traps repeatedly for collection, and the labor costs inherent in sorting material, makes litterfall methods uneconomical for extensive studies or for management purposes.

Another major category of LAI estimation methods is indirect noncontact methods. These methods all employ simple models of radiation transfer within forest canopies (such as the Beer–Lambert law), in combination with simplifying assumptions about the distribution of foliage elements in space as well as their angular distribution, to model the probability that a beam of light can pass unobstructed from the top of the canopy to a known sensor position. That probability typically depends not only on the total LAI of the stand, but on the angle of the beam. A variety of sensors have been employed to estimate transmission as a function of angle, and so to recover the unknown LAI and other parameters of the radiation model. These include commercial sensors that detect incoming radiation under the canopy, and either compare it to that at an open-sky location, or model the spatial distribution of gap sizes (such as the LAI-2000 Canopy Analyzer, LiCOR Inc., Lincoln, Nebraska; or the Sunfleck Ceptometer, Decagon Devices, Pullman, Washington, DC; and the TRAC instrument, 3rd Wave Engineering, Nepean, Ontario). Alternatively, hemispherical photography can be used, along with image processing, to estimate gap fraction as a function of zenith angle, which is then used to estimate LAI using a similar radiation transfer model (Rich, 1990; Jonckheere et al., 2004). All of these methods can be somewhat fussy in terms of the permissible sky conditions; hemispherical photography is also sensitive to the algorithms used to threshold sky versus nonsky pixels in the image. None is able to effectively separate photosynthetic tissue (i.e., foliage) from nonphotosynthetic tissue (e.g., stems and branches). Terrestrial laser scanning, which uses a laser beam as an active probe over a tripod-mounted instrument rather than a passive analysis of incoming light, may offer some advantages in this context (Henning and Radtke, 2006b; Strahler et al., 2008). However, the technology is still relatively new and expensive, and some studies suggest that estimates of LAI and other structural parameters may be sensitive to instrument characteristics and physical setup (e.g., Ducey et al., 2013a). Continued reductions in size and cost of terrestrial laser scanners, and improvements in algorithms, may yet make the approach attractive for applied research and management in the near future.

8.7. UNDERSTORY AND REGENERATION

Mosses, lichens, ferns, herbs, shrubs, and natural tree regeneration form the understory in many forest stands. Information on the amount and characteristics of regeneration and other understory vegetation and its biomass is required for forest management decisions, evaluating forest health, and for the estimation of total vegetation biomass and its carbon content. These components have not been measured historically in operational forest inventories to estimate timber quantities. Indeed, many older texts refer to the understory by the pejorative term "lesser vegetation." However, more recent work has emphasized the importance of the forest understory as an ecological indicator, and its disproportionate role in regulating a variety of ecosystem functions (Kerns and Ohmann, 2004; Dickie et al., 2011; Gonzalez et al., 2013). In managed forests, some understory vegetation also serves as competition for desirable tree regeneration and its abundance can influence silvicultural decisions (Wagner et al., 2006).

The most important characteristics of understory vegetation to measure depend on the goals and objectives of the inventory. However, some of the more commonly measured attributes include

- species composition;
- relative cover by species or species groups;
- density (number of stems or plants per unit area);
- frequency (the proportion of samples in which a species occurs);
- abundance (number of stems or plants per sample); and
- sizes (height, diameter), especially for tree regeneration.

Table 8.3 summarizes the sampling techniques commonly applied to understory vegetation. This section will review some of the more common techniques employed in estimation of these parameters. For a more complete treatment of understory vegetation measurement and sampling, refer to Kershaw and Looney (1985), Bonham (1989, 2013), or Sutherland (1996).

TABLE 8.3. Methods[a] for Sampling Understory Vegetation

| | Vegetation Type | | | | | |
Method	Saplings	Seedlings	Shrubs	Herbs and Grasses	Bryophytes	Fungi and lichen
Counts						
Fixed area plots	*	*	+	?	–	–
Point samples	+	?	?	?	–	–
Transects	*	*	*	*	?	?
Distance sampling	*	*	*	?	?	?
Cover						
Visual estimation	*	*	*	*	*	*
Frame quadrats	+	*	+	*	*	*
Point quadrats	–	+	+	*	*	*
Point samples	?	?	*	*	–	–
Transects	*	*	*	*	*	*
Biomass						
Harvesting	+	*	*	*	*	*
Reference units	+	+	+	?	?	–
Allometric relationships	*	*	*	*	?	?

Source: Adapted from Bullock (1996).

[a] *, Usual applicable; +, often applicable; ?, sometimes applicable; and –, generally not applicable.

8.7.1. Density and Frequency

Frequency is based on the presence or absence of a species in sample units (plots, transects, or points) and is defined as the number of times a species is present in a given number of sample units (Raunkiaer, 1934). Frequency is usually expressed as a percentage of the total number of sample units. Percent frequency is sometimes referred as *frequency index* (Bonham, 1989). Density is a quantitative expression of the number of plants per unit area. Depending upon the size of the plants, density is expressed as number of individuals per square foot or individuals per square meter for small plants and number of individuals per acre or individuals per hectare for larger plants.

The most common method for estimating understory vegetation frequency or density utilizes fixed-area square or circular plots that can vary in size depending on the class of vegetation to be studied. For example, Cain and de Oliveira Castro (1959) suggested the following plot areas: for the moss layer, $0.01–0.1\,m^2$; for the herb and small seedling layer, $1–2\,m^2$; for tall herbs and low shrubs, $4\,m^2$; for tall shrubs and low trees, $10\,m^2$; and for trees, $100\,m^2$. However, selection of an appropriate plot size for measurement is a subjective decision and should be based on the size and spacing of individuals of a species (Bonham, 1989). Curtis and McIntosh (1950) proposed that a plot should be no larger than one or two times the mean area per individual of the most common species. Bartlett (1948) determined the most efficient size of plots for density estimation corresponded to a 20% absence rate (frequency = 80%). A number of researchers have examined efficiency in terms of sampling error and time requirements for various plot sizes and shapes (e.g., Evans, 1952; Eddleman et al., 1964; Van Dyne et al., 1964; Hyder et al., 1965). Frequencies and densities can also be estimated using line transects (Section 9.2.2) and distance methods (Section 9.5). Density estimates for larger shrubs and small trees may be obtained using point sampling techniques (Section 9.3 and Chapter 11).

For frequency estimates, only the presence of a species on a plot needs to be noted. For density estimates, all individuals within the plot boundary need to be counted. Plot counts are expanded to per unit area counts using the ratio of unit area to plot area (see Chapter 9 for more details):

$$\text{Expansion factor} = \frac{\text{Unit area}}{\text{Plot area}} \tag{8.13}$$

For example, if the number of individuals on a $5\,\text{ft}^2$ plot is counted, the number of individuals/ft^2 is obtained by multiplying the count by 0.2 ($1\,\text{ft}^2/5\,\text{ft}^2$). If $100\,\text{m}^2$ plots are used, the number of individuals per hectare is obtained by multiplying the count by 100 ($10,000\,\text{m}^2/\text{ha} / 100\,\text{m}^2/\text{plot}$).

To estimate density, each individual plant on a plot needs to be identified and counted. One of the greatest difficulties in estimating density of understory vegetation is identification of individual plants (Bonham, 1989). Strickler and Stearns (1962) defined an individual as the aerial parts corresponding to a single root system. For single stem species, like most trees and annuals, this definition is simple to apply; however, for many grass species, clonal herbs and multistemmed shrubs, where the number of individual plant stems per root system vary considerably, this definition is not easily applied and may not even be relevant. While the only practical counting unit for density estimation is the individual, the value of carefully defining the individual depends on the purpose of the study, definition of the unit, and precision of the count (Bonham, 1989). The number of individual stems or shoots may be more highly correlated with other parameters of interest, such as biomass and cover, than the number of individual plants.

8.7.2. Cover and Competition

The simplest definition of cover is the percentage of ground surface covered by vegetative material. Cover is generally expressed as a fraction or percentage of total area. Cover of a species or life form is expressed as a percentage of total vegetation is referred to as *relative cover*.

Cover is generally measured as the vertical projection of vegetative material on to the ground surface (see Fig. 5.18 for an example of crown projection area). Cover is one of the most commonly measured vegetation parameters. Because cover is an expression of vertical projection, its measurement does not require identification of individual plants; thus, estimation of cover is often easier than estimation of density. Cover also has the advantage of being able to express measures of different life forms (e.g., mosses, grasses, forbs, shrubs, and trees) in comparable terms. If the vegetation occurs in distinct layers (e.g., trees, shrubs, and undergrowth), then depending upon the objectives of the sample, cover of each species can be measured separately by layer (Bonham, 1989).

When understory species have compact, regular crowns, cover may be estimated by measuring plant dimensions and applying formulas for regular geometric shapes. For example, crown area is often calculated by measuring crown diameter along one or more axes and applying the formula for a circle (see Section 5.5.2). More commonly, cover is

(a)

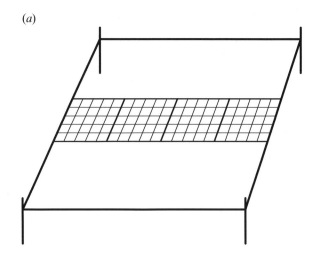

(b)

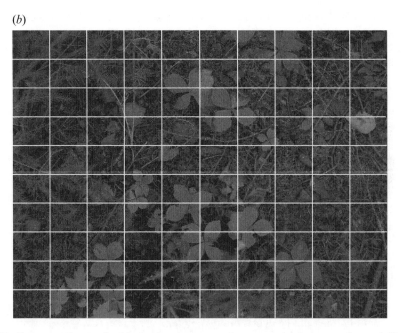

FIG. 8.7. Frame quadrat method for estimating vegetation cover: (*a*) frame quadrat and (*b*) photographic "frame" quadrat.

visually estimated using small fixed-area plots. A frame quadrat (Fig. 8.7*a*) can be used to facilitate visual estimation. A frame quadrat divides a larger fixed-area plot into smaller subplots. The example illustrated in Figure 8.7*a* has divided the larger plot into 100 equal area subplots. Cover can be visually estimated for each subplot or the subplot can be classified as either covered or not covered by a species. Cover for the larger plot is then obtained by summation of the cover values of the smaller plots.

Photographic frame quadrats (Fig. 8.7b) are an efficient method of estimating cover. Vertical photographs are obtained at a constant height above the vegetation or ground. A grid is superimposed on the image either by using a filter at the time of image capture or by manually overlaying a grid on the developed image. As with the normal frame quadrat, cover can be visually estimated for each subplot or the subplot can be classified as covered or not covered. For example, a subplot can be considered covered if 50% of its area is occupied by a given species. Using this definition, the number of subplots in Figure 8.7b covered by bunchberry (*Cornus canadensis*) is 16 and the percentage cover is estimated to be $100(16/100) = 16\%$. The availability of reasonably priced, high-quality digital cameras and powerful image analysis software makes photographic frame quadrats an easier and less time-consuming method of obtaining cover estimates.

Visual estimation of cover has the disadvantages of potential variation among different observations and the potential for high observer bias (Bonham, 1989). A number of intercept methods have been developed, which eliminate much of the potential observer variation and bias. Two commonly employed intercept methods are the point intercept technique and the line intercept technique.

The point intercept technique involves lowering a pin through the vegetation canopy and recording the number of hits (interceptions) by species (Fig. 8.8a). Percent cover is obtained by dividing the number of hits by the total number of pins. For example, in Figure 8.8a, a forb is hit by 4 out of 10 pins, therefore, the percent cover for forbs is $100(4/10) = 40\%$. Likewise, grass had one hit, representing 10% cover, and bare ground had five hits, representing 50% cover.

The point intercept technique can also be applied to vertical photographs (Fig. 8.8b). As with the photo frame quadrat, a dot grid can be superimposed on the photograph and the number of dots falling on a given species counted. In Figure 8.8b, 13 dots fall on bunchberry, thus the percentage cover is estimated as $100(13/100) = 13\%$.

The line intercept technique utilizes a transect and measures the length of the transect intercepted by the vertical projection of a species (Fig. 8.9a). The percentage cover is estimated as the ratio of interception length to total transect length:

$$\text{Percent cover} = 100 \frac{\Sigma I}{L} \qquad (8.14)$$

where I = interception length

　　　　　L = total transect length

For some species, such as trees and larger shrubs, the determination of intercepted distance can be difficult and subject to observer bias (Bonham, 1989). A point transect (Fig. 8.9b) is a modification of the line intercept technique. The point transect technique is similar to the point intercept technique. Points along a transect are established at predetermined intervals. Each point is assessed to determine whether it is covered or not covered. The percent cover is estimated as the number of covered points divided by the total number of points.

Bonham (1989) describes another technique for estimating cover using a modification of sampling with probability proportional to size (Section 11.1). An angle gauge (Fig. 8.10a)

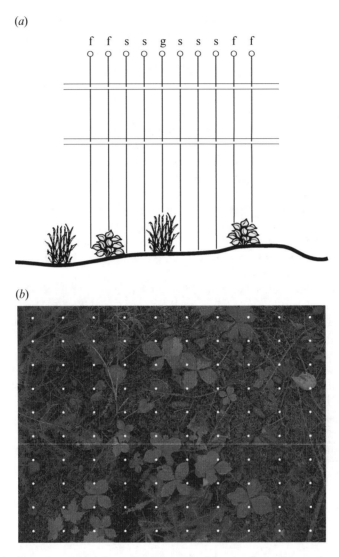

FIG. 8.8. Point intercept method for estimating cover: (*a*) point frame intersections (*f*, first hit on forb; *g*, first hit on grass; and *s*, first hit on soil) and (*b*) photographic point intercept method.

is built using two sticks. The gauge constant is determined from the ratio of stick length to crossbar length:

$$K = \frac{L}{l} \qquad (8.15)$$

where K = gauge constant
         L = stick length
         l = crossbar length

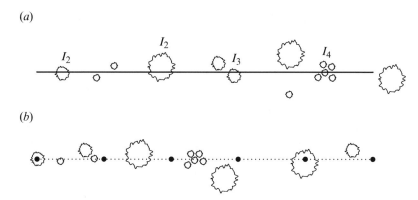

FIG. 8.9. Transect techniques for cover estimation: (*a*) line intercept technique and (*b*) point transect method.

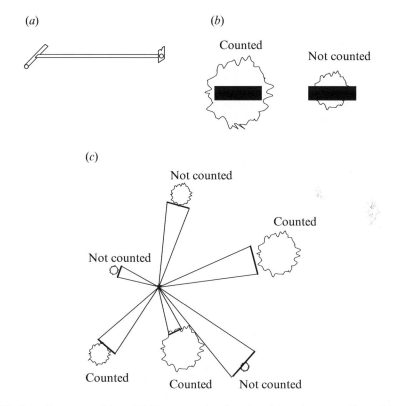

FIG. 8.10. Sampling cover with probability proportional to size: (*a*) angle gauge, (*b*) counting cover samples, and (*c*) cover point sample (adapted from Bonham (1989)).

This gauge constant, K, is the reciprocal of the gauge constant, k, used in horizontal point sampling (Section 9.3.1). Plants are counted if the diameter of their associated crown cross section is greater than the projected crossbar length (Fig. 8.10*b* and *c*). The percent cover represented by each plant counted is determined from the ratio of the area of a circle whose diameter equals the crossbar length to the area of a circle whose radius equals the stick length:

$$\text{Percent cover} = 100\frac{\pi(l/2)^2}{\pi L^2} = 100\frac{l^2}{4L^2} = 25\frac{l^2}{L^2} = \frac{25}{K^2} \qquad (8.16)$$

If $K = 5$, then each plant counted represents 1% cover. Total cover is obtained by counting all plants in a 360° search around the sample point. This angle gauge and approach is similar to the approach more recently devised by Stenberg et al. (2008) for estimating canopy cover in trees.

8.7.3. Biomass and Forage

Interest in *understory biomass* has increased because of its role in biodiversity, global climate change, carbon sequestering, wildlife habitat evaluation, and forest fuel assessment. Understory biomass may be directly measured using harvest techniques or indirectly measured using a variety of nondestructive measurements and regression equations relating biomass to these measures.

Harvest methods may employ either complete removal of vegetation within a sample unit or some sort of proportional harvesting scheme. When using complete removal methods, all vegetation within a predefined sample area is removed. Often, the plot boundaries are vertically projected and only vegetation within the plot space is removed. Portions of plants rooted within the plot boundary, but projecting out of the plot space, are discarded from the sample. Likewise, parts of plants not rooted within the plot boundary, but projecting into the plot space, are counted. The clipped vegetation may be weighed fresh or dried and may be separated into components (foliages, branches, stems) depending upon the requirements of the survey. In some surveys, clipped vegetation may be weighed fresh in the field and a fixed proportion of the material selected and dried. A ratio between the fresh weight and dried weight of the subsample is then used to estimate dried weight for the entire sample. Biomass per unit area is obtained by expanding the plot measurement to per unit area values using an appropriate expansion factor (eq. (8.13)).

Proportional harvesting schemes utilize either a fixed percentage removal or simply sample individual plants or individual plant parts. Sampling individual plant parts is often referred to as *reference unit sampling*. Reference unit sampling is commonly used on multistem shrub species. One stem is randomly harvested and its biomass determined. Total biomass for the plot is obtained by multiplying the reference unit weight by the number of reference units on the plot. See Bonham (1989, 2013) or Pitt and Schwab (1988) for detailed discussions of implementing reference unit sampling.

The most common approach to understory biomass determination is to predict biomass from nondestructive measurements of plants. A variety of regression models have been developed to predict biomass from plant measurements. Many of these equations are similar in form to those used for trees (Section 6.2). Regression equations have been developed that utilize stem diameter at the ground, stem diameter at some height above ground, crown area, plant height or combinations of these variables. Examples of biomass equations can be found in Brown (1976), Smith and Brand (1983), Wharton and Griffith (1998), and Chamberlain et al. (2013).

8.7.4. Regeneration Surveys

Information on regeneration is important in forest management to determine its current status, survival, and whether or not it is sufficient, to predict its future, to decide on silvicultural treatments, to forecast future stand yields, and to see if it complies with legislative requirements (Stage and Ferguson, 1984; Stein, 1984a).

Regeneration studies or inventories should first define what constitutes the population of interest. The population may consist of

- all tree species or only chosen ones and
- all individuals (seedlings, sprouts, saplings) or only healthy ones.

If all individuals are included, indications may be made of their health or condition, such as browsed by animals. The heights and/or diameters may be measured for the individuals in the regeneration population.

Natural tree regeneration consists of seedlings, sprouts, and saplings. The definitions and size characteristics of these components of regeneration vary in different studies, although in general terms seedlings are small plants of tree species originating from seed; sprouts are stems that have originated from a dead or cut tree stem or from roots; and saplings are young trees that have not attained a given minimum dbh or height. In the Forest Health Monitoring Guide Program (Tallent-Halsell, 1994), seedlings are individuals of tree species at least 1 ft (0.3 m) tall but less than 1.0 in. (2.5 cm) dbh. Saplings are individuals from 1.0 in. (2.5 cm) to 4.9 in. (12.4 cm) dbh. In a study of biomass in Maine (Wharton and Griffith, 1998), seedlings were defined as trees less than 1.0 in. dbh and saplings were trees at least 1.0 in. but less than 5.0 in. dbh. The *Forestry Handbook* (Wenger, 1984) defines a seedling as a woody plant of either a shrub or tree species that is less than 3.0 ft (90 cm) tall; a shrub as a woody plant greater than 3.0 ft (90 cm) tall that will not become a dominant or codominant in the canopy; and a sapling as a woody plant with a dbh_{ob} less than 4.0 in (10 cm) but greater than or equal to 1.0 in (2.5 cm). The same size definitions may be applied to sprouts although their origin should be noted.

Several methods have been developed to estimate the amount regeneration in a forest area. The most important methods are total stocking methods and the stocked-quadrat method.

Total stocking or *plot-count* methods use small fixed-area plots, such as a circular or square milacre plot. Tree counts on each plot are made by species, kind of regeneration, health, or condition. In more detailed studies, the heights and diameters may be recorded for each individual on a plot. Counts are then converted into per unit area values using appropriate expansion factors (eq. (8.13)). For example, if milacre (1/1000 acre) plots are used, the count is multiplied by 1000 to convert to per acre basis.

Counts of this kind are useful measures of the density of regeneration but are often insufficient. In addition to density, a complete characterization of regeneration should include some measure of spatial distribution. Loetsch et al. (1973) present a summary of methods to measure this characteristic. They describe an index of heterogeneity, which is the ratio of the variance to the population mean of the number of individuals per sample plot:

$$I_H = \frac{\sigma^2}{\mu} \tag{8.17}$$

where I_H = the index of heterogeneity

σ^2 = population variation of number of individuals per plot

μ = population mean of number of individuals per plot

If individuals are randomly distributed (i.e., Poisson distributed), $I_H = 1$. If $I_H < 1$ the distribution is more systematic (i.e., equally spaced). The smaller the value of I_H, the more regular the spacing. If $I_H > 1$, then the distribution is clustered or clumped.

Vertical point sampling (Section 11.3.2) is an overlooked alternative to using small fixed-area plots to obtain regeneration counts. Descriptions of the application of vertical sampling to regeneration assessment are found in Beers and Miller (1976) and Eichenberger et al. (1982).

In the *stocked quadrat* method, a sample plot is stocked if it has at least one seedling of the species of interest. The presence or absence of a tree on the plot, not the total number, is the primary focus of the stocked-quadrat method. The method emphasizes the evaluation of tree distribution rather than tree density. The basic idea is to divide an area into small squares of a size such that one tree per square represents full stocking at maturity (Fig. 8.11a). Stein (1984b) pointed out that the size of the quadrat can be determined as the reciprocal of the number of stems per unit area that constitute full stocking. Thus, if full stocking is 250 stems per acre, presence of trees should be checked on plots 1/250 acre in size; if full stocking is 2000 stems per hectare, plot size should be 1/2000 ha. The number of stocked quadrats in relation to the total number of plots gives the stocking percentage:

$$\%\text{Stocking} = 100 \frac{\text{Number of stocked quadrats}}{\text{Total number of quadrats}} \tag{8.18}$$

Plot size can also be determined from ideal spacing. For example, if full stocking is 2500 stems per hectare, then the ideal tree spacing is $\sqrt{10,000/2,500} = \sqrt{4} = 2\,\text{m}$. The corresponding plot size is (spacing)^2.

Establishing stocking plot in the ideal configuration shown in Figure 8.11a is time-consuming and nearly impossible in field conditions. As a result, most surveys use circular plots rather than square plots. However, if trees are arranged in a square or rectangular fashion, as is often the case in plantations or managed, young stands, the use of circular plots whose size equals (spacing)^2 (i.e., $r = \text{spacing}/\sqrt{\pi}$), can result in an underestimation of stocking because a circular plot can be placed in a fully stocked situation and yet contain no trees (i.e., not stocked, Fig. 8.11b and c). Using the "Law of the Unconscious Statistician" (Ross, 1980), one can show that the expected value of stocking obtained using plots with $r = \text{spacing}/\sqrt{\pi}$ is approximately 91% in a perfectly spaced, fully stocked stand. A solution to this would be to make the plot large enough so that no matter where it is placed in a perfectly spaced, fully stocked stand, it would always contain a tree. The minimum plot size so that this condition occurs is half the diagonal spacing ($r = \sqrt{2}/2$, Fig. 8.11d). While this solves the issue of underestimating fully stocked conditions, it results in increased overestimation in understocked situations because there is a high probability (approximately 57%) of a plot containing more than one tree in a perfectly spaced, fully stocked stand (Fig. 8.11e). Any plot radius between these two radii represents a tradeoff between underestimation at high stocking and overestimation at lower stocking.

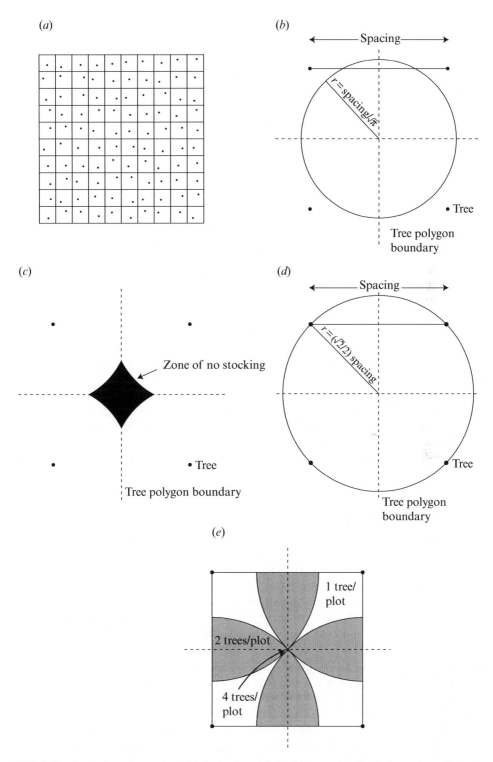

(*a*)

(*b*) Spacing

$r = $ spacing$/\sqrt{\pi}$

• Tree

Tree polygon boundary

(*c*) Zone of no stocking

• Tree

Tree polygon boundary

(*d*) Spacing

$r = (\sqrt{2}/2)$ spacing

Tree

Tree polygon boundary

(*e*) 1 tree/ plot

2 trees/plot

4 trees/ plot

FIG. 8.11. Stocked quadrat method: (*a*) the land area divided into quadrats such that each quadrat contains one tree, (*b*) a circular stocked quadrat plot with area equal to spacing^2 (i.e., $r = $ spacing$/\sqrt{\pi}$), (*c*) zone of no stocking when $r = $ spacing$/\sqrt{\pi}$, (*d*) a circular stocked quadrat plot with radius equal to half the diagonal spacing, and (*e*) sample geometry associated with (*d*).

Young stands often present very difficult conditions for field work. Naturally regenerated stands can have over 100,000 stems per hectare (40,000 per acre), and in plantations, smaller planted trees can be difficult to find when the level of competing vegetation is high. Larger plots can be very difficult to establish and time consuming to carefully search for acceptable trees, hence the reliance on small plots for regeneration surveys. The stocked quadrat method represents the extreme of this rationale, and, because plot size is small (targeting a single plot tree), stocked quadrat estimates are often highly variable. Forest managers, silviculturists, and other resource professionals often distrust density estimates because of this high variability (and often worry that a single plot with a large number of trees can "bias" density estimates). Methods such as the stocked quadrat were developed to place more emphasis on spatial distribution rather than density; however, these methods have complex sample geometries that often have unexpected results. Density estimates, where every acceptable tree is counted, are objective and will produce unbiased results. One of the methods described by Loetsch et al. (1973) can be used to assess spatial distribution (such as the index of heterogeneity shown in eq. (8.17)) or systematic sampling (Section 10.3) can be applied and the plots stratified by density classes and these strata mapped. However, small sample sizes, utilizing small plots, inevitably produce highly variable results.

8.8. SITE QUALITY

Within the context of a long tradition in forestry, *site quality* is usually interpreted to mean the average or potential productivity of a designated land area in terms of tree stem volume (e.g., Husch et al., 2003). However, as the goals of forest management have broadened, and as silviculture has become more sophisticated, there has been growing recognition of the need to characterize forest sites beyond the capacity for volume production. The diagnosis of site attributes that help predict not only growth, but also response to a variety of silvicultural treatments and natural stresses and disturbances, has become an important goal of forest site evaluation. The *site* is usually considered to be an area that is sufficiently homogenous in its environmental factors (such as climate, soil fertility, and soil moisture) that it is meaningful to treat it as a single unit for management or research purposes, although it must always be expected that sites will show some internal heterogeneity.

Site quality is usually evaluated on a stand basis. A common way of expressing relative site quality is to set up from three to five classes, or ordinal ranks, such as site I, site II, and site III, designating comparative productive capacities in descending rank. The characteristics of each class must be defined to enable any area to be classified. In many cases, such approaches use qualitative terms to define site classes, so that the ranking of a site may be quite simple in the field but may also be somewhat subjective. Thus, an important step forward has often been to introduce numerical definitions to improve the precision of site quality ranking. Attempts to refine classification systems so that they give an indication not only of productivity, but also of limiting or abundant site resources (e.g., water, nutrients) or response to management (e.g., prescribed fire) have proven useful in many regions.

Site quality can be evaluated in two general ways:

1. By the measurement of one or more of the individual site factors based on climate, terrain, landform, and soils. Such approaches are termed *geocentric* (Skovsgaard and Vanclay, 2008).

2. By the measurement of some characteristics of the plants (including trees or lesser vegetation) that are considered expressive of site quality. This approach assesses site quality from the effects of the environment on the vegetation. Such approaches are called *phytocentric*. When phytocentric approaches depend primarily on the dimensions and/or age of trees, they are also called *dendrocentric* (Skovsgaard and Vanclay, 2008).

Historically, dendrocentric approaches have tended to dominate operational forestry practice. However, multifactor systems that incorporate geocentric, phytocentric, and dendrocentric elements are becoming increasingly common in many regions.

8.8.1. Geocentric Approaches

Even when predicting tree volume or biomass productivity is the ultimate goal of a site quality system, a geocentric approach may have considerable appeal. Indeed, the earliest approaches to site classification in European forestry were geocentric and descriptive, and emphasized general descriptions of soils, species, and qualitative levels of productivity (Skovsgaard and Vanclay, 2008). Knowledge of the growth responses of forest trees to environmental factors is important to forest management. The forester can use this knowledge to encourage the establishment and growth of desirable species or genotypes, either by matching species and genotypes to sites with appropriate environmental characteristics, or by modifying the environment itself. In the past, foresters have been restricted in their inability to modify the environment on a given site in order to change stand density and structure. With the use of fertilization and irrigation practices, it is possible to modify site resources, especially soil fertility and moisture content. However, most application of site–growth relationships consists of encouraging the growth of a species on sites where its establishment is readily achieved or where its growth potential can be fully realized. An emerging area of work will be the need to anticipate changes in forest composition, tree growth, and silvicultural response with a variable and changing climate.

Unfortunately, despite its intuitive nature, the relationship between environmental factors and forest growth is notoriously challenging to measure. The factors of the site and the plants themselves are interacting and interdependent, making it difficult to assign cause and affect relationships. The environmental factors of the site can be grouped into edaphic, climatic, topographic, and competitive components. The most important elements of each group are shown below:

Edaphic. Soil depth and texture, nutrient levels, pH, moisture, drainage.
Climatic. Air temperature, precipitation, humidity, length of growing season, available light, wind.
Topographic. Altitude, slope angle, position and length, aspect.
Competitive. Other trees and vegetation.

A considerable amount of effort has been directed toward investigating the characteristics of the soil in an attempt to find a single environmental factor to serve as a reliable indicator of site quality. This approach has often led to simple, practical tools for site quality evaluation, but frequently leaves a sizable amount of the variation in tree growth or other key responses unexplained. To understand fully the growth of trees in relation to the environment, individual site factors cannot be studied in isolation. The interdependencies and influences of the other factors may be masked and, consequently, not recognized. Even if the primary interest is the study of one factor of the site, it must be done with recognition of the effects of other factors.

The usual form of a soil-site quality investigation has taken a dendrometric characterization of site quality (usually tree biomass, stem volume, or dominant tree height, at a given age) as the reference value that should be predicted from environmental factors. Statistically, such studies usually consist of a multiple regression analysis with biomass, volume, or height as the dependent variable and a number of soil and other environmental characteristics as the independent variables. Sometimes, site index (Section 8.8.3) is the dependent variable; a goal for such a study would be to predict site index on sites where it cannot be measured directly. A commonly used model has been

$$\log H = b_0 + b_1 \left(\frac{1}{A}\right) + b_2 (B) + b_3 (C) + \cdots + b_n (N)$$

where H = height

A = age

$B, C, \ldots, N$ = soil or other environmental factors

b_i = constants to be determined by regression analysis

A major difficulty in this type of analysis pertains to the numerical coding of qualitative variables to use in the regression analysis. Such a procedure can lead to difficulty because of the arbitrary assignment of numbers to discrete variables and subsequent computations, by assuming a continuous scale, equal units are implied that may not represent the actual relationship between the original discrete classes. The thoughtful assignment of numerical codes to qualitative variables will make the subsequent regression analysis more meaningful. For an example of this type of coding, refer to Beers et al. (1966) and Stage (1976). Alternatively, statistical techniques appropriate to ordinal variables can be used (e.g., classification and regression trees or random forest techniques).

Geocentric approaches are especially common in the ecological literature, where they are used to study forest productivity over large areas and to assist in drawing inferences about the effects of land use, management, or other disturbances on regional carbon and nutrient cycles. For example, Zarin et al. (2001) predict total aboveground biomass B (T/ha) for secondary forests in the Amazon basin as

$$B = -131.05 + 34.70 \ln\left[(\text{GST} \times \text{GSL})A\right]$$

for forests on sandy soils, and

$$B = 0.67 + 0.43(\text{GST} \times \text{GSL})A$$

for forests on nonsandy soils, where

GST = mean growing season temperature, °C

GSL = growing season length, as a fraction of the year

A = stand age, years

Thus, the combination of soil texture (sandy or nonsandy) with the climatic variable GST × GSL defines a simple, empirical site quality system, albeit one that has relatively large errors for site-specific predictions (Zarin et al., 2001). Nonetheless, it can be used to detect and benchmark influences such as that of repeated burning during the agricultural phase on subsequent forest regrowth (Zarin et al., 2005).

Advantages of geocentric approaches include potentially direct prediction of forest productivity from relevant environmental factors, including some (such as landform and climate) for which there is now readily available geospatial information in many parts of the world. However, detailed assessments of soil structure, chemistry, and soil moisture can be labor-intensive and expensive, and the challenges associated with determining which specific environmental factors are the best predictors of forest productivity cannot be overstated. Site quality systems based on simple, easily-obtained geocentric data are often associated with large prediction errors, and are often at their best in describing broad regional trends in productivity rather than in site-specific assessments for local management purposes.

8.8.2. Phytocentric Approaches Using Vegetation Composition

An alternative approach to site quality and site characterization recognizes that the vegetation on a site is growing in response to many of the site attributes (including climate, soil moisture and fertility, and past disturbance) that may influence future tree growth and response to management or disturbance. In European forestry practice, there is a long tradition of predicting potential forest yield on the basis of vegetation composition, with particular focus on understory plants (e.g., Cajander, 1926). The approach has also been widely used in the United States within specific regions or research contexts, though dendrometric approaches (especially site index; Section 8.8.3) remain dominant in practical forestry (Carmean, 1975). In Canada, however, multifactor systems that rely heavily on vegetation composition for fine-scale local diagnoses are widely used (e.g., Pojar et al., 1987).

One of the strongest advocates of site quality diagnosis based on plant community composition in North America was Rexford Daubenmire. Daubenmire (1976) argued that vegetation, and in particular the species composition of late-successional forests, reflected the sum of the environmental factors that governed plant growth and development. He also argued vehemently against the dominance of the site index paradigm (Section 8.8.3). Although it is tempting to view Daubenmire's approach, with a strong emphasis on "climax" forest communities, as somewhat quaint given the move away from climax concepts in current views of forest dynamics (e.g., Oliver and Larson, 1996), as well as the fact that many managed forest landscapes are comprised of stands that are too young (where even-aged management dominates) or too frequently disturbed (where uneven-aged management is the norm) for understory vegetation to come into complete equilibrium with site factors. However, investigations of species composition in stands of advanced development, coupled with examination of underlying soil relationships, have led to useful systems of site description that have proven to be practical in predicting site index, yield, and developmental trends in younger stands (e.g., Leak, 1976, 1978, 1982). Both existing and potential vegetation are viewed as important site descriptors in national efforts to map terrestrial ecological units in the United States (Brohman and Bryant, 2005; Winthers et al., 2005).

Phytocentric approaches that depend on species composition are attractive in their foundation on ecological linkages between plants and the physical components of the ecosystem. Such approaches may have value for a range of decisions in forest management and silviculture, and not just for predicting potential growth rates and yield levels. However, numerous objections have been raised to their widespread use. Perhaps the most serious of these are the need for some training of foresters and ecologists in the identification of nontree vegetation, and the fact that some species with potential value as indicators may not be evident throughout the year (e.g., when not flowering or when buried in snow). The need for expert and occasionally subjective judgment is a limitation to phytocentric, composition-based methods. However, other objections raised by authors such as Carmean (1975), including the possibility of making similar assessments using soils or topography, potential influence of overstory vegetation on the understory community, and deeper rooting among trees than among lower-stature plants (leading to responses to different environmental signals), do not seem as serious in practice (Daubenmire, 1976). If at least some potential indicator plants are robust to variation in shade and to recent disturbance, and if foresters and ecologists have sufficient training to pick these indicators out among the many species that may have little value as indicators, then evidence suggests that composition-based methods can be a powerful tool, especially when used in concert with other geocentric and dendrocentric approaches (e.g., Pojar et al., 1987).

8.8.3. Dendrocentric Approaches

When the emphasis or sole focus of site quality assessment is productivity of tree volume or biomass, its most direct measures are the quantity of wood grown on an area of land within a given period or mean annual increment of the stand. This has led to a long-term emphasis on dendrocentric approaches since the shift away from qualitative geocentric approaches early in the history of European forestry (Skovsgaard and Vanclay, 2008). In unthinned, even-aged stands, or in those that had been thinned only lightly and from below, standing volume in relation to age was used to establish site quality classes. Tesch (1980) suggests that this practice persisted in Europe in the late 1800s because so many forests were well-regulated with uniform density. However, as thinning practices became more diverse in Europe, and as foresters in North America, Australia, and New Zealand encountered unmanaged forests reflecting a wide range of stocking levels, the use of height as a proxy and correlate for volume production became much more popular. Initially, the approach used to classify sites by height paralleled that used to classify them by volume. For example, one might obtain measurements of age and height (or volume per hectare) for a large number of stands of a particular species in a region, and plot age on the x-axis and height (or volume per hectare) on the y-axis. Curves would be drawn by hand to reflect the upper envelope or boundary of the points and the lower envelope or boundary. Then, a series of curves would be drawn at equal spacings between the two outer curves and sharing the same shape to divide the sites into the desired number of classes. The curves were assumed to reflect the trajectories of stands through time within in the different site classes (Graves, 1906). Later, height curves came to be labeled with the height value reached by the curve at a standard "base" age. For example, if the base age is 100 years and a curve passes through a height of 25 m at age 100, then the value assigned to the curve is 25. That value is called the *site index*. Site index has the clear advantage of allowing site quality to be represented as a continuous value and not just a small number of discrete classes.

Once a set of curves has been developed for a particular species and region, site index can be evaluated in any stand (or, at least, any even-aged stand) by measurement of height and age of the appropriate trees in the stand, and simple comparison with the curves. Thus, it has become the dominant method of expressing site quality in many operational forestry settings.

The use of tree height to classify or index site productivity rests on a series of assumptions, reviewed by Skovsgaard and Vanclay (2008). Foremost among these is that height growth of the trees used for site evaluation reflects, and has reflected, potential growth of trees of that species on that site. This assumption is approximately true for dominant and codominant trees in an even-aged, monospecific stand: height growth is much less sensitive to competition than diameter growth across a reasonably wide range of stand densities. Thus, the height growth of those trees is assumed to reflect limitations imposed by the site and the biology of the trees themselves.

The development and practical use of site index curves depends on mensurational choices about the age and height to be used. Age at breast height is commonly used in site index curves for naturally regenerated stands, while total age is sometimes used for curves developed for plantations (where age may be known from records of planting). Husch (1956) has pointed out several advantages of using age at dbh instead of total age. Measuring age at dbh eliminates the necessity of adding arbitrary corrections to convert age at increment-boring level to total age; it measures tree age after the initial period of establishment and adjustment has passed; it also utilizes a standard and conventional point for age determination. The height of the dominant and codominant trees has usually been taken as representative of stand height for site index work. This choice is not without its limitations. It relies on subjective decisions about which trees are dominants and codominants. In addition, the crown position of a tree and its height growth may be affected by stand alterations, such as cuttings. Other stand measures such as the average height of a specified number of the largest trees (e.g., top height) or the height of the tallest tree have been suggested as alternative parameters. Unless otherwise specified, site index is generally defined as the average height that the dominant and codominant trees on an area will attain at key ages such as 50 or 100 years. For example, in imperial units, site index 70 on a 50-year basis means that the dominants and codominants reach an average height of 70 ft in 50 years. Site index 120 on a 100-year basis means an average height of 120 ft in 100 years. Site index curves are prepared for even-aged stands as shown in Figure 8.12 to allow site classification for a stand at any age.

To assess the site quality of an area, it is necessary to determine the average height of the dominant and codominant trees and their average age. The position of these coordinates is then located on the site index chart for the species. The site index for the stand can be read from the closest curve. For example, a stand with an average height of the dominants and codominants of 65 ft and 40 years of age would have a site index of 75 for white pine in the southern Appalachians.

It is important to understand that site index varies according to species. Site index charts are prepared for individual species or for typical forest types, such as the charts prepared by Schnur (1937) for mixed oak forest in the Central States of the United States. A single forest area may have different site index values, depending on the species and site index chart. To help solve this problem, regression formulas can be derived that relate the site index or height of one species with that of another. Examples of this type of study are those of Foster (1959), Deitschman and Green (1965), and Norman and Curlin (1968).

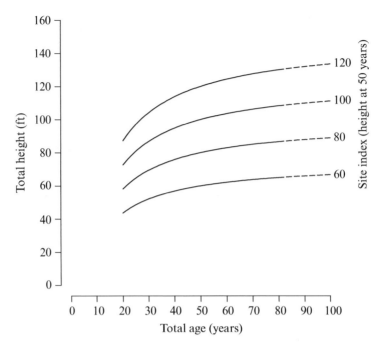

FIG. 8.12. Site index curves for natural stands of white pine in the southern Appalachians (adapted from Doolittle and Vimmerstedt (1960)).

The relationship of age and total height expressed by site index has enjoyed widespread popularity for several reasons. Height has been found closely correlated to the ultimate measure—volume. In addition, the two requisite measurements, height and age, are relatively quick and easy to determine. It is generally considered that height growth is only slightly affected by stand density, although some studies have shown that stand height and site index in some cases are related to stand density (Gaiser and Merz, 1951; Husch and Lyford, 1956). Finally, site index has been popular because it provides a numerical expression for site quality, rather than a generalized qualitative description.

In uneven-aged stands, height in relation to age cannot be used to express site quality. The height growth of a species in this type of stand is not closely related to age but more to the varying stand conditions by which it has been affected during its life. The same may also be true of trees that spent their early years in a subordinate stratum of a stratified, single-cohort mixture. The classic concept of site index has consequently been restricted to species normally occurring as dominants in even-aged stands. McLintock and Bickford (1957) considered several alternatives for evaluating site quality in uneven-aged stands in their study of site quality for red spruce in the northeastern United States. They concluded that the relationship between height and dbh of the dominant trees in a stand was the most sensitive and reliable measure of site quality. Site index according to this concept is then defined as the height attained by dominant trees at a standard dbh. The site index for a tree of any age can be read from the height–dbh curves for the range of site indices as shown in Figure 8.13. This chart utilizes a standard dbh of 14 in.

Although the use of height–diameter curves in a fashion analogous to height–age curves is appealing in its simplicity of measurement, tree diameter is believed to be even more

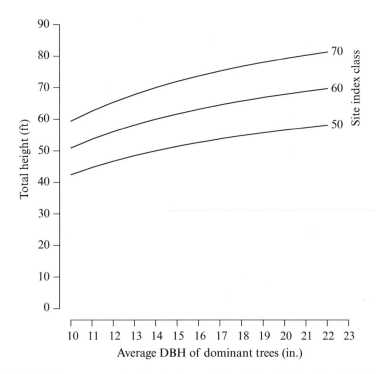

FIG. 8.13. Site index for uneven-aged stands of red spruce (adapted from McLintock and Bickford (1957)).

sensitive to competition than tree height. Thus, the stability of such a dendrometric approach for uneven-aged stands is open to question, and its adequacy for any particular management purpose should be tested against data on forest productivity. Geocentric or phytocentric composition-based approaches may be more suitable in landscapes dominated by uneven-aged stands or complex species mixtures.

8.8.4. Preparation of Site Index Curves

The preparation of site index curves for even-aged stands is based on average height and age measurements of the dominant and codominant trees on a series of sample plots. When temporary sample plots are used, there should be a sufficient number and distribution to cover equally the range of age and site classes found under natural conditions. For reliable relationships, a total of at least 100 plots is necessary, although more are desirable.

Site index curves can be prepared using anamorphic or polymorphic techniques. The anamorphic techniques assume that the curves of height over age for different sites all have the same form and shape. This assumption does not always hold so that anamorphic site index curves may not represent the true forms of curves for different site indexes. Polymorphic site index curves eliminate this limitation by showing different shapes of curves for different site indices.

In the past, graphical techniques have been used to construct the desired series of ana- morphic site index curves. In very brief form, this method graphed average height over age found on each sample plot on graph paper. A curve was then fitted freehand to these plots (see Husch (1963) for freehand curve fitting). The average site index for all plots is the height as read from this curve for the key age. Based on the average curve, curves for other site index classes are constructed. The assumption is made that these other curves will have a trend and shape similar to the average site index curve. Under this assumption, the other curves are spaced above or below the average curve in proportion to their height at the key age. For example, the average site index curve might have a height of 65 ft at a 50-year key age. The site index 60 curve would be $[(65-60)/65]100 = 7.7\%$ below the average site index curve. The ordinates of the site index 60 curve would be obtained by reducing the ordinate of the average curve at any age by 7.7%. The same procedure would be followed for all other site index curves.

In more modern times, regression techniques have been employed to remove the subjectivity involved in the freehand fitting of curves. One commonly used approach is described in the following paragraphs. Using field plot data involving total tree height and tree age (plot averages or individual tree data) obtained from selected dominant and codominant trees, site index curves similar to those shown in Figure 8.12 can be derived. The procedure for preparing anamorphic site index equations is as follows[‡]:

1. Transform the height and age data to logarithm of height H and reciprocal of age A and fit a simple linear regression, obtaining numerical values for the constants a and b in the model:

$$\log H = a + b\left(\frac{1}{A}\right) \tag{8.19}$$

   The resulting equation represents the average site index curve for the data.

2. Locate points needed to draw a specified site index curve by moving the Y-axis to the key age (by mathematical translation) and then, by definition, the Y-intercept is the log of site index S. For example, if the key age is 50 years, eq. (8.19) is translated to become

$$\log H = \log S + b\left(\frac{1}{A} - \frac{1}{50}\right) \tag{8.20}$$

   For example, if the regression coefficients for eq. (8.19) are $a = 1.950$ and $b = -4.611$, then the translated equation for key age of 50 years is

$$\log H = \log S - 4.611\left(\frac{1}{A} - \frac{1}{50}\right)$$

$$= \log S - \frac{4.611}{A} + \frac{4.611}{50}$$

$$= \log S - \frac{4.611}{A} + 0.09222.$$

   The curve for each site index is found by substituting in the desired site index value for S and computing the height for various ages. For example, the curve for site index 80 would be

[‡]This technique is referred to as the *guide curve* method (Clutter et al., 1983).

$$\log H = \log 80 - \frac{4.611}{A} + 0.09222$$

$$= 1.90309 - \frac{4.611}{A} + 0.09222$$

$$= 1.99531 - \frac{4.611}{A}.$$

3. Locate other site index curves by substituting the pertinent site index number into eq. (8.20) and proceeding as described in step 2.

It is often useful to calculate site index from the fitted equation rather than to read a graph. For this purpose, eq. (8.20) is rearranged by solving for the log site index to become

$$\log S = \log H - \frac{b}{A} + \frac{b}{A_S} \qquad (8.21)$$

where A_S is the key age. For the example used in step 2, the resulting equation is

$$\log S = \log H + \frac{4.611}{A} - \frac{4.611}{50}$$

$$= \log H + \frac{4.611}{A} - 0.09222.$$

For a given plot, one can substitute values of average tree height (as a logarithm) and average tree age and obtain logarithm of site index; then, by using antilogs, site index can be estimated.

Polymorphic site index curves fall into two broad classes: disjoint and nondisjoint. *Disjoint polymorphic curves* are a set of site index curves that, even though the form and shape of each individual site index curve differs, the individual lines representing the different site indices do not cross within the age range of interest. *Nondisjoint polymorphic curves* do cross one another within the age range of interest. Clutter et al. (1983) discuss the procedures for fitting polymorphic disjoint site index curves. Newberry and Pienaar (1978) is an example of fitting polymorphic nondisjoint curves. Cieszewski and Bailey (2000) describe a generalized approach that can be used to generate and fit polymorphic site index curves having a range of desirable properties. Their generalized algebraic difference approach is the foundation for a great deal of current work in site index curve development.

8.9. DENSITY AND STOCKING

Measures of stand density and forest stocking are both used to depict the degree to which a given site is being utilized by the growing trees or simply to indicate the quantity of wood on an area. However, a distinction is usually made between the two terms. Gingrich (1967) describes the distinction in this way:

Stand density is a quantitative measurement of a stand in terms of square feet of basal area, number of trees or volume per acre. It reflects the degree of crowding of stems within the area. *Stocking,* on the other hand is a relative term used to describe the adequacy of a given stand density in meeting the management objective. Thus, a stand with a density of 70 square feet of basal area per acre may be classified as overstocked, or understocked, depending on what density is considered desirable.

Stand density can be expressed, as above, in absolute units per unit land area according to such stand parameters as volume, basal area, crown coverage, number of trees, and so on, but often it is expressed on a relative scale as a percent of "normal" (either full or desirable) density or as a percent of the average density. When expressed in this form, it should be clear that "relative density" exists as a term transitional between "stand density" and "stocking" as above.

A major goal in describing stand density is usually to predict one or more attributes of the production ecology of the stand. Gingrich writes of "the degree of crowding"; stand density measures are often used to draw inferences about the degree of utilization of site resources, the productivity of a stand in relation to site potential, and losses due to competition-induced mortality and self-thinning. Stand density measures are also used to infer aspects of individual tree growth and form, including crown dimensions that impact lumber grades and yields (Long, 1985). Increasingly, stand density measures are also used to predict aspects of structure for wildlife habitat and stand vulnerability to insects and disease (e.g., Anhold and Jenkins, 1987; McTague and Patton, 1989; Moore and Deiter, 1992; Woodall et al., 2005). In this context, a density measure must strike a balance between simplicity in its input variables and transparency in its calculation, on the one hand, and the desire to capture potentially important ecological detail, on the other.

Many of the density measures used in practice share some algebraic similarities, and some can be derived from theories of growth and yield (e.g., Sterba, 1975) or from principles of stem mechanics (Dean and Baldwin, 1996). Some have mathematical forms that make it easier to deal with statistical aspects, such as the computation of confidence limits (Ducey and Larson, 1997, 1999). However, these differences do not explain the regional variation in their use. For example, Reineke's stand density index (SDI) (Section 8.9.2) is widely used in the western United States, while methods originally derived from CCF (Section 8.9.4) are used for the same purposes in the eastern United States. Diagrams derived from the −3/2 rule (Section 8.9.8) are common in Canada, while relative spacing (Section 8.9.5) is much more commonly applied in Europe. All of these density measures are tools to allow foresters and ecologists to gain insights into tree growth and stand dynamics from a simplified set of measurements; their practical intent is similar, and the primary difference between them is the choice of tree size parameter and algebraic form (Curtis, 1970).

8.9.1. Relative Density Based on Volume

For many forestry purposes, volume has been the ultimate expression of stand density. Relative density in terms of volume is determined by comparing the volume of an observed stand with the volume of some standard, such as the volumes of fully stocked stands of the same species for specified ages and site qualities as given in a yield table. For example, a 50-year-old stand of white pine with a site index of 58 in Massachusetts was measured and its volume was found to be 4500 board feet/acre. A normal yield table showed theoretically that the full stocking for this age and site was 6290 board feet. The relative density is then $100(4500/6290) = 71.5\%$. Note that relative density percentages using volume depend on the volume unit chosen. For example, board-foot density percentages will differ from cubic-foot density percentages for the same stand.

Density as measured by relative volumes has several disadvantages. Volumes as expressed for the standard or fully stocked stand may be based on different merchantability limits, different log rules, or different volume units from the stand under investigation, making valid comparison difficult. In addition, stand volume estimates are too expensive if only a measure of relative density is needed. For example, an advantage of basal area is that it is easily determined and is quite consistent for fully stocked stands of specified species, ages, and sites that have the same quadratic mean diameter. Just as relative density depends on the unit of volume used, even for the same stand, relative density using basal area will not necessarily be the same as that using volume.

8.9.2. Stand Density Index

The number of trees per unit area can be used as another measure of stand density. At any age, there can be a wide range in the number of trees per unit area so that frequency by itself is of little value. For a useful descriptive measure of stand density, number of trees must be qualified by tree size. The stand structure as shown in a stand table (Table 8.2) describes this, but is too cumbersome for practical use as a stand density description. A useful measure of density for even-aged stands based on number of trees is *Reineke's* SDI (Reineke, 1933). Reineke developed his SDI as part of an empirical investigation into the relationships between variables in normal yield tables for monospecific conifer stands in North America. SDI can be interpreted as the number of trees per unit area that a stand would have at a standard average dbh. In the imperial system, the standard dbh is generally 10 in. and in SI the standard dbh is usually taken as 25 cm.

Historically, the easiest way to calculate SDI for a stand was by referring to a SDI chart for the species. As shown in Figure 8.14, the chart consists of a series of lines representing the relationship between number of trees per acre and average stand diameter. The chart can be constructed by plotting the logarithm of number of trees per acre over the logarithm of the average stand dbh on rectangular coordinate paper. Alternately, as shown in Figure 8.14, the natural values can be plotted on logarithmic paper yielding the same results. The number of trees and the average stand diameters are obtained from a series of sample plots. Average stand diameter is taken as the diameter of the tree of arithmetic mean basal area (i.e., quadratic mean diameter). Reineke (1933) defines the maximum SDI relationship as

$$\log N = b \cdot \log \bar{D}_Q + a \tag{8.22}$$

where N = stand density (trees per unit area)

$\bar{D}_Q$ = quadratic mean diameter (diameter of the tree of average basal area)

Note that eq. (8.22) does not hold for all stands in general, only for those that are at normal stocking. Reineke (1933) found that the b constant was -1.605 for several species

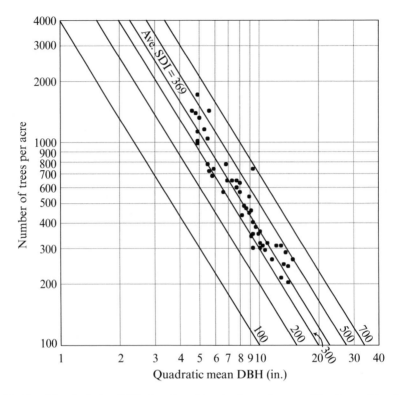

FIG. 8.14. Stand density index (SDI) chart for white pine in southeastern New Hampshire. Equation is $\log(N) = -1.598 \cdot \log\left(\bar{D}_Q\right) + 4.165$. Based on 53 sample plots (adapted from Husch and Lyford (1956)).

and was independent of site quality and age. Other investigators note that the linear relationship expressed by the equation holds for many species and that the slope (*b*) differs little, although the constant a (i.e., the intercept) varies considerably.

An average SDI relationship for the observations is obtained by fitting the stand data to eq. (8.22) using regression analysis. This line represents the average stocking of all plots (Where the maximum stocking is desired, either quantile regression or stochastic frontier analysis should be used in place of ordinary regression.) (Bi, 2004; Weiskittel et al., 2009a; Zhang et al., 2011). For the data shown in Figure 8.14, the ordinary regression line is $\log N = -1.598 \cdot \log \bar{D}_Q + 4.165$. The number of trees indicated by the intersection of this line and the ordinate at the standard dbh is the average SDI. In Figure 8.14, the 10-in. ordinate is taken as the standard. The average line intersects the 10-in. ordinate at $N = 369$, so the average SDI is 369. A series of parallel lines is then constructed to intersect the standard diameter ordinate at specified numbers of trees per acre. The numbers at these intersections are the SDI values for the set of parallel lines.

Any two stands falling on the same line have the same SDI (i.e., the same relative density). For example, in Figure 8.14, a stand with 300 trees/acre and a quadratic mean dbh of 8 in. and a stand with 900 trees/acre and a quadratic mean dbh of 4 in. would both have an SDI = 200 (i.e., the same relative density as a stand with 200 trees/acre and a quadratic mean dbh of 10 in.).

The SDI for any stand can be determined by plotting the position of the observed number of trees/acre and the quadratic mean dbh on the stand density chart for the species. The SDI is indicated by the closest line to the plotted point or can be found by interpolation between the index lines. Alternatively, the SDI can also be calculated from the formula

$$\log \text{SDI} = \log N - b \cdot \left(\log \bar{D}_Q - \log \bar{D}_I \right) \tag{8.23}$$

where   SDI = Reineke's SDI

   N = number of trees per unit area

   $\bar{D}_Q$ = observed quadratic mean diameter (diameter of the tree of average basal area)

   $\bar{D}_I$ = the standard diameter

The formula is derived from eq. (8.22) by setting $\bar{D}_Q$ equal to 10. Consequently, in this case $\log N = \log \text{SDI}$. For the example, a stand with 265 trees/acre and a $\bar{D}_Q = 14.6$ would have an SDI of

$$\begin{aligned}
\log \text{SDI} &= \log 265 - (-1.598) \cdot \left(\log 14.6 - \log 10 \right) \\
&= 2.423 + 1.598 (1.164 - 1) \\
&= 2.423 + 1.598 (0.164) \\
&= 2.686 \\
\text{SDI} &= 10^{2.686} \\
&= 485
\end{aligned}$$

Provided the maximum SDI for a given species is known, Long (1985) suggests some standard interpretations of SDI values. For example, initial crown closure is believed to occur at approximately 25% of maximum SDI, full site occupancy at 35%, and the onset of self-thinning at approximately 60%. These thresholds are only rough guidelines, however, and should be evaluated on a species-by-species basis.

8.9.3. Tree–Area Ratio

The tree–area ratio is a measure of density proposed by Chisman and Schumacher (1940) as a measure independent of stand age and site quality and appropriate for even- or uneven-aged stands. It is based on the concept that if the space on the ground Y, occupied by a tree of diameter at breast height d, can be expressed by the parabolic relation

$$Y = b_0 + b_1 d + b_2 d^2 \tag{8.24}$$

then the total area of growing space represented on a plot or unit of ground area can be found by summing over all the trees on the plot or unit area. Thus, a measure of growing space utilization can be obtained by adding over all trees (n) on a unit area, leading to

$$\text{Tree area} = b_0 n + b_1 \sum d + b_2 \sum d^2 \tag{8.25}$$

After the constants b_0, b_1, and b_2 are obtained by least squares, the contribution to tree area of a single tree of diameter d can be found by letting $n = 1$, $\Sigma d = d$, and $\Sigma d^2 = d^2$.

Using data obtained from sample plots adjusted to the appropriate unit area basis and setting the tree area value to 1, the constants in eq. (8.25) are obtained using regression analysis. If the data used to derive the coefficients b_0, b_1, and b_2 in eq. (8.25) came from stands that were deliberately chosen to be "fully" or "normally" stocked, then the application of the regression equation to other stands and substitution of n, Σd, and Σd^2 will provide a tree–area figure that will reflect the proportion of *full stocking* demonstrated by that stand. On the other hand, if the data used to derive the coefficients came from stands having a range of densities, substitution into the equation for a given stand will reflect the proportion of stocking compared to the *average stocking* of the basic data. Note that the "growing space" identified in the tree area ratio need not conform to crown area or any other physically occupied region; rather, it represents a proportion of full site occupancy as defined by the data selected for fitting the equation. Curtis (1971) suggested replacing the parabola of eqs. (8.24) and (8.25) by a power function, in which case the density measure becomes superficially similar to Reineke's SDI.

8.9.4. Crown Competition Factor

A measure of stand density, which in final form is similar to the tree–area ratio, although considerably different in derivation, is the CCF proposed by Krajicek et al. (1961). The CCF is considered independent of site quality and stand age and can be used in both even- and uneven-aged stands.

The development of a CCF formula for a given species or species group would proceed as follows, using the example given by Krajicek et al. (1961).

1. Measurements of crown width (CW) and dbh are taken on a satisfactory number of truly open-grown trees, carefully selected to ensure that they have developed in an undisturbed, competition-free environment.

2. For this measured sample, the relationship between CW and dbh (d) is found by least squares. For example, Krajicek et al. (1961) used the relationship $CW = b_0 + b_1 d$ and obtained the equation $CW = 3.12 + 1.829d$.

3. A formula for the crown area of individual trees expressed as a percent of unit area (A) is derived and is called *maximum crown area* (MCA) since it indicates the maximum proportion of an area that the crowns of trees of a given dbh can occupy:

$$MCA = 100 \left(\frac{\pi (CW)^2}{4} \right) \left(\frac{1}{A} \right) \tag{8.26}$$

In the imperial system, $MCA = 0.0018(CW)^2$, in SI $MCA = 0.0078(CW)^2$.

4. The regression relating CW to dbh is then substituted into eq. (8.26) to obtain an expression of MCA as a function of dbh:

$$\mathrm{MCA} = 100 \left(\frac{\pi \left(b_0 + b_1 d \right)^2}{4} \right) \left(\frac{1}{A} \right)$$

$$= \left(\frac{100\pi}{4A} \right) \left(b_0^2 + 2b_0 b_1 d + b_1^2 d^2 \right)$$

In Krajicek et al.'s example:

$$\mathrm{MCA} = 0.0018 \left(3.12 + 1.829d \right)^2$$
$$= 0.0175 + 0.02050d + 0.0060d^2$$

5. By adding the MCAs for all trees on a per unit area basis of forestland, an expression of stand density, called CCF, is obtained:

$$\mathrm{CCF} = \sum_{i=1}^{m} \left(n_i \mathrm{MCA}_i \right)$$

$$= \sum_{i=1}^{m} \left(n_i \left(\frac{100\pi}{4A} \right) \left(b_0^2 + 2b_0 b_1 d_i + b_1^2 d_i^2 \right) \right) \qquad (8.27)$$

$$= \left(\frac{100\pi}{4A} \right) \left[b_0^2 \sum_{i=1}^{m} n_i + 2b_0 b_1 \sum_{i=1}^{m} d_i n_i + b_1^2 \sum_{i=1}^{m} d_i^2 n_i \right]$$

where   CCF = crown competition factor

     m = number of dbh classes

     n_i = the number of trees per unit area in the ith dbh class

     d_i = the midpoint diameter of the ith dbh class

     b_j = the CW coefficients

     A = unit area (43,560 ft^2 or 10,000 m^2)

For the example above, the CCF for a specific stand is determined by

$$\mathrm{CCF} = \left(0.0175 \sum n_i + 0.0205 \sum d_i n_i + 0.0060 \sum d_i^2 n_i \right) \qquad (8.28)$$

In eq. (8.28), it is assumed the n_i are expressed on a per acre basis. If the n_i are for the total stand, then eq. (8.28) must be divided by stand area to obtain CCF on a per acre basis.

Any combination of MCAs that sum to 100 (CCF = 100) represents a closed canopy and reflects a situation where the tree crowns just touch and are sufficiently distorted to completely cover each acre of ground. However, the authors of this index state:

It should be emphasized that CCF is not essentially a measure of crown closure. Theoretically, complete crown closure can occur from CCF 100 to the maximum for the species (e.g., in oaks, approximately CCF 200). Instead of estimating crown closure, CCF estimates the area available to the average tree in the stand in relation to the maximum area it could use, if it were open grown.

8.9.5. Relative Spacing

Another expression of stand density is *relative spacing*. RS is the average spacing between trees, assuming square spacing, divided by the average height of the dominant trees:

$$RS = \frac{\sqrt{A/N}}{H_D} \tag{8.29}$$

RS yields numbers expressed as a fraction, such as 0.2 or 0.15, and the value declines as density increases. Thus, for even-aged stands, *RS* decreases as the stand grows older and the number of trees decreases due to mortality. Relative spacing was originally devised by Hart (1928) and has been widely used to formulate stocking guidance in Europe. In North America, Wilson (1946, 1951) advocated for Hart's ideas, which were nonetheless quickly dismissed as a "rule of thumb." However, relative spacing has a certain appeal in its geometric simplicity and can be a strong predictor of crown shape and form (Ducey, 2009b). RS is also closely related to the tree count in vertical point sampling (Section 11.3.2).

8.9.6. Density of Mixed-Species and Complex-Structure Stands

Most of the density measures described so far were originally developed for use in even-aged monocultures, and their application to mixed species stands, those with complex structure, and those with multiple cohorts or age classes has been challenging. Most efforts to quantify the density of such stands have focused on indices in which the total density is the sum of additive contributions of individual trees. For example, Long and Daniel (1990) suggested replacing Reineke's SDI with the power-form version of tree area ratio proposed by Curtis (1970) in uneven-aged stands:

$$ASDI = \sum_i N_i \left(\frac{D_i}{25}\right)^{1.6} \tag{8.30}$$

where the summation is over diameter classes or categories of trees, with N_i and D_i being the number of trees per hectare in each class and their diameter, respectively. This version of SDI ("additive SDI") forms the core of the mixed-species indices proposed by Woodall et al. (2005) and Ducey and Knapp (2010). Rivoire and Le Moguedec (2012) provide an extended analysis that incorporates site effects and yields a similar formulation but allows the exponents for individual species to vary. Alternatively, Sterba and Monserud (1993) employed a derivation that would give rise to Reineke's SDI in simple stands, but applied an adjustment for the shape of the diameter distribution. Their results give rise to a more complex and quite different formulation in mixed stands.

8.9.7. Point Density and Competition Indices

The previously described measures of density are usually employed to determine density of a stand in general or "on average." More specific measures of density have been developed to describe the degree of competition at a given point or tree in the stand. These measures have been referred to as either point density estimates or competition indices. The basic idea of these indices is to describe the degree to which growth resources (light, water, nutrients, and physical growing space) available to an individual tree are limited by neighboring trees. The increased interest in modeling individual tree growth and yield has produced a number of competition indices. These indices can be broadly classified into two categories: (1) distance independent measures and (2) distance dependent measures.

Distance independent measures describe the competitive status of a tree or class of trees relative to all trees in the stand. An example of a distance independent index is the basal area index proposed by Glover and Hool (1979):

$$G_i = \frac{\pi \left(D_i/2 \right)^2}{\pi \left[\left(\sum_{j=1}^{n} D_j / n \right) / 2 \right]^2} = \frac{D_i^2}{\bar{D}^2} \tag{8.31}$$

where G_i = the basal area index for the ith tree
          D_i = the diameter of the ith tree
          $\bar{D}$ = the mean plot or stand diameter

An alternative expression of this index is the ratio of a tree's basal area to the mean basal area (Daniels et al., 1986):

$$G_{Bi} = \frac{BA_i}{\overline{BA}} = \frac{D_i^2}{\left(\sum_{j=1}^{n} D_j^2 \right) / n} = \frac{D_i^2}{D_Q^2}.$$

Other indices have been calculated based on ratios of tree height to mean height, tree height to dominant height, and tree volume to mean volume. The main advantage of distance independent indices is that time-consuming measures of tree location are not required. The main disadvantage is that these indices measure a tree's status relative to average stand conditions rather than the immediate conditions surrounding the tree.

Distance-dependent indices attempt to describe a tree's competitive status based on the immediate conditions surrounding the tree. Distance-dependent indices fall into three broad classes (Tomé and Burkhart, 1989): (1) area overlap indices, (2) distance-weighted size ratios, and (3) area potentially available indices.

Area overlap indices are based on the idea that each tree has a potential area of influence over which it obtains or competes for site factors (Opie, 1968). All trees whose area of influence overlaps with a subject tree's area of influence are considered competitors (Fig. 8.15a).

Spurr's (1962) point density is an example of an area overlap index. Spurr (1962) adapted Bitterlich's point basal area to be used as a point density measure. All trees included in a basal area sweep are considered competitors. The estimate of point density is given by

$$PD = k \left(\frac{\frac{1}{2}\left(D_1/L_1 \right)^2 + \frac{3}{2}\left(D_2/L_2 \right)^2 + \cdots + \left(n - \frac{1}{2} \right)\left(D_n/L_n \right)^2}{n} \right) = k \left(\frac{\sum_{i=1}^{n} \left(i - \frac{1}{2} \right)\left(D_i/L_i \right)^2}{n} \right)$$

$$\tag{8.32}$$

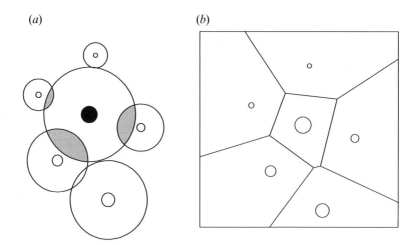

FIG. 8.15. Examples of distance-dependent point density measures: (*a*) area overlap concept and (*b*) area potentially available.

where   PD = point density

k = unit area conversion factor (k = 43,560/(144·4) = 75.625 in the imperial system and k = 10,000/(10,000·4) = 0.25 in SI)

D_i = diameter

L_i = distance from point to tree

n = number of trees in the basal area sweep

Trees are entered into the formula starting with the largest angle subtended from the viewing point (i.e., largest D/L), followed by the next largest, and so on. As Spurr (1962) points out, the tree with the largest angle is not necessarily the largest tree or the closest tree.

Distance-weighted size ratios are calculated as the sum of the ratios between the dimensions of each competitor and to the subject tree, weighted by a function of intertree distance (Tomé and Burkhart, 1989). An example is the competition index proposed by Hegyi (1974):

$$C_i = \sum_{j=1}^{n} \left(\frac{D_j / D_i}{L_{ij}} \right) \tag{8.33}$$

where C_i = the competition index for the subject tree

D_j = the diameter of the jth competitor

D_i = the diameter of the subject tree

L_{ij} = the distance from the subject tree to the jth competitor

n = number of competitors

Hegyi (1974) defined n as the number of trees within a fixed radius of the subject tree. Daniels (1976) modified the index by defining n as the number of trees within a fixed angle gauge sweep.

The *area potentially available index* utilizes polygons created by the intersections of the perpendicular bisectors of the distance between a subject tree and its competitors (Fig. 8.15b). The polygon area, as calculated from the coordinates of the vertices, is the area potentially available for tree growth (Brown, 1967). Moore et al. (1973) modified the index so the division was weighted by tree size:

$$I_{ij} = \left(\frac{D_i^2}{D_i^2 + D_j^2} \right) L_{ij} \qquad (8.34)$$

where I_{ij} = the distance from the subject tree to the weighted midpoint between the subject tree and the jth competitor

D_i = the diameter of the subject tree

D_j = the diameter of the jth competitor

L_{ij} = the distance between the subject tree and the jth competitor

As mentioned above, numerous indices have been developed. Each index is derived based on certain assumptions about how the competitive process is manifested. Daniels (1976) suggests that the utility of an index be judged based on correlation with observed tree growth and computational simplicity. Several studies have investigated the efficacy of these measures for predicting individual tree growth (e.g., Opie, 1968; Gerrard, 1969; Johnson, 1973; Daniels, 1976; Alemdag, 1978; Noone and Bell, 1980; Martin and Ek, 1984; Daniels et al., 1986; Tomé and Burkhart, 1989; Biging and Dobbertin, 1992). The results are variable and no single index has emerged as universally superior. The choice of an index depends primarily upon the use and data available.

8.9.8. Forest Stocking and Density Management Diagrams

In an early discussion of stocking and density, Bickford et al. (1957) accept the Society of American Forester's definition of stocking as "an indication of the number of trees in a stand as compared to the desirable number for best growth and management: such as well-stocked, over-stocked, partly stocked." Although they point out that there are varying shades of meaning for the silviculturist, economist, or forest manager, the most common connotation of "forest stocking" is in the sense of "best growth." That is, the terms "over-stocked" and "understocked" represent the upper and lower limits of site occupancy within which there exists a degree of stocking where forest growth will be optimum. Ideally, forest managers should be able to recognize this optimum stocking point under the complete range of stand composition conditions they encounter. However, field application of the kinds of density measures described here is made challenging by their mathematical complexity. Hence, considerable effort has also gone into representing the same relationships using a variety of charts and diagrams:

Basal Area Stocking Diagrams. Based on the idea of CCF, the work of Gingrich (1964, 1967) culminated in a stocking chart (Fig. 8.16) that has found considerable usage in applied forest management in the eastern and central United States.

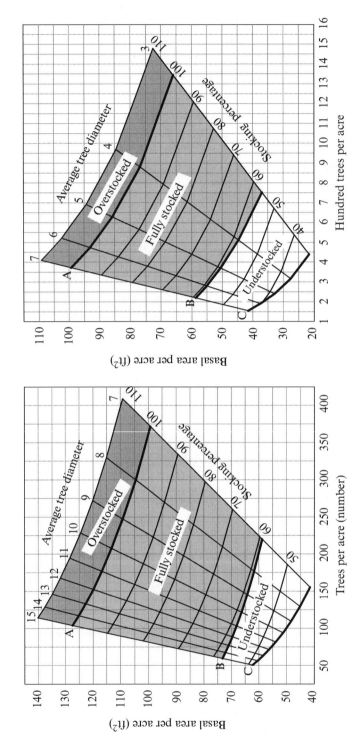

FIG. 8.16. Basal area stocking guide for upland hardwood stands. The area between curves A and B indicates the range of stocking where the trees can fully utilize the growing space. Curve C shows the lower limit of stocking necessary to reach the B level in 10 years on average sites (adapted from Gingrich (1967)).

The chart was developed using (1) a tree–area equation (Section 8.9.3) derived from data for stands that were deliberately chosen to be "fully stocked," and (2) an equation based on the CCF (Section 8.9.4), using data from open-grown trees. Thus, in Figure 8.16, the A-line (100% stocking) represents a normal condition of maximum stocking for undisturbed stands of upland hardwoods of average structure, and the B-line represents the lower limit of stocking for full site occupancy (Gingrich, 1967). The entire range of stocking between A and B is called "fully stocked" because the growing space can be fully utilized. It follows that stands having a combination of basal area per acre and trees per acre falling outside the range will demonstrate a gross increment less than the potential of the site, because of either overstocking or understocking.

Application of the stocking chart to determine a prescription of silvicultural treatment is in itself simple. For example, if stand measurements indicate a basal area per acre of 80 ft^2 and 2175 trees/acre, an average diameter of 9.2 in. is indicated in Figure 8.16. Following the 9.2 line down to the point where it crosses the B-line, one finds that a basal area of 65 ft^2 is the minimum basal area to maintain full stocking at that average diameter. Therefore, a silvicultural treatment that removes 15 ft^2 of basal area without materially changing the average stand diameter leaves a stand that will still fully utilize the site and produce wood efficiently. For greater detail on application, the reader is referred to the handbooks written by Roach and Gingrich (1962, 1968) and Roach (1977).

Stand Density Management Diagrams. Drew and Flewelling (1979) developed a stand density management diagram for Douglas-fir based on the maximum size–density relationship and the −3/2 rule of self-thinning. The maximum size–density relationship is a general principle of plant population biology that postulates that, for even-aged, pure species stands, the maximum mean tree size attainable for any density can be determined from the −3/2 power rule:

$$\overline{V} = a \cdot N^{-3/2} \tag{8.35}$$

where $\overline{V}$ = mean tree volume

   a = species-specific constant

   N = trees per unit area

The maximum size–density line represents maximum stocking. Relative density is defined as the ratio of actual density to the maximum density attainable in a stand with the same mean tree volume:

$$R = \frac{N}{N_{max}}$$

where R = relative density

   N = actual stand density

   N_{max} = the maximum density attainable:

$$N_{max} = \left(\frac{a}{\overline{V}}\right)^{2/3}$$

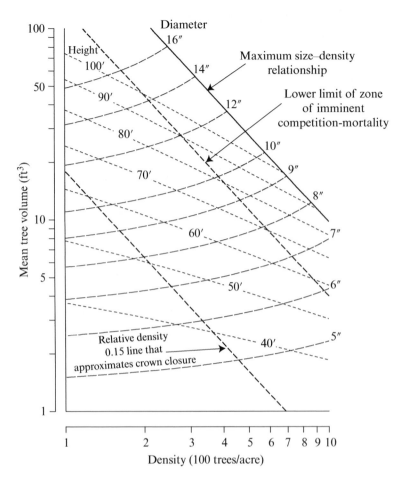

FIG. 8.17. Douglas-fir stand density management diagram (adapted from Drew and Flewelling (1979)).

In addition to the relative density lines, the lower limit of competition-induced mortality (Drew and Flewelling, 1977) and crown closure lines are drawn on their diagram. The lower limit of competition-induced mortality is approximately 55% of the maximum size–density relationship and crown closure is approximately 15% of the maximum. The resulting stand density management diagram is shown in Figure 8.17.

Similar diagrams utilizing either the −3/2 power rule, SDI (eq. (8.22)), or relative spacing (eq. (8.29)) have been developed for several species worldwide (e.g., McCarter and Long (1986) for lodgepole pine and Wilson (1979) for red pine). Like the basal area stocking diagrams, these diagrams are widely used as a means of assessing thinning needs and predicting stand development and yields. For a complete description of the application of density management diagrams, see Drew and Flewelling (1977, 1979) and McCarter and Long (1986).

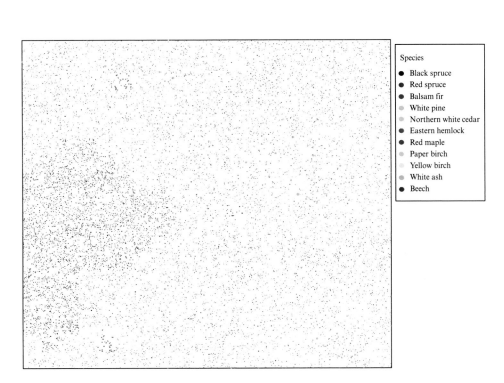

PLATE 1. Map of individual tree locations in a 25-ha mixed species forest in central New Brunswick, Canada.

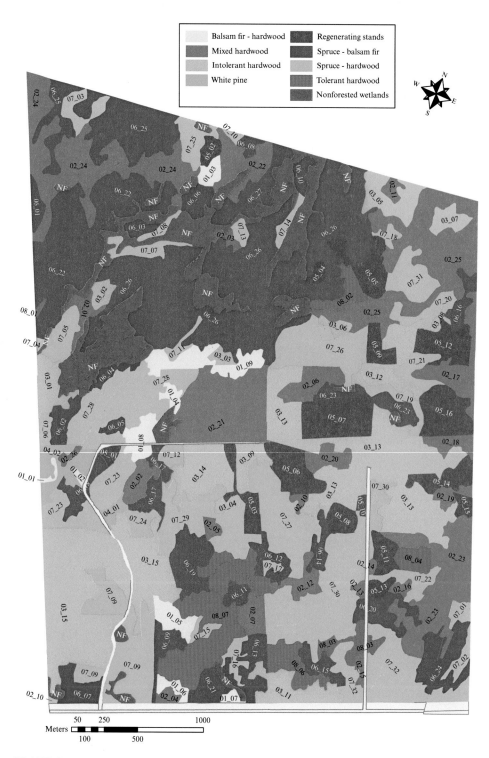

PLATE 2. Stand map of Noonan Research Forest, managed by University of New Brunswick, located in central New Brunswick, Canada.

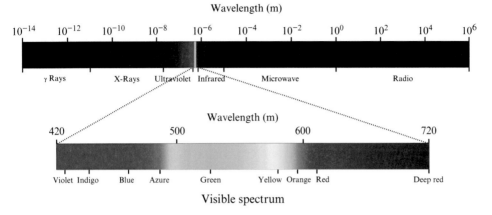

Electromagnetic spectrum

Wavelength (m)

10^{-14} 10^{-12} 10^{-10} 10^{-8} 10^{-6} 10^{-4} 10^{-2} 10^{0} 10^{2} 10^{4} 10^{6}

γ Rays   X-Rays   Ultraviolet Infrared   Microwave   Radio

Wavelength (m)

420   500   600   720

Violet Indigo   Blue   Azure   Green   Yellow Orange Red   Deep red

Visible spectrum

PLATE 3. The electromagnetic spectrum.

PLATE 4. One-meter resolution false-color infrared imagery from the U.S. National Agricultural Imagery Program, for a rugged area of northeastern Oregon.

PLATE 5. Thirty-meter resolution LANDSAT data for the same area as shown in Plate 4. Individual trees can no longer be resolved, but major land cover types are discernible.

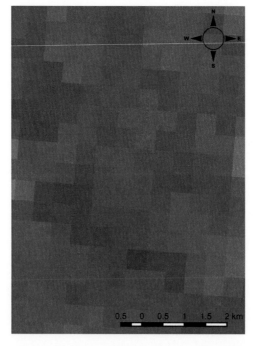

PLATE 6. MODIS data for the same area as shown in Plate 4. Although individual trees and stands are no longer visible, MODIS does provide frequent overflight of the entire globe.

PLATE 7. Point cloud from a portion of the Wild Upper Ammonoosuc valley on the White Mountain National Forest, New Hampshire, USA.

PLATE 8. False-color range image of a terrestrial laser scanner point cloud collected in a stand in southeastern New Hampshire. Hue indicates distance from the scanner, while value (dark vs. light) indicates intensity of the laser return.

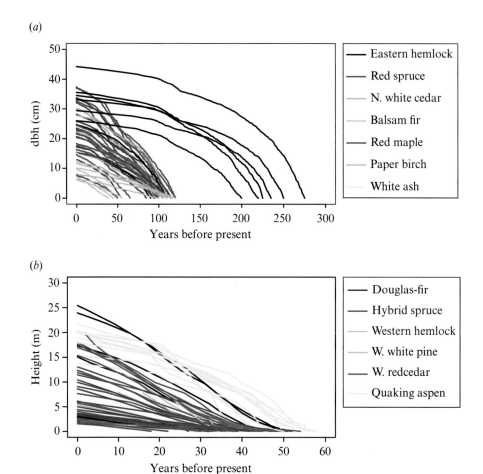

PLATE 9. Long-term, historic stand reconstruction data: (*a*) based on diameter growth trends in a mixed hardwood/conifer forest in central New Brunswick and (*b*) based on height growth trends in a mixed conifer forest from interior British Columbia.

9

SAMPLING UNITS FOR ESTIMATING PARAMETERS

To obtain information regarding parameters of a stand or forest area, the most common practice is to use sample plots. The *sample plot* is the unit for recording information and measurements. In most cases, this corresponds to the basic sampling unit. In cluster sampling, the basic sampling plot is subdivided into several subplots. A closely related concept is that of the *inclusion zone*. Each tree has its own inclusion zone, and a tree is included on a plot whenever the sample point (usually the plot center) falls within its inclusion zone.

For many years, the most common sample plot has been a unit of fixed area. This results in a probabilistic procedure for selecting trees around a sample point; the probability of selection is constant and equal for all individuals that make up the population because the inclusion zones for the individual trees all have the same area. The principles of fixed-area plot sampling are simple and apply directly to other types of material in the forest, including understory vegetation and wildlife sign.

It is also possible to use a sampling procedure in which the probability of selecting a tree is variable and depends on some dimension of the tree. In this case, the inclusion zone area also varies from tree to tree. This is the case for sampling with probability proportional to size, commonly known as PPS sampling. The best-developed examples of PPS sampling are for trees, but other examples—including those designed to assess stand-level attributes directly without detailed measurements of individual trees—may be of some use. PPS sampling is also well-developed for downed dead wood, which is treated in detail in Chapter 12. Nearest-neighbor methods, which typically measure only the n closest trees, are another example of sampling procedures where the probability of selecting trees varies, but it can be difficult to recover the probabilities and hence difficult to obtain unbiased estimates. When the detectability of trees or other objects in the population is imperfect, other methods, such as distance sampling, can be employed.

Forest Mensuration, Fifth Edition. John A. Kershaw, Jr., Mark J. Ducey, Thomas W. Beers and Bertram Husch.
© 2017 John Wiley & Sons, Ltd. Published 2017 by John Wiley & Sons, Ltd.

The choice of using plots of fixed or variable size depends on the practicality of their use in addition to theoretical considerations. In general, it is more efficient to use a sampling unit where trees are selected with a probability proportional to the variable of interest. For example, if one is interested in the estimation of volume of a stand, it will be more efficient to use PPS sampling since the selection of sample trees is proportional to their basal area, which is closely related to volume. On the other hand, if one is interested in determining the number of trees in a stand, it will be more efficient to use fixed-area plots. These same considerations apply when sampling for other attributes of a stand, including standing and downed dead wood, crown surface area, and so on.

9.1. THE FACTOR CONCEPT

In most analyses of forest inventory data, measurements are summarized and expressed on a per unit area basis (per acre for the imperial system and per hectare for SI). Sample measurements are scaled to per unit area measurements using a ratio of unit area to associated inclusion zone area:

$$\mathrm{TF}_i = \frac{\mathrm{Unit\,area}}{\mathrm{Inclusion\,zone\,area}_i} \qquad (9.1)$$

where TF_i = the expansion factor of the ith sample tree
Unit area = 43,560 ft^2 (1 acre) or 10,000 m^2 (1 ha)
Inclusion zone area$_i$ = size of the inclusion zone in ft^2 or m^2 associated with the ith tree

Each tree selected for measurement represents TF trees per unit area. For example, if 0.2-acre fixed area plots are used, the inclusion zones for all trees are also 0.2 acres, each tree sampled will represent $(1/0.2) = 5$ trees per acre as illustrated in Figure 9.1.

The expansion factor shown in eq. (9.1) is often referred to as the *tree factor* since it gives the number of trees per unit area each sample tree represents. Depending upon the

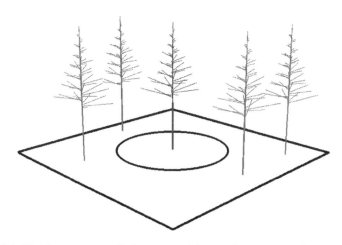

FIG. 9.1. The factor concept. Each tree on a 0.2-acre plot represents 5 trees per acre.

type of sampling unit used, the tree factor may be constant or may vary by tree size or some other factor. The number of trees per unit area is obtained by summing the tree factors for each tree on the plot. If tree factor is constant for all sample trees, then trees per unit area may be obtained by multiplying the constant tree factor by the number of trees per plot.

The expansion factors for other tree characteristics (XF_i), such as basal area or volume, are obtained by multiplying the value of the characteristic, X_i, by the tree factor:

$$XF_i = X_i \cdot TF_i \tag{9.2}$$

For example, the basal area factor for an individual tree is

$$BAF_i = BA_i \cdot TF_i = \left(c \cdot D_i^2\right) \cdot \left(\frac{\text{Unit area}}{\text{Inclusion zone area}}\right)$$

where BAF_i = the basal area factor of the ith tree
$\quad\quad BA_i$ = the basal area of the ith tree
$\quad\quad TF_i$ = the tree factor associated with the ith tree (eq. (10.1))
$\quad\quad\quad c$ = the basal area conversion factor (0.005454 in imperial system, 0.00007854 in SI)
$\quad\quad\quad D_i$ = the dbh of the ith tree

Per unit area estimates for a single plot are obtained by summing factors:

$$\frac{X}{\text{Unit area}} = XF_1 + XF_2 + XF_3 + \cdots + XF_n$$

$$= X_1\left(TF_1\right) + X_2\left(TF_2\right) + X_3\left(TF_3\right) + \cdots + X_n\left(TF_n\right) \tag{9.3}$$

$$= \sum_{i=1}^{n} XF_i = \sum_{i=1}^{n} X_i\left(TF_i\right)$$

For example, basal area per unit area is obtained by summing the BAFs for each sample tree. If TF is constant for each tree, then basal area per unit area may be obtained by summing the basal area of each tree and multiplying that sum by the constant tree factor:

$$BA \text{ per unit area} = \sum BAF_i = \sum TF \cdot BA_i = TF \cdot \sum BA_i.$$

Factors for other stand parameters are obtained in a similar manner.

The factors derived using this method lead to the following definitions:

Tree factor. The number of trees per unit area represented by each tree tallied.

Basal area factor. The number of units of basal area per unit area represented by each tree tallied.

Volume factor. The number of units of volume per unit area represented by each tree tallied.

Height factor. The number of units of height per unit area represented by each tree tallied.

The factor concept provides a simple and unified approach to forest inventory analysis. By focusing on the derivation of tree factor, all other factors are simply obtained by multiplying the value of that parameter by the tree factor. In the following sections, various sampling units will be described and the factors important to foresters derived for each type of sampling unit.

9.2. FIXED-AREA PLOTS

Fixed-area sampling units in a forest inventory or research are called plots or strips depending on their dimensions. The term *plot* is loosely applied to sampling units of small areas of square, rectangular, circular, or triangular shape. The term *strip* is generally used to refer to a rectangular plot whose length is many times its width. Strips often have a length that spans the entire area being surveyed, so different strips in the same inventory may have different lengths.

9.2.1. Circular Plots

Circular plots have been widely used since a single dimension, the radius, can be used to define the perimeter. When circular plots are used, the inclusion zone for each tree is a circle, centered on the tree, that has the same radius as the plot radius, and hence the same area as the plot area. The dimensions of commonly used circular sample plots are shown in Table 9.1. Circular plots have three advantages:

1. A circle has the minimum perimeter for a given area, which implies fewer decisions for trees near the plot boundary.
2. A circular plot has no predetermined orientation, which in some cases, especially in plantations, can be a cause of appreciable variability.
3. Methods of correcting slopover bias (Section 9.2.6) are simplest for circular plots.

TABLE 9.1. Dimensions of Commonly Used Fixed-Area Plots

Plot Area as a Fraction of Per Unit Area	Imperial Units			SI Units		
	Plot Area (ft^2)	Circular Radius (ft)	Square Side (ft)	Plot Area (m^2)	Circular Radius (m)	Square Side (m)
1/1000	43.56	3.7	6.6	10	1.78	3.16
1/500	87.12	5.3	9.3	20	2.52	4.47
1/250	174.24	7.4	13.2	40	3.57	6.32
1/100	435.6	9.8	20.9	100	5.64	10.00
1/50	871.2	16.7	29.5	200	7.98	14.14
1/25	1,742.4	23.6	41.7	400	9.28	20.00
1/20	2,178	26.3	46.7	500	12.62	22.36
1/10	4,356	37.2	66.0	1,000	17.84	31.62
1/5	8,712	52.7	93.3	2,000	25.23	44.72
1/4	10,890	58.9	104.4	2,500	28.21	50.00
1/2	21,780	83.3	147.6	5,000	39.89	70.71
1	43,560	97.8	208.7	10,000	56.42	100.00

The main disadvantage of a circular plot is the error that may arise if the plot boundary is not carefully observed and if ocular estimations of limiting trees are not well done. Modern ultrasonic and laser distance measurement tools make checking trees near the boundary of a circular plot especially fast and simple.

In general, small circular plots are more efficient than large ones, at least in theory. With large plots, the decision regarding trees near the plot boundary is often difficult, and it can be more time-consuming to move physically to each tree both for measurement and for checking borderline conditions. Consequently, it is advisable to use plots of a size that allows an easier control of trees near plot limits. An experienced person standing at plot center can do this with confidence if the radius of the plot is no greater than about 5 m. Of course, one can determine more precisely if a tree near the perimeter is in the plot by measuring the distance from plot center to the tree using a tape or electronic distance measurement device (Section 4.2). With a plot that is too large, it also becomes easy to miss trees that should have been tallied, leading to a downward bias. The chief drawback of small plots is an increase in sampling variance as plot size declines, requiring many more plots to achieve comparable confidence limits. This issue is addressed more fully in Section 10.2.1.

Distances should be measured in the horizontal plane, even if the plot is on sloping terrain. A circular plot of radius r on flat terrain has an area of $a = \pi r^2$. If the circular plot is located on terrain with a slope of α degrees and distances are measured parallel to the slope, its horizontal projection will generate an elliptical plot with major radius $r_{max} = r$ and minor radius $r_{min} = r \cdot \cos \alpha$. The area of the elliptical plot is less than the area of the circular plot in a horizontal plane, so using the ordinary tree factor would lead to a downward bias in the estimates. An alternative technique is described by Beers (1969). When distances have been measured along the slope, and slope has been recorded, calculation of the tree factor from the area of the ellipse will yield unbiased estimates, but the tree factors (and other factors that follow from it) will vary from plot to plot within an inventory as slope changes.

9.2.2. Square and Rectangular Plots

The advantage of square plots is the somewhat greater ease of deciding if trees are in or out of the plot because plot boundaries are straight lines. The plot limits can be easily established by measuring diagonals at right angles from the plot center. Using a compass or right angle prism to establish the direction of the diagonals, one measures and marks the distance from plot center to the corners of the square. The trees or plants are then measured systematically in the four triangles formed by the diagonals and periphery of the plot. Of course, the plot boundaries can be established by marking directly the corners using compass and tape; however, great care should be taken in laying out the plot corners and boundaries because errors that affect the plot area translate directly into biased estimates. The dimensions for commonly used square plots are shown in Table 9.1.

In rectangular plots, the width and length are not equal (the term length is generally applied to the longer dimension). Rectangular plots are usually established from the central axis of a given length. The width is measured from this axis and corners can be established. Rectangular plots are especially useful in natural forests with difficult topography and large altitudinal variation. In these situations, the axis of the rectangular plot should be oriented in advance to cross the maximum slope so that it may sample the maximum variability of the forest; however, the orientation should not be determined subjectively as the inventory is in progress because this creates an opportunity to introduce bias into the results.

A strip is a type of rectangular plot whose length is many times its width. For recording purposes, the continuous strip may be subdivided into smaller recording units. It is important to remember, however, that the entire strip is still considered the ultimate sampling unit and is the basis for the number of degrees of freedom in subsequent statistical computations. (If interrupted strips are used, e.g., alternate lengths of a strip tallied, the individual units of the strip tallied can be treated like plots in subsequent calculations in most practical work.)

An advantage of a continuous strip over plots is that a tally is taken for the entire strip traversed so that there is no unproductive walking time between sampling units; however, continuous strips have a disadvantage: the number of sampling units, and thus the degrees of freedom, is small. Comparing strip sampling to plot sampling of equal intensity, the size of the sampling unit is larger and the number of sampling units smaller. The larger sampling unit results in a reduction in variability, but the smaller number of sampling units counteracts this advantage. For this reason, sampling error of a strip sampling design is usually larger than for plot sampling, assuming the same sampling intensity (see Section 10.2.1). Difficult terrain can also create practical challenges in measuring on continuous strips; in a plot layout, one can simply navigate around obstacles, but in a strip survey one must continue measuring through them.

9.2.3. Subplots

With fixed-area plots, such as circular ones, the different size classes of trees or plants (e.g., dbh or height) are sampled in proportion to their frequency in the population. Normally, in forest stands there are more small trees than large ones (this is typical in natural forests—not necessarily in plantations). Since each tree has the same probability of selection, there will be many small trees in the sample. This has the disadvantage that it will be necessary to measure many more small trees than large ones in spite of the fact that many times they will be of less importance, especially if volume, biomass, or carbon content is the parameter of interest. To remedy this situation, it is possible to modify the sampling plan so that more large trees and fewer small ones are measured in the sampling process. This modification consists of using different sizes of plots for different size classes or attributes of trees or vegetation. The common approach is to use a large plot for big trees and small plots for small trees and lesser vegetation. This can be accomplished by nesting or subdividing the large plot into subplots of different sizes. For example, one could establish a circular plot of $1000 \, m^2$ (0.1 ha.) with radius of 17.84 m, for measuring trees greater than 25 cm dbh. Within this plot, a concentric plot of $500 \, m^2$ (0.05 ha.) having a radius of 12.62 m can be established at the same center for trees with a dbh up to 25 cm. Van Den Meersschaut and Vandekerkhove (2000) used a plot design of four concentric circular sample plots with areas of 16, 64, 255, and $1018 \, m^2$ to measure seedlings, shrubs, and trees (living and dead) of different size limits. They also established a 16×16 m plot for measuring lesser vegetation and lying deadwood. Concentric or nested plot designs are extremely common in national forest inventories and other permanent plot systems.

Other systems of subplots can be established, such as small circular or square subplots established at some fixed design within a large circular plot or square plot. Subplots of different shapes and locations can be used with any plot shape. The choice of geometric arrangement is at least partly practical; for example, subplots designed for assessing regeneration are often offset from the plot center, where any regeneration present might be subject to trampling during the initial phases of plot layout and measurement.

(a)   (b)

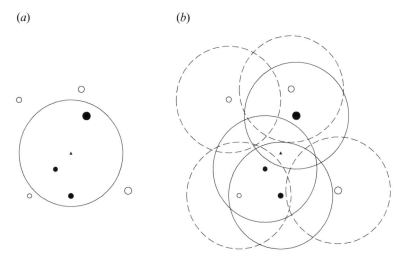

FIG. 9.2. Selection of trees in fixed-area plot sampling: (*a*) plot-centered approach and (*b*) tree-centered approach.

9.2.4. Selection of Plots and Trees

To respect the laws of probability and obtain unbiased estimates of the parameters of a stand or forest as well as the sampling variance, the centers of the sample plots should be selected at random from the total population of possible plot centers, that is, the tract or area of interest. A systematic sample (such as a grid) selected with a random start gives data that can provide an unbiased estimate of the parameters themselves, but not of the sampling variance. All trees or individual plants that are within the limits of a plot are usually measured to obtain information regarding the attributes of interest, such as dbh, height, or basal area. It is possible to use nested plots to obtain an unbiased subsample of trees for more time-consuming measurements, such as tree height, which are then extended to the rest of the population through regression.

With fixed-area plots, the tendency is to think of the tree selection process as the result of establishing a plot center and boundary, then measuring all trees contained within that boundary, as illustrated for a circular plot in Figure 9.2a. This view is often referred to as the *plot-centered approach* (Grosenbaugh, 1958; Beers and Miller, 1973; Oderwald, 1981). It is also possible to view the selection process by visualizing an imaginary fixed-area plot—the *inclusion zone*— around each tree (Fig. 9.2b). A plot center is established, and trees are included in the sample if this plot center is included in their inclusion zones. This view is often referred to as the *tree-centered approach* (Grosenbaugh, 1958; Beers and Miller, 1973; Oderwald, 1981).

The tree-centered approach aids in understanding the probability of a tree being selected for inclusion in a plot and provides a fundamental link between forest inventory practice and modern sampling theory in statistics. A tree is included in a plot if its imaginary plot or inclusion zone contains the established plot center (Fig. 9.2b). If random sampling is used, then the probability of selection, P_j, is equal to the inclusion zone area (which equals the plot area) divided by the area of the stand:

$$P_j = \frac{\text{Inclusion zone area}}{\text{Stand area}}$$

(9.4)

Comparing eq. (9.4) to eq. (9.1), we find that TF_j = Unit area / (Stand area $\times P_j$); there is a one-to-one relationship between probability of selection and the tree factor. Because the inclusion zone area is the same for all trees in a simple fixed-area plot design, the probability of being selected is constant for all trees, and so is the tree factor. If we consider the expected contribution of an individual tree to the estimate of trees per unit area, for a randomly located sample point, we have

$$
\begin{aligned}
E_j &= \left(1 - P_j\right)\cdot 0 + P_j \cdot TF_j \\
&= \left(\frac{\text{Inclusion zone area}}{\text{Stand area}}\right) \times \frac{\text{Unit area}}{\text{Stand area} \times \left(\text{Inclusion zone area}/\text{Stand area}\right)} \\
&= \frac{\text{Unit area}}{\text{Stand area}}
\end{aligned}
\tag{9.5}
$$

Now, if there is actually a total of N trees in the stand, the density of those trees per unit area is (N / Stand area) $\times$ Unit area. Summing the expected contribution of the N trees in the stand to the estimate of trees per unit area from a sample plot, we have

$$
\sum_{j=1}^{N} E_j = N\left(\frac{\text{Unit area}}{\text{Stand area}}\right)
\tag{9.6}
$$

which is equal to the actual, unobserved density. Thus, so long as the inclusion zone areas and probabilities are correct, we know that the density estimate for a fixed-area plot sample is unbiased. When nested plots are used to tally trees of differing sizes, trees of different sizes will have different plot areas and inclusion zone areas. These in turn yield different selection probabilities and hence tree factors. But the unbiasedness of the fixed area plot estimate still holds because the plot areas (inclusion zone areas) still cancel in eq. (9.5). Extending the proof of unbiasedness in eq. (9.6) to other quantities, such as basal area or volume per unit area, is straightforward.

9.2.5. Stand and Stock Tables

Plot measures are scaled to per unit area measures using the expansion factors described in Section 9.1. Average per unit area values and total values are obtained using the methods described in Chapters 3 and 10. While estimates of the averages per unit area and total values for various stand parameters are important, foresters typically desire more detailed summaries. Typically, foresters will summarize data into stand and stock tables. A *stand table* gives number of trees by species and dbh class. A *stock table* gives volumes (or weights) by similar classifications. In many cases, basal area and average height are included. Occasionally, tables may be subdivided by height class as well as dbh class. A *combined stand and stock table* gives density and volume by species and dbh class. Values within stand and stock tables are normally expressed on a per unit area basis; however, total values for the stand may be given.

The construction of stand and stock tables from fixed-area plot data will be illustrated using the following example. Four 20 m $\times$ 20 m plots were measured and the following total tally obtained:

dbh class	10	12	14	16	18	20	22	24	26	28	30	32	34	36	38	40	42+
No. tallied	9	6	17	8	9	4	8	7	5	6	2	5	3	1	3	0	2

Table 9.2 shows the combined stand and stock table. Column 1 is the dbh classes corresponding to the tally. The numbers of trees tallied are in column 5. Columns 2, 3, and 4 contain the per tree average values for height, basal area, and volume. Total height was measured on all trees shown in the above tally, and average height (column 2) was calculated for each dbh class. The basal area per tree (column 3) is found by calculating the cross-sectional area of a circle corresponding to the diameter class:

$$BA_{tree} = \pi \cdot \left(\frac{D}{2 \cdot 100}\right)^2$$

$$= \frac{\pi}{40,000} \cdot D^2$$

$$= 0.00007854 \cdot D^2.$$

Volume per tree (column 6) can be estimated using a variety of methods including local volume table, standard volume, or a volume equation. For this example, trees were assumed to be paraboloids (eq. (6.2)) and volume per tree is obtained by multiplying form (0.5) by basal area per tree (column 3) by average height (column 4). For example, the volume per tree for the 10 cm dbh class is

$$VOL_{10} = \frac{1}{2}\left(A_b \cdot H\right)$$

$$= \frac{1}{2}\left(BA \cdot H\right)$$

$$= \frac{1}{2}\left(0.0079 \cdot 11.0\right)$$

$$= 0.043 \text{ m}^3.$$

Tree factor (column 6) is constant for fixed-area plots and is obtained using eq. (9.1):

$$TF = \frac{\text{Unit area}}{\text{Plot area}}$$

$$= \frac{10,000}{(20 \cdot 20)}$$

$$= 25$$

The basal area factor (column 7) and the volume factor (column 8) are obtained by multiplying the tree factors (column 6) by the corresponding per tree values (eq. (9.2)). For the 10 cm class, the basal area factor is

$$BAF_{10} = BA_{10} \cdot TF_{10}$$

$$= (\text{Column } 3)(\text{Column } 6)$$

$$= 0.0079 \cdot 25$$

$$= 0.1964$$

TABLE 9.2. A Combined Stand and Stock Table for a Northern Hardwood Stand Located in Central New Brunswick

Column 1	Column 2	Column 3	Column 4	Column 5	Column 6	Column 7	Column 8	Column 9	Column 10	Column 11
dbh	Per Tree Averages			Number of Trees		Factors[a]		Per Hectare Averages		
Class (cm)	Height (m)	BA (m²)	Volume (m³)	Tallied	Tree (number)	BA (m²)	Volume (m³)	Trees (number)	BA (m²)	Volume (m³)
10	11.0	0.0079	0.043	9	25	0.1964	1.080	56.25	0.44	2.4
12	11.7	0.0113	0.066	6	25	0.2827	1.654	37.50	0.42	2.5
14	12.9	0.0154	0.099	17	25	0.3848	2.482	106.25	1.64	10.5
16	13.8	0.0201	0.139	8	25	0.5027	3.468	50.00	1.01	6.9
18	13.9	0.0254	0.177	9	25	0.6362	4.421	56.25	1.43	9.9
20	16.5	0.0314	0.259	4	25	0.7854	6.480	25.00	0.79	6.5
22	16.8	0.0380	0.319	8	25	0.9503	7.983	50.00	1.90	16.0
24	17.3	0.0452	0.391	7	25	1.1310	9.783	43.75	1.98	17.1
26	19.8	0.0531	0.526	5	25	1.3273	13.141	31.25	1.66	16.4
28	18.0	0.0616	0.554	6	25	1.5394	13.854	37.50	2.31	20.8
30	20.0	0.0707	0.707	2	25	1.7672	17.672	12.50	0.88	8.8
32	18.2	0.0804	0.732	5	25	2.0106	18.297	31.25	2.51	22.9
34	21.0	0.0908	0.953	3	25	2.2698	23.833	18.75	1.70	17.9
36	19.0	0.1018	0.967	1	25	2.5447	24.175	6.25	0.64	6.0
38	19.7	0.1134	1.117	3	25	2.8353	27.928	18.75	2.13	20.9
40	20.0	0.1257	1.257	0	25	3.1416	31.416	0.00	0.00	0.0
42	20.0	0.1385	1.385	2	25	3.4636	34.636	12.50	1.73	17.3
Total				95				593.75	23.17	203.0

Based on four 20 m × 20 m fixed-area plots.

[a] Factors are the value per unit area represented by each tree tallied.

and the volume factor is

$$VF_{10} = Vol_{10} \cdot TF_{10}$$
$$= (Column\ 4)(Column\ 6)$$
$$= 0.043 \cdot 25$$
$$= 1.080.$$

Average per unit area values for each parameter are obtained by multiplying corresponding factors by the number of trees tallied and dividing by the number of plots. Trees per hectare (column 9) is then obtained by multiplying the tree factor (column 6) by the number of trees tallied (column 5) and dividing by the number of plots:

$$Trees/ha = \frac{TF \cdot (No.\ tallied)}{(No.\ plots)}$$

For example, there were nine trees tallied in the 10 cm class; therefore, the trees per hectare for the 10 cm class becomes

$$Trees/ha = \frac{TF \cdot (No.\ tallied)}{No.\ plots}$$
$$Column\ 9 = \frac{(Column\ 6) \cdot (Column\ 5)}{No.\ plots}$$
$$= \frac{25 \cdot 9}{4}$$
$$= 56.25.$$

Similarly, basal area per hectare (column 10) is

$$BA/ha = \frac{BAF \cdot (No.\ tallied)}{No.\ plots}$$
$$Column\ 10 = \frac{(Column\ 7) \cdot (Column\ 5)}{No.\ plots}$$
$$= \frac{0.1964 \cdot 9}{4}$$
$$= 0.44$$

and volume per hectare (column 9) is

$$Volume/ha = \frac{VF \cdot (No.\ tallied)}{No.\ plots}$$
$$Column\ 11 = \frac{(Column\ 8) \cdot (Column\ 5)}{No.\ plots}$$
$$= \frac{1.080 \cdot 9}{4}$$
$$= 2.4.$$

9.2.6. Boundary Slopover

In Section 9.2.4, we found that if the inclusion probabilities for individual trees are correct, the estimate of trees per unit area (and related quantities, such as basal area and volume per unit area) from a single randomly located plot is an unbiased estimate of the actual density per unit area. Failure to compensate properly for slope, and careless work in laying out plots and checking boundary trees, can all make the inclusion probabilities inaccurate and hence create bias in the estimates. Another source of bias occurs whenever trees are suffi- ciently close to the boundary of the stand (or stratum, or other inventory unit) that their inclusion zone falls partly outside the area where plot centers can be located. In such cases, if no corrective action is taken, trees near the boundary have an actual inclusion probability that is less than their nominal inclusion probability, and the results of the inventory will be biased downward. This bias is called *boundary slopover* bias. Note that because bias is a matter of expectation, the bias is present even if no actual inventory plots fall near the boundary. Historically, this issue has not always been understood clearly, and several of the "fixes" proposed were actually counterproductive. Luckily there are now techniques to minimize or eliminate slopover bias that require very little field effort.

Edge areas are often of specific interest in and of themselves. When a forest area is bounded by nonforest or other land use classes, trees near the border will usually be some- what different than those in the interior of the forest. If the perimeter proportion of a forest area is large, it will be necessary to take precautions that this zone is adequately represented in the sample. It is not reasonable to assume that areas in the interior of the forest provide a fair representation of the entire forest area. In addition, very frequently, sample plots are located near the border of a forest area so that a portion of the plot falls outside the limits of the forest or stand or portions of the plot fall in different forest conditions or types. Where it is desired to maintain an estimate for a stand or stratum that is pure—that reflects only the contributions of trees from that stand or stratum, even if adjoining areas are under the same ownership—it can be desirable to treat stand or stratum boundaries as internal partitions of the total area that are equivalent to the boundaries that define the overall area.

One solution to the problem of boundary slopover is to allow sample points to fall outside the area being inventoried, in such a way that the inclusion zone for a tree standing immediately on the boundary still falls entirely within the area where plot centers can fall. When circular plots are used, this essentially entails conducting the inventory over a buff- ered version of the original tract, where the buffer width equals the plot radius. Iles (2003) discusses a practical modification of this method, called the *toss-back method*. In the toss- back approach, trees from the plots located outside the tract are "tossed back" onto the nearest plot inside the tract for computational purposes. Its principal disadvantage is the need to work outside the boundary, which can be difficult if the boundary is a water body, cliff, or busy road.

Over time, several other field methods have been developed to deal with boundary slopover. These methods may modify the location, size, or form of those plots whose centers are within the forest or given stand condition but a portion of the plot is outside. Not all of these methods are sound. Some of these historical methods are shown diagrammatically in Figure 9.3.

a) The plot is displaced so that it falls completely within the forest. This method is questionable and can potentially lead to significant bias if the stand has a large proportion of its area as edge. When edge is a relatively small proportion of the

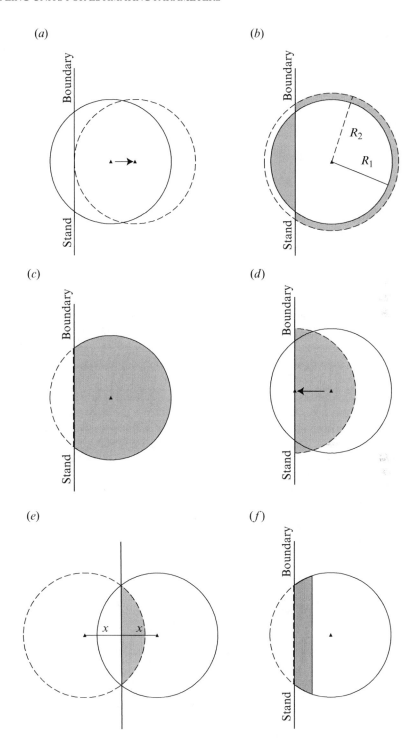

FIG. 9.3. Methods of boundary slopover correction from a plot-centered perspective: (*a*) plot shifting; (*b*) plot enlargement; (*c*) plot area adjustment; (*d*) half plot; (*e*) mirage method; (*f*) walkthrough method.

stand, the bias is small, but still present. By moving the plot away from the edge, edge trees are further undersampled, thus aggravating the very problem that boundary slopover bias creates (Gregoire and Scott, 2003). This method is relatively time-consuming, at its best has little effect, and can make the boundary slopover problem worse. It has little place in modern forest inventory.

b) The radius of the plot is increased so that the increased size is equivalent to the part of the original plot outside the stand. While an improvement over method a, edge trees are still undersampled using this method. It is difficult to calculate the increase in radius for anything other than a straight-line boundary.

c) The area of the plot within the stand is determined, and one measures the trees only in the part of the plot within the stand. This method can be extremely time-consuming since determining the portion of the plot within the stand can be complex. Ratio estimation (Section 10.8.2) is employed to obtain valid estimates of means and totals.

d) The plot is divided into two and a half-plot is established exactly on the stand edge. Trees within the half-plot are measured. Plot characteristics are determined as usual and the results doubled. In this method, edge trees are oversampled, thus producing biased results.

e) Use of a reflection or mirage method in which the part of the plot that is outside the border of the stand is "reflected" toward the interior of the plot that is within the stand. In practice, this is done by establishing a "mirage" sample point that is located by reflecting the original sample point through the boundary (i.e., the same distance from the edge as the current plot center, but on the opposite side of the boundary). Trees or plants inside the boundary, and on the mirage plot, are tallied. When inclusion zones are circular, all trees tallied on the mirage plot will already have been tallied on the original plot, so they are tallied twice; however, when inclusion zones are not circular (e.g., square plots), some mirage trees may be new tallies (Ducey et al., 2001). This method was originally proposed by Schmid-Haas (1969) and is relatively easy to implement in the field; however, it requires boundaries that are straight lines (or pieces of straight lines with well-defined corners). Like the toss-back method, it also requires working outside the tract. Gregoire (1982) provides a proof that the mirage method yields unbiased results, along with a lucid discussion of the slopover problem in general.

f) The walkthrough method (Ducey et al., 2004). In this method, one reflects the sample point through the trees that are near the border of the stand. Specifically, one imagines walking from the sample point to the tree in question, and continuing for the same distance and bearing straight through the tree and beyond. If the resulting point falls outside the stand, the tree is double-tallied. Strictly speaking, the walkthrough method reduces slopover bias, rather than eliminating it, because there are a small number of geometric situations in which walkthrough cannot correct the problem. Walkthrough is extremely fast in the field because for most trees a quick ocular check is sufficient to determine whether the tree should be double-tallied or not. Moreover, travel outside the boundary is not actually needed; if physical walkthrough would send the forester outside the tract, then the tree should be double-tallied, and this can be determined upon reaching the boundary. Flewelling and Strunk (2013) recently developed a generalization, the walking-to-and-fro correction, that corrects slopover bias in all cases; however, it does require more complex field work.

West (2013) studied the performance of the toss-back method, the mirage method, the walkthrough method, and walking-to-and-fro using computer simulation. He found that the differences in performance between these four methods were negligible. Thus, the choice between them can be driven by operational considerations, including time, cost, and the ability of field crews to execute them safely and correctly.

Cluster Plots. When clusters are used, some of the subplots may fall in nonforest areas or in different forest stands. For a description of several techniques to solve this problem, see Hahn et al. (1995). Cluster plots do pose significant challenges for mirage and walkthrough approaches (Valentine et al., 2006). The U.S. Forest Service FIA program uses the approach outlined by Bechtold and Patterson (2005). In this method, the areas in the different condition classes of a subplot are calculated and their proportions estimated. The attributes of interest of any tree characteristic are measured on these condition class areas and the population total and its variance can be estimated for each of the condition classes.

9.3. SAMPLING TREES WITH VARIABLE PROBABILITY

Up to now, we have considered plots where their size or area is fixed and constant. As illustrated in Figure 9.2, the inclusion zones around each tree are of the same size, regardless of tree size. This means that the probability of selecting any tree is constant regardless of tree size, unless nested plots are used.

It is also possible to devise selection procedures where the probability of selecting a tree varies continuously according to some characteristic of that tree; for example, its dbh or height. In this case, we can visualize inclusion zones around each tree where the size of the inclusion zone depends upon the size of the tree characteristic. Trees of large size will have larger zones and small trees will have smaller zones. In other words, the probability of selecting a tree is proportional to its size.

9.3.1. Horizontal Point Samples

The idea of selecting individuals in a sample in proportion to some characteristic has been recognized for a long time in statistics, but it was not applied in forest mensuration until an Austrian forester, Walter Bitterlich, invented an ingenious method of determining stand basal area (Bitterlich, 1947). Bitterlich called his method *angle count sampling* because an angle was used to select trees and basal area was obtained by counting selected trees (Bitterlich, 1947, 1984). Originally, it was not realized that this was a form of sampling with variable probability until another forester, the American, Lewis Grosenbaugh, recognized it (Grosenbaugh, 1955, 1958). He pointed out that the Bitterlich method was really an application of selecting samples with probability proportional to size (commonly called PPS). The development and use of this type of sampling has been an important landmark in forest mensuration.

The most used application of sampling with probability proportional to size in forestry is that originally developed by Bitterlich, commonly called *horizontal point sampling.**

---

* Horizontal point sampling is the term frequently used in North America, Bitterlich (1947) referred to the method as angle count sampling, it is also known as Bitterlich sampling, prism sampling, plot-less sampling, and variable plot sampling. We use the term horizontal point sampling because it accurately describes the process of projecting a horizontal angle and sweeping 360° about a sample point.

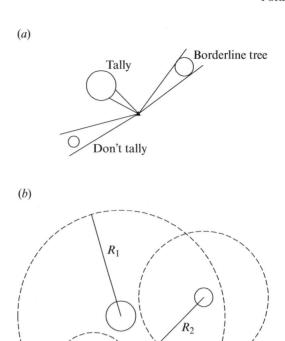

FIG. 9.4. Selection of trees in horizontal point sampling: (*a*) point-centered approach and (*b*) tree-centered approach.

This is a type of sampling where the probability of selecting a tree is proportional to its basal area. In this application, a series of sampling points, similar to the centers of circular fixed-area plots, are located in the forest area. The observer occupies the point and projects a horizontal angle with an instrument toward each tree at the dbh level. As shown in Figure 9.4*a*, all trees that appear as large as or larger than the projected angle are counted. Trees that are smaller than the projected angle are not counted.

The circles represent the cross sections of trees at dbh level and the lines indicate the angle projected from the sampling point. One can measure any attribute of the selected trees, as in fixed-area plots (dbh, height, crown, health, etc.). These measures are scaled to per unit area values using the factor concept discussed in Section 9.1. From eq. (9.1), we know that the tree factor is determined using

$$TF_i = \frac{\text{Unit area}}{\text{Inclusion zone area}_i}.$$

In simple fixed-area plots, the inclusion zone area was constant, and with nested plots, the inclusion zone area could take one of a small number of values. In horizontal point sampling, the inclusion zone area varies continuously as a function of tree cross-sectional area.

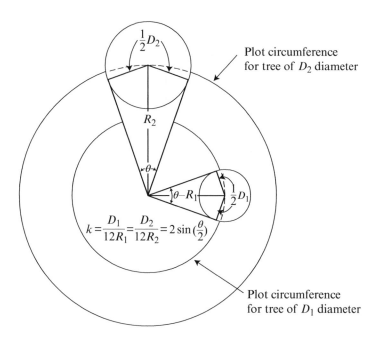

FIG. 9.5. Gauge constant k where D is tree diameter in inches and R is plot radius in feet.

This is illustrated in Figure 9.4b using the tree inclusion zones. Each of the three trees has a different "plot radius" (inclusion zone radius) or limiting distance (R_i) that is proportional to each tree's dbh. To determine the tree factor, we must determine each tree's inclusion zone radius. (Later, we will show a shortcut formula.) It is helpful to consider a tree at the borderline condition, where the gauge angle is precisely tangent to the breast height cross section (Fig. 9.5), and to keep in mind the following:

1. The angle gauge is projecting a fixed horizontal angle θ.
2. At any sampling point, a series of concentric circular plots is conceptually established, a different plot radius or limiting distance being associated with every different tree diameter.
3. The radius of each concentric plot is determined by each different tree diameter and is not influenced by the actual spatial location of the tree; therefore, the actual distance from the sampling point to the tree is not needed.
4. At the borderline condition, the ratio of tree diameter to plot radius is a constant. Thus, for a given angle θ, tree diameter in inches, D, and plot radius in feet, R, we can define the gauge constant k to be

$$k = \frac{D}{12R} = 2\sin\frac{\theta}{2} \tag{9.7}$$

Therefore, the plot (inclusion zone) radius is

$$R = \frac{D}{12k} \tag{9.8}$$

Since plot area $= \pi R^2$, in imperial units the tree factor is determined to be

$$
\begin{aligned}
\mathrm{TF} &= \frac{43{,}560}{\pi R_i^2} = \frac{43{,}560}{\pi \left(D_i / 12k \right)^2} \\
&= \frac{43{,}560 k^2}{\left(\pi / 144 \right) D_i^2}
\end{aligned}
$$

By dividing the top and bottom by 4, we obtain

$$
\mathrm{TF}_i = \frac{\left(1/4\right) 43{,}560 k^2}{\left(1/4\right) \left(\pi / 144\right) D_i^2} = \frac{10{,}890 k^2}{0.005454 D_i^2} = \frac{10{,}890 k^2}{\mathrm{BA}_i} \tag{9.9}
$$

In SI units, tree diameter, D, is measured in centimeters and plot radius, R, is measured in meters. For a given angle, θ, the gauge constant is

$$
k = \frac{D}{100R} = 2 \sin \frac{\theta}{2} \tag{9.10}
$$

and the plot radius becomes

$$
R = \frac{D}{100k} \tag{9.11}
$$

and the tree factor is

$$
\begin{aligned}
\mathrm{TF} &= \frac{10{,}000}{\pi R_i^2} = \frac{10{,}000}{\pi \left(D_i / 100k \right)^2} \\
&= \frac{10{,}000 k^2}{\left(\pi / 10{,}000 \right) D_i^2}
\end{aligned}
$$

finally, by dividing top and bottom by 4, we obtain

$$
\mathrm{TF}_i = \frac{\left(1/4\right) 10{,}000 k^2}{\left(1/4\right) \left(\pi / 10{,}000\right) D_i^2} = \frac{2{,}500 k^2}{0.00007854 D_i^2} = \frac{2{,}500 k^2}{\mathrm{BA}_i} \tag{9.12}
$$

As discussed in Section 9.1, other factors are obtained by multiplying the tree factor by the value of the tree characteristic. In horizontal point sampling, the basal area factor is of particular interest. In the imperial system, the basal area factor is

$$
\mathrm{BAF}_i = \left(\mathrm{TF}_i \right) \left(\mathrm{BA}_i \right) = \left(\frac{10{,}890 k^2}{0.005454 D_i^2} \right) \left(0.005454 D_i^2 \right) = 10{,}890 k^2 \tag{9.13}
$$

and for SI:

$$
\mathrm{BAF}_i = \left(\mathrm{TF}_i \right) \left(\mathrm{BA}_i \right) = \left(\frac{2500 k^2}{0.00007854 D_i^2} \right) \left(0.00007854 D_i^2 \right) = 2500 k^2 \tag{9.14}
$$

By examining eqs. (9.13) and (9.14), an outstanding aspect of horizontal point sampling is observed. The basal area factor, BAF, is dependent only on k (likewise, tree factor can be expressed in terms of BAF and tree basal area: $TF_i = BAF / BA_i$). Since k is constant for a given sample, this implies that the basal area per unit area represented by each tree tallied is constant. Thus, basal area per unit area is estimated by counting the number of trees tallied and multiplying by eq. (9.13) or (9.14). *No tree measurements are required to estimate basal area per unit area.*

The value of BAF depends on the value of k, which in turn depends on the angle chosen. Generally, the gauge constant k is chosen such that $10{,}890k^2$ or $2{,}500k^2$ is a convenient number. Once the BAF is fixed, the diameter to plot radius ratio is fixed, which, in turn, implies a plots radius for every tree diameter (as a result, tree factor can be represented as $TF_i = BAF / BA_i$). Consequently, all trees of the same diameter that are located less than their "plot–radius distance" from a given sampling point will be "on the plot" (Fig. 9.6). The plot is completed by rotating 360° about the plot center and determining all "in" trees.

By fixing BAF, the diameter to inclusion zone radius, the gauge constant, and the horizontal angle can all be calculated. For example, if we want the value of BAF to be 10, in imperial units, then

$$10 = 10{,}890k^2$$
$$k^2 = 0.000918$$
$$k = 0.030303$$

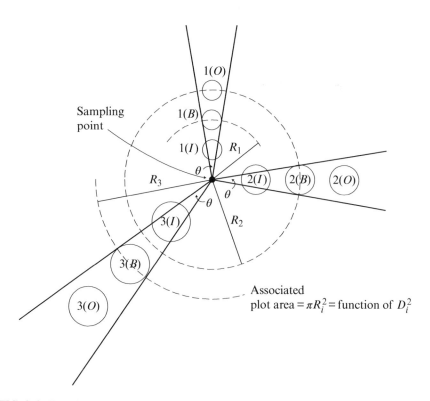

FIG. 9.6. Tree size and plot radius. *I*, designates "in" trees; *B*, border trees; and *O*, out trees.

The angle corresponding to k is

$$k = 2\sin\left(\frac{\theta}{2}\right)$$

$$\sin\left(\frac{\theta}{2}\right) = \frac{k}{2}$$

$$\theta = 2\sin^{-1}\left(\frac{k}{2}\right)$$

$$\theta = 2\sin^{-1}\left(\frac{0.030303}{2}\right)$$

$$= 2\left(52'\right)$$

$$= 1°44'$$

The reciprocal of the diameter to plot radius, referred to as the horizontal distance multiplier (Husch et al., 1983) is

$$R = \frac{D}{12k}$$

$$\frac{R}{D} = \text{HDM} = \frac{1}{12k}$$

$$k = \sqrt{\frac{\text{BAF}}{10,890}}$$

$$\text{HDM} = \frac{1}{12\sqrt{\text{BAF}/10,890}} = \frac{33\sqrt{10}}{12\sqrt{\text{BAF}}}$$

Suppose we want to use a value of BAF=4 for SI units. Then,

$$4 = 2500k^2$$

$$k^2 = 0.0016$$

$$k = 0.04$$

and

$$\theta = 2\sin^{-1}\left(\frac{0.04}{2}\right)$$

$$= 2\left(1°9'\right)$$

$$= 2°18'$$

and

$$R = \frac{D}{100k}$$

$$\frac{R}{D} = \text{HDM} = \frac{1}{100k}$$

$$k = \sqrt{\frac{\text{BAF}}{2500}}$$

$$\text{HDM} = \frac{1}{100\sqrt{\text{BAF}/2500}} = \frac{1}{2\sqrt{\text{BAF}}}$$

TABLE 9.3. Diameter : Plot Radius Ratios, Gauge Constants, and Angles for Common Imperial and SI BAFs

	Imperial Units				SI Units		
BAF	$D:R^a$ (In. : ft)	Gauge Constant (k)	Angle	BAF (m²/ha/tree)	$D:R^a$ (cm : m)	Gauge Constant (k)	Angle
5	1 : 3.9	0.0214	1°14′	1	1 : 0.50	0.0200	1°9′
10	1 : 2.8	0.0303	1°44′	2	1 : 0.35	0.0283	1°37′
15	1 : 2.2	0.0371	2°8′	3	1 : 0.29	0.0346	1°59′
20	1 : 1.9	0.0429	2°27′	4	1 : 0.25	0.0400	2°18′
25	1 : 1.7	0.0479	2°45′	5	1 : 0.22	0.0447	2°34′
30	1 : 1.6	0.0525	3°	6	1 : 0.20	0.0490	2°48′

[a] In imperial units, the diameter : plot radius ratio gives the plot radius in feet per inch of diameter. For example, a 10-in. tree when using a BAF = 10 would have a corresponding plot radius of 10(2.8) = 28 ft; in SI, plot radius is given in meters per centimeter of diameter.

Table 9.3 gives the diameter : plot radius ratios, gauge constants, and horizontal angles for some common imperial and SI BAFs. Types of angle gauges and their calibration and use in forest inventory are discussed in Section 11.1.

For field use and for quick tabulation of results, it is convenient to have a table of plot radii and tree factors for a range of dbh and basal area factors. An example table with values in SI and imperial units is shown in Table A.5.

9.3.2. Stand and Stock Tables

Stand and stock tables can be constructed from horizontal point data using a process similar to constructing stand and stock tables from fixed-area plot data (Section 9.2.5). An example will be used to illustrate the construction of stand and stock tables from horizontal point data. Four horizontal points were sampled from a northern hardwood stand in central New Brunswick, Canada, using a 2M (m²/ha/tree) BAF prism. The following tally was obtained:

dbh class	10	12	14	16	18	20	22	24	26	28	30	32	34	36	38	40	42+
No. tallied	1	2	3	2	3	6	6	6	5	3	6	5	1	1	3	0	2

Table 9.4 shows the combined stand and stock table. Columns 1 through 5 are similar to those in Table 9.2, the stand and stock table based on fixed area plots (Section 9.2.5). Total height was measured on each sample tree and Lorey's average height (Section 8.4.1) calculated (column 2). Column 5 is the number of trees tallied.

The basal area factor (column 7) is constant for horizontal points and was 2 for this sample. Because inclusion zone size varies with tree diameter, the tree factor (column 6) varies and is determined using

$$TF_{dbh} = \frac{BAF}{BA_{dbh}}$$

For this example, the tree factor will be

$$TF_{dbh} = \frac{2.0}{0.00007854 (dbh)^2}$$

TABLE 9.4. A Combined Stand and Stock Table for a Northern Hardwood Stand Located in Central New Brunswick

Column 1	Column 2	Column 3	Column 4	Column 5	Column 6	Column 7	Column 8	Column 9	Column 10	Column 11
dbh	Per Tree Averages			Number of		Factors[a]		Per Hectare Averages		
Class (cm)	Height (m)	BA (m^2)	Volume (m^3)	Trees Tallied	Tree (number)	BA (m^2)	Volume (m^3)	Trees (number)	BA (m^2)	Volume (m^3)
10	11.0	0.0079	0.043	1	254.6	2.0	11.0	63.66	0.50	2.8
12	12.0	0.0113	0.066	2	176.8	2.0	12.0	88.42	1.00	6.0
14	14.2	0.0154	0.099	5	129.9	2.0	14.2	162.40	2.50	17.8
16	16.0	0.0201	0.139	3	99.5	2.0	16.0	74.60	1.50	12.0
18	16.3	0.0254	0.177	2	78.6	2.0	16.3	39.30	1.00	8.2
20	17.7	0.0314	0.259	3	63.7	2.0	17.7	47.75	1.50	13.3
22	18.0	0.0380	0.319	6	52.6	2.0	18.0	78.92	3.00	27.0
24	18.7	0.0452	0.391	6	44.2	2.0	18.7	66.31	3.00	28.1
26	19.0	0.0531	0.526	6	37.7	2.0	19.0	56.50	3.00	28.5
28	19.0	0.0616	0.554	5	32.5	2.0	19.0	40.60	2.50	23.8
30	19.3	0.0707	0.707	3	28.3	2.0	19.3	21.22	1.50	14.5
32	19.0	0.0804	0.732	6	24.9	2.0	19.0	37.30	3.00	28.5
34	19.3	0.0908	0.953	5	22.0	2.0	19.3	27.54	2.50	24.1
36	19.3	0.1018	0.967	1	19.6	2.0	19.3	4.91	0.50	4.8
38	19.4	0.1134	1.117	5	17.6	2.0	19.4	22.04	2.50	24.3
40	19.7	0.1257	1.257	1	15.9	2.0	19.7	3.98	0.50	4.9
42	20.6	0.1385	1.385	3	14.4	2.0	20.6	10.83	1.50	15.5
Total				63				846.29	31.50	283.8

Based on four 2M BAF points.

[a]Factors are the value per unit area represented by each tree tallied.

For example, the tree factor for the 10 cm dbh class is

$$TF_{10} = \frac{2.0}{0.00007854(10)^2} = \frac{2.0}{0.007854} = 254.6$$

The volume factor is obtained by multiplying the volume per tree (column 4) by the tree factor (column 7):

$$VF_{dbh} = V_{dbh} \cdot TF_{dbh}$$

and volume factor for the 10 cm class is

$$\begin{aligned} VF_{10} &= V_{10} \cdot TF_{10} \\ &= 0.043 \cdot 254.6 \\ &= 11.0. \end{aligned}$$

By comparing average height (column 2) with the volume factor (column 8), an interesting and powerful feature of horizontal point sampling can be observed. In this example, volume per tree was estimated using the formula for a paraboloid (eq. (6.2)), which is a constant form factor equation (i.e., $V_i = f \cdot BA_i \cdot H_i$), by substituting this formula for volume into the volume factor equation, we get

$$\begin{aligned} VF_{dbh} &= V_{dbh} \cdot TF_{dbh} \\ &= \left(f \cdot BA_{dbh} \cdot H_{dbh}\right)\left(\frac{BAF}{BA_{dbh}}\right) \\ &= \left(f \cdot BAF\right)H_{dbh} \end{aligned}$$

Thus, if volume can be estimated using a constant form factor equation, then height is the only tree variable that needs to be measured. A further discussion of volume estimation using horizontal point sampling is found in Section 11.1.9.

Trees per hectare (column 9) are obtained by multiplying the number of trees tallied (column 5) by the tree factor (column 6) and dividing by the number of points:

$$Trees/ha = \frac{TF \cdot \left(No.\ tallied\right)}{\left(No.\ plots\right)}$$

For the 10 cm dbh class, the number of trees per hectare is

$$\begin{aligned} Trees/ha &= \frac{TF \cdot \left(No.\ tallied\right)}{No.\ plots} \\ Column\ 9 &= \frac{\left(Column\ 6\right) \cdot \left(Column\ 5\right)}{No.\ plots} \\ &= \frac{254.6 \cdot 1}{4} \\ &= 63.66 \end{aligned}$$

Similarly, basal area per hectare (column 10) is obtained by multiplying the BAF (column 7) by the number of trees tallied (column 5) and dividing by the number of points and volume per hectare (column 9) is obtained by multiplying the volume factor (column 8) by the number of trees tallied (column 5) and dividing by the number of points. In this particular example, because form $= \frac{1}{2}$ and BAF $= 2$, the $VF_i = H_i$ and volume per hectare is simply the sum of the heights of the "in" trees.

9.3.3. Boundary Slopover Bias

As with fixed-area plots, when trees are located near stand or stratum boundaries, part of their inclusion zones may slop over the boundary. The main difference with horizontal point sampling is that the inclusion zone radius or limiting distance is a continuous function of the diameter of the individual trees; thus, it may require careful attention to determine whether a particular sample point has fallen close enough to the edge to require field correction. Any of the methods discussed in Section 9.2.6 can be applied to horizontal point samples. The toss-back method requires an estimate of the maximum inclusion zone radius, which in turn depends on the BAF and the diameter of the largest tree that might be encountered. That radius should be estimated conservatively (i.e., overestimated) to ensure that all inclusion zones are fully captured in the buffered tract. The mirage method, originally developed by Schmid-Haas (1969) and later described by Beers (1977), was considered the easiest and (except for the flawed approach of moving sampling points away from the boundary) was the most commonly employed method of boundary overlap correction in the field for many years. The walkthrough method of Ducey et al. (2004) is also very fast when applied to horizontal point sampling, and the walking-to-and-fro approach of Flewelling and Strunk (2013) is also applicable. The *direct weighting procedure* (Beers, 1966) can also be used when work outside the boundaries is difficult. Just as with fixed-area plots, the choice of boundary slopover correction should depend on applicability (whether the boundaries are straight, and whether measurements can be done safely outside the tract), and also on field efficiency.

9.3.4. Other Forms of Sampling Proportional to Size

Other forms of sampling standing tree stems with probability proportional to size include horizontal line sampling, vertical point sampling, and vertical line sampling. None of these techniques are as widely applied as horizontal point sampling, but they may have application for some problems. For a complete explanation of sampling theory and estimation procedures for these alternate forms, see Grosenbaugh (1958) and Husch et al. (1983).

Vertical Point Sampling. Vertical point sampling was developed by Hirata (1955). In its original development, vertical point sampling is implemented by projecting a vertical angle ϕ from a point location. Trees are selected for sampling if the angle to the tip of the tree is greater than the projected angle (Fig. 9.7a). The gauge constant q is

$$q = \frac{H}{R} = \tan \phi$$

and the associated inclusion zone radius is given by

$$R_i = \frac{H_i}{q}.$$

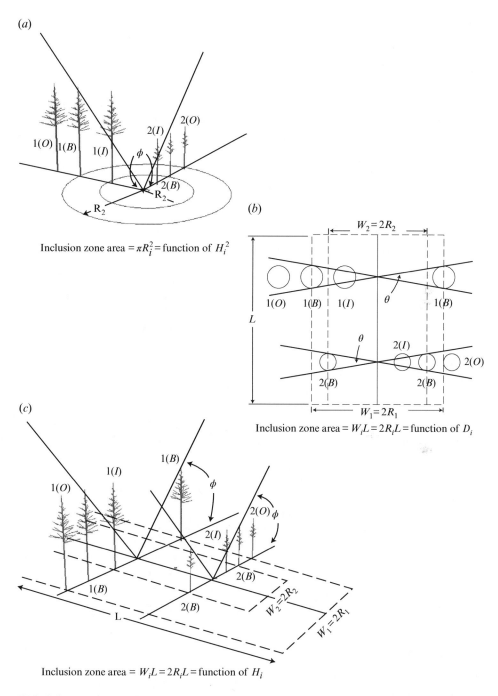

(a)

Inclusion zone area $= \pi R_i^2 =$ function of H_i^2

(b)

Inclusion zone area $= W_i L = 2 R_i L =$ function of D_i

(c)

Inclusion zone area $= W_i L = 2 R_i L =$ function of H_i

FIG. 9.7. Other forms of PPS sampling: (a) vertical point sampling, (b) horizontal line sampling, and (c) vertical line sampling.

Since inclusion zone radius is proportional to height, plot area is proportional to height squared, so vertical point sampling selects trees with probability proportional to height squared.

Horizontal Line Sampling. In horizontal line sampling, as originally developed by Strand (1957), a horizontal angle θ is projected perpendicular to a line of length L (Fig. 9.7b). The line may span the entire tract of interest or it may be a short fixed-length segment centered on a sample point. Trees are tallied on both sides of the line. The gauge constant k and the radius R_i are the same as in horizontal point sampling, and indeed the same instruments are often used. The associated plot area is $L \cdot 2R_i$. Since R_i is proportional to individual tree diameter, so is plot area; thus, horizontal line sampling selects trees with probability proportional to diameter. Whereas horizontal point sampling gives a "variable radius plot," horizontal line sampling gives a "variable width strip."

Vertical Line Sampling. In vertical line sampling, which was also originally developed by Strand (1957), a vertical angle ϕ is projected perpendicular to a line of length L (Fig. 9.7c). Trees are tallied on both sides of the line. The gauge constant q and the radius R_i are the same as in vertical point sampling. The associated plot area is $L \cdot 2R_i$. Just as in horizontal line sampling, the sample is a variable-width strip; but vertical line sampling yields a probability proportional to tree height.

The tree factors and factors for other characteristics can be derived for each PPS sampling technique using the same process shown for horizontal point sampling and the inclusion zone areas (Fig. 9.7). When using horizontal point sampling, basal area per unit area is estimated directly without additional tree measurements. With horizontal line sampling, the sum of diameters per unit area is estimated directly. With vertical point sampling, the sum of squared heights per unit area is estimated directly, and with vertical line sampling the sum of heights per unit area is estimated directly. Factor formulas for various stand characteristics are given in Table A.6 for all four of the main forms of PPS sampling.

9.4. OTHER EXAMPLES OF VARIABLE PROBABILITY SAMPLING

Although PPS designs focused on tree stems and using a relatively simple geometric approach based on fixed angles are the most developed and most widely used designs in forest inventory, they are by no means the only such approaches. Nor are they the only approaches that can be used effectively in forestry. An exhaustive survey of all possible variable probability approaches could consume an entire book in itself. Here, we briefly summarize a selected group to give an indication of the additional possibilities.

9.4.1. Point Intercept Sampling

Point intercept sampling was introduced to applied ecology by Levy and Madden (1933) for the analysis of range vegetation, and its use has quickly spread to other situations. In point intercept sampling, individual points, a line of points, or a grid of points are laid out and a physical or optical probe is run vertically. The material that contacts the probe is tallied. The use of a dot grid to estimate area on maps and aerial photographs (Section 4.4.2) is a special case of point intercept sampling, and a primary use of point intercept sampling is to estimate vegetative cover (both in aggregate and by species). If the material tallied can be assigned to individual items, such as individual plant crowns, it is easy to see that point intercept sampling includes those items with probability proportional to their horizontal coverage. If the horizontally projected area of those items can be measured, then it is possible to estimate the number of such items in addition to their coverage.

9.4.2. Line Intercept Sampling

In line intercept sampling, a sample line (ideally with its position and orientation chosen at random) is run from a sample point, and the material that crosses the line is tallied. Line intercept sampling was developed by Canfield (1941) for sampling range vegetation, and it was quickly adapted to other material. In line intersect sampling, the proportion of the sample line that is covered by the material gives a direct estimate of the proportion of the ground (or other horizontal plane) that is covered. When the material being sampled consists of lines or curves, line intersect sampling includes that material with probability proportional to its length; thus, a simple count of objects encountered multiplied by a length factor (see Chapter 12) yields an estimate of the cumulative length per unit area of material. This approach has been widely used to estimate road density from aerial photos, and the density of animal tracks both from aerial photographs and field surveys. A primary use of line intercept sampling in forests is sampling downed dead wood (Section 12.2). Brown (1974) details the closely related *plane intercept sampling* method, which is perhaps the most widely used method worldwide for estimating coarse and fine fuels. Line intercept sampling has also been used to estimate the area and characteristics of canopy gaps (Battles et al., 1996).

9.5. DISTANCE-BASED SAMPLING UNITS

A variety of other sampling methods have been developed in the history of ecology, and those that use the distance from a sample point to one or more nearest neighbors have held an enduring fascination. Although they have not been widely used in forest inventory, especially in North America, they have been used in other contexts. However, they also present some significant challenges, especially in populations that show strong spatial pattern, and at present no estimating equations exist for these methods that can be shown to be unequivocally unbiased. A different set of sampling approaches, called *distance sampling*, uses the distances to sampled objects to deal with the problem of objects that have not been detected in the survey. Although originally developed in the context of wildlife surveys, these methods are beginning to find use in assessing forests as well.

9.5.1. Nearest-Neighbor Methods

While nearest-neighbor estimators have been widely used in plant ecology (Pielou, 1977), they rarely have been used in forest inventories, especially in North America. Some authors refer to these techniques as "distance sampling" but that term properly belongs to the family of techniques presented in Section 9.5.2. In the most common approach to nearest-neighbor sampling, once a sample point is located, the distances to the k nearest neighbors are measured along with their attributes. For example, when $k=4$, the distances to the

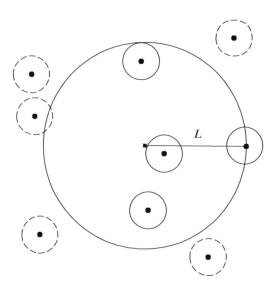

FIG. 9.8. Nearest-neighbor sampling.

closest four trees to the sample point are needed; the distance to the fourth tree describes a circular area that is thought of as being akin to a plot. The point-centered quarter method (Cottam et al., 1953), in which the distance to the nearest neighbor in each of four 90-degree quadrants is used, can be considered a special case of this more general family (a cluster of samples with $k=1$, each with a wedge-shaped rather than circular plot). Sheil et al. (2003) describe an approach that can be thought of as giving variable-length strips.

Intuitively, the distance to the kth nearest neighbor describes a sample area; knowing the number of objects in that area, one can calculate a density. Unfortunately, as we shall see below, the unbiased estimation of density or other attributes from such data is fraught with difficulty. Nonetheless, nearest-neighbor methods have been suggested for a variety of purposes in a variety of forested contexts. Dennis (1984) discusses a variety of distance-based estimators for assessing regeneration. Payandeh and Ek (1986) assessed the relative performance of five different distance–density estimators. Jonsson et al. (1992) describe a forest inventory system using density-adapted circular plot sizes. Lessard et al. (1994) compared nearest-neighbor sampling with point and fixed-area plots in northern hardwood stands in Michigan for a variety of values of k. They found that, if biased-corrected estimators are utilized, nearest-neighbor sampling produced estimates comparable to point and fixed-area plot samples. They found that nearest-neighbor methods could be cost-competitive with fixed-area plots and horizontal point sampling. Lynch and Rusydi (1999) found similar results for teak plantations in Indonesia.

With nearest-neighbor sampling, the number of trees k per plot is fixed and the distance L to the kth nearest tree is measured. In Figure 9.8, this is illustrated for $k=4$. The distance L_i is the plot radius for the ith plot, and, if circular plots are used, plot size is $a_i = \pi L_i^2$. Since the kth tree lies exactly on the plot boundary (Fig. 9.8), the plot is considered to have $k-\frac{1}{2}$ trees and the estimate of density for the ith plot was suggested by Prodan (1968b) as

$$N_i = \frac{A(k-0.5)}{\pi L_i^2}$$

where $A = 43,560\,\text{ft}^2$ (imperial units) or $10,000\,\text{m}^2$ (SI units). Mean density is estimated as

$$\bar{N} = \frac{\sum_{i=1}^{n} N_i}{n} = \frac{1}{n}\sum_{i=1}^{n}\frac{A(k-0.5)}{\pi L_i^2} = \frac{A(k-0.5)}{n\pi}\sum_{i=1}^{n}\frac{1}{L_i^2} \tag{9.15}$$

where $\bar{N}$ = mean density (trees per unit area)
      A = unit area ($43,560\,\text{ft}^2$ or $10,000\,\text{m}^2$)
      k = fixed number of trees per plot
      n = number of plots
      L_i = distance (ft or m) to kth nearest tree on the ith plot

The estimated density from eq. (9.15) is biased except perhaps when a completely random spatial pattern is assumed for the population (Payandeh and Ek, 1986). Moore (1954) suggested the use of $(k-1)/k$ as a bias correction. In this case, the mean density is estimated as

$$\bar{N} = \left(\frac{k-1}{k}\right)\frac{1}{n}\sum_{i=1}^{n}\frac{Ak}{\pi L_i^2} = \left(\frac{k-1}{k}\right)\frac{Ak}{n\pi}\sum_{i=1}^{n}\frac{1}{L_i^2} = \frac{A(k-1)}{n\pi}\sum_{i=1}^{n}\frac{1}{L_i^2} \tag{9.16}$$

Equation (9.16) suggests that the kth tree is only used to establish the plot radius (Jonsson et al., 1992). Estimation of other parameters, such as basal area or volume, is made using

$$\bar{X} = \left(\frac{k-1}{k}\right)\frac{1}{n}\sum_{i=1}^{n}\frac{AX_i}{\pi L_i^2} = \left(\frac{k-1}{k}\right)\frac{A}{n\pi}\sum_{i=1}^{n}\frac{X_i}{L_i^2} \tag{9.17}$$

where X_i = the sum of the values of X (basal area, volume, etc.) for the k trees of the ith plot: $X_i = \sum_{j=1}^{k} X_{ij}$.

The amount of bias present in these estimates is dependent upon the spatial distribution and the number of trees measured per plot. Prodan (1968b) used $k=6$ trees. Lessard et al. (1994) and Lynch and Rusydi (1999) found that $k \geq 5$ generally produce acceptable results. Payandeh and Ek (1986) recommended using $k=7$; however, the inability to say definitively that the estimates are unbiased, and the sensitivity of bias to spatial pattern in the population, has proven disconcerting. Fundamentally, the challenge is that the "plot areas" implied in eqs. (9.15) through (9.17) are not the areas of the inclusion zones of the trees. The inclusion zones for nearest-neighbor sampling are kth order Voronoi polygons and cannot be recovered either from the tree sizes or from the sample point-to-tree measurements (Kleinn and Vilčko, 2006). A brute-force approach would require mapping the trees in the neighborhood of the sample trees, and this would undermine the apparent simplicity and speed that make nearest-neighbor approaches seductive (Iles, 2009). Fehrmann et al. (2012) present a triangulation-based method that would reduce the field work, while Magnussen et al. (2008, 2012) present alternative estimating equations that achieve better performance at a considerable cost in complexity; however, a completely general and practical solution has yet to be found. Some authors have found the possibility of bias in nearest-neighbor methods completely unacceptable. For example, Schreuder (2004, 3) wrote, "The approach is biased and can be hard to implement in practice. In addition it is not fulfilling a need that cannot be met by traditional sampling designs." In fact, the field efficiencies that nearest-neighbor methods are trying to achieve can be simply achieved by controlling the size of the sampling unit during the design phase to obtain a targeted number of trees per sample point and avoids all the uncertainties in the estimates from nearest-neighbor methods (Section 9.6).

9.5.2. Distance Sampling

In all of the other sampling approaches presented in this chapter, it is assumed that if a tree (or other object) is supposed to be included in the sample, it will be. For example, if a tree is within the plot radius of a fixed-area plot, it will be found and tallied. This contrasts sharply with the situation in direct assessment of wildlife populations as animals often hide, flee, or otherwise evade detection during inventories. Wildlife ecologists have developed a very sophisticated set of sampling approaches that use distances to detected organisms to model and account for imperfect detection. This collection of sampling approaches, known as *distance sampling*, can be implemented around sample points or along sample lines, and involves making the following assumptions:

1. Objects falling on the point or line are detected with certainty.
2. Objects are detected at their initial point, before moving in response to the approach or presence of the observer.
3. The distances to objects can be measured accurately, or assigned accurately to discrete groupings.
4. The success or failure in detecting one object is independent of the success or failure in detecting others.

For a complete treatment of the theory and application of distance sampling, see Buckland et al. (2001, 2004).

Distance sampling has only recently begun to be used in forestry applications, but it has appeal in certain circumstances. For example, if a particular type of tree or kind of understory vegetation is rare and difficult to detect, it could be extremely time-consuming to ensure 100% detection on plots that are of a sufficient area to provide an adequate sample. Kissa and Sheil (2012) demonstrate the use of distance sampling along a line for rare but important species in diverse tropical forests. In some cases, field crews may not be sufficiently diligent at finding large, nearly borderline trees, as in the case with horizontal point sampling with a small basal area factor. Ritter et al. (2013) develop distance sampling approaches that can be used with horizontal point sample data; Ritter and Saborowski (2012) provide similar developments for downed dead wood sampling. Alternatively, as interest shifts to new technologies such as terrestrial LiDAR, it may be impossible to guarantee 100% detection of trees in automated processing of digital data. Distance sampling can help correct for nondetection in such data (Ducey and Astrup, 2013). Given the range of potential applications, continued development and adoption of distance sampling approaches in forestry seems likely.

9.6. SELECTING APPROPRIATE SAMPLING UNITS

When selecting a sampling unit, one must consider both the type of sampling unit and its size and shape. In general, the most efficient sampling unit is the one that samples proportional to the variance of the stand parameters of interest. If density (trees per unit area) is the primary variable of interest, fixed-area plots are generally more efficient than PPS sampling. Fixed-area plots sample proportional to frequency and, since tree factor is constant for a given plot size, no additional individual tree measurements are required. If basal area or volume is the primary variable of interest, horizontal point samples are generally more efficient than fixed area plots. Horizontal point samples sample proportional to tree basal area. No additional individual tree measurements are required to estimate basal area. In many cases, estimates of volume can be obtained from basal area and measurements of average height and form (Section 11.1.9).

Unbiased estimates of timber quantities and other stand parameters can be obtained from any sampling unit type and size, although the precision and cost of the survey or study may vary significantly. Smaller sampling units are frequently more efficient than larger ones (Section 10.2.1). In a fairly homogeneous forest, precision for a given sampling intensity tends to be greater for small sampling units than for large ones because the number of independent sampling units is greater. However, the size of the most efficient unit is also influenced by the variability of the forest. When small sampling units are measured in heterogeneous forests, high coefficients of variation will be obtained. In such cases, large sampling units are more desirable.

For a given sampling intensity (the total area of all sampling units), the smaller the sampling units the greater will be the precision because there will be more units. On the other hand, the greater the number of units, the higher will be the cost of sampling. In general, the cost of sampling will be greater for a large number of small sampling units than for fewer units of larger size. In addition, each sampling unit should be large enough so that it will reasonably represent the composition and structure of the forest. If the sampling unit is too small, the probability that it will not be representative of the population increases.

The guiding principle in the choice of sampling unit size should be to have a sample unit large enough to include a representative number of trees, but small enough so that the time required for measurement is not excessive. For dense stands of small trees, units should be relatively small; for widely spaced stands of large trees, units should be relatively large.

When using horizontal point samples, the desired BAF can be determined using an estimate of mean basal area per unit area and the desired number of trees m to be included at each point:

$$\text{BAF} = \frac{\text{BA/Unit area}}{m} \tag{9.18}$$

For example, if a forest averages $125\,\text{ft}^2/\text{acre}$ of basal area and we want to count 10 on average, the appropriate BAF would be $125/10 = 12.5$. A similar process, based on average density, can be used to determine the size of fixed-area plots. The relative efficiency of different-sized plots can be measured by

$$e = \frac{\left(s_{\bar{x}}\right)_1^2 t_1}{\left(s_{\bar{x}}\right)_2^2 t_2} \tag{9.19}$$

where $(s_{\bar{x}})_1$ = standard error in percent for one sampling unit as the basis for comparison
        $(s_{\bar{x}})_2$ = standard error in percent for the other sampling unit to be compared
          t_1 = cost or time for base sampling unit
          t_2 = cost or time for compared sampling unit

The equation gives the efficiency of sampling unit 2, relative to sampling unit 1. If e is less than 1, then sampling unit 1 is more efficient than plot 2. If e is greater than 1, sampling unit 2 is more efficient than plot 1. Equation (9.19) ignores travel time between sample plots or points; a discussion of the rationale for this assumption, and its implications, can be found in Ducey et al. (2013b). Other procedures for investigating optimum sampling units are given by Mesavage and Grosenbaugh (1956), Freese (1962), Tardif (1965), O'Regan and Arvanitis (1966), Zeide (1980), Oderwald (1981), and Shiver and Borders (1996).

Zeide (1980) presented a procedure for determining optimal plot size in one-stage systematic or random sampling designs. The optimal plot size was proposed to be the one that minimizes total time required for location and measurement to achieve a given precision. Zeide then developed the following formula for calculating optimal plot size:

$$\text{Optimal plot size} = P_1 \left(\frac{t}{m} \right)^2 \tag{9.20}$$

where P_1 = size of plot used in preliminary sample to assess time and variation
       t = average travel time between two neighboring plots of size P_1 (the distance between these plots should be for the number of plots that provides the desired precision)
     m = average plot measurement time for plot size P_1

The formula indicates that the greater the distance between plots the larger they must be. When plot size is optimal, equal amounts of time will be spent on travel and on plot measurement.

10

SAMPLING DESIGNS IN FOREST INVENTORIES

On a large forest, sampling can provide all the necessary information in less time and at a lower cost than a 100% inventory. Indeed, it is impractical on a large forest to measure all the trees or other characteristics of interest. Since fewer measurements are needed, sampling may produce more reliable results than a complete tally because fewer, better-trained personnel can be used, and better field supervision of the work can be exercised, and results can be obtained in a more timely fashion. In addition, the idea of precision is in the forefront throughout a sampling inventory, whereas a complete tally may give the illusion of the acquisition of information without error. This chapter is concerned with the application of sampling theory and techniques to forest resource evaluation applicable to both timber and nontimber parameters. The reader is referred to Cochran (1977) for a basic treatment of sampling theory and techniques. Freese (1962), Schreuder et al. (1993), Shiver and Borders (1996), and Gregoire and Valentine (2008) are additional useful references for sampling in forest inventory.

10.1. BASIC CONSIDERATIONS

In Chapter 3, a comprehensive overview of basic statistical concepts and methods was given. A few of these basic concepts will be briefly reviewed here in the context of forest inventory. In forest inventory work, sampling consists of making observations on portions of a population (the forest and its characteristics) to obtain estimates that are representative of the parent population.

An observation is a recording of information, such as the volume of a fixed-area plot or the basal area of a horizontal point sample. The sampling units on which observations are

Forest Mensuration, Fifth Edition. John A. Kershaw, Jr., Mark J. Ducey, Thomas W. Beers and Bertram Husch.
© 2017 John Wiley & Sons, Ltd. Published 2017 by John Wiley & Sons, Ltd.

made may be stands, compartments, administrative units, fixed-area plots, strips, or points. The aggregate of all possible sampling units constitutes its population. The group of sampling units chosen for measurement constitutes the sample. For purposes of selecting sampling units, a sample frame can be prepared that can be a diagram or list of all possible sampling units in the population. Some populations are finite; that is, they consist of a fixed number of sampling units such as the 100 nonoverlapping 50 m by 50 m square plots shown in Figure 3.1. Other populations, such as all possible points (as used in horizontal point sampling), are infinite; that is, they consist of an unlimited number of sampling units. In the statistical treatment of data, it is important to know whether the population is infinite or finite; however, as a practical matter, very large finite populations can be treated as infinite.

The essential problem in sampling is to obtain a sample that is representative of the population. If the sample is representative of the population, then useful statements can be made about the characteristics of the population (volume or weight per unit area, number of trees per unit area, total volume in the area, etc.) on the basis of the characteristics observed in the sample observations. The characteristics of the population are referred to as *parameters*. The exact values of the parameters would be known if the entire population was measured. In most situations, however, complete enumerations are too time consuming and costly, and sampling is used to estimate parameter values. Such estimates are calculated from a sample and are called *statistics*. A statistic therefore is a summary value calculated from a sample to estimate a population parameter.

A sampling design is determined by the kind of sampling units used, the number of sampling units employed, and the manner of selecting and distributing these sample units over the forest area, as well as the procedures for making measurements and analyzing the results. The specifications for each of these elements can be varied to yield the desired precision at a minimum specified cost.

Basic inventory designs generally fall into the following categories:

1. Probability sampling
   a. Simple random sampling (SRS)
   b. Systematic sampling (SYS)
   c. Stratified sampling (STS)
   d. Multistage sampling
   e. Multiphase sampling
   f. Sampling with varying probabilities
2. Nonrandom or selective sampling

This chapter concentrates on sampling designs using fixed-area plots (Section 9.2), although the basic sampling designs are applicable to sampling units of any type. Chapter 11 will discuss sample design issues unique to sampling units with varying probability, and Chapter 12 will discuss sampling issues related to snags and down woody debris.

10.1.1. Errors in Forest Inventories

The *precision* of a forest inventory based on sampling is indicated by the size of the sampling error and excludes effects of nonsampling errors. *Accuracy* of an inventory refers to the size of the total error and includes effects of nonsampling errors. In forest inventory, as in any sampling procedure, we are primarily concerned with accuracy. We try to achieve accuracy by designing and executing an inventory for an acceptable precision and by minimizing nonsampling errors.

Sampling errors result from the fact that the sample is only a portion of the population and may not produce estimates identical to the population parameter. The sampling error is expressed as the standard error of the mean (Section 3.5). In SRS, this error can be estimated using

$$s_{\bar{x}} = \frac{s}{\sqrt{n}} \tag{10.1}$$

where s is the standard deviation of the sample (eq. (3.12)) and n is the sample size. For more complex sample designs, the formula for standard error is modified to reflect design considerations and other restrictions on randomization (these are discussed more fully in each section).

The primary objective of a sample design is to allocate and locate sampling units such that the best unbiased estimate of the mean, $\bar{x}$, with the smallest possible standard error, $s_{\bar{x}}$, given a fixed level of effort (sample size), is achieved. Understanding how $s_{\bar{x}}$ is influenced by sampling designs is critical to choosing the most efficient sampling design for the inventory situation.

Nonsampling errors are errors not connected with the statistical problem of selecting the sampling units, and may, therefore, occur whether the entire population, or a sample of the population, is measured. Thus, nonsampling errors are always present, while sampling errors are present only when sampling methods are employed. Nonsampling errors arise from defects in the sampling frame, mistakes in the collection of data due to bias or negligence, and mistakes in the recording or processing stages. When nonsampling errors are large, the total error will be reduced only slightly by taking a large sample (to decrease sampling error). Indeed, when the nonsampling errors are large, attention should be focused on reducing them before the sample size is increased to reduce sampling errors.

Prodan et al. (1997) compiled a detailed categorization of nonsampling errors:

1. *Design errors.* These are errors that produce bias in the estimations due to the non-observance of the probability of selection (subjective distribution) or the independence between sample units. The systematic distribution of sampling units may cause bias if the distribution coincides with some specific characteristic of the area such as all sampling units falling in river bottoms, ridge tops, or on some given contour. Iles (2003) points out that, while this could occur, it is unlikely; however, it is important that such considerations be taken into account when developing a sample design.

2. *Operational errors.* The principal operational errors are the erroneous location of samples, inaccurate establishment of sample unit boundaries, and errors in the measurement of tree dimensions or nontimber elements. These errors can be minimized with appropriate training of field crews and the exercise of quality control.

3. *Errors in the functions used to quantify parameters.* For example, inappropriate mathematical models or errors in their coefficients and constants. (For example, errors in or inappropriate functions to estimate tree volumes.)

4. *Errors in determination of areas.* These are errors that may arise in the preparation of maps and the determination of areas from them.

5. *Errors in management and processing of data.* Examples are the incorrect codification and registering of measurements, errors in their transfer from field forms, and errors that may occur in the use of programs for the processing of data.

10.1.2. Confidence Limits

Because forest inventory estimates include sampling error, they often are expressed as a range. This range may be expressed as a function of $s_{\bar{x}}$: $\bar{x} \pm s_{\bar{x}}$, or more formally as a confidence interval (Section 3.5.2)- bounded limits of a random interval that has a probability (i.e., confidence) of including the population parameter (i.e., the mean). The confidence interval is expressed by

$$CI = \bar{x} \pm t \cdot s_{\bar{x}} \tag{10.2}$$

The value of t for a chosen probability level $(1-\alpha)$ can be found from Student's t distribution (Table A.4) using $n-1$ degrees of freedom, where n is the size of the sample.

If $\bar{x}$ is the mean volume per unit area and $s_{\bar{x}}$ is the standard error of the mean from a sample-based inventory, then the confidence interval for the total timber volume estimate for a forest area A (assuming A is known without error) is

$$CI(\text{total}) = A \cdot \bar{x} \pm A \cdot t \cdot s_{\bar{x}} \tag{10.3}$$

As an example, assume a 5,000-acre forest area has been inventoried. Based on a sample of 144 fixed-area plots, the mean volume per acre was estimated as 11,500 board feet, with a standard deviation of 6,000 board feet. The 95% confidence interval for the per-acre volume would be calculated as follows:

$$s_{\bar{x}} = \frac{s}{\sqrt{n}} = \frac{6000}{\sqrt{144}} = 500,$$

and

$$\begin{aligned}
CI(\text{per acre}) &= 11,500 \pm (1.9759)(500) \\
&= 11,500 \pm 988 \\
&= 10,512 \text{ to } 12,488 \\
&= 10,500 \text{ to } 12,500 \text{ board feet/acre.}
\end{aligned}$$

The value of $t = 1.9759$ was determined from Table A.4 using the column for probability $= 0.05$ $(1 - 0.95)$ and, since 144-1 was not included in the table, the row corresponding to degrees of freedom nearest 143 (i.e., 150) was used.

Assuming the value for the area, $A = 5000$, is known without error, then the confidence interval and its limits for the total timber volume estimate is

$$\begin{aligned}
CI(\text{total}) &= (5,000)11,500 \pm (5,000)(1.9759)(500) \\
&= 57,500,000 \pm 4,939,750 \\
&= 52,560,250 \text{ to } 62,439,750 \\
&= 52,600,000 \text{ to } 62,400,000 \text{ board feet.}
\end{aligned}$$

Equation (10.3) assumes the total forest area, A, is known without error. This is seldom the case. Usually, the forest area is estimated by some procedure, such as repeated planimeter, dot grid, or GPS measurements, and the standard error computed. Then, considering both the sampling error of volume per unit area and the sampling error of the area estimate, s_A, the confidence interval can be approximated using (Goodman, 1960)

$$\text{CI(total)} = A\bar{x} \pm t\sqrt{(As_{\bar{x}})^2 + (\bar{x}s_A)^2} \tag{10.4}$$

Assuming that the estimate of forest area in the example has a standard error of 75 acres, then

$$\text{CI(total)} = (5,000)(11,500) \pm 1.9759\sqrt{[(5,000)(500)]^2 + [(11,500)(75)]^2}$$
$$= 57,500,000 \pm 5,225,464$$
$$= 52,300,000 \text{ to } 62,700,000 \text{ board feet.}$$

In the above calculation, we used the t value associated with the degrees of freedom for the inventory. Degrees of freedom for area determination are often quite small or not clearly specified. The approach used above produces a conservative (wider) confidence interval. For more detailed approaches for determining degrees of freedom and appropriate t values for two samples with unequal sample sizes and unequal variance, refer to Zar (2009).

The estimate of a forest area, A, may be determined as a proportion, p, of a total area using a procedure such as dot counts on aerial photos. In this case, one assumes the total area A_T is known without error; however, the estimate of the proportion, p, will have a standard error s_p. The approximate confidence interval for the total volume would then be

$$\text{CI(total)} = A\bar{x} \pm t\sqrt{(As_{\bar{x}})^2 + (\bar{x}A_T s_p)^2} \tag{10.5}$$

where A, the forest area, is given by $p \cdot A_T$.

Assuming that the forest area of 5000 acres in the previous example was determined from a dot count showing 62.5% in forest of a total land area of 8000 acres, and that the estimate of $s_p = 0.007$, then

$$A = 0.625 \cdot 8000 = 5000 \text{ acres}$$

and

$$\text{CI(total)} = 5,000(11,500) \pm 1.9759\sqrt{[(5,000)(500)]^2 + [(11,500)(8,000)(0.007)]^2}$$
$$= 57.500.000 \pm 5,101,0013$$
$$= 52,400,000 \text{ to } 62,600,000 \text{ board feet.}$$

The width of the confidence interval $ts_{\bar{x}}$ can be considered a measure of the error associated with an estimated mean at the desired level of probability. By dividing this quantity by the mean, the percent error of an estimated mean $E\%$ at the desired probability level is obtained:

$$E\% = 100\left(\frac{ts_{\bar{x}}}{\bar{x}}\right) \tag{10.6}$$

This is also called the percent sampling error and expresses the precision of the inventory. For the example, we have been discussing in this section, E (in percent), for the 0.95 probability level would be

$$E\% = 100\left(\frac{(1.9759)(500)}{11,500}\right) = 8.6\%$$

Instead of expressing a timber estimate by its mean and confidence interval, an analogous expression called the *reliable minimum estimate* (RME) can be used (Dawkins, 1957). The RME estimates the minimum quantity expected to be present with its probability:

$$\text{RME} = \bar{x} - ts_{\bar{x}} \qquad (10.7)$$

The value of t for this probability level is obtained from one side of the distribution. In using a table of t values where the sign is ignored (i.e., "two-tailed" table as shown in Table A.4), the appropriate value would be obtained using the column for double the probability level required. Thus, the t value for a probability level of 0.05 would be read under the column headed 0.10, recognizing the appropriate degrees of freedom. For the example we have been following, the RME would be

$$\text{RME}(\text{per acre}) = 11,500 - (1.645)(500)$$
$$= 11,500 \pm 820$$
$$= 10,300 \text{ board feet/acre.}$$

10.1.3. Precision Level and Intensity

The choice of precision level for a forest inventory depends on the level of sampling error one is willing to accept in the estimates. Thus, one might ask: What would occur if decisions on investment, forest management, and so forth, were based on estimates with sampling errors of ±1%, ±5%, or ±10%? In most forest inventory work, this is not done. Precision levels used are ones traditionally employed in similar inventories. The difficulty in attacking this problem comes from the lack of methods to quantify effects of inventory sampling errors of different sizes on forest management decisions (Hamilton, 1978; Husch, 1980).

The precision level can be expressed in relative terms as a percent (eq. (10.6)) or as a standard error of the mean (eq. (10.1)). In the following sections, the formulas for the determination of sample size utilize these expressions of acceptable precision.

The *intensity* of sampling expresses the percentage of the total area actually included in the sample. Sample intensity can be calculated for fixed-area plot sampling, but not for variable probability sampling. For example, if 200 0.2-acre plots are measured in a 1000-acre forest, the sample intensity, I, is

$$I = \frac{(200)(0.2)}{1000}(100) = 4.0\%.$$

For horizontal point sampling, attempts to calculate sample intensity based on the largest inclusion radius or average inclusion radius are incorrect and do not give true percentages sampled. Expressing a *sample rate* such as number of sample points per unit area or area per sample point is more meaningful.

10.2. SIMPLE RANDOM SAMPLING (SRS)

SRS is the fundamental selection method. All other sampling procedures are modifications of SRS that are designed to achieve greater economy or precision. SRS requires that there be an equal chance of selecting all possible combinations of n sampling units from the population. Selection of each sampling unit must be free from deliberate choice and must be completely independent of the selection of all other units.

In SRS, the entire forest area is treated as a single population of N units. If nonoverlapping, fixed-area sampling units are used, the population size, N, has definable limits. If points are used (or fixed-area plots overlap), N can be considered infinite. From the population, a sample of n sampling units with equal probability of selection is randomly chosen.

Simple, or unrestricted, random sampling in forest inventory yields an unbiased estimate of the population mean and the information required to assess sampling error; however, SRS has the following disadvantages:

1. Requirement of devising a system for randomly selecting plots or points.
2. Difficulties in locating widely dispersed field positions of selected sampling units.
3. Time-consuming and expensive nonproductive traveling time between units.
4. Possibility of a clumpy distribution of sampling units may result in atypical estimates of the mean, standard deviation, and other measures.

Forest inventory using SRS requires the establishment of a sampling frame, such as maps, aerial photos, or satellite images, from which to draw the sample. Consider the 25 ha forest used in Chapter 3, Plate 1 is a stem map of the individual trees by species. A sampling frame could be devised using two methods:

1. The forest is divided up into a grid of equal-sized square or rectangular plots, the plots are numbered and randomly selected by number (Fig. 10.1a).
2. The center of each plot is selected by randomly generating plot coordinates within the boundary of the forest (Fig. 10.1b).

The first approach is rarely carried out in practice, and circular plots, which do not completely tessellate an area, often are used for many logistical reasons (Section 9.2.1). The second approach is more common. For the forest in Plate 1, this is straightforward since it is a square and an SRS can be designed by randomly selecting pairs of numbers from the uniform distribution, $U[0,500]$. More complex shapes would require more sophisticated algorithms. Most geographic information systems (GIS) software have algorithms that can perform this task. Graphically oriented statistical software, such as R (R Development Core Team, 2013), also have routines that can be used.

After the number of sampling units, n, has been determined, they are chosen from the sample frame using any accepted procedure for random selection. The selection can be with or without replacement. If sampling with replacement is followed, since there is a possibility of the same sampling unit being selected more than once, the population can be considered infinite. For large finite populations, the calculation of mean and standard errors can be done as though dealing with an infinite population since the finite population correction factor $(N-n)/N$ (eq. (3.15)) approaches unity. Most sampling with fixed-area

(*a*)   (*b*)

(*c*)

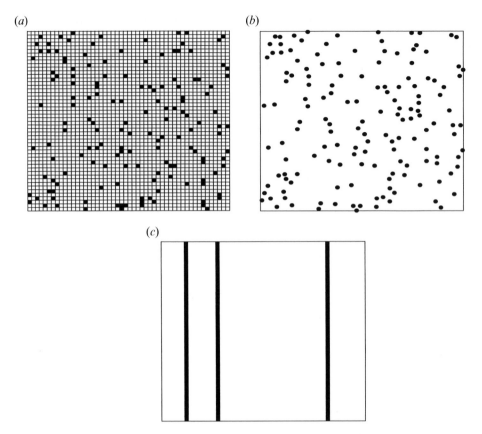

FIG. 10.1. Simple random sampling: (*a*) random sampling from a grid of 10 m by 10 m square plots, (*b*) random sampling by generating random plot center coordinates, and (*c*) random sampling using 10 m by 500 m strips.

plot or strip sampling is conducted without replacement. For the second approach described above, sampling without replacement would require discarding sample points that overlap with previously selected plots. When using variable probability sampling, the population is infinite, and the sampling is carried out with replacement.

Returning to the 25 ha forest as an example, we want to conduct a 6% inventory. Using 100 m² plots, $n = 0.06 \cdot 250{,}000/100 = 150$ plots. Figure 10.1*a* shows an example 6% random sample based on approach 1, and Figure 10.1*b* shows the same sample (using the same random number seed) using approach 2. With approach 2, some of the plots slop over the boundary and would require correction using one of the methods described in Section 9.2.6. Using 10 m wide strip plots, $n = 0.06 \cdot 250{,}000/(10 \cdot 500) = 3$ strips. Figure 10.1*c* shows one of the many potential SRSs that could be selected for the 10 m by 500 m strips.

Analysis of the resulting data from the sampling units is summarized below. The formulas are applicable whether dealing with plots, strips,* or points. Separate tallies for each sampling unit in a forest inventory must be recorded to calculate sampling error. More details on these formulas and their use are given in Chapter 3.

---

* Implementation of strip sampling will be discussed in more detail in Section 10.3.3.

Given:

n = number of sampling units measured

N = total number of sampling units in the population (the forest)

X_i = the quantity of X measured in the ith sampling unit

$\bar{x}$ = the mean of X per sampling unit; an estimate of the population mean

s = standard deviation of the sample

$s_{\bar{x}}$ = standard error of the mean

$\hat{X}$ = estimated total of X for population

CV = coefficient of variation (eq. (3.13))

E = allowable error of X

$E\%$ = allowable error as a percent of the mean

t = Student's t for desired probability level

Then,

$$\bar{x} = \frac{\sum\limits_{i=1}^{n} X_i}{n} \tag{10.8}$$

$$s^2 = \frac{\sum\limits_{i=1}^{n}\left(X_i - \bar{x}\right)^2}{n-1} = \frac{\sum\limits_{i=1}^{n}X_i^2 - \left(\sum\limits_{i=1}^{n}X_i\right)^2 \big/ n}{n-1} \tag{10.9}$$

$$s_{\bar{x}}^2 = \frac{s^2}{n}\left(\frac{N-n}{N}\right) \tag{10.10}$$

For an infinite population, the finite correction factor, $([N-n]/N)$, would be omitted. The estimate of the population total is

$$\hat{X} = N\bar{x} \tag{10.11}$$

The number of sampling units needed to yield an estimate of the mean with a specified allowable error and probability can be calculated from

$$n = \frac{Nt^2 s^2}{NE^2 + t^2 s^2} \tag{10.12}$$

For an infinite population:

$$n = \frac{t^2 s^2}{E^2} \tag{10.13}$$

Expressing the standard deviation and allowable error in percentages:

$$n = \frac{Nt^2\left(\text{CV}\right)^2}{N\left(E\%\right)^2 + t^2\left(\text{CV}\right)^2} \tag{10.14}$$

For an infinite population:

$$n = \frac{t^2 (CV)^2}{(E\%)^2} \qquad (10.15)$$

Note that in Equations (10.12) through (10.15), s and CV refer to the standard deviation and coefficient of variation of a preliminary sample measured to give an indication of the variability of the population. E is an arbitrarily chosen level (often E will be set by common practice, contract specifications, or, in some cases, by legislation) and the t-value depends on the required probability (also arbitrarily chosen or otherwise preset) and degrees of freedom. The t-value is determined from Student's t distribution (Table A.4) with $n-1$ degrees of freedom, where n refers to the number of sampling units in the preliminary sample. To be correct, n should be the number of sampling units sought; however, since this is unknown, the n of the preliminary sample can be used and the equation solved iteratively (Section 3.6). If a preliminary sample has not been taken, and if the expected sample size is large (over 30), the t-values for an infinite number of degrees of freedom may be used. A detailed explanation of the implementation of these formulas and an example is found in Section 3.6.

If the amount of money for a survey is fixed, the number of sampling units must be determined within this restriction. If the total cost for a survey is given as c_t, then

$$c_t = c_o + nc_1 \qquad (10.16)$$

where c_o=overhead cost of a survey including planning, organization, analysis, and compilation

c_1=cost per sampling unit

n=number of sampling units

The number of sampling units is indicated as

$$n = \frac{c_t - c_o}{c_1} \qquad (10.17)$$

For small inventories, such as the 25 ha forest, overhead and analysis costs are generally minimal, perhaps representing one day of effort ($500–$1000), while for larger forest inventories, these costs may be quite substantial, especially if field verification is included. Costs per sample unit vary greatly depending upon how large the sampling unit is and what measurements are required. Temporary inventory plots may range in cost from $10 to $100 while permanent sample plots may range in cost from $250 to well over $1000. If preliminary information on the mean and variance of the population is known from experience or a preliminary sample, it is then possible to estimate the precision that can be obtained for a given cost.

Table A.7 contains the plot summary data associated with the $100\,m^2$ circular plot SRS shown in Figure 10.1b. The tree expansion factor (Section 9.1) for these plots is: TF=10,000/100=100. Table 10.1 summarizes the inventory data for trees, basal area, and total volume using eqs. (10.8) through (10.11). Sampling errors of about 8% were achieved for all three stand parameters of interest.

TABLE 10.1. Summary Statistics and Total Estimates for the Simple Random Sample of the 25 ha Forest Shown in Figure 10.1b.

Parameter	Per Hectare Estimates			Total Estimates		
	Trees (#/ha)	Basal Area (m²/ha)	Total Volume (m³/ha)	Trees (#)	Basal Area (m²)	Total Volume (m³)
$\bar{X}$	934	22.8	148	23,350	570	3700
s	475	10.1	71	11,864	252	1770
CV (%)	51	44	48			
$s_{\bar{x}}$	39	0.8	6	967	21	144
Confidence width[a]	76	1.6	11	1,914	41	285
Error (%)	8	7	8			

[a]95% confidence level was used for both confidence width and error calculations.
Note: finite correction factor was not used in these summaries.

10.2.1. Influence of Plot Size on Simple Random Sample Designs

When designing a sample for a forest inventory, one must consider tradeoffs among the types of sample units used, the sizes of the sample units, the numbers of sample units, and the allocation of sample units across the forest. Frequently, the type of sample unit, their size, and allocation are chosen based on experience and/or common practice in the region, and sample design primarily focuses on sample sizes (though this too may be set by common practice).

One of the primary reasons for reliance on common practice is a lack of comparison data. Collection of inventory data based on different plot types, sizes, numbers, and allocations requires considerable time and effort. Datasets like those shown in Plate 1 are rare, and simulation systems that produce realistic stand structure have only recently become available (e.g., Valentine et al., 2000; Kokkila et al., 2002; Goreaud et al., 2004; Kershaw et al., 2010). Simulating different sample designs for forests generated using these simulators can provide important insights to consider in sample design.

Figure 10.2 shows the average standard errors (uncorrected and corrected), based on 100 simulated samples with sample size varying from 5 plots to 6000 plots and plot size varying from 25 to 625 m² (note that maximum intensity simulated was 60%). From Figure 10.2a and b, it is clear that, for the same number of larger plots, larger plots produce lower standard errors than smaller plots; however, if standard error is viewed as a function of sample intensity (percent area sampled), there is little difference in standard errors across plot sizes (Fig. 10.2c and d).

Similarly, the results shown in Table 10.1 are for one sample out of a number of possible random samples. The forest could be resampled using the same sample design or the sample design could be altered by changing number of plots and/or the size and shape of the plots. As discussed in Section 9.2, smaller fixed-area plots should be, theoretically, more efficient than larger plots. This concept can be illustrated by holding the sample intensity constant at 6% and increasing plot size. Figure 10.3 shows the effect of changing plot size on the estimated means and standard errors for the three stand parameters commonly estimated in forest inventories. Plot size has no appreciable influence on the estimated mean values (Fig. 10.3a, b, and c), and, because sample intensity was held constant,

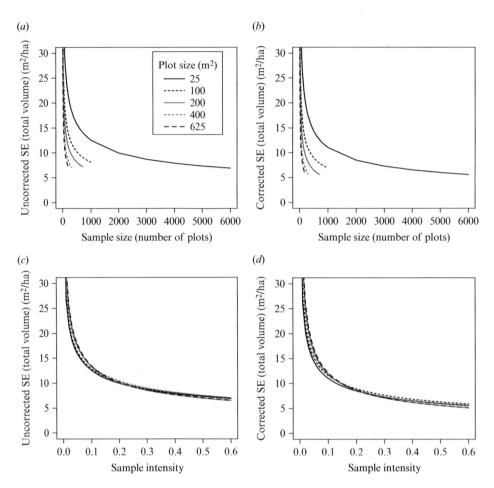

FIG. 10.2. Influence of plot size on resulting standard error estimates: (*a*) uncorrected (infinite population) standard error by plot size and sample size, (*b*) corrected (finite population) standard error by plot size and sample size, (*c*) uncorrected (infinite population) standard error by plot size and sample intensity, and (*d*) corrected (finite population) standard error by plot size and sample intensity.

plot size has little influence on the range of estimated mean values. This makes sense considering that sample intensity, and, thus, the area over which we are averaging, is held constant.

Standard error, on the other hand, was much more influenced by plot size (Fig. 10.3*d*, *e*, and *f*). Standard errors were consistently smaller for the smaller plot sizes, initially increased as plot size increased, then leveled off and remained more or less constant for the larger plot sizes. The primary factor influencing this trend was sample size. As plot size increases, the number of plots decreases from 1000 plots for the smallest plot size (15 m^2) to 2 plots for the largest plot size (7500 m^2). Given that the range of variability is almost constant across the range of plot sizes, the standard errors arise by dividing a relatively constant variance (standard deviation) by a smaller number of plots ($\sqrt{n}$) as plot size increases.

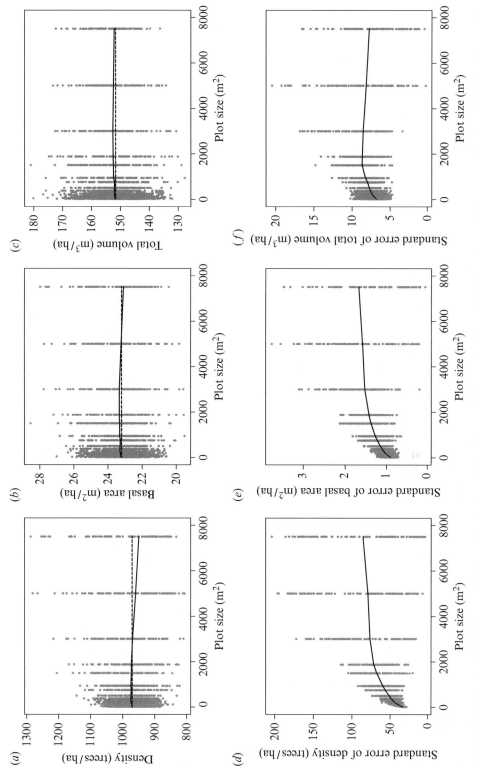

FIG. 10.3. Influence of plot size on: (*a*) mean density (trees/ha) estimates, (*b*) mean basal area (m²/ha) estimates, (*c*) mean total volume (m³/ha) estimates, (*d*) standard error of mean density, (*e*) standard error of basal area, and (*f*) standard error of volume.

317

There are many lines of reason that may lead one to choose large plots or small plots. One could argue that the influence of missing a plot tree or including a nonplot tree is larger on a small plot (small plots have large expansion factors, hence inclusion errors represent larger trees per unit area) than on a large plot; however, one could also argue that the likelihood of missing a tree on a large plot is greater and that, because of the increased perimeter, there will be more borderline trees and the ability to accurately measure the borderline condition decreases with increasing plot size. Iles (2003) suggests that for most forest inventories targeting a plot size that yields 4–5 trees per plot is probably adequate. For the forest area in Plate 1, the mean density is 970, thus a plot size of roughly $50\,m^2$ ($10{,}000{\cdot}[5/970]$) would yield this targeted trees/plot. Of course, the final plot size chosen will depend on several factors and should ideally balance statistical considerations and the possibility of nonsampling errors with logistical and cost considerations.

10.3. SYSTEMATIC SAMPLING (SYS)

In SYS, sampling units are spaced at fixed intervals throughout the population. SYS designs, which are widely used, have a number of advantages: (1) they provide reliable estimates of population means and totals by spreading the sample over the entire population; (2) they are usually faster and cheaper to execute than designs based on random sampling because the choice of sampling units is mechanical, eliminating the need for a random selection process; (3) travel between successive sampling units is easier since fixed directional bearings are followed, and the resulting travel time is usually less than that required for locating randomly selected units; and (4) the size of the population need not be known since units are chosen at a fixed interval after an initial point has been selected. In addition, mapping can be carried out concurrently on the ground since the field party traverses the area in a systematic grid pattern. In forest inventory work, the systematic distribution of sampling units can be used with fixed-area plots or strips, or points and lines in PPS sampling (Chapter 11).

Since the units for a SYS are located at some regular interval, there will be a fixed set of possible nonoverlapping samples. The number of possible samples is set by the sampling interval (distance between plots) and the plot size. For the mean of a SYS to be unbiased, some form of random selection must be incorporated in the sampling process. The only randomization possible is the selection of one of the fixed sets of systematic samples. The set selected will depend on the selection of the initial sampling unit in the population.

10.3.1. Systematic Plot Sampling

If sampling units such as plots or points are used, the sample is systematic in two dimensions; that is, the sampling units are selected at intervals in two directions perpendicular to each other. The following discussion of SYS uses fixed-area plots; the principles, however, apply to variable probability sampling as well.

The typical goal of SYS is to allocate samples at regular intervals across the area of interest, although, as Iles (2003) points out, SYS does not only apply to geographic layout and can be used with any ordered list of sampling units (see Iles (2003, chapter 5) for several examples of SYS from ordered lists). If the number of plots is a square number, for example 100, and the area to be sampled is also square, like the 25 ha area shown in Plate 1, then the plots can be located in a square grid. For example, with 100 plots there would be a grid of 10 by 10 plots, and, given that the area is 500 m by 500 m, the plot spacing would be 50 m by 50 m (500/10). The same reasoning could be applied to Figure 10.1a, where the area was divided into a grid of nonoverlapping 10 m by 10 m plots. In this case, the area is divided into 50 plots by 50 plots, and the spacing, in terms of numbers of plots, would be 5 plots by 5 plots (50/10). Spacing for square plot arrangement can be found using

$$D_{sq} = \sqrt{\frac{A \cdot c}{n}} \tag{10.18}$$

where D_{sq} = distance (spacing) between square arranged systematic plots (m or ft)

A = area of the forest to be sampled (ha or acre)

c = conversion factor for area units to distance units (10,000 m²/ha for SI and 43,560 ft²/acre for imperial units)

n = number of plots

D_{sq} also can be determined from plot size and sampling intensity, $D_{sq} = \sqrt{a/P}$, where a is plot area in squared units of distance (m² or ft²) and P is sample intensity (percent of area sampled).

In general, the number of plots and/or the forested areas to be sampled are seldom square. If individual forest stands are being sampled, the boundaries can be quite irregular. While square spacing of plots can be used in any situation, the process can produce slight variation in sample sizes depending upon the selected starting point and the grid orientation relative to the boundaries of the area to be sampled. Thus, for practical reasons, the square-grid arrangement of sampling units is often dropped, resulting in the interval between lines being different than between sampling units. This modification when used with fixed-area plots is called line-plot sampling; however, some foresters also consider the square-grid arrangement of plots as line-plot sampling. Numerous possible line-plot distributions can be devised, depending on the size of the plot, the distance between plots on a line, the distance between lines, and the overall shape of the area being sampled. A line-plot design can be drawn up using the following relationships:

$$n = \frac{AP}{a} \tag{10.19}$$

$$P = \frac{a}{D_l D_p} \tag{10.20}$$

where A = total stand area (ha or acre)

P = proportion of area covered by plots

a = plot area in squared units of distance (m², ft², or chains²)

n = number of plots

D_l = spacing of lines in a given unit (m, ft, or chains)

D_p = spacing of plots on lines in same units as D_l

Once the line and plot spacing are determined, the sample plots could then be located using one of two methods:

1. A random starting point within the forest area is selected and the plot grid centered on this starting point.
2. A convenient corner of the area is selected as a starting area, and the starting point is selected by randomly selecting a starting point within this area. The starting area should be defined by D_l and D_p (or D_{sq}).

Considering the forest depicted in Plate 1, the random sample shown in Figure 10.1 was a 6% cruise; using 0.01 ha (100 m²) plots, the number of plots in this sample was

$$n = \frac{25 \text{ ha} \cdot 0.06}{0.01 \text{ ha}} = \frac{250{,}000 \text{ m}^2 \cdot 0.06}{100 \text{ m}^2} = 150 \text{ plots}$$

Since 150 is not a square number, a systematic sample based on a 6% sample intensity would be most conveniently designed using line-plot sampling. The roots of 150 are 2, 3, 5, and 5. We can combine these roots in any combination to determine the number of lines and number of plots per line. Often, it is useful to consider the ratio of the long and short axes of the area being sampled. For example, if our area was twice as long as it was wide, we might consider 6 lines (2·3) with 25 plots (5·5) per line; however, the forest we are working with is square (500 m by 500 m), so we may want the number of lines and number of plots per line as close to equal as possible. For the examples illustrated here, we selected 15 lines (3·5) with 10 plots (2·5) per line to achieves this. Given that our area is 500 m by 500 m, our line spacing can be determined by

$$D_l = \frac{500 \text{ m}}{15} = 33.3 \text{ m}$$

The plot spacing along lines could be found the same way ($D_p = 500 \text{ m}/10 = 50 \text{ m}$) or by using eq. (10.20) and solving for D_p:

$$D_p = \frac{a}{D_l P} = \frac{100 \text{ m}^2}{33.33 \text{ m} \cdot 0.06} = \frac{100 \text{ m}^2}{2 \text{ m}} = 50 \text{ m}$$

Using method 1, described above for locating the systematic grid, we would generate a starting x–y grid point within the forest area. The easiest way to do this would be to generate two uniform random numbers between 0 and 500, $\{U[0,500], U[0,500]\}$, and then distribute the lines at 33.3 m intervals about the x coordinate and the plots along the lines at 50 m intervals about the y coordinate. Figure 10.4a illustrates one possible systematic sample using method 1. Using method 2, we would start in the 33.3 m by 50 m southwest (lower-right-hand) corner and generate a random starting point, $\{U[0,33.3], U[0,50]\}$, and the lines are spaced at 33.3 m intervals and the plots are space along the lines at 50 m intervals starting at this point. Figure 10.4b shows one example of a systematic sample generated using this approach.

In Figure 10.4, the selection grid was oriented parallel to the forest boundary; however, this need not be the case, and the grid can be randomly rotated (Fig. 10.5). Randomly orienting the selection grid has some design advantages in that it adds another element of randomness to the sample. It may help ameliorate situations where there is a periodic

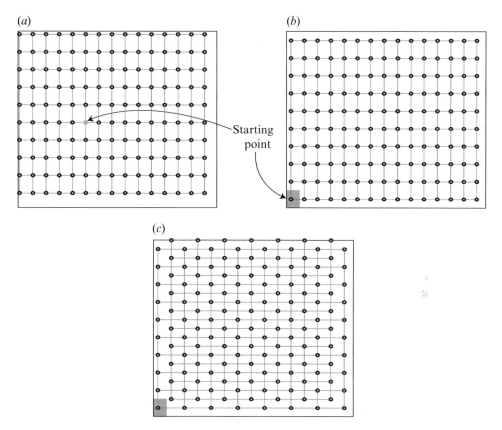

FIG. 10.4. Systematic sampling layout: (*a*) random start within the forest, (*b*) random start from selected corner, and (*c*) random start from selected corner with offset grid.

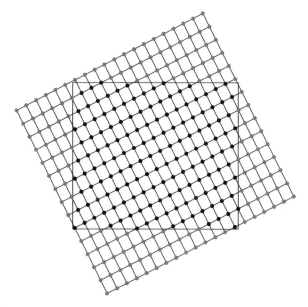

FIG. 10.5. Systematic sampling layout with a randomly rotated selection grid.

pattern in the population (e.g., systematic variation in forest density following mechanized harvesting along a "fishbone" pattern of skid trails) and reduces the variability associated with the chance that either a great many plots may fall close to a linear edge, or that none will. The use of a random orientation may result in slight variation in sample sizes. An alternative to grid rotation is to use an offset grid. In this case, spacing of plots along adjacent lines is offset by half of the plot spacing (Fig. 10.4c).

Prior to the advent of personal computers and GIS systems, design of systematic samples was done by hand using maps or photos of the area of interest. Often, the selection grid would be drawn on a clear overlay, randomly placed on the map or photo, and pin pricks used to transfer plot locations from the clear overlay to the map or photo (Spurr, 1952). GIS, such as Quantum GIS (QGIS, 2013), have routines that can largely automate the generation of systematic (and random) samples. Statistical graphics packages, such as R (R Development Core Team, 2013), also can be used to automate the design of samples.

When selection grids are rotated, square or rectangular plots can be oriented using cardinal directions, the line direction, or parallel to one of the area's boundaries. Using circular plots eliminates the need to consider plot orientation. Plot orientation is generally not an issue as long as consistent orientations are used; however, in plantations where trees are found in regular rows, substantial variability can be introduced into the results if plot orientation corresponds to the orientation of the plantation rows.

10.3.2. Sampling Error for a SYS Inventory

With SYS, there is a fixed number of possible samples that is determined by the sampling interval and plot size chosen. While there are an infinite number of possible starting points, once a starting point is selected, the sampling interval and plot size determines the number of possible nonoverlapping systematic samples that can be placed on the forest given the starting point. In the case where the plots completely tessellate the forest area, as shown in Figure 10.1a, the number of possible samples is finite. In this case, the 33.3 m by 50 m starting area includes five rows by all of three columns and a part of the fourth column. While many different starting points could be randomly selected, the resulting selection grid would only produce $4 \cdot 5 = 20$ different independent samples. In cases where plots do not completely tessellate the area (such as when circular plots are used) or when plots are centered about the selection grid, there are an infinite number of possible systematic samples; however, the outcomes of these samples are still bounded by the spacing imposed.

SYS with a random start will result in unbiased estimates of the mean and total population, and the formulas used with SRS can be applied (eqs. (10.8) and (10.11)) to estimate the mean and total. For example, Table A.8 contains the plot summaries for the systematic sample illustrated in Figure 10.4c. The mean and total estimates of this sample are shown in Table 10.2.

With a typical SYS, as described above, the starting point is the only random choice, once the starting point is selected all other sample points are fixed. Variance measures differences between randomly selected observations and their mean, thus variance requires at least two independent observations. The sampling interval divides the population into clusters or sets of n sampling units. A SYS (the entire set of units) consists of a single selection from this population of clusters, thus there is only one independent observation; therefore, there are no differences available to measure. At least two of the clusters or sets would need to be selected in order to estimate sampling error. Various methods to approximate the

TABLE 10.2. Summary Statistics and Total Estimates for the Systematic Sample of the 25 ha Forest Shown in Figure 10.4c Using the Variance and Standard Error Formulas for Simple Random Sampling

Parameter	Per Hectare Estimates			Total Estimates		
	Trees (#/ha)	Basal Area (m²/ha)	Total Volume (m³/ha)	Trees (#)	Basal Area (m²)	Total Volume (m³)
$\bar{X}$	950	23.0	150	23,750	575	3740
s	425	10.9	77	10,610	271	1930
CV (%)	45	47	52			
$s_{\bar{x}}$	35	0.9	6	870	22.2	158
Confidence width[a]	68	1.8	12	1,712	43.8	312
Error (%)	7	8	8			

Finite correction factor was not used in these summaries.
[a] 95% confidence level was used for both confidence width and error calculations.

sampling error of a systematic sample have been suggested (e.g., Schaeffer et al., 1990; Johnson, 2001). Johnson (2001) provides examples of some of the common methods applied to systematic forest inventory. Here, we are going to consider two methods.

If the total population of sampling units in a forest were randomly distributed, exhibiting no pattern of variation, then SYS would be equivalent to SRS and the random sampling formulas would be applicable for estimating sampling error (Schaeffer et al., 1990). Table 10.2 shows estimates of sampling error based on the formulas for SRS. Osborne (1942) has shown that the sampling error computed in this way estimates the maximum sampling error, which may considerably overestimate the actual sampling error because, theoretically, the reduced number of sample combinations resulting from SYS should yield lower sampling errors than SRS. In the example presented here, the sampling error from SYS (Table 10.2) is only slightly smaller than what was obtained using SRS (Table 10.1) for trees/ha and about the same for basal area and volume per hectare.

Schaeffer et al. (1990) agree with Osborne (1942) for random and ordered populations and suggest that, for most field sampling situations, the sampling error for SYS will be close to the estimate obtained using the formula for SRS (eq. (10.10)); however, for populations with a periodic distribution, the true sampling error could be substantially larger than that estimated using SRS formulas. In a forest, components are rarely, if ever, arranged completely independent of each other, but, instead, may show some level of systematic or periodic variation from place to place, especially in intensively cultured landscapes (Shiver and Borders, 1996). If sampling units are systematically selected, then variation in the observed values may no longer be ascribable to randomness if the interval between sampling units happens to coincide with a periodic pattern of population variation. Schaeffer et al. (1990) suggest methods for increasing the sampling error based on the correlation between sampling units; however, Iles (2003) points out that, while coincidence between sampling interval and the periodic pattern within a population could occur, it is highly unlikely to occur and there are no documented examples of it ever occurring. One should keep in mind that there are no statistically valid estimates of sampling error based on a single SYS, and, while estimates based on SRS are frequently used, these estimates generally overestimate the true sampling error.

The larger the forest area inventoried, the greater the variation that can be expected and the more likelihood that a systematic sample will give a better estimate of the mean than a completely random sample. Even for a stratified population (Section 10.5), a systematic sample will probably yield a better estimate of the mean if the strata are large and variable. As the homogeneity of the defined strata increases, estimates from a random and systematic sample will tend to agree.

Shiue (1960) proposed a method of SYS that maintains the advantages of SYS and provides a reasonable means of estimating sampling error. In this method, several systematic samples are taken, with the initial sampling unit chosen randomly for each start. Using a line-plot procedure, the first sample of systematically located plots constitutes the first cluster. Another systematic sample would be the second cluster, and so on. Based on these clusters, estimates of the mean plot volume and its sampling error can be computed using the formulas for cluster analysis (Section 10.6). To avoid a large t value and to maintain a small confidence interval for a given probability, at least five random starts were recommended (Shiue, 1960). Schaeffer et al. (1990), Iles (2003), and Gregoire and Valentine (2008) propose similar approaches to obtaining estimates of sample error from systematic samples.

10.3.3. Systematic Strip Sampling

By using strips as the sampling unit, the systematic distribution is accomplished by first dividing the forest area into N strips of uniform size. Sampling units would then be taken at intervals of every kth strip to form the sample of n strips. Figure 10.6 shows the 25 ha forest area divided into 50 strips of uniform length and clustered into 10 groups. The selection of n strips at a sampling interval of k strips can be carried out in two ways.

1. A random selection of a number from 1 to N can be made and the corresponding strip chosen as the initial sampling unit. Sampling units at the interval k are then taken in both directions from this initial strip.
2. Randomly select a number between 1 and k for the first strip. All subsequent strips are taken at intervals of k strips.

Both procedures will yield the probable number of systematic samples. The first procedure should be used if possible because it assures that the *a priori* probability of selection is the same for every strip ($1/k$). If the value of N is not an exact multiple of k, then some strips will not have the same probability of selection if the second method is used. The first procedure will yield an unbiased estimate of the mean, whereas the second procedure may give a slightly biased result depending upon the estimator used. The second procedure, however, must be used if the size of the population is not known, as may occur in sampling a forest where no map is available.

Often, a forest is irregular in shape rather than square or rectangular. If the area is then divided into strips of equal width, the strips will differ in length and consequently in area. Thus, the possible systematic samples that could be obtained by taking every kth strip may differ in intensity. Zenger (1964) has described a method that draws a systematic sample in such a way that the probability of selection of a strip is proportional to its length.

Applying strip sampling, the field party starts from a baseline, or one side of the tract, and runs a straight strip on a compass bearing or GPS track across the tract stopping at the other side. The party then offsets the determined interval and runs back to the baseline or boundary. The procedure continues until all the strips in the sample have been measured.

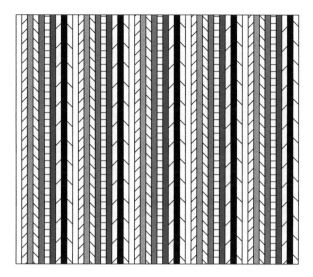

FIG 10.6. Forest area divided into strips for systematic strip inventory. There are 10 clusters of 5 strips per cluster.

The measurements of the variables occurring on the strips are tallied and represent the sample of the entire forest. Separate tally sheets may be kept for different forest types or classes so that separate estimates may be made for them.

It is desirable to orient strips at right angles to the drainage pattern in order to increase the likelihood of having the strip intersect all stand conditions. In North America, the traditional width of strip is one chain (66 ft) or less. The width of the strip and the interval between strips determine the intensity or percentage of the total area tallied. The intensity of a systematic strip design is expressed by

$$P = \left(\frac{W}{D}\right)$$ (10.21)

where P = proportion of area covered by strips

W = width of strip in a given unit

D = distance between strips in same units as W

Equation (10.21) is used principally to determine the spacing of strips when the cruise intensity and strip width are known. For example, if a 10% cruise ($P=0.10$) is to be conducted with 1-chain wide strips, the spacing of strips would be $D=W/P=1/0.10=10$ chains. For the 6% cruise of the 25 ha forest illustrated in Plate 1 using 10 m wide strips, the spacing of strips would be $D=10/0.06=167$ m. For a given intensity, narrow strips at a closer spacing will give more uniform distribution and better coverage of the stand area than fewer strips that are wider but more distantly spaced, although the costs may be greater.

When all the strips are of equal size, the estimate of the population mean is often calculated from eq. (10.8) and the sampling error from eq. (10.10) under the assumption that the strips constitute a random sample. When sampling units are not of equal size, as would occur in a strip sample of an irregular-shaped forest area, ratio estimation can be used (Section 10.8). Thus, an estimate of the area of each sampling unit X_i and the observation, such as volume, on the units V_i are required to obtain the ratio estimate as described in Section 10.8.1.

10.4. SELECTIVE OR OPPORTUNISTIC SAMPLING

Selective (or opportunistic) sampling is an example of nonrandom sampling used in forest inventory work. In selective sampling, an observer selects a sample of plots that appears to be representative of the forest and measures them. Selective sampling may give good approximations of population parameters, but is not generally recommended. Because human choice is often prejudiced by individual opinion, selective sampling may result in bias. In addition, selected samples will not yield a measure of reliability of the estimate because probability theory is based on the laws of chance. Selective sampling was used in the pioneer days of American forestry but is now rarely justified in normal operational forest inventories.

While selective sampling is rarely used to obtain inventory estimates, selective sampling does have a legitimate role in some resource assessments. For example, growth and yield studies may focus data collection on fully stocked, undisturbed portions of selected stands. Such studies provide cost-effective means to develop normal yield curves and establish boundaries for growth and yield models; however, they do not represent a sample of the forest, so care must be exercised when extrapolating results from these data or inferring regional processes or changes based on remeasured permanent plots located in this manner. Similarly, studies placed in remnant, nonharvested areas may provide important site documentation for ecological processes. Again, care must be exercised in extrapolating these results to regional processes because these sites are not a sample of the region and often are not harvested because of unique site characteristics that may not be representative of the sites across regions.

An area of application where selective sampling has provided some success is in the detection of rare or invasive species. Normal sampling designs like systematic or random samples rarely provide the information needed to determine if rare species are present (Goff et al., 1982). Such designs are good for estimating dominant vegetation components, but are seldom intensive enough to provide reliable detection data for rare species. Goff et al. (1982) proposed a subjective sampling technique they called the *timed meander method*. The method, as described by Goff et al. (1982), is as follows:

1. Vegetation units (generally stands) are delineated.
2. Starting at a convenient edge of the stand, the observer meanders through the area so as to thoroughly cover the stand (Fig. 10.7a).
3. At the starting point, the observer notes the starting time.
4. New species are identified and noted on a field sheet as the observer meanders through the stand.
5. Periodically, the observer notes the time.

Early in the survey the time intervals might be 1 minute as many new species are initially noted. As the search continues, fewer new species will detected and the time interval can increase to 10 or more minutes. The survey ends when the observer feels they have covered the area. The use of a GPS and its tracking capabilities could reduce the subjectivity of observer coverage assessment. Goff et al. (1982) suggested several useful ways of summarizing and using timed meander data. The cumulative species effort curve (Fig. 10.7b) was suggested as one way of assessing effectiveness of the survey—as the number of new species plateaus, the observer can be sure that most species present on the site has been detected. The timed meander method has become a common approach for assessing species richness (Elzinga et al., 2001).

(*a*)   (*b*)

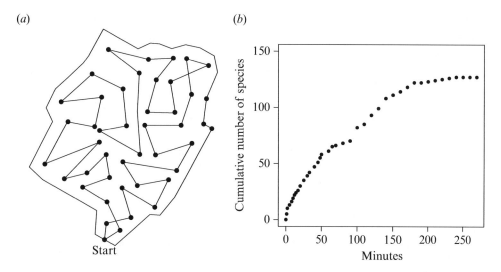

FIG. 10.7. Timed meander sampling: (*a*) schematic illustration of timed meander path through a stand and (*b*) species–effort curve for Consumers Power Company field unit 1, Ottawa, MI (adapted from Goff et al., 1982).

While the timed meander method provides an efficient assessment of species richness, it does not provide any quantitative assessment of spatial or temporal patterns (Huebner, 2007). Detection and monitoring of invasive species requires a sampling system that provides both detection and assessment of patterns and importance over time.

10.5. STRATIFIED SAMPLING (STS)

In many cases, a heterogeneous forest may be divided by stratification into subdivisions called *strata*. In forest inventory work, the purpose of stratification is to reduce variation within forest subdivisions and increase precision of population estimates. STS in forest inventory has the following advantages over simple random or SYS:

1. Separate estimates of the means and variances can be made for each of the forest subdivisions.
2. For a given sampling intensity, stratification often yields more precise estimates of the forest parameters than does a simple random or stratified sample of the same size. This will be achieved if the established strata result in a greater homogeneity of the sampling units within a stratum than for the population as a whole.

On the other hand, the disadvantages of stratification are that the size of each stratum must be known or at least a reasonable estimate be available, and that the sampling units must be taken in each stratum if an estimate for that stratum is needed.

Stratification is achieved by subdividing the forest area into strata on the basis of some criteria such as topographical features, forest types, density classes, or volume, height, age, or site classes. Stratification also can be arbitrary, and occasionally this is used when inventorying large, remote areas, or when aerial photos provided little basis for stratification; however, for best results, stratification should be based on some attribute of interest. If possible, the basis of stratification should be the same characteristic that will be estimated in the sampling procedure. Thus, if volume per unit area is the parameter to be estimated, it is desirable to stratify the forest on the basis of volume classes. The different strata into which a forest may be divided can be irregular in shape, of many sizes, and of varying importance. Stratification permits sampling intensity and precision to be varied for the different strata. Aerial photographs and satellite imagery are of tremendous assistance in stratification for forest inventory.

For example, consider the 25 ha forest area in Plate 1. Careful examination of the species distribution shows that most of the area is dominated by red maple and paper birch with a mixture of other species including balsam fir, beech, white pine, and others (IH stratum). On the left-hand side (west), there is an area dominated by black spruce (BS stratum) and there are three very small areas dominated by eastern hemlock (EH stratum). Using these species compositions, the area can be divided into the three strata as shown in Figure 10.8.

Within each of the M strata into which a forest is divided, a number of sampling units are selected. Sampling units may be selected randomly or systematically. Systematically selected sampling units have the same constraints of error estimation as discussed in Section 10.3. The analysis of the data obtained from stratified random sampling is summarized below for the situation where the plots or strips are all of a uniform size, and assuming sampling without replacement.

Given

M = number of strata in the population

n = total number of sampling units measured for all strata

n_j = total number of sampling units measured in the jth stratum

N = total number of sampling units in the population

N_j = total number of sampling units in the jth stratum

a_j = area in jth stratum

A = total area in forest

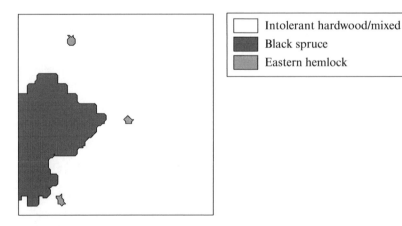

FIG. 10.8. Stratification of the forest shown in Plate 1 into 3 strata based on species composition.

X_{ij} = quantity of X measured on the ith sampling unit of the jth stratum

$\bar{x}_j$ = mean of X for the jth stratum

$\bar{x}$ = estimated mean of X for the population

P_j = proportion of the total forest area in the jth stratum $= \dfrac{a_j}{A}$

$\hat{X}$ = estimated total of X for the population

s_j^2 = variance of X for the jth stratum

$s_{\bar{x}}^2$ = estimated variance for the mean of the population

s_X^2 = estimated variance of $\hat{X}$

E = allowable standard error in units of X

The estimate of the mean per stratum is

$$\bar{x}_j = \frac{\sum_{i=1}^{n} X_{ij}}{n_j} \tag{10.22}$$

The estimate of the mean for the population (i.e., total forest) is

$$\bar{x} = \frac{\sum_{i=1}^{n} N_j \bar{x}_j}{N} = \sum_{j=1}^{M} P_j \bar{x}_j \tag{10.23}$$

The estimate of the total for X for the entire population is

$$\hat{X} = \sum_{j=1}^{M} N_j \bar{x}_j = N\bar{x} \tag{10.24}$$

The variance for each stratum in the population, s_j^2, is calculated as described for SRS (eq. (10.9)). The variance of the mean for the population is then calculated from

$$s_{\bar{x}}^2 = \frac{1}{N^2} \sum_{j=1}^{M} \left[\frac{N_j^2 s_j^2}{n_j} \left(\frac{N_j - n_j}{N_j} \right) \right] \tag{10.25}$$

If the population consists of strata sufficiently large that the finite correction factor is insignificant, or information is available on the relative sizes of strata, then we can estimate variance from

$$s_{\bar{x}}^2 = \sum_{j=1}^{M} P_j^2 \frac{s_j^2}{n_j} \tag{10.26}$$

The standard error squared of the total estimate of X is then

$$s_X^2 = N^2 s_{\bar{x}}^2 \tag{10.27}$$

The n_j sampling units actually chosen and measured per stratum should not be used to estimate the relative size of a stratum. Proportions in the strata should be estimated "a priori" to the actual sampling. Proportions are best estimated on the basis of area.

10.5.1. Estimation of Number of Sampling Units

To estimate the number of sampling units needed, it is necessary to have preliminary information on the variability of the strata in the population and to choose an allowable error and probability level. With this information, the intensity of sampling can be estimated. The total number of sampling units can then be allocated to the different strata either by proportional or optimum allocation.

In *proportional allocation*, the number of sampling units in a stratum is proportional to the area of the stratum to the whole forest. In *optimum allocation* (sometimes referred to as Neyman allocation), the number of sampling units per stratum is proportional to the area-weighted stratum standard error. Optimum allocation will give the smallest standard error for a stratified population when a given total number of sampling units is measured. If we wish to get the most precise estimate of the population mean for the expenditure of money, then optimum allocation should be used. Allocation can also include the costs of sampling if they differ across strata.

Using proportional allocation, strata variances are not required for allocating sample sizes across strata (n_j); however, advance knowledge of the variability in the different strata is desirable to determine total sample size (n). When no information on variability of individual strata is available, it is necessary to estimate total sample sizes for the entire population as though SRS was being employed (i.e., using eqs. (10.12) through (10.15)). Under proportional allocation, the determination of total sample size, n, for a given precision when information on variability per strata is available is

$$n = \frac{Nt^2 \sum_{j=1}^{M} P_j s_j^2}{NE^2 + t^2 \sum_{j=1}^{M} P_j s_j^2} \tag{10.28}$$

and

$$n_j = P_j n \tag{10.29}$$

If the population can be considered infinite, then

$$n = \frac{t^2 \sum_{j=1}^{M} P_j s_j^2}{E^2} \tag{10.30}$$

Under optimum allocation, the determination of sample size n with specified precision is shown in eq. (10.31) for the simplest case when costs per sampling unit are the same for all strata. The sampling intensity is changed in each stratum, according to its variability, to achieve a given precision with the smallest possible number of sampling units:

$$n = \frac{Nt^2 \left(\sum_{j=1}^{M} P_j s_j \right)^2}{NE^2 + t^2 \sum_{j=1}^{M} P_j s_j^2} \tag{10.31}$$

and

$$n_j = n\left(\frac{P_j s_j}{\sum\limits_{j=1}^{M} P_j s_j}\right) \qquad (10.32)$$

If the population can be considered infinite, then

$$n = \frac{t^2 \left(\sum\limits_{j=1}^{M} P_j s_j\right)^2}{E^2} \qquad (10.33)$$

For the determination of sample size by optimum allocation when the costs per sampling unit vary by stratum, or when the total cost of the inventory is fixed, see Cochran (1977), Schreuder et al. (1993), or Shiver and Borders (1996).

For the 25 ha forest and the stratification shown in Figure 10.8, if a 6% sample intensity is required and 100 m² plots are used, the total number of sampling units would be 150. The strata areas are 20.65 ha for the IH stratum, 4.25 ha for the BS stratum, and 0.1 ha for the EH stratum. Using proportional allocation and applying eq. (10.32) would yield 124 plots in the IH stratum ($150 \cdot [20.65/25] = 123.9 = 124$), 26 plots in the BS stratum ($150 \cdot [4.25/25] = 25.5 = 26$), and 1 plot in the EH stratum ($150 \cdot [0.1/25] = 0.6 = 1$); however, because more than one sampling unit is needed to calculate sampling error and because there are three distinct elements in the EH stratum, three plots should be allocated to the EH stratum. The remaining 147 plots would then be allocated using eq. (10.32): 122 for the IH stratum ($147 \cdot [20.65/24.9]$) and 25 for the BS stratum ($145 \cdot [4.25/24.9]$).

SRS or SYS within each strata is used to select sampling units. Figure 10.9 illustrates this for a stratified random sample based on the allocation determined above. Boundary slopover correction (Section 9.2.6) should be applied to plots that are not all contained within a single strata. Alternatively, ratio estimation (Section 10.8.2) could be applied using the partial plots (Loetsch and Haller, 1964; Cochran, 1977; Shiver and Borders, 1996; Husch et al., 2003).

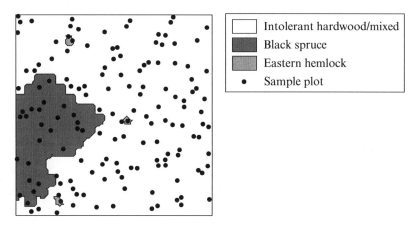

FIG. 10.9. Stratified random sample layout using proportional allocation and an overall 6% sample intensity.

10.5.2. Sample Efficiency

Table A.9 contains the plot summary data for the STS illustrated in Figure 10.9. Table 10.3 shows the strata level and forest level per ha and total summaries for the sample. Table 10.4 compares the sample estimates of the three sampling approaches with the true population mean. Overall, the three approaches produced very similar results. The STS and SYS designs had smaller overall standard errors for trees/ha. In relationship to the true population mean (Table 10.4), STS also produced better estimates than SYS, which was better than SRS (though none of the estimates are statistically different). In almost all sampling situations, the true population mean is never available; therefore, sampling error is often the only means of comparing performance of different sample designs. The relative efficiency of each sample design can be calculated using eq. (9.19). In this case, sample size and sampling unit size are constant, so we can assume that costs are constant for the three sample designs and only standard error would need consideration.

The results presented here are only for one sample of the many possible for each sample design. Different samples of each design from this population would yield different results, and, while the results shown here conform to the expectation that STS and SYS are more efficient than random sampling, any individual set of samples may show different results and multiple samples from the same population would be required to fully understand the design implications. Simulation of samples, such as carried out here, is an excellent way to explore sample design. For example, Figure 10.10 compares 500 random samples using SRS and STS to the 20 possible combinations arising from SYS. Again the results are very similar and most likely due to the dominance of the IH strata relative to the other strata

TABLE 10.3. Summary Statistics and Total Estimates for the Stratified Random Sample of the 25 ha Forest Shown in Figure 10.9 Using the Variance and Standard Error Formulas for Simple Random Sampling

Parameter					Per ha Estimates		Forest Total Estimates		
Strata	Area (ha)	No. Plots	Mean	$s_{\bar{x}_s}$	Confidence Width[a]	Error (%)	Total	$s_{\hat{x}}$	Confidence Width[a]
Trees (#/ha)									
BS	4.25	25	1428	89	184	13	6,069	379	782
EH	0.10	3	667	88	379	57	67	9	39
IH	20.65	122	874	40	80	9	18,044	829	1642
Overall	25.00	150	967	36	75	7	24,179	912	1802
Basal area (m²/ha)									
BS	4.25	25	29.4	1.4	3.0	10	125	6.2	12.7
EH	0.10	3	50.6	3.9	16.7	33	5	0.4	1.7
IH	20.65	122	21.4	1.2	2.3	11	442	24.2	47.9
Overall	25.00	150	22.9	1.0	2.0	9	572	25.0	49.4
Total volume (m³/ha)									
BS	4.25	25	172	9	18	11	733	38	79
EH	0.10	3	369	31	132	36	37	3	13
IH	20.65	122	142	9	18	13	2,938	187	370
Overall	25.00	150	148	8	15	10	3,707	191	377

Finite correction factor was not used in these summaries.
[a] 95% confidence level was used for both confidence width and error calculations.

TABLE 10.4. Comparison of Population Values (per ha) to Sample Estimates from Simple Random Sampling, Systematic Sampling, and Stratified Sampling

	Mean			Standard Error ($S_{\bar{x}}$)		
Sample Type	Trees (#/ha)	Basal Area (m²/ha)	Total Volume (m³/ha)	Trees (#/ha)	Basal Area (m²/ha)	Total Volume (m³/ha)
True population	971	23.2	151			
Simple random	934	22.8	148	39	0.8	6
Systematic	950	23.0	150	35	0.9	6
Stratified random	967	22.9	148	36	1.0	8

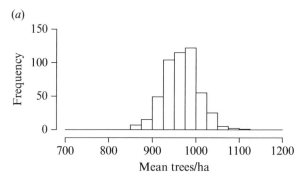

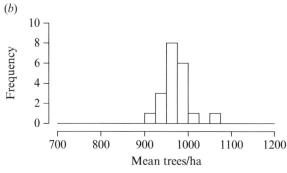

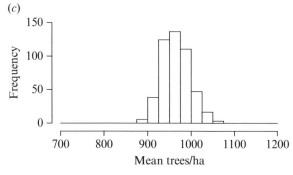

FIG. 10.10. Comparison of distribution of means arising from: (*a*) 500 simple random samples, (*b*) 20 systematic samples, and (*c*) 500 stratified random samples. The number of sampling units (sample size) was 150 for each simulated sample.

(STS has greater reductions in sampling errors when strata are larger and the differences between strata are greater than what was used here for this example); however, as observed in the detailed examples provided, the simulated samples also show that SRS produces the largest sampling errors with SYS and STS being lower.

10.6. CLUSTER SAMPLING

A cluster consists of a group of smaller measurement units that, taken together, form the sampling unit. The typical cluster consists of a number of subplots (measurement units) located around the center in a fixed configuration (Fig. 10.11). The group of subplots forms the sampling unit. Clusters can assume many different configurations depending on the number of measurement units, the distance between units, and the geometric distribution.

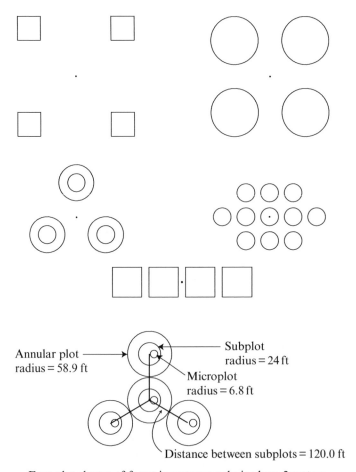

Four plot cluster of forest inventory analysis phase 2 system

FIG. 10.11. Typical cluster configurations.

Clusters are frequently used in forest inventories of large areas, such as regional or national inventories, especially in remote areas with difficult access or long distances between plots. Instead of a distribution of single plots, which require much time moving from one to another, clusters can be used. When using clusters, fewer sample locations are used and a number of measurement units are measured at each location instead of a single plot. The intent is to reduce the time for movement from one sample point to another and reduce costs of fieldwork.

Cluster sampling is classified into two categories: single stage and two stage. In *single-stage cluster sampling*, a sample of cluster locations is selected from a set of locations. At each location, a fixed set of measurement units (subplots) are measured. In *two-stage cluster sampling*, a sample of cluster locations is selected from a set of locations. At each location, a subsample of measurement units is selected from a list of measurement units and this subset measured. It is also possible to have a situation where the number of measurement units per location is not the same and where individual measurement units are not of equal size (e.g., PPS sampling may be used as the measurement unit). The analysis discussed below focuses on single-stage clusters of equal size. For a discussion of the analysis of two-stage cluster sampling and clusters of unequal size, see Cochran (1977).

Given n=the number of clusters (locations) sampled, M=the number of subunits per cluster, and X_{ij}=the value of the parameter (e.g., basal area or volume) for the jth subunit of the ith cluster. The total for the ith cluster is

$$x_i = \sum_{j=1}^{M} X_{ij} \qquad (10.34)$$

and the mean subunit of the ith cluster is

$$\bar{x}_i = \frac{x_i}{M} = \frac{\sum_{j=1}^{M} X_{ij}}{M} \qquad (10.35)$$

The estimate of the sample mean of the clusters is

$$\bar{x} = \frac{\sum_{i=1}^{n} x_i}{n} \qquad (10.36)$$

And the estimated mean of the subunits is

$$\bar{x} = \frac{\sum_{i=1}^{n}\sum_{j=1}^{M} X_{ij}}{nM} = \frac{\sum_{i=1}^{n}\bar{x}_i}{n} = \frac{\bar{x}}{M} \qquad (10.37)$$

The estimated variation among subunits has two components and can be estimated using (Cochran, 1977)

$$s_{\bar{x}}^2 = \frac{(n-1)s_b^2 + n(M-1)s_w^2}{nM-1} \qquad (10.38)$$

where the variation between clusters is

$$s_b^2 = \frac{\sum_{i=1}^{n}\left(\overline{x}_i - \overline{\overline{x}}\right)^2}{n-1} \tag{10.39}$$

and the variance within clusters is

$$s_w^2 = \frac{\sum_{i=1}^{n}\sum_{j=1}^{M}\left(X_{ij} - \overline{x}_i\right)^2}{n(M-1)} \tag{10.40}$$

The standard error for the clusters would be $s_{\overline{x}} = \sqrt{s_b^2/n}$ and the standard error for the subunits would be $s_{\overline{\overline{x}}} = \sqrt{s_{\overline{x}}^2 / (Mn)}$.

To illustrate the use of cluster sampling, we will return to the 25 ha example forest (Plate 1) and the systematic sample from Section 10.3. A 6% sample using 100 m² plots resulted in a sample size of 150 plots. Because SYS only has one random choice (the starting point), estimation of sampling error is not possible from a single systematic sample. Shiue (1960) suggested using multiple systematic samples, each with their own random start. Each systematic sample can be considered a cluster and cluster sampling analysis can be used to estimate sampling error for SYS with multiple starts. Shiue (1960) recommended at least five random starts; here, we used 6 random starts, thus each systematic sample (cluster) is composed of 25 plots (150/6 = 25). Using square grid spacing, plots within each systematic sample are spaced at 100 m spacing ($D_{sq} = \sqrt{25 \cdot 10,000 / 25} = 100$ m). Sampling without replacement was used, so no overlap in the clusters (systematic samples) was allowed. Figure 10.12 shows the map of the six systematic sample clusters, and Table A.10 contains the plot-level sample data.

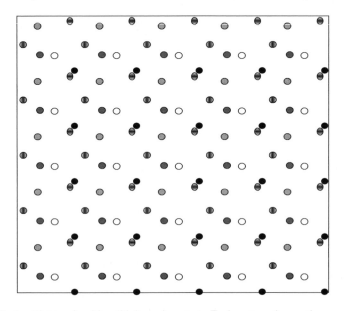

FIG. 10.12. Systematic sample with multiple random starts. Each systematic sample represents a cluster. The different plot color patterns represent the six different clusters.

TABLE 10.5. Cluster Summaries for Volume/ha for Systematic Sampling with Multiple Random Starts

Cluster	x_i	$\bar{x}_i$	$\left(\bar{x}_i - \bar{\bar{x}}\right)^2$	$\dfrac{\sum_{j=1}^{M}\left(X_{ij} - \bar{x}_i\right)^2}{M-1}$
1	3190.0	127.6	334.79	5584.01
2	3495.6	139.8	36.90	7102.33
3	3676.8	147.1	1.38	6326.36
4	3783.7	151.3	29.69	5847.80
5	3677.0	147.1	1.39	6270.73
6	4061.7	162.5	274.52	7057.75
Total	21885.0	875.4	678.66	38188.98

All plot-level values are scale to per hectare at the subunit. Individual plot-level summaries are shown in Table A.10.

Table 10.5 shows the summary statistics for each cluster for total volume. Column 2 in Table 10.5 is the total for each cluster (eq. (10.34)), and column 3 is the average subunit for each cluster (eq. (10.35)). The mean cluster is obtained by dividing the total of column 2 by the number of clusters (eq. (10.36)):

$$\bar{x} = \frac{\sum_{i=1}^{n} x_i}{n} = \frac{21885.0}{6} = 3647.5.$$

Because each subunit had been scaled to volume (m^3) per hectare, the units for the mean cluster is sum of m^3/ha, which is not a very useful value. The mean subunit, which is expressed in m^3/ha, is obtained by either dividing the total of column 2 by the total number of subunits (nM) or by dividing the total of column 3 by the number of clusters (eq. (10.37)):

$$\bar{\bar{x}} = \frac{\sum_{i=1}^{n}\sum_{j=1}^{M} X_{ij}}{nM} = \frac{21885.0}{6 \cdot 25} = \frac{875.4}{6} = 145.9\,m^3/ha.$$

The variance between clusters is obtained by dividing the total of column 4 by the number of clusters minus 1 (eq. (10.39)):

$$s_b^2 = \frac{\sum_{i=1}^{n}\left(\bar{x}_i - \bar{\bar{x}}\right)^2}{n-1} = \frac{678.66}{6-1} = 135.73,$$

and the variance within clusters is obtained by dividing the total of column 5 by the number of clusters (eq. (10.40)):

$$s_w^2 = \frac{\sum_{i=1}^{n}\sum_{j=1}^{M}\left(X_{ij} - \bar{x}_i\right)^2}{n(M-1)} = \frac{\sum_{i=1}^{n}\left(\frac{\sum_{j=1}^{M}\left(X_{ij} - \bar{x}_i\right)^2}{M-1}\right)}{n} = \frac{38188.98}{6} = 6364.83.$$

Finally, the variance among subunits is (eq. (10.38))

$$s_{\bar{x}}^2 = \frac{(n-1)s_b^2 + n(M-1)s_w^2}{nM-1} = \frac{(6-1)135.73 + 6(25-1)6364.83}{6\cdot 25-1} = 6155.80,$$

and the standard error is $s_{\bar{x}} = \sqrt{s_{\bar{x}}^2/nM} = \sqrt{6155.80/150} = 6.41\,\mathrm{m}^3/\mathrm{ha}$. Comparing this estimate to the estimate obtained for a single larger systematic sample (Table 10.2), we see, as is often the case, that the estimated standard error from clusters is slightly larger than the standard error we would have obtained from a single larger sample (Freese, 1962).

Shiue (1960) suggested that the standard error derived from the variance between clusters (eq. (10.39)) would be a better measure of standard error for SYS ($s_{\bar{x}} = \sqrt{s_b^2/n} = \sqrt{135.73/6} = 4.76\,\mathrm{m}^3/\mathrm{ha}$) than the standard error based on subunits. Schaeffer et al. (1990) suggested this estimate was a minimum bound on the error.

Those familiar with analysis of variance will note the similarity between cluster analysis and the formulas for analysis of variance. Equation (10.38) is the formula for total sums of squares, and eq. (10.40) is the formula for the error (residual) sum of squares; however, eq. (10.39) differs from the equation for treatment sums of squares by a factor equal to the number of replicates (in cluster analysis, this is the number of subunits per cluster). In this example, the estimate from eq. (10.39) would differ by a factor of 25 from the estimate based on treatment sums of squares, and while some authors have considered this an erroneous formulation, Cochran (1977) points out that, in terms of standard error at the subunit level, the simpler formulation presented here produces almost equivalent results to that using the analysis of variance formulation (6.41 m³/ha vs. 6.46 m³/ha).

10.7. MULTISTAGE SAMPLING

In multistage sampling, a population consists of a list of sampling units (primary stage), each of which is made up of smaller units (second stage), which in turn could be made up of smaller units (third stage), and so on. A sample would be chosen from the primary units. A subsample of the secondary units would then be taken in each of the selected primary units, and the procedure would be continued to the desired stage. In general, this procedure is called multistage sampling.

Two-stage sampling, which is discussed in this section, is the most common application in forest inventory and indicates that the sampling stops at the secondary stage. For example, a forest to be inventoried might consist of numerous compartments, or stands, that could be considered the primary units in a sampling design. Plots chosen in the selected compartments would then form the secondary units. Similarly, an inventory design using plots on randomly chosen lines or strips is a form of two-stage sampling. Another frequently used two-stage sampling design in forest inventory employs clusters of plots or sampling points at randomly chosen locations. Multistage sampling in forest inventory is not restricted to fixed-area sampling units and can be employed with variable probability sampling procedures. Thus, either a series of compartments or primary locations could be selected from the forest, and a number of secondary points chosen for selection of trees using a variable probability procedure.

Often, primary units are grouped based on similar characteristics, and referred to as strata (Section 10.5) and a subset of primary units is selected from each strata. This is referred to as *stratified two-stage sampling* (Freese, 1962). If all primary units within all strata are selected at the first stage, the design is equivalent to STS. In stratified two-stage sampling, two-stage sampling estimation procedures are used on each stratum and STS estimation procedures are used for the overall forest (Freese, 1962).

Multistage sampling has the principal advantage of concentrating measurement work close to the location of the chosen primary sampling units rather than spreading it over the entire forest area to be inventoried. This is advantageous when it is difficult and costly to locate and get to the ultimate sampling unit, while it is comparatively easy and cheap to select and reach the first-stage unit. Stratified two-stage sampling also is advantageous when the strata are composed of many smaller stands or compartments.

To permit calculation of unbiased estimates of means and standard errors, random selection of sampling units at all stages should be used. It also is possible to select primary units with probability proportional to their sizes. Systematic selection of primary units across sizes ordered from smallest to largest also is common (Iles, 2003). Secondary units can be selected on a random basis; however, frequent use is made of systematic selection in two-stage sampling. A common design employed in forest inventory utilizes randomly chosen primary sampling units with systematic selection of secondary units within each primary unit.

In two-stage sampling, m sampling units are selected from the M primary units of the population as a first stage. From each of the selected m units, n secondary units are then chosen from the population of N secondary units within each primary sampling unit. The most common cases encountered using two-stage sampling in forest inventory are

1. All primary units of equal size containing equal numbers of secondary units of equal size.
2. Primary units of unequal sizes containing varying numbers of secondary units of equal size.
3. Primary units of unequal sizes containing varying number of secondary units of variable sizes.

Case 1 is a special case of cluster sampling (Section 10.7) and is rarely encountered in most forest inventory situations. Even in the most industrialized forested landscape, stand or compartment sizes vary. Formulas for the estimates associated with case 1 were presented in earlier editions of this textbook (Husch et al., 1983, 2003) and are also found in Freese (1962). Cases 2 and 3 frequently occur in forest inventory since forests are often divided on the basis of age, species composition, or stand structure, which are forest characteristics that often conform to topographic or other edaphic factors that vary in size and shape. Case 2 applies when using fixed-area plots for secondary units. Case 3 arises when using variable probability sampling at the secondary stage or when fixed-area secondary units slop over forest or strata boundaries. If slop over correction is not employed, then the analysis requires the use of ratio estimation.

TABLE 10.6. Number of Stands, Total Area, and Range of Stand Sizes by Stand Type for the 1500 ha Noonan Forest in New Brunswick, Canada

Strata	Number of Stands	Stand Area (ha)			
		Total	Minimum	Mean	Maximum
Balsam fir -hardwood (BFIH)	9	35.7	0.8	4.0	12.5
Mixed hardwood (HDWD)	24	249.4	0.8	10.4	51.0
Intolerant hardwood (INHW)	15	222.5	0.3	14.8	74.2
White pine (PINE)	2	12.5	2.3	6.2	10.1
Regenerating stands (REGEN)	15	106.4	1.1	6.6	17.0
Spruce-balsam fir (SPBF)	27	371.5	0.9	13.8	135.7
Spruce-hardwood (SPHW)	32	334.1	0.2	10.4	47.7
Tolerant hardwood (TOHW)	6	64.2	0.3	10.7	37.6
Total forested	131	1396.3			
Nonforested wetlands (NF)		115.0			
Total		1511.3			

An example for case 2 is developed here for stratified two-stage sampling. Plate 2 shows a map of the 1500 ha (3700 acres) Noonan Research Forest in central New Brunswick, Canada. The forest is composed of 131 forested stands across 8 strata (stand types) and nonforested wetlands (Table 10.6). The goal of the inventory is to estimate total volume by strata type and for the overall forest. A total sample size of 150 plots was targeted. To simplify the two-stage calculations, the same number of plots per stand, $m_{kj} = 3$, was used. With 3 plots per stand, a total of 50 stands could be selected for sampling. Proportional allocation (on the basis of total area) was used to allocate the number of sample stands (n_k) within each stratum. A minimum of two stands was required to calculate a standard error. The white pine stratum only has two stands, so both stands were sampled (in this stratum, the design was a simple random sample); the remaining 48 primary units (stands) were proportionally allocated to the other strata using eq. (10.29). The final numbers of stands sampled were 2 in BFIH; 9 in HDWD; 8 in INHW; 2 in PINE; 4 in REGEN; 13 in SPBF; 11 in SPHW; and 2 in TOHW (note: because of rounding, 51 stands were sampled). Stands within a stratum were sorted by size (hectares) and an initial stand selected. The remaining stands were selected systemically across the sorted list of stands starting with the initially selected stand. Within each selected stand, three 100 m² circular plots were randomly selected and measured. The resulting plot data are shown in Table A.11.

The mean and standard error for each stand (primary unit) is estimated using the formulas for a simple random sample:

$$\overline{x}_{kj} = \frac{\sum_{i=1}^{m_{kj}} X_{kji}}{m_{kj}}$$

and

$$s\left(\overline{x}_{kj}\right) = \sqrt{\frac{\sum_{i=1}^{m_{kj}} X_{kji}^2 - \frac{\left(\sum_{i=1}^{m_{kj}} X_{kji}\right)^2}{m_{kj}}}{m_{kj}\left(m_{kj} - 1\right)}}$$

where X_{kji} = the value of the ith secondary unit in the jth primary in the kth stratum

m_{kj} = the number of secondary units in the jth primary in the kth stratum

The mean for the kth stratum is obtained using

$$\bar{x}_k = \frac{\sum_{j=1}^{n_k} M_{kj} \bar{x}_{kj}}{\sum_{j=1}^{n_k} M_{kj}} \tag{10.41}$$

where $\bar{x}_{kj}$ = the mean of the jth primary in the kth stratum

M_{kj} = the number of secondary units in the jth primary in the kth stratum

M_{kj} is an expression of size of the jth primary and can be expressed as the number of plots (A_{kj}/a), the area of the stand (A_{kj}), or as a relative measure of size (e.g., portion of stratum, A_{kj}/A_k). If there is correlation between the mean of a primary and its area, then eq. (10.41) will result in slightly biased results. Freese (1962) presents an alternative formula for this case.

The standard error for the estimate shown in eq. (10.41) is

$$s(\bar{x}_k) = \bar{x}_k \sqrt{\left(\frac{n_k}{n_k - 1}\right)\left(\frac{\sum_{j=1}^{n_k} M_{kj}^2}{\left(\sum_{j=1}^{n_k} M_{kj}\right)^2} + \frac{\sum_{j=1}^{n_k} \hat{X}_{kj}^2}{\left(\sum_{j=1}^{n_k} \hat{X}_{kj}\right)^2} - \frac{2\sum_{j=1}^{n_k} M_{kj} \hat{X}_{kj}}{\sum_{j=1}^{n_k} M_{kj} \sum_{j=1}^{n_k} \hat{X}_{kj}}\right)\left(1 - \frac{n_k}{N_k}\right)} \tag{10.42}$$

where $\bar{x}_k$ = the mean of the kth stratum

n_k = the number of primaries (stands) selected in the kth stratum

N_k = the total number of primaries (stands) in the kth stratum

M_{kj} = an expression of size of the jth primary in the kth stratum (as described above)

$\hat{X}_{kj}$ = the total for the jth primary in the kth stratum ($\hat{X}_{kj} = M_{kj} \cdot \bar{x}_{kj}$)

While initially this formula looks complex, when divided into components it is quite manageable (Table 10.7), and the calculations are easily completed using a spreadsheet program. Five summations are required: sum of the sizes of the primaries sampled (column 5), sum of the squared sizes (column 8), sum of the totals of the primaries sampled (column 9), sum of the squared totals (column 11), and the sum of the cross product between size and total (column 12). Once the means and standard errors are estimated for each stratum, the forest-level mean and standard error are estimated using the formulas for STS (eqs. (10.23) and (10.26)). These estimates are slightly biased for small numbers of sampled primaries. Freese (1962) presents alternative unbiased estimators; however, these estimators require knowledge of the total number of secondary units, but are preferred when the bias associated with the above estimators is expected to be large.

Sample size determination involves tradeoffs between the numbers of primaries sampled and the numbers of secondary units per primary. Because the number of primary units

TABLE 10.7. Strata Estimates and the Component Calculations for a Stratified Two-Stage Sample of Noonan Forest

Strata	$\bar{x}_k$ (m³/ha)	n_k	N_k	$\sum_{j=1}^{n_k} M_{kj}$ (ha)	$\sum_{j=1}^{N_k} M_{kj}$ (ha)	$\left(\sum_{j=1}^{n_k} M_{kj}\right)^2$ (ha²)	$\sum_{j=1}^{n_k} M_{kj}^2$ (ha²)	$\sum_{j=1}^{n_k} \hat{X}_{kj}$ (m³)	$\left(\sum_{j=1}^{n_k} \hat{X}_{kj}\right)^2$ (m³)²	$\sum_{j=1}^{n_k} \hat{X}_{kj}^2$ (m³)²	$\sum_{j=1}^{n_k} M_{kj}\hat{X}_{kj}$ (ha/m³)	$s(\bar{x}_k)$ (m³/ha)	$\hat{X}_k$ (m³)	$s(\hat{X}_k)$ (m³)
BFHW	158	2	9	3.5	35.7	12.25	8.01	552	304,240	304,240	1,500	61.9	5,631	2,213
HDWD	101	9	24	75.2	249.4	5661.06	2066.85	7,575	57,385,305	18,749,133	194,997	7.4	25,111	1,833
INHW	64	8	15	94.1	222.5	8862.34	4770.22	6,068	36,824,374	10,804,603	203,660	16.2	14,341	3,608
PINE	320	2	2	12.5	12.5	15,525	107.89	3,991	15,929,260	12,966,032	37,168	49.7[a]	3,991	621
REGEN	10	4	16	20.1	106.4	402.00	163.85	193	37,227	19,840	304	8.5	1,024	906
SPBF	144	13	27	141.6	371.5	20036.40	4554.76	20,419	416,953,204	96,947,416	636,530	15.1	53,591	5,593
SPHW	144	11	32	131.7	334.1	17339.62	4176.43	19,018	361,681,947	96,052,763	609,515	15.9	48,253	5,327
TOHW	198	2	6	43.0	64.2	1849.00	1446.14	8,532	72,797,211	59,939,575	294,256	8.6	12,743	552
Total	118	51	131		1396.3							14.7	164,685	20,526

[a] $s(\bar{x}_{\mathrm{PINE}})$ is based on standard error for simple random sampling since all primaries are sampled.

342

is finite (and usually small relative to the number of secondary units) and because variance between primaries is often greater than variance within primaries, it is often statistically more efficient to sample more primaries than sample additional secondary units within primaries. In the case where two-stage sampling involves primaries that are all the same size (case 1 described above), sample size determination is similar to the approach described for cluster sampling (Section 10.6). In the case of stratified two-stage sampling, an overall number of primaries can be determined using an approach similar to sample size determination for stratified random sampling (Cochran, 1977; Schaeffer et al., 1990), and the number of primaries per stratum allocated using optimal allocation. However, any attempt to determine sample size requires some knowledge of the variation within and between primary units.

One of the main reasons two-stage sampling is utilized is because it is often much more expensive to move between primaries than within primaries. Costs are often an important consideration in determining the number of primary units versus secondary units within primaries. Both Freese (1962) and Iles (2003) discuss methods for incorporating differences in costs of sampling primaries versus secondary units into sample size calculations.

10.8. SAMPLING WITH COVARIATES

Section 3.9 introduced the concept of using covariates to improve estimation. In situations where the parameter of interest is difficult, costly, or destructive to measure, sampling with covariates, which are easier and less costly to measure, but correlated well with the parameter of interest, can be an effective sampling scheme. There are typically two cases that arise in forest inventory: regression sampling and double sampling.

Unfortunately, there exists some ambiguity in the term "regression" in the sampling context. It always pertains to situations where the covariate mean or total is known without error (as opposed to "double sampling" where the mean or total is estimated with some error), or the case where the model assumed has an intercept value (as opposed to "ratio" estimation where the intercept is zero). By careful use of the words *sampling* and *estimation*, the ambiguity can perhaps be minimized. Thus, we might consider the following classification using the notion presented by Freese (1962):

1. Regression sampling (true mean of covariate, X, known)
   - Regression estimation: $\bar{y}_R = \bar{y} + b(\mu_x - \bar{x})$, where $\bar{y}$, b, and $\bar{x}$ are estimated from the sample and μ_x is the known population mean.
   - Ratio estimation: $\bar{y}_R = \hat{R}\mu_x$, where $\hat{R}$ is an estimated ratio obtained from the sample by either $\hat{R} = \bar{y}/\bar{x}$ (ratio of means) or $\hat{R} = \sum\left(\dfrac{y}{x}\right)/n$ (mean of ratios), dependent on certain variance assumptions.

2. Double sampling (true mean of covariate, X, unknown, and must be estimated from a sample)
   - Regression estimation: $\bar{y}_{Rd} = \bar{y}_S + b(\bar{x}_L - \bar{x}_S)$, where $\bar{y}_S$, b, and $\bar{x}_S$ are determined from a small sample and $\bar{x}_L$ is determined from the large sample.
   - Ratio estimation: $\bar{y}_{Rd} = \hat{R}\bar{x}_L$, where $\hat{R}$ is either a ratio of means or a mean of ratios obtained from the small sample and $\bar{x}_L$ is obtained from the large sample.

For more details regarding sample sizes needed and formulas for the estimation of means and sampling errors of the various estimators, a text such as Cochran (1977) or Schreuder et al. (1993) should be consulted.

For ratio estimation, the choice of using either the ratio of the means or the mean ratio is not arbitrary. If the covariate, X, is selected with equal probability, and the question of interest focuses on the population total (or its areal mean, as in the case of many forestry applications), then the ratio of means is usually appropriate (Thompson, 2012, section 5.6). The mean of ratios is more informative for describing the relationship between variables at the level of individual observations. If the covariate is selected with variable probability, then an alternative estimator may be more appropriate (e.g., the generalized ratio estimator; Brewer, 1963). As Iles (2003) points out, it is important to document selection probabilities so that appropriate computational methods can be selected.

10.8.1. Regression Sampling

As the classification scheme above shows, regression sampling arises in a situation where the covariate is completely enumerated (such that the population mean is known without error) and a subsample of the covariate and the parameter of interest is then obtained. This rarely occurs in most forest inventory applications, thus double sampling is more generally encountered.

Regression Estimation. Using this method, an estimate of the mean of the sampling units is adjusted by means of a regression coefficient. The regression coefficient indicates the average change in the parameter of interest per unit change in the covariate for the sample and the population. For example, suppose we had a 100% cruise for the forest shown in Plate 1 and had dbh measurements for all of these trees. The basal area (m²) per hectare could be calculated for this forest:

$$\mu_{BA} = \frac{\sum_{i=1}^{24,273} 0.00007854 \cdot dbh_i^2}{25} = \frac{580.3561}{25} = 23.2 \, m^2 \, / \, ha.$$

Utilizing the relationship between basal area and volume, a regression estimator could be used to estimate total volume (m³/ha) based on a smaller subsample of basal area and volume. Table 10.8 shows the sample values for thirty 100 m² plots randomly sampled from the 25 ha forest.

The regression estimator for mean volume per hectare is

$$\bar{v}_{reg} = \bar{v} + b\left(\mu_{BA} - \bar{x}_{BA}\right) \tag{10.43}$$

where $\bar{v}_{reg}$ = adjusted estimate of volume (m³/ha) for the population

         $\bar{v}$ = sample mean volume (m³/ha)

         μ_{BA} = the true population mean basal area (m²/ha)

         $\bar{x}_{BA}$ = the sample mean basal area (m²/ha)

         b = the regression coefficient

TABLE 10.8. Basal Area (m²/ha) and Total Volume (m³/ha) for Thirty 100 m² Plots Randomly Sampled from the Forest Shown in Plate 1

Plot	BA (m²/ha)	Volume (m³/ha)	Plot	BA (m²/ha)	Volume (m³/ha)	Plot	BA (m²/ha)	Volume (m³/ha)
1	40.4	268	11	23.9	154	21	20.5	150
2	24.4	137	12	17.1	106	22	12.7	89
3	30.3	227	13	18.9	104	23	11.4	82
4	18.8	155	14	15.5	91	24	26.7	184
5	26.9	201	15	19.5	117	25	14.5	78
6	10.1	57	16	15.1	84	26	5.5	36
7	29.7	199	17	28.0	159	27	21.7	133
8	27.3	185	18	7.1	42	28	15.7	97
9	15.5	106	19	43.2	304	29	31.5	189
10	30.3	191	20	230	144	30	50.2	343

The regression coefficient is estimated using

$$b = \frac{SS_{XV}}{SS_{XX}}$$

(10.44)

where

$$SS_{XV} = \sum_{i=1}^{n} X_i V_i - \frac{\sum_{i=1}^{n} X_i \sum_{i=1}^{n} V_i}{n}$$

$$SS_{XX} = \sum_{i=1}^{n} X_i^2 - \frac{\left(\sum_{i=1}^{n} X_i\right)^2}{n}$$

X_i = basal area (m²/ha) of the ith sampling unit
V_i = volume (m³/ha) of the ith sampling unit

From Table 10.8, the required summations can be obtained:

$$\sum_{i=1}^{n} X_i = 675.3$$

$$\sum_{i=1}^{n} V_i = 4413$$

$$\sum_{i=1}^{n} X_i^2 = 18304.66$$

$$\sum_{i=1}^{n} V_i^2 = 805,346$$

$$\sum_{i=1}^{n} X_i V_i = 120,929$$

The estimate for the regression coefficient, b, is then

$$b = \frac{SS_{XV}}{SS_{XX}} = \frac{120,929 - \dfrac{675.3 \cdot 4413}{30}}{18304.36 - \dfrac{(675.3)^2}{30}} = \frac{21599.2}{3102.81} = 6.9612.$$

The regression adjusted volume (m³/ha) is then estimated as

$$\bar{V}_{reg} = \bar{v} + b\left(\mu_{BA} - \bar{x}_{BA}\right) = 147 + 6.9612\left(23.2 - 22.5\right) = 147 + 6.9612 \cdot 0.7 = 147 + 4.87 = 152.$$

To estimate standard error of a regression estimator, $s_{\bar{V}_{reg}}$, an estimate of the standard deviation about the regression line is needed (Freese, 1962):

$$s_{v|x} = \sqrt{\frac{SS_{VV} - b \cdot SS_{XV}}{n-2}} \tag{10.45}$$

where $\quad SS_{VV} = \sum_{i=1}^{n} V_i^2 - \dfrac{\left(\sum_{i=1}^{n} V_i\right)^2}{n}$

and the estimate of standard error is

$$s_{\bar{V}_{reg}} = s_{V|X}\sqrt{\frac{1}{n} + \frac{\left(\mu_{BA} - \bar{x}_{BA}\right)^2}{SS_{XX}}\left(1 - \frac{n}{N}\right)} \tag{10.46}$$

where $\quad N =$ population size

If the sample size, n, is small relative to the population size, N, the finite population correction factor, $\left(1 - (n/N)\right)$, can be omitted. So, for the example in Table 10.8, the standard deviation for the regression is estimated to be

$$s_{V|X} = \sqrt{\frac{SS_{VV} - b \cdot SS_{XV}}{n-2}} = \sqrt{\frac{156319.2 - 6.9612 \cdot 21599.2}{30 - 2}} = \sqrt{\frac{5962.85}{28}} = \sqrt{212.96} = 14.6,$$

And the standard error for the regression estimate is

$$s_{\bar{V}_{reg}} = s_{V|X}\sqrt{\left(\frac{1}{n} + \frac{\left(\mu_{BA} - \bar{x}_{BA}\right)^2}{SS_{XX}}\right)\left(1 - \frac{n}{N}\right)}$$

$$= 14.6\sqrt{\left(\frac{1}{30} + \frac{(23.2 - 22.5)^2}{3102.81}\right)\left(1 - \frac{30}{2500}\right)} = 14.6 \cdot 0.1819 = 2.66.$$

For most practical applications, a simpler formulation of standard error may be used:

$$s_{\bar{V}_{reg}} = s_{V|X}\sqrt{\frac{1}{n}\left(1 - \frac{n}{N}\right)} \tag{10.47}$$

which yields

$$s_{\bar{v}_{reg}} = 14.6 \sqrt{\frac{1}{30}\left(1 - \frac{30}{2500}\right)} = 14.6 \cdot 0.1814 = 2.65.$$

A comparison of the standard error for the regression estimator and the standard error for the volume sample alone will help to appreciate the advantage of using regression sampling and the impact of the covariate on estimation. For the sample shown in Table 10.8, the standard error for volume based on the 30 sample plots would be 13.4 m³/ha versus the 2.66 m³/ha obtained here. Even with respect to the samples summarized in Table 10.4, the standard error obtained here is around one-half to one-third of the value obtained from a sample with five times the number of observations. Of course, the estimate here relies on knowing the true population mean basal area per hectare. While measuring the variables required to estimate volume is time consuming, measuring the dbh of all 24,273 trees is probably more time consuming. Except for very small tracts or where the value of the product merits such effort (such as veneer quality black walnut), regression sampling would seldom be an efficient sampling strategy.

Ratio Estimation. One situation where regression sampling might arise frequently is when using fixed-area sampling units in a forest or stand with irregular boundaries. In this situation, the possibility exists of having fractional plots or irregular length strips; therefore, it is necessary to measure two variables for each plot or strip: the area of the sampling unit which falls in the forest area, and the quantity of the variable, such as volume of timber, on this area. The mean volume per unit area for the forest must then be calculated using ratio estimation.

In the majority of cases involving plot sampling, ratio or regression procedures have been bypassed in the interest of simplifying field work, saving time, and reducing later calculations. Thus, if a plot falls so that it straddles a forest or stratum boundary, instead of measuring both the area and quantities of that portion of the unit in the forest, a number of boundary slopover corrections have been proposed (Section 9.2.6). Older, more questionable procedures involve moving the plots so that they fall entirely within or without the forest or assigning them to the stratum in which the center point falls. To avoid the possibility of bias arising from these procedures, ratio or regression methods are preferred.

When the ratio estimation procedure is employed, the sample should consist of 30 or more units so that the inherent bias of the method becomes negligible. In addition, the method is only truly applicable when the linear relationship between quantity and area per sampling unit passes through the origin. When the linear regression line does not pass through the origin (e.g., volume per unit area can be zero when the plot area is nonzero), regression estimation is preferable.

The ratio estimator is given by

$$\bar{r} = \frac{\bar{v}}{\bar{x}} = \frac{\sum\limits_{i=1}^{n} V_i / n}{\sum\limits_{i=1}^{n} X_i / n} = \frac{\sum\limits_{i=1}^{n} V_i}{\sum\limits_{i=1}^{n} X_i} \tag{10.48}$$

where $\bar{r}$ = estimate of the mean quantity per unit area
         $\bar{v}$ = mean of the observed quantities on the selected sampling units
         $\bar{x}$ = mean area of the sampling units selected
         V_i = quantity measured on the ith sampling unit
         X_i = area of the ith sampling unit
         n = number of units in the sample

The estimate of the total quantity is given by

$$\hat{V} = X\bar{r} \tag{10.49}$$

where $\hat{V}$ = estimate of the total quantity (e.g., volume) of the forest
         X = known total area of the forest

If the area, X, is known without any sampling error, then the standard error of the total estimate of quantity is

$$s_{\hat{V}} = \sqrt{X^2 s_{\bar{r}}^2} \tag{10.50}$$

If the total area of the forest is an estimate, $\hat{X}$, and has a sampling error, then the standard error is

$$s_{\hat{V}} = \sqrt{\hat{X}^2 s_{\bar{r}}^2 + \bar{r}^2 s_{\hat{X}}^2} \tag{10.51}$$

where $s_{\hat{V}}$ = standard error of the total quantity
         $s_{\bar{r}}$ = standard error of the ratio $\bar{r}$
         $s_{\hat{X}}$ = standard error of the total area estimate

The standard error of the ratio, $\bar{r}$, is calculated using

$$s_{\bar{r}} = \sqrt{\frac{s^2}{n}\left(1 - \frac{n}{N}\right)} \tag{10.52}$$

where

$$s^2 = \frac{\bar{r}^2}{n-1}\left(\frac{\sum_{i=1}^{n} V_i^2}{\bar{V}^2} + \frac{\sum_{i=1}^{n} X_i^2}{\bar{X}^2} - \frac{2\sum_{i=1}^{n} V_i X_i}{\bar{V}\bar{X}}\right) \tag{10.53}$$

and
    N = number of sample units in the population
The number of sampling units for a given precision can be estimated by solving eq. (10.52) for n and using a preliminary estimate of s^2. Student's t can be incorporated to vary the probability level.

10.8.2. Double Sampling

Double sampling is a form of multiphase sampling limited to two phases. In double sampling, an estimate of the principle variable is obtained by utilizing its relationship with a covariate. The method is useful when information on the principal variable is costly and difficult to obtain, whereas the covariate can be more easily and cheaply observed. The aim of double sampling is to reduce the number of measurements of the costly, principal variable without sacrificing the precision of the estimate.

The general procedure in double sampling is that in a first phase a large random sample is taken of the covariate X that will yield a precise estimate of its population mean or total. In a second phase, a random subsample is selected from the previous sample, and on these sampling units measurements are taken of the principal variable Y. Note that the first and second phases are mutually dependent since the measurements in the second phase are taken from a portion of the sampling units of the first phase. Thus, we have a small sample on which both the covariate, X, and the principal variable, Y, have been measured. With these data, a regression or ratio can be developed between the two variables that can be utilized with the large sample of the auxiliary variable to make an estimate of the mean and total for the principal variable.

The relationship of Y and X may have one of numerous forms. For illustrative purposes, a simple linear relationship will be demonstrated; however, it is well to bear in mind that in many instances a curvilinear relationship may be required.

Regression Estimation. In double sampling, the corrected estimate is obtained from a regression of the form

$$\overline{y}_{\text{reg}} = \overline{y}_S + b\left(\overline{x}_L - \overline{x}_S\right) \tag{10.54}$$

where $\overline{y}_{\text{reg}}$ = regression estimate of the mean of Y (principal variable) from double sampling

$\overline{y}_S$ = estimate of the mean of Y from the second, small sample

$\overline{x}_S$ = estimate of the mean of X from the second, small sample

$\overline{x}_L$ = estimate of the mean of X (auxiliary variable) from the first, large sample

b = linear regression coefficient

The regression coefficient is estimated using eq. (10.44):

$$b = \frac{SS_{XY}}{SS_{XX}}$$

where

$$SS_{XY} = \sum_{i=1}^{n_S} X_i Y_i - \frac{\sum\limits_{i=1}^{n_S} X_i \sum\limits_{i=1}^{n_S} Y_i}{n_S}$$

$$SS_{XX} = \sum_{i=1}^{n_S} X_i^2 - \frac{\left(\sum\limits_{i=1}^{n_S} X_i\right)^2}{n_S}$$

X_i = quantity X measured on the ith sampling unit of the second, small sample

Y_i = quantity Y measured on the ith sampling unit of the second, small sample

n_S = number of sampling units in the second, small sample

The standard error of the regression estimate is then the sum of the variance of Y about its mean and the variance of Y about the regression line and is obtained using

$$s_{\bar{y}_{reg}} = \sqrt{s_{Y|X}^2 \left(\frac{1}{n_S} + \frac{(\bar{x}_l - \bar{x}_s)^2}{SS_{XX}}\right)\left(1 - \frac{n_S}{n_L}\right) + s_Y^2 \left(\frac{1}{n_L}\right)\left(1 - \frac{n_L}{N}\right)} \tag{10.55}$$

where s_Y^2 = the sample variance of Y from the second, small sample $\left(s_Y^2 = \dfrac{SS_{YY}}{n_S - 1}\right)$

$$SS_{YY} = \sum_{i=1}^{n_S} Y_i^2 - \frac{\left(\sum_{i=1}^{n_S} Y_i\right)^2}{n_S}$$

$s_{Y|X}^2$ is the variance about the regression[†] (the square of eq. (10.45)): $s_{Y|X}^2 = \dfrac{SS_{YY} - b \cdot SS_{XY}}{n_S - 2}$

n_L = the number of observations from the first, larger sample

N = the total number of sampling units in the first phase of double sampling (large sample)

Note that in eq. (10.55) $(1 - (n_L / N))$ is the finite population correction factor that is dropped if the population is considered infinite or if n_L is small relative to N.

In the above example for regression sampling (Section 10.8.1), the population mean basal area (m²/ha) was required. While substantial improvements in standard error were achieved, the effort required to obtain the population mean would be large. The data shown in Table 10.8 are a subsample of plots from Table A.7. Using the basal area estimates from Table A.7 as the large sample, and the data from Table 10.8 as the small sample, the regression estimates for double sampling would be

$$\bar{x}_L = \frac{\sum_{i=1}^{150} X_i}{150} = \frac{3422.82}{150} = 22.8\,\text{m}^2 \,/\,\text{ha}$$

$$\bar{x}_S = \frac{\sum_{i=1}^{30} X_i}{30} = \frac{675.32}{30} = 22.5\,\text{m}^2 \,/\,\text{ha}$$

[†] This equation can also be written in terms of SS_{XX}: $s_{Y|X}^2 = \dfrac{SS_{YY} - b^2 SS_{XX}}{n_S - 2}$, the two equations are algebraically equivalent, though this second formulation is found more commonly in statistical literature (Cochran, 1977; Schaeffer et al., 1990).

$$\bar{y}_S = \frac{\sum\limits_{i=1}^{30} Y_i}{30} = \frac{4413}{30} = 147 \ \text{m}^3 \, / \, \text{ha}$$

$$b = \frac{SS_{XY}}{SS_{XX}} = \frac{120{,}929 - \left(675.3 \cdot 4{,}413/30\right)}{18304.36 - \left(\left(675.3\right)^2/30\right)} = \frac{21599.2}{3102.81} = 6.9612$$

The regression adjusted estimate of mean volume (eq. (10.54)) is then

$$\bar{y}_{reg} = \bar{y}_S + b\left(\bar{x}_L - \bar{x}_S\right) = 147 + 6.9612\left(22.8 - 22.5\right) = 147 + 2.1 = 149 \ \text{m}^3 \, / \, \text{ha}.$$

The estimated variance due to regression is

$$s^2_{Y|X} = \frac{SS_{YY} - b^2 \cdot SS_{XY}}{n_S - 2} = \frac{156319.2 - 6.9612 \cdot 21599.2}{30 - 2} = \frac{5962.85}{28} = 212.96,$$

the sample variance of Y is

$$s^2_Y = \frac{SS_{YY}}{n_S - 1} = \frac{156{,}319}{30 - 1} = 5{,}390,$$

and, finally, the standard error of the regression estimate is

$$S_{\bar{y}_{reg}} = \sqrt{s^2_{Y|X}\left(\frac{1}{n_S} + \frac{\left(\bar{x}_L - \bar{x}_S\right)^2}{SS_{XX}}\right)\left(1 - \frac{n_S}{n_L}\right) + s^2_Y\left(\frac{1}{n_L}\right)\left(1 - \frac{n_L}{N}\right)}$$

$$= \sqrt{212.96\left(\frac{1}{30} + \frac{\left(22.8 - 22.5\right)^2}{3102.81}\right)\left(1 - \frac{30}{150}\right) + 5390\left(\frac{1}{150}\right)\left(1 - \frac{150}{2500}\right)}$$

$$= \sqrt{5.6839 + 33.7773} = \sqrt{39.4612} = 6.28 \ \text{m}^3 \, / \, \text{ha}.$$

Again, to appreciate the efficiency of double sampling and the regression estimator, the standard error obtained here must be compared to the estimate obtained from SRS. The estimated standard error from the original 150 plot sample was 5.8 (6 as reported in Table 10.4) and was 6.3 here. In the original SRS, 1401 trees were measured for dbh and height to calculate volume, in the double sample, the same 1401 trees were measured for dbh, but 272 trees were required to have height measurements. Using a conservative estimate of 2 minutes per tree to measure heights, the double sampling scheme would save almost 38 hours of field work with very little increase in standard error.

Ratio Estimation. Double sampling can also be carried out using ratio estimates, remembering that the ratio estimate is a conditioned regression in which the relationship between the two variables X and Y is such that a zero value of X means a zero value of Y. In the example above, it would be expected that volume would be 0 when basal area is 0, thus a ratio estimator might be more appropriate than the regression estimator. The mean of ratios or the ratio of means may be used to estimate the adjustment ratio. The mean ratio estimator is

$$\bar{r}_m = \frac{\sum_{i=1}^{n_S} \dfrac{Y_i}{X_i}}{n_S} \tag{10.56a}$$

and the ratio of means estimator is the same as shown in eq. (10.48):

$$\bar{r} = \frac{\bar{v}}{\bar{x}} = \frac{\left(\sum_{i=1}^{n_S} V_i\right)/n_S}{\left(\sum_{i=1}^{n_S} X_i\right)/n_S} = \frac{\sum_{i=1}^{n_S} V_i}{\sum_{i=1}^{n_S} X_i} \tag{10.56b}$$

The ratio adjusted estimator is then

$$\bar{y}_{\text{rat}} = R_d \cdot \bar{x}_L \tag{10.57}$$

where R_d = the estimated ratio from eqs. (10.56a) $(\bar{r}_m)$ or (10.56b) $(\bar{r})$
For the example in Table 10.8, the mean ratio is

$$\bar{r}_m = \frac{\sum_{i=1}^{n_S} \dfrac{Y_i}{X_i}}{n_S} = \frac{193.7709}{30} = 6.459$$

and the ratio of the means is

$$\bar{r} = \frac{\sum_{i=1}^{n_S} Y_i}{\sum_{i=1}^{n_S} X_i} = \frac{4412.57}{675.319} = 6.534.$$

The ratio-adjusted mean volume estimates are then $\bar{y}_{\text{rat}} = \bar{r}_m \cdot \bar{x}_L = 6.459 \cdot 22.8 = 147 \, \text{m}^3/\text{ha}$
for the mean ratio and $\bar{y}_{\text{rat}} = \bar{r} \cdot \bar{x}_L = 6.534 \cdot 22.8 = 149 \, \text{m}^3/\text{ha}$ for the ratio of means.

The standard error for the mean ratio estimator is given by

$$s_{\bar{y}_{\text{rat}}} = \sqrt{\bar{x}_L^2 \left(\frac{s_r^2}{n_S}\right)\left(1 - \frac{n_S}{n_L}\right) + \left(\frac{s_Y^2}{n_L}\right)\left(1 - \frac{n_L}{N}\right)} \tag{10.58a}$$

and the standard error for the ratio of means estimator is given by

$$s_{\bar{y}_{\text{rat}}} = \sqrt{\left(\frac{\bar{x}_L}{\bar{x}_S}\right)^2 \left(\frac{s_Y^2 + \bar{r}^2 \cdot s_X^2 - 2\bar{r} \cdot s_{XY}}{n_S}\right)\left(1 - \frac{n_S}{n_L}\right) + \left(\frac{s_Y^2}{n_L}\right)\left(1 - \frac{n_L}{N}\right)} \tag{10.58b}$$

where s_r^2 = the sample variance of the ratios ($r_i = Y_i / X_i$) from the second, smaller sample
s_X^2 = the sample variance of X from the second, smaller sample

s_Y^2 = the sample variance of Y from the second, smaller sample

s_{XY} = the covariance (eq. (3.21)) of X and Y from the second, smaller sample

The sample variance of r is

$$s_r^2 = \frac{\sum\limits_{i=1}^{n_S} r_i^2 - \dfrac{\left(\sum\limits_{i=1}^{n_S} r_i\right)^2}{n_S}}{n_S - 1} = \frac{1265.627 - \dfrac{(193.771)^2}{30}}{30 - 1} = \frac{14.0551}{29} = 0.485$$

and the standard error for the mean ratio estimator is

$$s_{\bar{y}_{\text{rat}}} = \sqrt{(22.8^2)\left(\frac{0.484}{30}\right)\left(1 - \frac{30}{150}\right) + \left(\frac{5390}{150}\right)\left(1 - \frac{150}{2500}\right)}$$

$$= \sqrt{6.709 + 33.777} = \sqrt{40.487} = 6.36 \, \text{m}^3 \, / \, \text{ha}.$$

The covariance between X and Y is

$$s_{XY} = \frac{SS_{XY}}{n_S - 1} = \frac{21599.2}{30 - 1} = 744.8$$

and the standard error for the ratio of means estimator is

$$s_{\bar{y}_{\text{rat}}} = \sqrt{\left(\frac{22.8}{22.5}\right)^2 \left(\frac{5390 + 6.534^2 \cdot 106.993 - 2 * 6.534 \cdot 744.8}{30}\right)\left(1 - \frac{30}{150}\right) + \left(\frac{5390}{150}\right)\left(1 - \frac{150}{2500}\right)}$$

$$= \sqrt{1.027 \cdot 7.494 \cdot 0.8 + 35.933 \cdot 0.94} = \sqrt{39.934} = 6.32 \, \text{m}^3 / \text{ha}.$$

Thus, the ratio of means estimator provides a slightly better estimator than the mean ratio estimator in terms of standard error, but is larger than the standard error for the regression estimator. Comparing the estimated mean volumes to the true population mean ($152 \, \text{m}^3/\text{ha}$), we see that the regression estimator and the ratio of means estimators give almost identical and closer results ($149 \, \text{m}^3/\text{ha}$) than the mean of ratios estimate ($147 \, \text{m}^3/\text{ha}$). The estimate from the SRS sample was $148 \, \text{m}^3/\text{ha}$. Again, these results are for only one sample of numerous possible samples, results from other samples will give slightly different results.

10.9. LIST SAMPLING

List sampling with varying probabilities can be applied where the listed items are individual elements having different sizes, or where the listed items are clusters having different numbers of elements, or some other appropriate expression of size. In the usual forestry context, we can think of individual trees as the elements, and compartments or stands as clusters. This type of variable probability sampling applied to individual listed trees is usually impractical because of the necessity of listing each element prior to sampling.

Therefore, we will discuss the technique for compartments (clusters of trees) having varying, but known areas.

The sampling method is carried out first by listing the compartments in any order, along with their measure of size, say area. A sample size is decided on, and sampling is performed in such a way as to give the larger compartments greater chance of being selected. The selected compartments are then visited and measurements are taken on the variable of interest. The analysis of the data, subsequently shown, yields unbiased estimates of means, totals, and their variances, as long as sampling is performed with replacement (i.e., once chosen, a compartment is "put back" in the population and can be drawn again). In application, sampling is often without replacement; in this case, bias (often negligible) in the variance estimate will be incurred if eq. (10.60) is used.

To illustrate the procedure, we will consider a 150-acre mixed oak–hickory–maple forest located in central Indiana. As part of an appraisal exercise, all trees 9 in. dbh and greater were measured for species, dbh, and height to a 4 in. top diameter. Board foot volume was estimated using Beers (1964a) composite volume tables. The forest is made up of 25 compartments having areas designated by X_i. We want to choose n compartments with selection probability proportional to X_i and measure the variable Y_i on each of the chosen compartments. The first step is to list the compartments as shown in Table 10.9, obtaining a column of cumulative areas in the process. It is also helpful to show a column of associated numbers that indicates the set of consecutive integers from the one above the previous cumulative total to, and including, the cumulative total of the compartment in question. These numbers will be used to select compartments during the sampling process.

For this example, we decide that $n=7$ compartments is a sufficient sample. Seven random numbers are appropriately drawn from the range of integers 1 through 150, the total area of the compartments. A compartment is chosen as part of the sample if the random number falls within the interval indicated in the column of associated numbers. After the compartments are selected, they are visited and measurements are taken regarding the variables of interest, Y_i. Table 10.10 shows the random integers, selected compartments, compartment size, and total board foot volume.

A mean ratio estimation approach is used in this case, and the estimate of the mean value of Y per unit area is

$$\bar{r}_m = \frac{1}{n}\sum_{i=1}^{n}\frac{Y_i}{X_i} = \frac{1}{n}\sum_{i=1}^{n}r_i \tag{10.59}$$

where Y_i = the quantity measured on the ith compartment, for example, total compartment volume

X_i = size of the ith compartment, for example, compartment area

n = number of compartments chosen

$\bar{r}_m$ = mean ratio of Y per unit value of X (e.g., mean volume per unit area)

r_i = ratio of Y_i to X_i

For the data shown in Table 10.10, we have

$$\bar{r}_m = \frac{\sum_{i=1}^{n}r_i}{n} = \frac{98{,}656}{7} = 14{,}093 \text{ board feet/acre.}$$

TABLE 10.9. List of Compartments and Individual and Cumulative Areas for Use in List Sampling with Varying Probabilities, and the Total Board Foot Volume in Each Compartment

Compartment	Area (Acres) X_i	Cumulative Total of X_i	Associated Sample Numbers	Total Board Foot Volume
1	12	12	1–12	115,523
2	5	17	13–17	56,295
3	13	30	18–30	323,237
4	3	33	31–33	40,546
5	8	41	34–41	65,828
6	2	43	42–43	24,168
7	6	49	44–49	138,700
8	3	52	50–52	54,578
9	4	56	53–56	56,119
10	11	67	57–67	90,793
11	5	72	68–72	56,201
12	4	76	73–76	45,012
13	9	85	77–85	41,432
14	6	91	86–91	66,935
15	7	98	92–98	149,009
16	10	108	99–108	165,950
17	4	112	109–112	35,195
18	6	118	113–118	78,040
19	2	120	119–120	34,877
20	8	128	121–128	132,343
21	7	135	129–135	137,511
22	4	139	136–139	55,409
23	4	143	140–143	60,272
24	3	146	144–146	44,318
25	4	150	147–150	67,196
Total	150			2,135,487

TABLE 10.10. Random Integers, Selected Compartment, and Associated Area and Board Foot Volume for List Sample

Random Integer Drawn	Compartment Selected	Compartment Area (Acres)	Compartment Volume (bd. ft)
7	1	12	115,523
25	3	13	323,237
58	10	11	90,793
81	13	9	41,432
108	16	10	165,950
133	21	7	137,511
141	23	4	60,272
Total		66	934,718

The standard error of $\bar{r}_m$ can be estimated from the simple random sample variance (eq. (3.10)) and the formula for standard error (eq. (3.14)):

$$s_{\bar{r}_m} = \sqrt{\frac{s_r^2}{n}} = \sqrt{\frac{\displaystyle\sum_{i=1}^{n} r_i^2 - \frac{\left(\displaystyle\sum_{i=1}^{n} r_i\right)^2}{n}}{n(n-1)}} \tag{10.60}$$

For the data in Table 10.10, we have

$$\begin{aligned} s_{\bar{r}_m} &= \sqrt{\frac{\displaystyle\sum_{i=1}^{n} r_i^2 - \frac{\left(\displaystyle\sum_{i=1}^{n} r_i\right)^2}{n}}{n(n-1)}} = \sqrt{\frac{1,688,577,109 - \frac{(98,656)^2}{7}}{7(7-1)}} \\ &= \sqrt{\frac{298,142,150}{42}} = \sqrt{7,098,623} = 2,664 \text{ board feet/acre} \end{aligned}$$

The estimate of the total volume for the population, $\hat{Y}$, is then

$$\hat{Y} = \bar{r}_m X \tag{10.61}$$

with estimated standard error of

$$s_{\hat{Y}} = s_{\bar{r}_m} \cdot X \tag{10.62}$$

So, finally, for the data in Table 10.10, we have

$$\hat{Y} = \bar{r}_m X = 14,094 \cdot 150 = 2,114,061 \text{ board feet.}$$

and

$$s_{\hat{Y}} = s_{\bar{r}_m} \cdot X = 2,664 \cdot 150 = 399,649 \text{ board feet.}$$

Comparing the estimated total of 2,114,061 to the known total shown in Table 10.9: 2,135,487, we see that there is close agreement (the values differ by about 1%). Thus, list sampling provided an extremely reliable estimate of total volume and only required measurement of 3920 of the 9160 trees on the tract.

Instead of a complete tally within each compartment as implied above, a more practical procedure would be to take a subsample of secondary sampling units within each of the chosen compartments. This practice implies a two-stage sampling, with secondary sampling units chosen from primary units of unequal size (Section 10.7).

List sampling with varying probabilities has been used in other forestry applications. For example, Scott (1979) described its use for mid-cycle updating of permanent inventory plots. The initial plot volumes were used as the listed item from which a sample of n plots was chosen for remeasurement to obtain an estimate of current volume.

10.10. 3P SAMPLING

The necessity of listing the units prior to sampling acts as a severe deterrent to list sampling in many forestry applications, especially those where individual trees are the potentially listed items. Grosenbaugh (1963b), making use of a principle similar to that proposed by Lahiri (1951) to overcome the prior listing requirement, proposed a type of sampling that utilizes the PPS concept, but the element of size considered in his original application was the timber cruiser's on-the-spot estimate of tree volume. The name used by Grosenbaugh for this technique is "sampling with probability proportional to prediction" or "3P sampling."

The purpose of the present treatment of 3P sampling is to make the reader aware of the general concept and application of the method and show the calculations for a simple cruising example. For details, reference should be made to the works of Grosenbaugh (1963b, 1965, 1967a, 1967b, 1979), Bell and Dilworth (1993), Mesavage (1965, 1971), and Schreuder et al. (1993). This type of sampling has been applied to timber sales where each tree in the population (all marked trees) is assessed for a "crude" prediction of volume or value, and a subsample of these trees is selected for more detailed measurements. For this purpose, it appears to be very efficient.

Before listing the steps in a simplified application of 3P sampling, it is worthwhile to explain the basic sampling concept that leads to the name "probability proportional to prediction." The analogy given by Mesavage (1965) is convenient for this purpose:

> ...[S]uppose we have 20 cards numbered one to twenty. If we stipulate that the predicted volume of a sample tree must be equal to or greater than the number on a card subsequently drawn at random, a tree with a prediction of 1 would have only 1 chance in 20 to qualify [as a sample tree to be carefully measured for volume], whereas one with a predicted volume of 15 would have 15 chances in 20. The probability of selection is thus seen to be proportional to prediction.

To illustrate the steps in conducting a 3P sampling for the purpose of estimating the total volume of timber marked for sale, we will use the data from the 150-acre example forest used in list sampling (Table 10.9). The steps might be as follows:

1. Designate a sample size n, the number of trees to be carefully measured for volume. This can be done using the minimum sample size formula for SRS (eq. (10.13)), assuming an infinite population, and noting that the coefficient of variation figure used should be based on the ratios of actual to estimated tree volume (defined in step number 8). This value is typically somewhat smaller than the usual coefficient of variation based on tree volume, which partially explains the usual high efficiency of 3P sampling. Alternatively, for sample size determination, one can make use of a crude guide such as that suggested by Mesavage (1965) for a large timber sale— for trained cruisers 100 or so trees are usually sufficient for an accuracy of 1.5% and, for inexperienced cruisers, approximately 200 trees are needed to achieve the same accuracy. It is worth noting here that it is the consistency (precision) of the cruiser's estimates that leads to high accuracy using 3P, and individual or volume table bias is of little consequence. Thus, the experienced cruiser, though possibly biased, is probably less erratic in estimates than the beginner and, therefore, likely to have a more efficient sample. For the example here, we are going to arbitrarily choose $n = 400$.

2. Estimate the sum of volumes for the N trees making up the sale. Thus,

$$\hat{X} = \text{estimated} \sum_{i=1}^{N} X_i$$

where X_i=cruiser's estimate of tree volume (this can also be an entry from a volume table utilizing the cruiser's estimate or measurement of tree diameter and possibly height)

Note that X, the actual sum of estimated volumes, is known only after the inventory is completed. From the list sample example, the estimated volume was 14,094 bd. ft/acre, so our estimate of total volume is $\hat{X} = 150 \cdot 14,094 = 2,114,061$ board feet.

3. Designate the maximum individual tree volume expected as K. Thus,

$$K = \text{maximum } X_i$$

K, then, is used as the upper limit of the set of integers running from 1 through K, which will act as the means by which each tree will be checked for qualification as a sample tree to be measured in detail. The maximum tree volume observed was $K = 2086$ bd. ft.

4. Adjust the set of integers to ensure that you obtain close to the sample size desired. That is, define

$$n' = \frac{\hat{X}}{K + Z} \tag{10.63}$$

where n' = expected sample size

$\hat{X}$ = estimate of the total volume of all trees

Z = number of "rejection symbols" to be randomly mixed with the set of integers 1 through K

Thus, for our example $n' = 400$, and if $\hat{X} = 2,114,061$ bd. ft and maximum tree volume $K = 2,086$ bd. ft, $Z = (\hat{X}/n') - K = (2,114,061/400) - 2,086 = 3,199$; otherwise, we would likely obtain $(\hat{X}/K) = 1022$ sample trees rather than the desired 400. In designing a bias–free 3P inventory, it is also worth following the guidelines suggested by Grosenbaugh (1963b) and cited by Johnson (1972), as summarized here.

a. $n' K$ must be less than $\hat{X}$.

b. $(Z/K)^2$ must be greater than $(4/n') - (4/N)$, where N equals the anticipated total number of trees in the timber sale.

So, for our example, we have $400 \cdot 2,086 = 834,400 < 2,114,061$ and $(3,199/2,086)^2 = 2.35 > (4/400) - (4/9,200) = 0.0096$.

5. Visit each of the N trees comprising the sale. At each tree, follow this procedure:

a. Estimate directly or indirectly, using a volume table, the tree volume (or value) of X_i. For the example shown here, volume was estimated using 2-in. diameter classes and an average height of 60 ft.

b. Record the estimate.

c. Draw a number (or symbol) at random from the set of integers 1 through K having the Z interspersed rejection symbols. A device invented and described by Mesavage (1967) facilitates this operation, though most handheld calculators or smartphones can be programmed to implement this.

6. If the volume estimate X_i is greater than or equal to the random integer, measure the tree for accurate volume determination. This volume is then recorded as Y_i, the actual volume of the ith tree.

7. If the volume estimate X_i is less than the random integer, or if instead of a number, a rejection symbol is drawn, nothing more is required from the tree and the crew moves on to the next marked or to-be-marked tree. At the completion of the inventory, one should have approximately the sample size, n', originally prescribed, but minor variations are possible because of vagaries associated with random numbers in the selection procedure.

8. After completion of the inventory, the total volume of the N marked trees can be estimated from the formula:

$$\hat{Y} = X \left(\frac{\sum\limits_{i=1}^{n} \dfrac{Y_i}{X_i}}{n} \right) = X \left(\frac{\sum r_i}{n} \right) = X \cdot \bar{r}_m \tag{10.64}$$

where

$$X = \sum_{i=1}^{N} X_i \quad \text{and} \quad r_i = \frac{Y_i}{X_i}$$

and n equals the number of sample trees on which Y_i has been measured. The estimated total marked volume is equal to the sum of the estimates of tree volumes obtained from the complete population, times the mean ratio of actual to estimated volume that is obtained from the n sample trees. For our example, $n = 585$ trees were sampled, and

$$X = \sum_{i=1}^{N} X_i = 2,090,359 \text{ board feet.}$$

$$\bar{r}_m = \frac{\sum\limits_{i=1}^{N} r_i}{n} = \frac{604.1397}{585} = 1.0327$$

and the adjusted total volume is

$$\hat{Y} = X \cdot \bar{r}_m = 2,090,359 \cdot 1.0327 = 2,158,714 \text{ board feet.}$$

The standard error of the estimate $\hat{Y}$ can be estimated, although it was pointed out by Ware (1967) that no exact expression for its true value exists. The following approximation cited in Grosenbaugh's early work on 3P sampling has most often been applied, perhaps because it is the same as that used for SRS:

$$s_{\hat{Y}} = \sqrt{\frac{\sum_{i=1}^{n}\left(\frac{Y_i X}{X_i} - \hat{Y}\right)^2}{n(n-1)}} = X\sqrt{\frac{s_r^2}{n}} = X \cdot s_{\bar{r}_m} \tag{10.65}$$

where s_r^2 = variance (eq. (3.12)) of the ratios (Y_i/X_i)
        $s_{\bar{r}_m}$ = the standard error of the mean ratio

So, for our example, we have

$$s_{\hat{Y}} = X\sqrt{\frac{s_r^2}{n}} = X\sqrt{\frac{\sum_{i=1}^{n} r_i^2 - \frac{\left(\sum_{i=1}^{n} r_i\right)^2}{n}}{n(n-1)}} = 2,090,359\sqrt{\frac{635.8269 - \left((604.1397)^2 / 585\right)}{585(585-1)}}$$

$$= 2,090,359\sqrt{\frac{11.9214}{341,640}} = 12,348 \text{ board feet.}$$

The estimate obtained here, 2,158,714 bd. ft is slightly larger than the true population value of 2,135,487 bd. ft, however, is only an error of about 1%, comparable to the error obtained from list sampling, but only required 585 trees to be measured for detailed volume estimation. The approach utilized here used approximations of volume based on a dbh measurement and average height. In practice, the volume estimate could have been based on a visual estimate. As long as the estimates were fairly consistent, results would be comparable to what was illustrated here. Like list sampling, the above example implies that every tree in the population needs visited, an alternative approach, referred to as point 3P sampling (Schreuder et al., 1993) will be discussed in Section 11.2.2, which combines horizontal point sampling and 3P sampling.

11

INVENTORY OF STANDING TREES USING SAMPLING WITH VARYING PROBABILITY

The sample designs discussed in Chapter 10 are equally applicable to inventories using variable probability. Often, one will encounter the misconception that fixed-area plots are measures and variable probability points are estimates. This misconception arises from the tendency to take a plot-centered viewpoint rather than a tree-centered viewpoint (Section 9.2.4). It is the size of a tree's inclusion zone that determines whether or not a tree is "in" at a particular sample point. A plot is simply a convenient method to select trees from the population when sampling with equal probability, and the inclusion zone (plot) radius can be thought of as the "gauge" by which we determine "in" or "out" trees. Similarly, the angle we project, horizontally or vertically, is the gauge we use to select "in" or "out" trees with variable probability sampling. Both methods result in measured values, which are estimates of the forest conditions around the sample point. While these measured values will differ slightly (or in some cases greatly), these differences are no more unexpected than differences obtained using fixed-area plots of different sizes. While the process is the same, and both result in measured values, there are, however, several issues unique to sampling with variable probability. This chapter focuses on these issues.

Three types of variable probability sampling are widely used in forest inventory applications:

1. *Sampling with probability proportional to size (PPS sampling)*. The most frequent application of PPS sampling is horizontal point sampling (HPS). Sampling units and basic theory associated with HPS are presented in Section 9.3.1. In this chapter, issues involved with designing and implementing an inventory using HPS are discussed. Horizontal line and vertical point and vertical line sampling (VLS) also are presented.

Forest Mensuration, Fifth Edition. John A. Kershaw, Jr., Mark J. Ducey, Thomas W. Beers and Bertram Husch.
© 2017 John Wiley & Sons, Ltd. Published 2017 by John Wiley & Sons, Ltd.

2. *List sampling.* In list sampling, the probability of selecting an individual is proportional to some listed quality associated with the individual. List sampling requires complete enumeration of groups of individuals prior to sample selection. List sampling is similar to cluster sampling and is often discussed in this context in traditional sampling texts. List sampling is presented in Section 10.9.

3. *Sampling with probability proportional to prediction (3P sampling).* 3P sampling, a term peculiar to forestry literature, is similar to list sampling. 3P sampling does not require a complete enumeration of individuals (or groups) prior to sampling, although each individual (or group) is eventually examined for some attribute. 3P sampling is presented in Section 10.10.

11.1. HORIZONTAL POINT SAMPLING (HPS)

In this application, a series of sampling points is chosen, much as one would select plot centers for fixed-area plots (Chapter 10). The observer occupies each sampling point, sights with an angle gauge at breast height on every tree visible from the point, and tallies all trees that are greater than the projected angle of the gauge. An estimate of the basal area per unit land area is then the number of trees counted times the basal area factor (BAF), as described in Section 9.3.1.

Any variable associated with the selected trees may be measured, just as in the case of fixed-sized plots. These variables are then scaled to per unit area values by multiplying the value of the attribute by the tree factor (Section 9.1). The unique feature of HPS is that no tree measurements are needed to obtain an unbiased estimate of basal area per unit area. This feature arises because, in HPS, the tree factor is inversely proportional to tree basal area:

$$\mathrm{TF}_i = \frac{\mathrm{BAF}}{\mathrm{BA}_i}$$

where BAF = the basal area factor of the angle gauge (eq. (9.13) or (9.14)) and BA_i = tree basal area. Thus, when tree basal area (BA_i) is multiplied by TF, a constant is obtained.

11.1.1. Angle Gauges for HPS

For HPS, one needs an angle gauge that will accurately project a small horizontal angle, generally under 5°. A tree that appears larger than the projected angle is considered "in," and a tree that appears smaller than the projected angle is considered "out" (Fig. 9.6). Basically, there are two different ways of projecting the angle:

1. By prolonging two lines of sight from the eye through two points whose lateral separation w is fixed, both of which are in the same horizontal plane and both of which are at the same fixed distance L from the eye (Fig. 11.1a).

2. By deviating the light rays from the tree through a fixed angle (Fig. 11.1b).

Types of Instruments. Instruments based on the first principle include the stick-type angle gauge and the Spiegel Relaskop (Fig. 5.7). Electronic devices, like those shown in

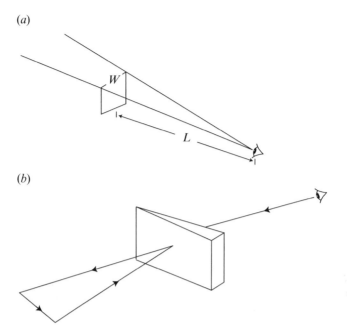

FIG. 11.1. Projecting a horizontal angle (*a*) by prolonging two lines of sight and (*b*) by deviating the light rays through a fixed angle.

Figure 5.8, also use this principle for projecting angles. To construct a stick-type angle gauge, a device that clearly illustrates the principle behind all of these instruments, one simply provides a stick with a peep sight to position the eye and a cross-arm of predetermined width, which is placed on the stick at a predetermined distance from the peep sight (Fig. 11.1*a*). To determine the width of the cross-arm, one must know the gauge constant k for the BAF to be used and make a decision on the stick length L. Then, cross-arm width will be $w = kL$.

The Spiegel Relaskop uses the same basic principle. Instead of using a stick on which to position the cross-arm, the Spiegel Relaskop utilizes the principle of the reflector sight to image the scale (cross-arm) that is on a wheel only a short distance from the eye, at a suitable viewing distance. The Spiegel Relaskop is available in both imperial and SI scales. In the imperial scale, sighting angles for BAFs of 5, 10, 20, and 40 ft²/acre are provided. The instrument automatically corrects for slope (Section 11.1.6) and is easy to use. It is relatively expensive (approximately US$2000) and sighting visibility is generally poor in low-light conditions. The electronic devices cost around the same amount as the Spiegel Relaskop and rely on laser distance measuring (Section 4.2.4), and, while their optics have improved, they often have issues of poor sighting visibility in low-light conditions. Smartphone apps such as iBitterlich (http://www.taakkumn.com/taakkumnRoom/iBitterlich.html*) for iPhone platforms and Bitterlich (http://www.deskis.ee/relaskoop/index.html†) for Android platforms use the camera focal point as the vertex of the angle and superimpose a bar or a

*Last accessed August 31, 2016.
†Last accessed August 31, 2016.

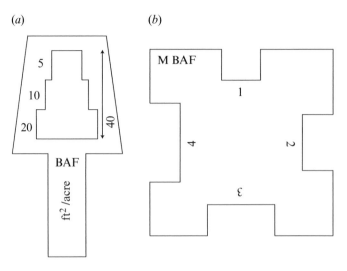

FIG. 11.2. Common angle gauges: (*a*) angle gauge and (*b*) metric relascope.

gap onto the camera image. The electronic devices and smartphone apps are both angle gauges of the first type.

Simple angle gauges also can be stamped from light-weight metal. Two examples of the shapes commonly used are shown in Figure 11.2. These gauges are held a fixed distance L from the eye and the width of the metal or the opening in the metal creates the angle projection. Usually, these gauges come with a chain of length L attached. The end of the chain is held next to the user's eye and the chain pulled tight to ensure the viewing distance is constant.

The thin prism or optical wedge is the only instrument of the second type that has been widely used by foresters (Fig. 11.1*b*). Briefly, a *prism* is a device made of optical glass or plastic in which the two surfaces are inclined at some angle A (the refraction angle), so that the deviation produced by the first surface is further increased by the second. The chromatic dispersion is also increased, but is not a cause of appreciable error unless a telescopic device is used in conjunction with the prism. The cruising prisms that foresters have used in HPS generally have a refracting angle of less than 6°. Such prisms are made in square, rectangular, and round shapes. Optical prisms can be combined to produce larger BAFs. The combined BAF can be found using (Marshall et al., 2004)

$$\text{BAF}_{\text{combined}} = \left(\sqrt{\text{BAF}_1} + \sqrt{\text{BAF}_2}\right)^2$$

For example, a 3 M prism and a 4 M prism can be held together to produce a $\left(\sqrt{3} + \sqrt{4}\right)^2 = (3.732)^2 = 13.9$ M prism.

Figure 11.3*a* illustrates the projection of a fixed angle with a thin prism. As shown in the figure, the ray that is tangent to points *a*, *b*, and *c* on the sides of the tree to the observer's right is refracted to *E*. Thus, when observers at *E* sight through the prism to points *a*, *b*, and *c*, they will see these points as if they were at *a′*, *b′*, and *c′*. Of course, all visible points on each tree cross section will be displaced so that each cross section will appear to be displaced as shown in the figure. To use the prism as a gauge, the observer looks *through* the prism at the right side of the trees—that is, at points *a*, *b*, and *c* and, at the same time, *over*

(*a*)

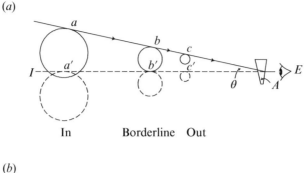

In   Borderline   Out

(*b*)

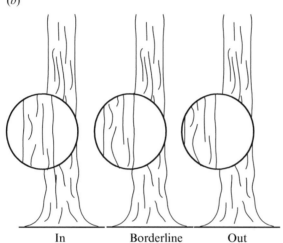

In   Borderline   Out

FIG. 11.3. Representation of image deflection using a prism: (*a*) deflection through the cross section of prism and trees and (*b*) deflection as seen by the observer when using a round prism.

the prism at the left side of the trees on the line of sight to *I*. One observes if the right side is, or is not, refracted past the left side. The actual "picture" that will be obtained when a round prism is used is shown in Figure 11.3*b*. Similar "pictures" are obtained with rectangular prisms.

Prism Calibration. Prisms for use in forest inventories are available as calibrated or noncalibrated. Although the latter are generally less expensive than calibrated prisms, they are more likely to deviate from the stated prism factor. In either case, checking or initial calibration (i.e., determination of BAF) is imperative before fieldwork commences.

Precise techniques for prism calibration have been described by Beers and Miller (1964) using a collimeter and by Stage (1962) using a projector and screen. Many field sampling situations may not warrant this precision, and the following procedure, illustrated in Figure 11.4, is suggested:

1. A flat "target" of width *w* is placed at some distance from the observer holding the prism.
2. The observer moves the prism toward or away from the target until the perpendicular line of sight to the target results in the completely offset picture—obtained

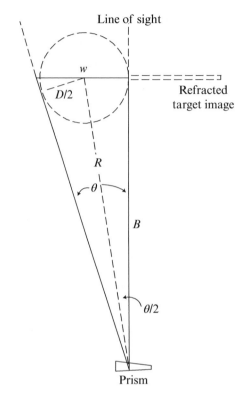

FIG. 11.4. The geometry of prism calibration. Target of width w located at distance B, which will provide the "completely offset" picture. Dotted lines pertain to a hypothetical borderline tree located properly in the generated gauge angle θ.

     when one side of the target seen over the prism precisely lines up with the other side seen through the prism.

3. The perpendicular distance B from the prism to the target is carefully measured in the same units as the target width.

4. The gauge angle θ is then calculated or determined from a table of trigonometric functions since

$$\tan\theta = \frac{w}{B} \tag{11.1}$$

$$\theta = \arctan\frac{w}{B} \tag{11.2}$$

5. The gauge constant k is then determined from the formula

$$k = 2\sin\frac{\theta}{2} \tag{11.3}$$

    The BAF is determined using eq. (9.13) for imperial units or 9.14 for SI units. Miller and Beers (1975) give additional details regarding prisms and their calibrations.

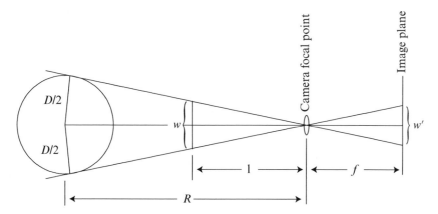

FIG. 11.5. Gauge constant for photographic horizontal point sampling. (Adapted from Decourt, 1956.)

11.1.2. Photographic Methods

In HPS, the basal area can be estimated directly by counting trees that subtend the projected angle; as a result, HPS is readily adaptable to photography. DeCourt (1956) was the first to recognize this. DeCourt observed that, by substituting the camera focal point for the human eye, the gauge constant, as expressed in Figure 9.5 and eq. (9.10), could be expressed in terms of the tree size on the image and focal length (Fig. 11.5):

$$\frac{D}{100R} = \frac{w}{l} = \frac{w'}{f} \tag{11.4}$$

where D = diameter (cm)

R = inclusion zone (plot) radius (m)

w = gauge opening (cm)

l = distance of gauge from eye (cm)

w' = tree diameter on image (mm)

f = focal length (mm)

(If imperial units are being used, the relationship is $D/(12R)$ and R is in ft, and D, w, l, w', and f are in inches.)

In HPS, the entire 360° circle is swept, searching for "in" trees. A single photographic image represents only a fraction of the circle, defined by the field of view, FOV°. DeCourt (1956) modified his BAF to account for this:

$$\text{BAF} = 2500 \left(\frac{w'}{f} \right)^2 \left(\frac{360°}{\text{FOV}°} \right) \tag{11.5}$$

DeCourt (1956) provides methods for determining f and w' for any camera, lens, and photo enlargement combination. Once f and w' are determined, they hold for all photographs taken with the same camera, lens, and enlargement ratio.

The main advantage of DeCourt's photographic system was that much of the measurement work could be done indoors under better work conditions eliminating fatigue and weather-related field errors. The method could also be applied to historical photos provided f, w', and FOV° could be determined. Even though the method is easily implemented, it has never been widely used.

The primary reason for lack of wide application has been the costs associated with film and developing. The advent of affordable, high-resolution digital cameras and image processing has resulted in a revival of interest in DeCourt's method. Stewart et al. (2004) applied DeCourt's methods to digital imagery and further refined the method by developing correction factors for trees near the edges of the FOV. Fastie (2010) and Dick (2012) applied DeCourt's method to panoramic photographs obtained by stitching together a series of individual photographs that sweep the entire 360° circle. Kershaw (2009) further refined the methods by coupling the camera with visible laser to aid tree delineation and measurement.

Because the panoramas sweep the entire circle, the angle gauge can be easily expressed in terms of pixels: $w_p = (\theta / 360) N_p$ where w_p is the width of the angle gauge in pixels, θ is the projected angle, and N_p is the pixel width of the 360° panorama. As Fastie (2010) pointed out, the method is reliable provided that (1) tree trunks are not obscured by other trees and shrubs, (2) stitching is error free, and (3) depth of field is sufficient to provide well-focused images. Dick (2012) suggested an overlap of approximately 20% between adjacent pairs of photos to minimize stitching errors.

The greatest drawback of DeCourt's method is obscured or hidden trees, which produces an undercount of "in" trees and a negative bias in the estimates of basal area per unit area and other stand-level parameters. Dick (2012) showed that the number of obscured or hidden trees could be controlled by adjusting the BAF, with greater than 80% of the trees visible when the BAF was greater than or equal to $5\,m^2/ha/tree$. The problem of obscured or hidden trees is not unique to photographic methods and often requires a significant amount of field time checking these trees. In the field, the observer can move around and easily check for obscured and hidden trees. On a photograph, this is not possible. Dick (2012) proposed using double sampling to correct for obscured and hidden trees. Kershaw (2010) proposed distance sampling techniques to correct for these trees. The correction techniques proposed by Ducey and Astrup (2013) for terrestrial LiDAR scanners could also be applied to photographic point samples. One of the greatest advantages of the panoramic approach is that the photographs provide a permanent, visual record of the plot that can be referred to in the future and potentially have additional measurements extracted.

11.1.3. Sample Size

In HPS, the sampling is performed *with replacement*.* (This is contrary to fixed-area plot sampling that is usually perceived as *sampling without replacement*.) Thus, it can be assumed that sampling is being performed in a population that is infinite. Consequently, for most situations, the following formula can be used to determine the number of points to visit:

---

* Since horizontal point sampling is based on sampling with replacement, any tree in the forest population may qualify and be tallied from two or more sample points. Provided the points have been chosen in an unbiased manner, tallying a qualifying tree from two such locations will cause no bias in the subsequent estimates of mean stand parameters. In fact, failure to do so will result in biased estimates. If the occurrence of such double-tally trees is high, however, excessive sampling is probably present and the inventory should be modified by either reducing the number of points (increasing spacing if systematic sampling is used), increasing the angle gauge, or both.

$$n = \frac{t^2 \text{CV}^2}{E^2} \tag{11.6}$$

where n=number of points required for desired precision E, with the probability level implied by the value of t

t=Student's t

CV=coefficient of variation (in percent) for the forest to be sampled

E=allowable error or desired precision (in percent) for the average volume (or basal area, etc.)

If prior experience or previous data are not available to estimate the value of the coefficient of variation, preliminary inventories might be necessary to establish reasonable approximations. When small forest areas are sampled (several hundred acres; ±40 ha or less), impractical results may be obtained if the formula is used. For example, for a 40-acre tract where the coefficient of variation is expected to be approximately 70% and a desired precision of 10% is required at the 0.95 probability level $(t \approx 2)$, the substitution into the formula leads to

$$n = \frac{(2)^2 (70)^2}{10^2} = 196 \text{ points}$$

Locating 196 sample points in a 40-acre woods is unnecessarily intensive. One should note that forest area does not appear in the formula used and that such a problem does not exist when dealing with fixed-area plot sampling since sampling is performed from a *finite* population and a correction factor involving forest area is part of the appropriate formula to determine number of plots.

To circumvent these impractical results for small areas, various approaches have been proposed (Shiver and Borders, 1996; Avery and Burkhart, 2002). Often, the procedure is to determine how many fixed-area plots of certain size might be required and then to locate somewhat more (or less) sample points depending upon local experience regarding comparative precision of plots versus points.

Another approach involves the use of a "rule of thumb" to facilitate the decision. For example, the following has provided reasonable results in the mixed hardwood forests of central United States.

If area in acres is	Number of points
Less than 10	10
11–40	1 per acre
41–80	20+0.5(area in acres)
81–200	40+0.25(area in acres)
Over 200	Found from eq. (11.6)

11.1.4. Choosing a Suitable Gauge Constant

The choice of an appropriate gauge constant for HPS is influenced primarily by the nature of the stands to be sampled, and secondarily by the objectives and conduct of the inventory. The choice is frequently based on local experience and other general guidelines such as: for small trees and open stands, use a small gauge constant; for large trees and dense stands, use a large gauge constant. Since small trees do not always occur in open stands and large trees do not always occur in dense stands, a compromise frequently must be made after assessing both average tree size and stand density. As discussed in Section 9.6, a more definitive approach is to first determine the average number of trees m we wish to tally per location, then after estimating the average stand basal area (in the case of HPS), find the BAF by division:

$$BAF = \frac{BA / unit\ area}{m}.$$

After the BAF is determined, the corresponding gauge constant and gauge angle can be calculated as discussed in Section 9.3.1.

Iles (2003) suggests that the desired average count should be between four and seven trees per point; however, others have suggested average tree counts between 6 and 16 (e.g., Beers and Miller, 1964). The choice of the desired average number of sample trees is somewhat arbitrary. However, several general concepts must be considered before making a decision about a suitable gauge constant.

1. For small average tree counts, there exists a great likelihood of getting many small individual counts (say less than 5), and conversely for large average tree counts, the likelihood is great for getting large individual counts (say more than 20).

2. Small or large tree counts can lead to inefficient samples. For very small counts, the forest is "undersampled" and the inventory is economically inefficient since the at-location costs are frequently much less than the between-location costs. Therefore, measuring a few more trees at each location might materially increase the overall statistical precision but not materially affect cost. For very large tree counts, the forest is "oversampled" since once a certain level of statistical precision is reached, the precision will be increased negligibly by measurement of additional trees. In this case, the measurement of additional trees is a waste of time.

3. Biases in the estimates of forest parameters may occur if tree counts are very high. The small-angle gauges required for large tree counts result in large inclusion zones, and the potential for "in" trees to be very far from the sample point. This increases the number of potential borderline trees and increases the possibility of overlooking borderline trees. As distance from the sample point increases, there is an increased difficulty in viewing all tree stems and the number of obscured or hidden trees increases. There also is a tendency to be less careful when there are many trees to be scrutinized.

4. In regard to low tree counts, the effect of a mistaken action will lead to a much larger bias (in percentage) than when the tree count is high.

5. In deciding on what gauge constant to use, one should remember that the gauge constant, gauge angle, and BAF all vary inversely to the expected tree count. Therefore, for HPS a large BAF implies small tree counts and a small BAF implies large tree counts.

One should note that the tradeoffs between large or small angle gauges are not any different from the tradeoffs between small and large fixed area plots. The logistics of plot

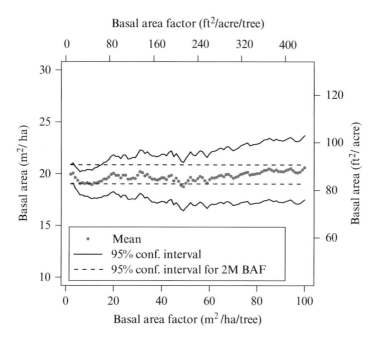

FIG. 11.6. Influence of basal area factor on mean basal area (m²) per hectare. Based on 700 sample points collected in central New Brunswick, Canada.

establishment, measurement, and checking borderline trees are essentially the same, and the implications associated with missing an "in" tree are similar.

While choice of angle gauge influences the average number of trees counted per plot, it does not influence the estimates of mean stand-level parameters (Fig. 11.6). The estimated mean basal area (m²/ha) does not vary appreciably across the range of BAFs of 2–100M (m²/ ha/tree), and in most cases is within the 95% confidence interval of the 2M estimate. While the estimate of the mean remains the same, standard error and confidence width increase. The increase in standard error is minor and regular until about 20M BAF, after which larger BAFs have larger and more irregular increases in standard error. At 20M, the sample points used in this example average one tree per plot; as BAF increases beyond 20, there is an increasing number of plots with no trees tallied. The excess number of plots with zero observations produces zero-inflated variances and may require alternative approaches for summarization and analyses (for more details, see Mullahy, 1986; Chin and Quddus, 2003; Rose et al., 2006).

11.1.5. Proper Use of Gauges

Basic to the field application of HPS is the requirement that once the necessary factors or constants are developed there exists a precise method for determining whether a given tree qualifies as a sample tree (i.e., whether it is "in" or "out"). Proper use of the angle gauge is very important for the determination of whether a tree qualifies or not. Improper use of the gauge will result in undercounting or overcounting of "in" trees depending on whether the misuse increases or decreases the angle. Consistent misuse can result in serious biases in the estimates obtained from the field data. Proper use depends upon the type of angle gauge employed.

When using a stick-type angle gauge, the Spiegel Relaskop, or the angle gauges illustrated in Figure 11.2, the angle generated has its vertex at the eye of the user; therefore, the user must stand over the sample point so that the vertex of the angle is positioned vertically above the point. Detailed directions for using the Spiegel Relaskop are provided by the manufacturer. While many of the electronic angle gauges project angles from a focal point within the device, some assume a distance to eye, manufacturers' users guides should be consulted for proper use of these devices.

Since the wedge prism is often misused, certain precautions should be observed. Once the observer occupies the sampling point or location, the prism must be properly positioned before each tree is sighted critically; otherwise, bias is likely to occur. The following rules should be followed:

1. Since the gauge angle originates with the prism, the prism must be held vertically above the sample point when sighting; this implies that the observer (not the prism) moves in a small circle about the point. The smartphone apps mentioned above (Section 11.1.1) use the camera focal point as the vertex of the angle, and, therefore, the phone must be held over the sample point in the same way that the optical prism is held over the point.

2. Each tree is sighted at breast height through and over the prism.

3. The line of sight should be perpendicular to the prism bisector (i.e., the plane bisecting the refracting angle of the prism).

4. The distance from the eye to the prism is immaterial, provided a clear picture is obtained. Ten inches (25 cm) is the normal viewing distance.

With the prism vertically above the designated sampling point, the prism bisector can be oriented perpendicular to the line of sight by the following technique. If the prism is properly oriented at right angles to the line of sight,

a. A rotation of the prism in the vertical plane perpendicular to the line of sight will *reduce* the amount of horizontal deflection of the tree (i.e., make the images overlap more)

b. A rotation of the prism by (1) "tipping" or (2) "swinging" will *increase* the amount of horizontal deflection of the tree (i.e., make the images overlap less)

The four positions of the prism (unrotated, rotated, tipped, and swung) are shown in Figure 11.7.

11.1.6. Checking Questionable Trees

In fixed-area plot sampling, an occasional tree must be checked for inclusion in the plot by lying off the plot radius or plot width. Similarly, in horizontal point or line sampling an occasional tree will be questionable and must be checked; however, the inclusion zone radius is not fixed for all trees as it is for fixed-area plot sampling. It depends on tree diameter and often requires questionable trees to be checked. Any tree that is questionably "in" or "out" (i.e., borderline; Fig. 11.3), as determined by the angle gauge, should be measured for dbh and the inclusion zone radius (or plot half-width) associated with the tree calculated by

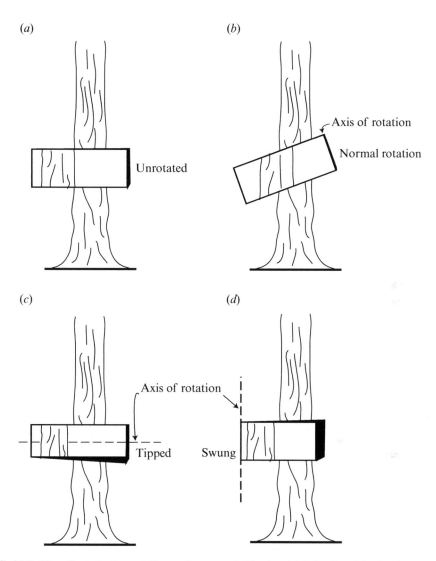

FIG. 11.7. The effect of prism rotation on the amount of horizontal deflection: (*a*) prism in unrotated position on a borderline tree; (*b*) prism rotated in the vertical plane perpendicular to the line of sight—reduced horizontal deflection; (*c*) prism tipped in the vertical plane parallel to the line of sight—increased horizontal deflection; and (*d*) prism swung in the horizontal plane—increased horizontal deflection.

multiplying the diameter by the horizontal distance multiplier (diameter:inclusion zone radius ratios) in Table 9.3 or by using the horizontal distance multipliers (Section 9.3.1 and Table A.5). For example (using imperial units), if the BAF is 10.0, then

$$\text{HDM} = \frac{33\sqrt{10}}{12\sqrt{10}} = 2.75$$

A 10-in. tree would then have an associated inclusion zone radius of 27.5 ft. If the measured horizontal distance to the center of a questionable tree is greater than the calculated inclusion zone radius, the tree is "out"; otherwise, it is "in." A table showing plot radii by dbh classes can be prepared (see Table A.5). A more efficient method of checking questionable trees makes use of a specially calibrated tape described by Beers and Miller (1964). Diameters are marked on the tape at the appropriate plot radius for the BAF being used, thus avoiding table references or calculations.

Truly borderline trees represent about 5% of the tallied trees (Iles and Fall, 1988). Experienced cruisers often are very good at correctly assessing "in" or "out" on most borderline trees. Inexperienced cruisers may need to check borderline trees more frequently. Even though the number of borderline trees is small, and the resulting biases arising from errant calls even smaller (around 0.5%), it is still a good idea to check these trees (Iles and Fall, 1988). Iles and Fall (1988) give several reasons for checking borderline trees including (1) the extra effort required to check borderline trees is not onerous, (2) many borderline trees are also visually obscured and would need checked anyway, and (3) a dbh measurement is generally required as part of the cruise data collected on "in" trees. For more details on the effects of borderline trees and the rationale for checking borderline trees, see Iles and Fall (1988).

11.1.7. Slope Correction

Since the determination of forest stand parameters is on a horizontal land area basis, the projected angle must be corrected if the terrain is not level. Specifically, the gauge constant k must be reduced to a value k_r where $k_r = k \cdot \cos S$, and where S is the slope angle in degrees.

For stick angle gauges based on the principle of projecting a horizontal angle by prolonging two lines of sight, as in Figure 11.1, slope correction can be achieved by reducing the cross-arm width w to the value w_r, where $w_r = w \cdot \cos S$, or by increasing stick length L to the value L_r, where $L_r = L/\cos S$. The Spiegel Relaskop uses an ingenious method of varying "cross-arm" width. It has a "strip" scale that varies in width directly as the cosine of the slope angle. This scale, mounted on a weighted wheel, is seen "projected" on the target (the tree) and rotates by gravity to the appropriate position as the line of sight is raised or lowered.

Slope correction can be achieved when the wedge prism is used by rotating the prism in a plane perpendicular to the line of sight through an angle equal to the slope angle. Such rotation properly reduces the gauge angle.*

Instead of making instrument adjustments, a simple procedure for correcting for slope is to measure the dbh of questionable trees and determine the horizontal distance (correcting slope distance using the slope angle) from the observation point to the tree center. If this horizontal distance to the center of the questionable tree is greater than the calculated inclusion zone radius, the tree is "out"; otherwise, it is "in."

11.1.8. Leaning and Hidden Trees

Moderate tree lean has little effect on field procedures. If, however, a tree leans severely to the right or left of the observer and appears to be "just in," a corrected gauge picture should be obtained. This is accomplished by rotating the angle gauge so that the vertical axis of the

---

* See Husch et al. (1983) for a more detailed discussion of various other methods for correcting for slope.

gauge parallels the axis of the leaning tree. Thus, if a prism is used, the sides of the prism should parallel the sides of the tree; if a stick-type gauge is used, the sides of the blade should parallel the tree sides; if a Spiegel Relaskop is used, the sides of the instrument should parallel the tree sides.

A leaning tree that appears "out" with the unrotated gauge need not be checked since the adjustment described above will lead to a further separation of the actual and refracted image. When a leaning tree is viewed with an unrotated prism, the effective gauge angle is reduced, and rotation of the prism is necessary to obtain the full angle.

On questionable trees and those that lean severely toward or away from the observer, a tape-check, as described earlier, should be made. When this is done, the tree center is commonly assumed to be a point vertically above the center of the tree cross section at the ground line.

In dense or brush-choked stands, precautions are necessary to avoid missing or double-counting trees. Trees hidden behind other trees can be detected by swaying from side to side after each obvious tree is examined. There is no error associated with viewing trees from another other point as long as the distance from the tree to the observer remains the same as the original distance to the sampling point. Moving from the sample point is also warranted if trees are close enough together that two or more stems are superimposed while being viewed with the angle gauge.

When limbs or underbrush blocks the breast height point, one can check a visible point higher on the stem for qualification. If a tree qualifies for inclusion at some point above breast height, it will qualify at breast height unless the tree leans toward the observer and the distance is critical. In any of the above situations, a tape-check should be made if a tree is questionable.

11.1.9. Volume Estimation

Volume per unit area can be estimated from horizontal point data using a number of methods. As described in Section 9.3.2, average volume per tree can be estimated for each diameter class and volume per unit area for each diameter class obtained by multiplying trees per unit area and volume per tree:

$$\left(\text{Volume per unit area}\right)_i = \left(\text{Trees per unit area}\right)_i \cdot \left(\text{Volume per tree}\right)_i$$

An estimate of volume per unit area for the stand or point is obtained by summing the volumes per unit area for each diameter class. A mathematically equivalent method is to calculate the volume factor (Section 9.1) for each diameter class:

$$\text{VF}_i = \text{TF}_i \cdot V_i = \frac{\text{BAF}}{\text{BA}_i} V_i = \frac{\text{BAF}}{cD_i^2} V_i \tag{11.7}$$

where VF_i = volume factor for the ith dbh class

TF_i = tree factor for the ith dbh class

V_i = average volume per tree for the ith dbh class

D_i = midpoint diameter of the ith dbh class

c = the cross-sectional area conversion factor (0.005454 for imperial units and 0.00007854 for SI units)

Volume per unit area for each diameter class is then obtained from

$$\left(\text{Volume per unit area}\right)_i = \left(\frac{\text{No. of trees tallied}_i}{\text{No. of plots}}\right) \cdot \text{VF}_i,$$

and the total for the stand or point is obtained by summing the values for each dbh class.

Another method of estimating volume per unit area is *diameter obviation* (Beers, 1964b). From eq. (11.7), it is noted that the volume factor can be written as

$$\text{VF}_i = \frac{\text{BAF}}{cD_i^2}V_i.$$

If volume per tree V_i can be estimated using a constant form factor equation, $V_i = bD_i^2 H$, then the volume factor becomes

$$\text{VF}_i = \frac{\text{BAF}}{cD_i^2}bD_i^2 H_i = \left(\frac{b \cdot \text{BAF}}{c}\right)H_i \tag{11.8}$$

where VF_i = the volume factor for the *i*th tree (or dbh class)

D_i = the dbh of the *i*th tree

H_i = the height of the *i*th tree

b = the constant form factor

c = the cross-sectional area conversion factor.

Volume per unit area is then estimated using

$$\left(\text{Volume per unit area}\right) = \left(\frac{1}{n}\right)\sum_{i=1}^{m}\left[\left(\frac{b \cdot \text{BAF}}{c}\right)H_i\right] = \left(\frac{b \cdot \text{BAF}}{c \cdot n}\right)\sum_{i=1}^{m}H_i \tag{11.9}$$

where m = number of tree tallied

n = number of points

Thus, volume is estimated from a sum of heights and dbh is not required unless it is used to place trees in diameter classes. A third method of volume estimation, which utilizes the volume:basal area ratio (VBAR) (Hummel, 1953), is discussed in Section 11.2.1.

11.1.10. Special-Purpose Modifications of HPS

Since the invention of HPS by Bitterlich (1947) and the development of its theoretical foundations by Grosenbaugh (1958), a variety of modifications of the basic approach have been developed, often with the goal of estimating stand attributes through a simple tree count. A comprehensive review of many early adaptations can be found in Bitterlich (1984).

HPS selects trees with probability proportional to basal area, or equivalently, dbh^2. Likewise, horizontal line sampling (HLS) selects trees with probability proportional to dbh^1, while fixed area plots select trees with equal probability, that is, proportional to dbh^0. Zöhrer (1978) suggested that these approaches could be viewed as special cases of a more general one, in which the exponent on dbh could be chosen to be any number, not just an integer. For arbitrary values of the exponent, the tree would no longer be at either a constant distance or a constant angle when in the borderline condition. Zöhrer (1978, 1979) envisioned using a precise dendrometer, such as a Tele-relaskop, in conjunction with a lookup table, to sample trees with probability proportional to a local volume table; careful measurement of distance and diameter, in conjunction with an appropriate equation, would suffice to determine the inclusion or exclusion of trees that were nearly borderline. This idea was further developed by Kuusela (1979), who used a variable-angle gauge for rapid determination of inclusion for individual trees. Ducey and Valentine (2008) suggested a modified prism technique to enable estimating the additive version of Reineke's Stand Density Index (Section 8.9.2), which has exponent 1.6, while Ducey (2009a) presented a method for tallying trees with probability proportional to the exponent in an allometric equation for total biomass. Though theoretically sound, these methods have not seen much practical use. In part, this may be due to fussy or clumsy field techniques, but it may also reflect the inherent efficiency of ordinary HPS. For example, Ducey (2009a) found that the gains in statistical efficiency were minimal when using an "optimal" exponent that matched that of a generalized biomass equation, in comparison to ordinary point sampling.

Other inventive applications have included the use of an angle gauge to measure the aggregate crown attributes of a stand. Stenberg et al. (2008) suggest the use of a very wide angle gauge to estimate canopy cover. Trees are included in the tally if their crowns are wider than the gauge when viewed from the sample point, and a simple count multiplied by a constant crown area factor gives an estimate of canopy cover. The geometry of the method is inexact and it can be challenging in dense forests, but it could give rapid estimates of acceptable quality in many situations.

11.2. SUBSAMPLING IN HPS

As discussed in Sections 10.7 and 10.8, subsampling can be used effectively in sample designs to reduce the number of measurements required while maintaining acceptable estimates of standard errors. Though not necessarily unique to HPS, there are several subsampling designs that have been developed for use with HPS (Beers and Miller, 1964; Schreuder et al., 1993; Stamatellos, 1995). Two of the more commonly applied methods are presented here.

11.2.1. Big BAF Sampling

Big BAF sampling is so named because two angle gauges are used: a small BAF gauge for counting trees and estimating basal area; and a big BAF gauge for selecting trees to measure for volume (or other attributes such as biomass or carbon) determination. The method utilizes the VBAR (Flury, 1897; Hummel, 1953, 1955) to estimate volume from basal area. Flury (1897), working in spruce and beech stands in Switzerland, observed a high correlation between average stand height and VBAR and developed a series of VBAR tables that could be used to estimate volume from estimates of stand basal area and average stand height. Grosenbaugh (1952b), in the first publication in North America on HPS, recognized the utility of VBAR for estimating volume when using HPS and recommended ways to utilize it in HPS.

The VBARs are obtained by dividing volume estimates from a volume table or equation by basal area:

$$VBAR_i = \frac{V_i}{BA_i}$$ (11.10)

where, $VBAR_i$ = the volume to basal area ratio for the ith tree

V_i = the volume per tree for the ith tree

BA_i = the basal area of the ith tree

The basic rationale behind big BAF sampling is derived from the observation that estimates of basal area often have higher standard errors than do estimates of VBAR (Bruce, 1961). As a result, a larger sample of points is used to estimate basal area per unit area and a smaller subsample of trees is used to estimate mean VBAR (Iles, 2003; Marshall et al., 2004). While VBAR can be used with any sampling unit type, it is particularly suited to use with HPS. As discussed above, and in Section 9.3.1, basal area per unit area is obtained by counting "in" trees using an angle gauge and multiplying by the BAF. Since standard errors are high and counting "in" trees relatively quick, a small BAF angle gauge is used to obtain a sufficient tree count. The subsample of trees selected for estimating VBAR (often referred to as the measure trees) can be selected using any probabilistic sampling method; however, using a large BAF angle gauge is most practical (Marshall et al., 2004). So at each sample point, two sweeps are made: the first with a small BAF angle gauge, counting "in" trees; and the second with a large BAF angle gauge to select trees for measurement for volume determination. The trees selected using the large-angle gauge are a subset of the trees counted using the small-angle gauge.

Table 11.1 shows a sample tally for count and measure* trees from a mixed intolerant hardwood–spruce–fir stand in central New Brunswick using a 2 M BAF for counting and a 16 M BAF for measuring. The mean VBAR is estimated using the mean ratio approach (eq. (10.56a)):

$$\overline{VBAR} = \frac{\sum_{i=1}^{n_S}\left(\dfrac{V_i}{BA_i}\right)}{m}$$ (11.11)

where V_i = volume of the ith measure tree

BA_i = basal area of the ith measure tree

m = the number of measure trees

For the data shown in Table 11.1, the $\overline{VBAR}$ is

$$\overline{VBAR} = \frac{1}{20}\sum_{i=1}^{20}\frac{V_i}{BA_i} = \frac{1}{20}(149.511) = 7.476$$

---

* It is merely coincidental that the number of count points (20) and number of measure trees (20) are identical. This does not imply that each count point has a measure tree.

TABLE 11.1. An Example Big BAF Tally from a Mixed Intolerant Hardwood–Spruce–Fir Stand in central New Brunswick, Canada

(a) Count Data

Plot	Count	Basal Area (m²/ha)	Plot	Count	Basal Area (m²/ha)
1	12	24	11	8	16
2	6	12	12	15	30
3	9	18	13	10	20
4	15	30	14	22	44
5	12	24	15	15	30
6	12	24	16	9	18
7	9	18	17	10	20
8	18	36	18	4	8
9	19	38	19	1	2
10	5	10	20	16	32

(b) Measure Data

Tree Number	dbh (cm)	Height (m)	BA (m²)	Volume (m³)	VBAR	Tree Number	dbh (cm)	Height (m)	BA (m²)	Volume (m³)	VBAR
1	35.3	19.4	0.09787	0.871	8.900	11	22.7	14.4	0.04047	0.282	6.968
2	39.6	20.3	0.12316	1.136	9.224	12	31.7	17.2	0.07892	0.637	8.071
3	18.9	15.3	0.02806	0.188	6.700	13	18.5	16.8	0.02688	0.185	6.882
4	19.6	18.1	0.03017	0.228	7.557	14	25.2	18.6	0.04988	0.375	7.518
5	21.6	17.5	0.03664	0.261	7.123	15	21.4	14.7	0.03597	0.255	7.089
6	17.3	16.9	0.02351	0.172	7.316	16	17.5	12.2	0.02405	0.145	6.029
7	45.5	22.1	0.16260	1.604	9.865	17	14.2	14.4	0.01584	0.095	5.997
8	41.0	21.4	0.13203	1.270	9.619	18	15.6	12.1	0.01911	0.115	6.018
9	25.4	15.8	0.05067	0.381	7.519	19	15.8	15.0	0.01961	0.122	6.221
10	25.6	15.4	0.05147	0.350	6.800	20	29.2	20.0	0.06697	0.542	8.093

Counts use a 2 M BAF and measures use a 16 M BAF.

The mean volume per unit area is then obtained by multiplying $\overline{\text{VBAR}}$ by $\overline{\text{BA}}$. Thus, for the data in Table 11.1, we have

$$\text{Volume} = \overline{\text{VBAR}} \cdot \overline{\text{BA}} = 7.476 \cdot 22.7 = 169.69 = 170 \, \text{m}^3/\text{ha}.$$

The standard error for this estimate can be estimated using eq. (10.58a), but is generally estimated using Bruce's formula (Marshall et al., 2004):

$$\%\text{SE}(\text{Volume}) = \sqrt{\%\text{SE}^2(\text{BA}) + \%\text{SE}^2(\text{VBAR})} \tag{11.12}$$

%SE is the standard error expressed as a percentage of the mean:

$$\%\text{SE} = 100 \frac{s_{\bar{x}}}{\bar{x}} = 100 \left(\frac{s}{\sqrt{m}} \right) \left(\frac{1}{\bar{x}} \right) = 100 \left(\frac{s}{\bar{x}} \right) \left(\frac{1}{\sqrt{m}} \right) = \frac{\text{CV}}{\sqrt{m}}$$

where $s_{\bar{x}}$ = standard error (eq. (3.14))
        s = sample standard deviation (eq. (3.12))
        CV = coefficient of variation (eq. (3.13))

The standard errors for both BA and VBAR are estimated from the sample data. For the BA samples shown in Table 11.1, we have

$$s_{\overline{\text{BA}}} = \sqrt{\frac{\displaystyle\sum_{i=1}^{n} \text{BA}_i^2 - \left[\left(\displaystyle\sum_{i=1}^{n} \text{BA}_i \right)^2 \Big/ n \right]}{n(n-1)}} = \sqrt{\frac{12,468 - \left(454^2/20\right)}{20(20-1)}} = \sqrt{\frac{2,162.2}{380}} = 2.385$$

and the %SE(BA) = $100 \cdot (2.385/22.7) = 10.51\%$. For VBAR, standard error is estimated using the individual V:BA ratios ($r = V/\text{BA}$) for the m measure trees:

$$s_{\overline{\text{VBAR}}} = \sqrt{\frac{\displaystyle\sum_{i=1}^{m} r_i^2 - \left[\left(\displaystyle\sum_{i=1}^{m} r_i \right)^2 \Big/ m \right]}{m(m-1)}} = \sqrt{\frac{1143.786 - \left(149.513^2/20\right)}{20(20-1)}} = \sqrt{\frac{26.07404}{380}} = 0.2619 \, \text{m}^3/\text{m}^2$$

and the %SE(VBAR) = $100(0.2619/7.476) = 3.50\%$. The combined standard error for the volume estimate is

$$\%\text{SE}(\text{Volume}) = \sqrt{(10.51)^2 + (3.50)^2} = 11.08\%.$$

Sample size determination requires assessing tradeoffs between the number of sample points at which counts are made versus the size of the subset of count trees to measure. This of course requires assessing the influence of different combinations of count BAFs and measure BAFs (Iles, 2003; Marshall et al., 2004). Bruce (1961) demonstrated that the greatest reduction in combined %SE is obtained by concentrating sampling effort on the factor with the greatest %SE. In big BAF sampling, generally more gain is achieved when more sample points are counted than when a larger proportion of count trees are measured;

however, a number of factors need to be considered including costs of moving between sample points, costs of counting trees, costs of measuring trees, and the required levels of precision. Marshall et al. (2004) proposed a system for determining sample sizes and the relative proportion of count points to measure trees that incorporates the expected standard errors of each component, the costs of acquiring data, and the desired combined standard error. The proportion of sample points from which counts are obtain to the number of measure trees is estimated using (Bell et al., 1983; Bell and Dilworth, 1993)

$$P_{n(\text{BA}):m(\text{VBAR})} = \frac{\text{CV}(\text{BA})\sqrt{\$\text{VBAR}}}{\text{CV}(\text{VBAR})\sqrt{\$\text{BA}}}$$
(11.13)

where $P_{n(\text{BA}):m(\text{VBAR})}$ = proportion of sample points from which counts are obtain to the number of trees that are measured

$\text{CV}(\text{BA})$ = coefficient of variation for basal area (or tree count)

$\text{CV}(\text{VBAR})$ = coefficient of variation of VBAR

$\$\text{VBAR}$ = cost of measuring trees (dbh, height, form, grade, etc.) to determine VBAR

$\$\text{BA}$ = cost of travelling to, establishing, and count "in" trees at a sample point

Once $P_{n(\text{BA}):m(\text{VBAR})}$ is determined, then the number of measure trees can be determined using (Marshall et al., 2004)

$$m(\text{VBAR}) = \frac{\text{CV}^2(\text{BA}) + P_{n(\text{BA}):m(\text{VBAR})} \cdot \text{CV}^2(\text{VBAR})}{P_{n(\text{BA}):m(\text{VBAR})} \cdot \%\text{SE}^2(\text{Volume})}$$
(11.14)

where $m(\text{VBAR})$ = the minimum number of trees to measure for VBAR

$\%\text{SE}(\text{Volume})$ = the desired combined %SE for the inventory

$\text{CV}(\text{BA})$, $\text{CV}(\text{VBAR})$, and $P_{n(\text{BA}):m(\text{VBAR})}$ are as defined above.

The number of sample points is then obtained using

$$n(\text{BA}) = P_{n(\text{BA}):m(\text{VBAR})} \cdot m(\text{VBAR})$$
(11.15)

where $n(\text{BA})$ = the number of sample point from which counts are obtained

For the data shown in Table 11.1, assuming the average $\$\text{VBAR}=\$3.65/\text{tree}$ and the average $\$\text{BA}=\$5.00/\text{sample point}$. The $P_{n(\text{BA}):m(\text{VBAR})}$ for this data would be

$$P_{n(\text{BA}):m(\text{VBAR})} = \frac{\text{CV}(\text{BA})\sqrt{\$\text{VBAR}}}{\text{CV}(\text{VBAR})\sqrt{\$\text{BA}}}$$

$$= \frac{47.0^2 \cdot \sqrt{3.65}}{15.7^2 \cdot \sqrt{5.00}}$$

$$= 7.68$$

If we desired a combined %SE(Volume) of 7.5%, then the minimum number of measure trees would be

$$m(\text{VBAR}) = \frac{\text{CV}^2(\text{BA}) + P_{n(\text{BA}):m(\text{VBAR})} \cdot \text{CV}^2(\text{VBAR})}{P_{n(\text{BA}):m(\text{VBAR})} \cdot \%\text{SE}^2(\text{Volume})}$$

$$= \frac{47.0^2 + 7.68 \cdot 15.7^2}{7.68 \cdot 7.5^2}$$

$$= \frac{4095.4}{529.2} = 9.5 = 10 \text{ trees}$$

and the minimum number of sample points is

$$n(\text{BA}) = P_{n(\text{BA}):m(\text{VBAR})} \cdot m(\text{VBAR})$$

$$= 7.68 \cdot 10 = 76.8 = 77 \text{ sample points}$$

Given that the BA/ha averages about 22.7 m²/ha, if we wanted to average about 5 trees per sample point, a BAF=4 M (m²/ha/tree) would be appropriate, and if we wanted to measure about 10 trees across the 77 sample points, a BAF of approximately 175 M (m²/ha/tree) would be required (BAF=22.7·77/10=174.9). While this may appear to be extreme, it emphasizes the observation that standard errors of VBARs tend to be low, standard errors of BA based on HPS counts tends to be high, and the greatest impact on combined standard errors is achieved by focusing on the component with the highest standard error (Bruce, 1961; Bell et al., 1983; Marshall et al., 2004).

While the above discussion focused on total volume, basal area ratios can be computed for merchantable volume, specific product volumes (saw timber, pulpwood, etc.), weight, biomass, carbon content, or any other tree and stand characteristic of interest. Standard errors in total VBAR tend to be low, standard errors for other ratios may be higher, and, while subsampling will most likely still be more efficient than measuring all count trees, the results might not be as dramatic as illustrated here for total volume.

11.2.2. Point 3P Sampling

Sampling with 3P was discussed in Section 10.10. Unlike list sampling (Section 10.9), 3P sampling does not require the sampling frame (list) to be completely enumerated prior to the sample, but does require each tree in the population to be visited at the prediction phase. The requirement of a census at the prediction phase has been a perceived deterrent to, and frequent criticism of, widespread application of 3P sampling, especially at large spatial scales. Grosenbaugh (1965) recognized this limitation and proposed a multistage sample design that combines HPS with 3P sampling (Grosenbaugh, 1971, 1974, 1979).

The basic approach to point-3P sampling is similar to the approach used for big BAF sampling. In the "point" phase of the sampling scheme, an angle gauge (usually small) is used to select trees at sample points for estimating basal area per unit area. In the "prediction" phase, each tree selected for basal area estimation also has the attribute of interest (usually height or volume, but it can be any individual tree attribute) predicted (visually estimated) in the same way that 3P predicts all trees in the population. A random number generator is used to select a random number across the expected range of the attribute of interest, and, if the predicted value is larger than the generated number, the tree is selected for detailed determination of the attribute of interest. As with 3P sampling, a certain number of rejection symbols can be mixed in the random number generator to help maintain the desired number of measure trees.

The most common implementation of point-3P sampling is for volume estimation, although any stand attribute can be estimated. Volume can be directly estimated during the prediction phase, though it is more common to estimate height. The trees available for prediction are selected using HPS, thus they are selected with probability proportional to basal area. By subsequently estimating height and selecting measure trees based on this prediction, the measure trees are selected with probability proportional to (basal area)·(height), which is closely related to volume. Other tree attributes may be measured on the "in" trees, such as dbh and species, as needed for development of stand and stock tables and other data summaries.

The equations presented here focus on estimating volume. Height is predicted for all HPS sample trees, and 3P sample techniques (Section 10.10) are used to select measurement trees. The estimator for average volume per unit area is derived from eq. (11.9), which shows that volume per unit area can be estimated for HPS samples using $\text{BAF} \cdot \Sigma H$ multiplied by a constant (average stand form). The estimator for average volume per unit area for point-3P sampling is given by (Grosenbaugh, 1971, 1979; Rennie, 1976)

$$\bar{V} = \text{BAF} \cdot \overline{\Sigma \text{Hp}} \cdot \left(\overline{\frac{V}{\text{BA} \cdot \text{Hm}}} \right) \tag{11.16}$$

where $\bar{V}$ = mean volume per unit area

   BAF = basal area factor

   $\overline{\Sigma \text{Hp}}$ = mean sum of predicted heights per sample point

$$\overline{\Sigma \text{Hp}} = \frac{\sum\limits_{i=1}^{n} \left(\sum\limits_{j=1}^{m_i} \text{Hp}_{ij} \right)}{n}$$

   Hp_{ij} = predicted height the jth tree on the ith sample point

   $\overline{\dfrac{V}{\text{BA} \cdot \text{Hm}}}$ = the mean ratio of observed volume to (basal area)(measured height) of the z individual trees selected for measurement

$$\overline{\frac{V}{\text{BA} \cdot \text{Hm}}} = \frac{\sum\limits_{k=1}^{z} \left(\dfrac{V_k}{\text{BA}_k \text{Hm}_k} \right)}{z} \tag{11.17}$$

   V_k = volume of the kth measurement tree

   BA_k = basal area of the kth measurement tree

   Hm_k = measured height on the kth measured tree

   and z = number of trees selected for detailed measurement in the 3P step

The volume:(basal area)(measured height) is a ratio estimator and is used in a similar manner to the way VBAR is used in big BAF sampling. The standard error of the mean volume per unit area is approximated using Bruce's (1961) formula (Grosenbaugh, 1971, 1979; Rennie, 1976):

$$s_{\bar{V}} = \bar{V} \sqrt{\frac{s^2\left(\sum Hp\right)}{n\left(\overline{\sum Hp}\right)^2} + \frac{s^2\left(\dfrac{V}{BA \cdot Hm}\right)}{z\left(\dfrac{V}{BA \cdot Hm}\right)^2}} \tag{11.18}$$

where $s^2\left(\sum Hp\right)$ is the sample variance of the sum of predicted heights per point

$s^2\left(\dfrac{V}{BA \cdot Hm}\right)$ is the sample variance of the ratio of volume:(basal area)(measured

height) of the z individual trees selected for detailed measurement

To illustrate the calculations, we will use the same 150-acre tract used in the list sampling (Section 10.9) and 3P (Section 10.10) sections. Tree locations were randomly simulated based on a map of the 25 compartments, and 30 HPS points were systematically located across the forest ignoring compartment boundaries. A 10 F BAF angle gauge (i.e., each tree represents 10 ft^2/acre) was used to select trees for HPS, and heights for the HPS sample trees were "predicted" using a height–diameter copula (Kershaw et al., 2010). The 3P selection phase incorporated a 30% rejection rate in order to keep sample sizes for measure trees near the desired size. The resulting HPS data are shown in Table 11.2, and the data for the measured trees are shown in Table A.12.

The mean sum of predicted heights ($\overline{\sum Hp}$) is

$$\overline{\sum Hp} = \frac{\sum\limits_{i=1}^{n}\left(\sum\limits_{j=1}^{m_i}Hp_{ij}\right)}{n} = \frac{15,594.7}{30} = 519.8$$

and the associated sample variance is

TABLE 11.2. Count and Sum of Predicted Heights (Hp, ft) by Plot for the Thirty 10 F BAF Sample Points Collected in a Mixed Oak–Hickory–Maple Forest in Central Indiana

Plot	Count	Sum of Hp	Plot	Count	Sum of Hp	Plot	Count	Sum of Hp
1	7	391.9	11	14	799.1	21	4	216.5
2	7	387.9	12	18	1071.2	22	10	567.1
3	8	470.8	13	6	354.2	23	5	303.9
4	7	431.7	14	7	422.7	24	11	607.8
5	10	573.0	15	6	358.1	25	11	655.3
6	7	413.6	16	9	508.6	26	9	468.7
7	10	583.3	17	12	700.1	27	18	1051.9
8	3	169.5	18	4	247.5	28	6	363.2
9	6	330.0	19	13	753.5	29	7	407.6
10	14	861.5	20	14	777.5	30	6	347.0

$$s^2\left(\sum\mathrm{Hp}\right)=\frac{\sum\limits_{i=1}^{n}\left(\sum\limits_{j=1}^{m_i}\mathrm{Hp}_{ij}\right)^2-\left[\left(\sum\limits_{i=1}^{n}\left(\sum\limits_{j=1}^{m_i}\mathrm{Hp}_{ij}\right)\right)^2\Big/n\right]}{n-1}$$

$$=\frac{9{,}636{,}556-\dfrac{\left(15{,}594.7\right)^2}{30}}{30-1}$$

$$=\frac{9{,}636{,}556-8{,}106{,}489}{29}=52{,}760.92$$

The mean ratio of volume:(basal area)(measured height) $\left(\overline{\dfrac{V}{\mathrm{BA}\cdot\mathrm{Hm}}}\right)$ is

$$\overline{\frac{V}{\mathrm{BA}\cdot\mathrm{Hm}}}=\frac{\sum\limits_{k=1}^{z}\dfrac{V_k}{\mathrm{BA}_k\cdot\mathrm{Hm}_k}}{z}=\frac{214.50}{78}=2.75$$

and the associated sample variance is

$$s^2\left(\frac{V}{\mathrm{BA}\cdot\mathrm{Hm}}\right)=\frac{\sum\limits_{k=1}^{z}\left(\dfrac{V_k}{\mathrm{BA}_k\cdot\mathrm{Hm}_k}\right)^2-\left[\left(\sum\limits_{k=1}^{z}\dfrac{V_k}{\mathrm{BA}_k\cdot\mathrm{Hm}_k}\right)^2\Big/k\right]}{k-1}$$

$$=\frac{642.91-\dfrac{\left(214.5\right)^2}{78}}{78-1}$$

$$=\frac{642.91-589.87}{77}=0.6887$$

The estimated mean volume per acre is

$$\frac{V}{\mathrm{acre}}=\mathrm{BAF}\cdot\overline{\sum\mathrm{Hp}}\cdot\left(\overline{\frac{V}{\mathrm{BA}\cdot\mathrm{Hm}}}\right)=10\cdot519.8\cdot2.75=14{,}295\,\mathrm{board\,feet}\,/\,\mathrm{acre}$$

and the total estimated volume for the 150-acre tract is 2,144,269 board feet. The standard error for the volume per acre estimate is

$$s_{\bar V}=\bar V\sqrt{\frac{s^2\left(\sum\mathrm{Hp}\right)}{n\left(\overline{\sum\mathrm{Hp}}\right)^2}+\frac{s^2\left(\overline{\dfrac{V}{\mathrm{BA}\cdot\mathrm{Hm}}}\right)}{z\left(\overline{\dfrac{V}{\mathrm{BA}\cdot\mathrm{Hm}}}\right)^2}}$$

$$=14{,}295\sqrt{\frac{52{,}760.92}{30\left(519.8\right)^2}+\frac{0.6887}{78\left(2.75\right)^2}}$$

$$=14{,}295\sqrt{0.006508+0.001168}=1252.4$$

The standard error for the entire 150-acre tract is 187,864.8 board feet. While the standard error for point-3P sampling is substantially larger (approximately 10 times) than the standard errors for list or 3P sampling, the total number of trees predicted and/or measured is substantially lower. For 3P sampling (Section 10.10), all 9106 trees had to be visited and predicted and 585 trees were measured. For list sampling (Section 10.9), 3920 trees were measured. For the example used here, 269 trees were counted on the 30 HPS, thus only 269 trees had to have height predicted, and the 3P phase only selected 78 trees. So while errors less than 1% were obtained using list and 3P sampling, errors around 8% were obtained from point-3P sampling with about 10% of the prediction and measurement efforts. The standard errors for the volume estimate from point-3P sampling are generally comparable to the standard errors obtained from similar sized normal HPS samples, where all "in" trees are measured for volume. For this particular example, the standard error would have been 1444, which is slightly larger than the point-3P standard error. As with big BAF sampling, most of the sampling error is associated with variation in counts between sample points rather than the associated ratio estimator.

Point-3P sampling was widely used in the southeastern United States in the 1970s (Steber and Space, 1972; Rennie, 1976). More recently, the method has not been widely applied, despite its potential for greatly reducing field inventory effort. Additional examples of point-3P sampling can be found in Grosenbaugh (1971, 1979), Steber and Space (1972), Van Hooser (1972, 1973), and Rennie (1976). Comparisons of point-3P sampling to other forms of multiphase HPS sampling can be found in Wood and Wiant (1992) and Stamatellos (1995).

11.3. OTHER VARIABLE PROBABILITY SAMPLING TECHNIQUES

Because of the ability to obtain basal area by counting "in" trees, and the close relationship between basal area and stem volume, HPS has been widely developed and implemented in forestry (Bitterlich, 1984). Other forms of variable probability sampling are often over-looked, or viewed as trivial variants, because they sample proportional to tree dimensions that are less useful for volume determination. As forest management continues to expand the resources and attributes of stands and forests that are incorporated into management plans, forest inventories need to adapt to estimate these resources. These additional variables may have allometric relationships with tree dimensions other than cross-sectional area and may be more efficiently sampled using one of the alternate forms of PPS sampling. The basic theory for these alternative sampling techniques and a summary of examples of how they have been utilized are presented here.

11.3.1. Horizontal Line Sampling (HLS)

HLS was originally proposed by Strand (1957). Grosenbaugh (1958) outlined the basic theory, and the concept was further expanded by Beers and Miller (1976). HLS generally utilizes the same angle gauges as HPS; however, instead of sweeping the angle gauge 360° around the sample point, the angle is projected perpendicular on both sides along a transect line (Fig. 9.7b). As a result, the inclusion zones for trees using HLS are rectangular (Fig. 11.8a). The area of the inclusion zone is a function of transect length and radius (half strip-width):

$$\text{Inclusion area} = L \cdot 2R_i$$

where L = transect length (m or ft)

R_i = inclusion zone (plot) radius (half strip-width) of the ith tree (m or ft)

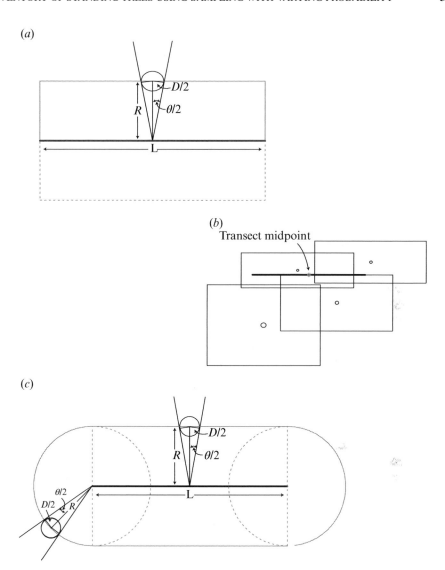

FIG. 11.8. Horizontal line sampling: (*a*) inclusion zone for a single tree, (*b*) inclusion zones for several trees along the transect, and (*c*) inclusion zone for modified HLS.

The inclusion zone radius is the same radius as for HPS: $R_i = D_i / (12k)$ for imperial units, and $R_i = D_i / (100k)$ for SI units. The gauge constant (k) is the same for HLS as for HPS: $k = 2\sin(\theta / 2)$. The tree factor, in imperial units, is

$$\mathrm{TF}_i = \frac{43{,}560}{L \cdot 2\,\dfrac{D_i}{12k}} = \frac{261{,}360k}{L \cdot D_i} \tag{11.19}$$

and in SI units is

$$TF_i = \frac{10,000}{L \cdot 2 \dfrac{D_i}{100k}} = \frac{50,000k}{L \cdot D_i} \tag{11.20}$$

The tree factor varies inversely proportionally to tree diameter, thus, the constant sampling factor in HLS is the diameter factor:

$$DF = TF_i \cdot D_i = \frac{261,360k}{L \cdot D_i} D_i = \frac{261,360k}{L} \tag{11.21}$$

for imperial units and

$$DF = TF_i \cdot D_i = \frac{50,000k}{L \cdot D_i} D_i = \frac{50,000k}{L} \tag{11.22}$$

for SI units. The count of "in" trees multiplied by DF gives an estimate of the sum of diameters per unit area.

The basic procedure for implementing HLS, as described in Beers and Miller (1976), is

1. Locate the center point of the transect and lay out the transect line along the prescribed azimuth. Transects may span the entire width of the area or be shorter lengths. Typical lengths are one chain (66 ft) or 20 m. Transects may be randomly oriented or systematically oriented along a prescribed grid. Transects that span an area do not need the center point located. Orientation should be predetermined.

2. Starting at one end of the transect, carefully sight each tree perpendicular to the azimuth of the transect line to determine whether the tree is in or not. (Note: If a tree is in when sighted at a nonperpendicular angle, it will be in at the perpendicular angle.) Both sides of the transect should be sighted. Near the end points of the transect, care needs to be exercised to make sure that perpendicular sightings are made. Trees whose perpendicular bisectors are beyond the ends of the transect are not in, even if their sighted distance indicates that they are within the limiting distance.

3. For each "in" tree, measure the required tree attributes.

Figure 11.8b illustrates HLS using inclusion zones. Working left to right, the first tree is not in because the inclusion zone is beyond the transect; the second tree is in because the inclusion zone intersects with the transect and contains the transect midpoint; the third tree is in, but border line; and the fourth tree is not in because it extends beyond the end of the transect line. Even though the fourth tree's inclusion zone intersects the transect, it does not include the transect midpoint, therefore, the tree is out.

One of the drawbacks of HLS is the time required to accurately determine if trees near the ends of transects are in based on perpendicular bisectors. To avoid this time-consuming task, and to increase sample sizes of snags, Ducey et al. (2002) proposed a modified form of HLS that incorporates an HPS at the ends of the transects (Fig. 11.8c). At each end of the transect, a half-sweep is made and all trees that are in based on the angle gauge are included in the sample. The inclusion zone becomes a "sausage"-shaped zone and the associated area is

$$a_i = 2(kD_i)L + \pi k^2 D_i^2 \tag{11.23}$$

where k is the gauge constant (eq. (9.7)). The modification is a practical solution because it both eliminates the error-prone and time-consuming task of determining the status of trees near the ends of the transects and does not require any additional field equipment since the angle gauge used for HLS is the same as that used for HPS.

Because most forest inventories focus on volume estimation, HLS sampling has not been extensively used since HPS is much more efficient for volume estimation. HLS has been applied effectively where stand structure heterogeneity is large and where the focus of the inventory is on estimation of relatively rare components of the stand or forest. For example, Rice et al. (2014) found that HLS was particularly effective in partially harvested stands having a regular pattern of parallel skid trails. Ducey et al. (2002), using the modified approach described above, found HLS to be efficient for inventories of standing dead trees that would not be sampled adequately by ordinary HPS with typical sample sizes (Kenning et al., 2005). HLS may be useful for other trees that ordinarily constitute a small fraction of the total overstory, such as especially high-value timber trees or trees with special wildlife habitat attributes. Birth (1977) describes how HLS can be used to obtain Reineke (1933) relative density measures.

11.3.2. Vertical Point and Line Sampling

Sampling with variable probability using vertical techniques requires projecting an angle vertically rather than horizontally. Projection of vertical angles is more difficult than projecting horizontal angles, and is one of the contributors to the lack of widespread application of vertical sampling. In addition, vertical sampling selects trees with probability proportional to height, in the case of VLS, or height squared, in the case of vertical point sampling (VPS). The relationship between height and volume is much more variable than between diameter or basal area and height. As a result, more vertical samples would be required to obtain the same levels of precision compared to horizontal samples.

VLS is implemented in the same way HPS is implemented. A sample point is located in the field, and a 360° sweep is conducted about the sample point. All trees whose heights extend above the line of sight projected by the vertical angle are counted as in on the sample point. "In" trees are then measured for the additional attributes required for the inventory.

The gauge constant for vertical sampling is defined as $q = H/R = \tan\phi$, where $H =$ tree height, $R =$ plot radius (or strip half-width), and ϕ is the vertical angle. The inclusion zone area for VPS is then

$$a_i = \pi \left(\frac{H_i}{\tan\phi} \right)^2 = \pi \left(\frac{H_i}{q} \right)^2 \tag{11.24}$$

In imperial units, the tree factor is

$$\mathrm{TF}_i = \frac{43,560}{\pi \left(H_i/q \right)^2} = \frac{43,560 q^2}{\pi H_i^2} \tag{11.25a}$$

and in SI units, the tree factor is

$$TF_i = \frac{10,000}{\pi \left(H_i/q\right)^2} = \frac{10,000q^2}{\pi H_i^2} \tag{11.25b}$$

The tree factor varies inversely proportional to height squared; therefore, the squared height factor (SHF) is the constant sampling factor, and a count of "in" trees multiplied by SHF yields the sum of squared heights per unit area.

While sum of squared heights per unit area is not a very useful quantity to use directly, it can be used as a covariate in a double sampling scheme if a field-efficient method of obtaining VPS estimates is available. Ducey and Kershaw (2011) point out that if an ordinary camera is pointed vertically above a sample point, and the number of trees in the frame is counted, the trees selected constitute a vertical point sample. Maynard et al. (2014) provide methods for slope correction and supporting software. The chief difficulty with VPS using a camera is the ability to see and distinguish individual trees in the photograph when a dense understory is present.

For VLS, the inclusion zone is given by

$$a_i = 2R_iL = 2\left(\frac{H_i}{\tan\varphi}\right)L = 2\left(\frac{H_i}{q}\right)L \tag{11.26}$$

The tree factor, in imperial units, is then

$$TF_i = \frac{43,560}{2\left(H_i/q\right)L} = \frac{21,780q}{H_iL} \tag{11.27a}$$

and in SI units is

$$TF_i = \frac{10,000}{2\left(H_i/q\right)L} = \frac{5,000q}{H_iL} \tag{11.27b}$$

From eq. (11.27), it can be seen that the tree factor varies inversely proportional to height, thus the height factor (HF) is the constant sampling factor in VLS. A count of in trees multiplied by the HF yields an estimate of the sum of heights per unit area.

VLS is implemented in the same manner as HLS. Transects may span the entire length of the stand or be some prescribed length. All trees along the transect are sighted perpendicular to the transect azimuth, and trees whose heights extend above the line of sight projected by the vertical angle are counted as in. As with HLS, care must be exercised near the ends of the transect to insure that the trees being sighted do not extend beyond the ends of the transects. The modified procedure of Ducey et al. (2002) proposed for HLS could be applied to VLS to avoid this time-consuming task.

The primary application of VLS has been regeneration surveys. Beers (1974b) and Beers and Miller (1976) describe how VLS can be used in a regeneration survey. Their approach built upon the approach proposed by Bickerstaff (1961). Lappi et al. (1983) develop a simple height relaskop for regeneration surveys. In an alternative application of VLS, Schnell et al. (2013) show how VLS can be effectively used to estimate stand-average crown ratio.

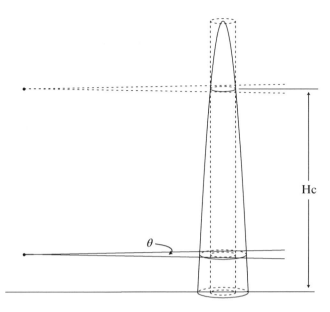

FIG. 11.9. Critical height sampling.

11.3.3. Critical Height Sampling

Bitterlich (1956) developed the idea of individual tree volume contribution. The concept of critical height sampling was originally derived by Kitamura (1962), who showed the volume at a given sample point could be estimated by multiplying the BAF by the sum of critical heights:

$$V = \text{BAF} \cdot \sum_{j=1}^{m} \text{Hc}_j \qquad (11.28)$$

where critical height, Hc, is defined as the height on each tree where the tree's stem diameter would be exactly borderline given the projected horizontal angle (Fig. 11.9). This same estimator was later independently derived by Iles in the early 1970s (Iles, 1979). Critical height sampling has the advantage that volume is determined without the need for volume tables or volume equations that can introduce bias in the estimates of volume per unit area (Bitterlich, 1976; Iles, 1979; McTague and Bailey, 1985; Van Deusen and Meerschaert, 1986).

One challenge with implementing critical height sampling is accurately determining the height of the borderline diameter. Bitterlich (1976) describes a method using the Spiegel Relaskop (Fig. 5.7*a*) or the Telerelaskop to indirectly determine critical height. McTague and Bailey (1985) use taper equations to estimate heights and demonstrate how unbiased volume estimates can be obtained. Van Deusen and Meerschaert (1986) further develop this idea and show how merchantable volume can be obtained. Lynch and Gove (2013) point out that, for trees near the sampling point, critical height will be very high on the tree stems, making measurement from the sample point problematic. To overcome this issue, they combined ideas of importance sampling (Section 6.5.2) with critical height sampling and introduce an antithetic variable ($1 - u$, where u is a uniform random variate based on ratios of upper stem cross-sectional area to basal area) and use this variable to select the point on the stem at which height is measured. The antithetic variable results in heights for stems near the sample point being low on the stem and higher on the stems for trees farther away, thus facilitating measurement from plot center. This problem could also be circumvented by simply moving from the sample point to obtain the needed height measurements, which is often required because of visibility issues.

Critical height sampling is a very efficient method to obtain volume estimates (Bitterlich, 1976; Iles, 1979) because it samples directly proportional to volume. Unfortunately, critical height sampling has not seen wide operational implementation. The lack of implementation is primarily due to the perceived difficulty in measuring upper stem diameters to determine critical height. Lack of familiarity with instruments such as the Relaskop, or some of the newer electronic stem measurement tools, contributes to this perception.

12

INVENTORY OF DOWNED DEAD MATERIAL USING SAMPLING WITH VARYING PROBABILITY

Over the past several decades, downed dead wood has emerged as a structural and functional component of key interest for management and conservation in forest ecosystems. Early concern over downed dead material focused on its role as logging waste (Warren and Olsen, 1964), a role that has reemerged as downed wood is considered a possible bioenergy resource in some regions. Downed wood represents fuel in many fire-prone systems and can harbor bark beetles and other agents of tree mortality (Brown et al., 2003). The negative connotations of downed dead wood are evident in the term "coarse woody debris," by which it is often known in the scientific literature. However, a great deal of interest also focuses on its contributions to ecosystem functions, including biogeochemical cycling, tree regeneration, and biodiversity (Harmon et al., 1986; Jonsson et al., 2005; Stokland et al., 2012). In the context of carbon inventories, dead wood is defined as all nonliving aboveground biomass derived from trees, including biomass that is standing or lying on the ground (IPCC, 2003). Standing dead trees can usually be inventoried using the same designs as standing live trees, albeit occasionally with modifications that reflect the relative rarity of standing dead trees in many forest ecosystems. However, downed dead material presents a series of challenges that have given rise to custom inventory approaches.

Downed dead wood also serves as a model for the creative application of geometry to develop variable-probability sampling methods. Although many downed dead wood inventories still use fixed-area plots (Section 12.1), line intersect sampling (Section 12.2) saw early development for downed wood inventory. As ecological interest in downed wood burgeoned at the end of the twentieth century, a recognition of the shortcomings of existing inventory approaches led to a rapid development and diversification of inventory techniques. Some of the challenges of measuring volume, weight, and nutrient content of individual pieces of downed wood have been described in Section 6.7.2.

Forest Mensuration, Fifth Edition. John A. Kershaw, Jr., Mark J. Ducey, Thomas W. Beers and Bertram Husch.
© 2017 John Wiley & Sons, Ltd. Published 2017 by John Wiley & Sons, Ltd.

12.1. FIXED-AREA PLOTS

The use of fixed-area plots is among the oldest sampling methods used for downed wood. Fixed-area plots are the most common design used for fallen wood inventory in forest inventories at the national scale, though many of the largest countries have shifted to line intersect sampling (Woodall et al., 2009). Downed wood inventory using fixed-area plots is quite simple and familiar, at least in principle, because it seems similar to the use of plots for standing trees and other vegetation. Sample points are selected randomly or systematically in the area of interest, and a geometric shape of known area, such as a circle, square, or rectangle, is laid out around each sample point. The downed wood on each plot is tallied and measured for whatever characteristics are necessary to calculate volumes or weights and to characterize the wood in terms of its decay class and other values. Harmon and Sexton (1996) emphasize square and rectangular plots, but the choice of plot shape and size should depend on field efficiency and accuracy of work as well as statistical variability, just as with standing trees (Section 9.2).

The simplicity of fixed-area plots ends there as questions arise as to what it means for downed wood to be on a plot (Ståhl et al., 2001). Ståhl et al. (2001) identify two main approaches to determining what fallen wood is included on a plot, and Gove and Van Deusen (2011) have identified additional alternatives that have actually been reported in the scientific and management literature:

1. A piece is tallied if some predefined, clearly identifiable location on the piece is inside the plot boundary. The most common predefined location is the pith at the large or basal end of the piece (Gove and Van Deusen, 2011). However, the mid-point of the piece could also be used (e.g., Vanderwel et al., 2008). Either protocol is acceptable so long as the location is defined in advance. Under this approach, if a piece is tallied on the plot, then the entire piece is treated as included in the inventory. For example, the entire piece would be measured for volume, and the volume of the entire piece would be multiplied by the reciprocal of plot area (i.e., piece factor, measured in hectares; see Section 9.1) to determine the contribution of that piece to volume per hectare. This approach is most directly analogous to the use of fixed-area plots for standing trees in terms of its field approach and computational details; the inclusion zone for a piece is the same as it would be for a standing tree, except that it is centered on the predefined point rather than at the location of a standing stem (Fig. 12.1). Estimating the number of pieces per hectare, volume per hectare, and similar attributes is essentially identical to the methodology used for standing trees. However, some downed wood may physically lie on the plot but may not be tallied because the center of the large end (or other predefined point) is outside the plot, while other wood is included in the estimates even though it physically crosses outside the plot boundary. Wood in the former condition is analogous to the parts of living trees that are rooted outside a fixed-area plot but overhang it. Wood in the latter is akin to the parts of trees rooted inside the plot that extend over and outside the plot boundary.

2. Other protocols include only those pieces and portions of pieces of downed wood that lie strictly within the plot boundary and exclude those portions that fall across the plot border. Gove and Van Deusen (2011) call this the "chain saw protocol": one can imagine cutting the downed wood at the plot boundary with a saw, discarding

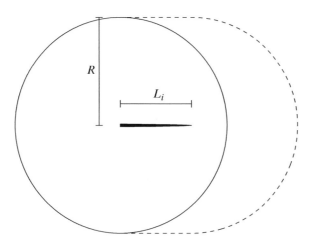

FIG. 12.1. A piece of downed wood and its inclusion zone when a circular plot has been used with the large-end protocol (solid line) and with the sausage protocol (dashed line).

the portions outside the plot, and only including those portions that remain in any estimates. This approach may be easiest to implement with square or rectangular plots because pieces are cut off by straight lines, though circular plots can also be used. Ståhl et al. (2001) point out that this approach can require tedious measurements on individual pieces to find what portion of each piece should be included. The total volume of wood per hectare, and similar quantities such as weight, can be estimated without bias using this protocol. However, simply multiplying the number of pieces tallied by the reciprocal of plot area leads to an estimate of the number of pieces per hectare that is biased upward. This protocol seems most useful when it is important to identify the downed wood physically lying on a given area of ground, as is often the case with biodiversity studies (Stokland et al., 2012) and certain biogeochemical investigations (Harmon and Sexton, 1996). However, it is not attractive when estimating the number of pieces per hectare for a larger inventory area is of interest.

3. Gove and Van Deusen (2011) also identify cases in the published scientific literature in which entire pieces of downed wood are used to develop estimates of pieces per hectare and volume per hectare (as in the first protocol), but pieces are tallied if any portion falls on the area of the plot. They call this the "sausage protocol" because of the shape of the inclusion zone that results. Using the ordinary formulas for fixed-area plots leads to an upward bias in estimates of pieces per hectare, volume per hectare, and similar quantities because the inclusion zone for a piece is larger than the plot area (Fig. 12.1). Gove and Van Deusen (2011) present alternative equations when the sample plots are circular, their whole-log inclusion zone is given as

$$a_s = 2RL_i + \pi R^2 \qquad (12.1)$$

where a_s = whole-log inclusion zone area (m^2 or ft^2)
  R = fixed-area plot (inclusion zone) radius (m or ft)
  L_i = length of piece (m or ft)

Note that this is always larger than the area of the ordinary inclusion zone for a circular plot, which equals πR^2. Thus, the correct piece factor under this protocol is always smaller than the conventional plot-based piece factor since the piece factor is unit area divided by inclusion zone area.

Gunn et al. (2014) give similar equations when pieces can be assumed to be linear and the plots are square or rectangular. For a rectangular plot of length l and width w, the expected inclusion zone area (taken as an integral over possible angles between the piece and the long axis of the plot, with the angle assumed random; Gunn et al., 2014) is

$$E\left[\text{Inclusion zone area}_i\right] = lw + \frac{2(l+w)}{\pi}L_i \qquad (12.2)$$

This expected inclusion zone area is also larger than the nominal one for a rectangular plot, which equals lw. Gunn et al. (2014) give an alternative equation that can be used if piece orientation relative to the plot is measured. Likewise, unfortunately, many studies that use this field protocol have not used proper equations to expand the plot tallies to areal estimates.

It is not always clear in inventories that use fixed-area plots, which of the three protocols outlined above has been prescribed, but some combinations of field protocols and estimating equations are highly biased (Gove and Van Deusen, 2011). Likewise, Woodall et al. (2009) note that protocols provided in their survey of national forest inventories lacked the basic information to match field procedures with estimating equations. Clearly specifying the field protocol is a critical element for fixed-area plot inventories of downed wood.

Depending on the protocol, fixed-area plot sampling requires measurements to characterize the entire volume and overall decay class of every piece of fallen wood on the plot (or portion of each piece, under the "chain saw protocol") whenever estimates of volume and related quantities are needed. Total length, one or more diameters or cross-sectional areas, as required by the volume formula, and an estimated overall decay class are usually needed (Section 6.7). However, any inaccuracies in the volume equations will carry through to final inventory estimates, and these can be substantial for downed wood (Fraver et al., 2007). All of the standard techniques for correcting boundary slopover, including the mirage and walkthrough methods, can be used with fixed-area plots.

Table 12.1 outlines the calculation of pieces/ha and volume/ha for a hypothetical 17.84 m radius, or 0.10 ha, circular plot, that happens to have fallen in an area with a fairly dense population of downed wood. Ten pieces impinge at least partially on the plot area; whether the large end of the piece falls on the plot, and the cubic volume of the entire piece as well as the portion on the plot, is given for each piece. Only those pieces with large ends within the plot would be included in the tally under the large-end protocol; however, the entire volume of each piece is used to estimate volume/ha. Under the chain saw protocol, all pieces touching the plot are used, but only the portion of the volume within the plot should be used, and no calculation of pieces/ha can be performed. Under the sausage protocol, all the pieces are used and the entire volume is used, but the piece factors must be adjusted. The inclusion zone area is calculated using eq. (12.1), and the piece factor is calculated based on the inclusion zone area. In this example, all three protocols give similar answers, though they are not identical. For all three protocols, the estimate of pieces/ha is the sum of the piece factors, and the estimate of volume/ha is the sum of the volume factors. Over a large population of samples, all three protocols are

TABLE 12.1. Example Calculation of Volume Per Hectare for an 0.10-ha Circular Plot Under Three Different Protocols

Piece	Total Length (m)	Total Volume (m³)	Volume on Plot (m³)	Large End on Plot?	Large End Protocol Piece Factor	Large End Protocol Volume Factor	Chain Saw Protocol Volume Factor	Sausage Protocol Inclusion Area (m²)	Sausage Protocol Piece Factor	Sausage Protocol Volume Factor
1	3	0.03	0.03	Y	10	0.3	0.3	1107.0	9.0	0.3
2	6	0.12	0.12	Y	10	1.2	1.2	1214.1	8.2	1.0
3	12	0.38	0.25	N	0	0	2.5	1428.2	7.0	2.7
4	4	0.21	0.21	Y	10	2.1	2.1	1142.7	8.8	1.8
5	5	0.16	0.12	Y	10	1.6	1.2	1178.4	8.5	1.4
6	10	0.49	0.19	N	0	0	1.9	1356.8	7.4	3.6
7	11	0.78	0.78	N	0	0	7.8	1392.5	7.2	5.6
8	2	0.04	0.04	Y	10	0.4	0.4	1071.4	9.3	0.4
9	1	0.01	0.01	Y	10	0.1	0.1	1035.7	9.7	0.1
10	20	2.55	1.55	Y	10	25.5	15.5	1713.6	5.8	14.9
Total					**70**	**31.2**	**33.0**		**80.9**	**31.7**

unbiased; the key is to ensure that the approach for computing density and volume estimates matches the actual field methods used.

Just as with live trees, there is a common tendency to treat estimates derived from fixed-area plots as "real" and those derived from other designs as somehow different and inferior. In the case where the population of interest is the downed wood on a single large, fixed-area plot, as it sometimes is in biodiversity studies (Stokland et al., 2012), then the enumeration of the downed wood on the plot is not a sample but a census, and the distinction is meaningful. Otherwise, it reflects a fundamental misunderstanding: fixed-area plots are just one possible sampling design and one that has its shortcomings. In many forested systems, the number of short, small-diameter pieces of downed wood is very high, and the time expended in measuring such pieces is out of all proportion to their contribution to estimates of volume, weight, and carbon or nutrient content. Such pieces also usually contribute little unique habitat value (though see Nordén et al., 2004, for a counterexample). Thus, a large proportion of overall field effort is consumed by small pieces that are of little consequence to final estimates, while large pieces are sampled so infrequently that the plot-to-plot variance of the resulting estimates is very large (Ståhl et al., 2001). Moreover, field errors of nondetection are a serious concern with fixed-area plots. The entire plot must be searched for small, hidden, or well-decayed pieces in a time-consuming and error-prone fashion (e.g., Jordan et al., 2004; Kenning, 2007). These challenges have led many practitioners to reject fixed-area sampling as a default method for downed wood inventory.

12.2. LINE INTERSECT SAMPLING

Line intersect* sampling and closely related techniques have been used widely for ecological sampling, with applications dating back at least to Canfield (1941). The basic mathematical underpinnings of line intersect sampling date to work by Buffon (1777). Line intersect sampling, or LIS as it is commonly abbreviated, was originally adapted in a simplified way to downed dead wood inventory by Warren and Olsen (1964), and extensions and adaptations rapidly followed (Van Wagner, 1968; Bailey, 1970; Brown, 1974). In line intersect sampling for downed wood, a sample line is run within the stand or tract of interest. The pieces of downed wood that cross the line are tallied, and additional measurements are taken on those pieces. Running multiple lines allows for the construction of confidence limits, in much the same way that using multiple plots or strips in fixed-area sampling does.

Many variations of line intersect sampling have been proposed. For example, in some versions, the sample lines run from border to border in the stand or tract, in the same way that a strip sample of standing trees would. In others, the sample lines are shorter segments centered on a randomly (or systematically) located sample point. In still others, a cluster of lines is used (often forming a cross, a Y, or even a box so that measurements are taken on a closed traverse with little "wasted" walking time). The combination of sampling geometry with a wide array of possible implementations led Ståhl et al. (2001) to write, "it [LIS] is somewhat difficult to describe in a simple way from a theoretical point of view." Certainly, the theoretical treatments of LIS support that assertion: the authoritative work by Kaiser

---

*The term line intercept sampling, as used in Section 9.4.2, is synonymous with line intersect sampling as used here. Original authors such as Canfield (1941) used the term intercept; more recent authors cited throughout Chapter 12 use the term intersect.

(1983) is nearly impenetrable, as is the treatment by de Vries (1979, 1986). Moreover, the literature on LIS is full of confused, inaccurate, or blatantly wrong statements (Gregoire and Valentine, 2003). However confused the literature may be, LIS itself need not be completely confusing. In this chapter, we will focus on a relatively simple type of LIS protocol that is readily adaptable to many forest inventory situations, so that the assumptions, field protocols, and estimating equations can be kept simple and clear. As we shall see, LIS is actually a PPS design (see Section 9.4.2), and samples pieces of downed wood with probability proportional to their length. Moreover, the point on a piece where the sample line crosses is randomly selected along the length of the piece, and this offers opportunities to estimate properties such as volume per unit area with simplified measurements, and without any need to assume a specific taper or volume equation.

12.2.1. Assumptions of Line Intersect Sampling

For the moment, let us suppose that a sample point has been randomly or systematically located within the stand or stratum that is being inventoried. A sample line is run for a fixed, predetermined length, centered on the sample point. All the pieces of downed wood that cross the line are tallied; those that do not cross the line are ignored. The sample layout and the corresponding inclusion zones for a group of pieces are depicted in Figure 12.2.

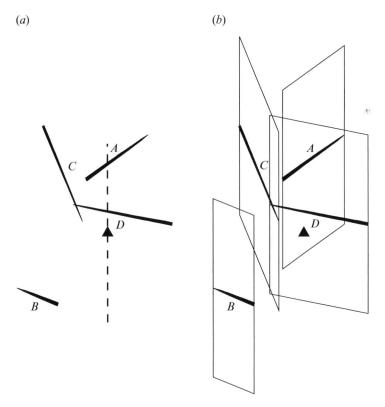

(a)   *(b)*

FIG. 12.2. An LIS sample point, its sample line, and several pieces of downed wood (*a*). The line crosses pieces *A* and *D* so those pieces are tallied. The corresponding inclusion zones are shown in (*b*).

In principle, it should be straightforward to calculate the inclusion zone area for any piece of downed wood as

$$a_i = S \cdot w_i \tag{12.3}$$

where a_i = inclusion zone area (m² or ft²)
  S = length of the sample line (m or ft)
  w_i = the projection of the piece perpendicular to the sample line in the same units as S

However, measurement of w_i is fussy in the field, and estimates are especially sensitive to it as downed wood becomes nearly parallel to the line. Therefore, nearly all practical work using LIS for downed wood inventory uses an alternative approach, which considers the *expected* value of w_i and of the inclusion zone area, taking the angle of intersection between the sample line and the downed wood as uniformly and randomly distributed. It turns out that these expected values are

$$E[w_i] = \frac{2L_{horizontal,i}}{\pi}$$

$$E[a_i] = \frac{2L_{horizontal,i}}{\pi} S$$

where $L_{horizontal,i}$ is the piece length, measured horizontally and in the same units as S. By analogy to the tree factor from sampling standing trees, we may define a piece factor, PF_i, as the number of pieces per unit area that a tallied piece represents. When S and $L_{horizontal,i}$ are in meters, and the stand area is in hectares, we have

$$PF_i = \frac{10,000\pi}{2SL_{horizontal,i}} \tag{12.4}$$

Likewise, when S and $L_{horizontal,i}$ are in feet, and the stand area is in acres, we have

$$PF_i = \frac{43,560\pi}{2SL_{horizontal,i}} \tag{12.5}$$

For example, suppose that we are using a sample line that is 100 m long, and we tally a piece that is 10 m in length. Then, PF_i for that piece equals

$$PF_i = \frac{10,000\pi}{2(100)(10)} = 5\pi \approx 15.71 \text{ pieces / ha}$$

If we tally a second piece that is 20 m in length, its PF_i is

$$PF_i = \frac{10,000\pi}{2(100)(20)} = 2.5\pi \approx 7.85 \text{ pieces / ha}$$

Formally speaking, the only assumption that has been added in moving from eq. (12.3) to eqs. (12.4) and (12.5) is that the angle of the intersection between the piece and the sample line is random. de Vries (1979, 1986) and many other researchers have based this on a further assumption that the orientation of the pieces themselves is random. Such an assumption is unrealistic in many, if not most, situations. Directional felling during harvesting, windthrow along the direction of prevailing winds, and even the tendency of fallen pieces to roll with the topography create populations of downed wood that show patterns of orientation. However, if the sample lines themselves are randomly oriented, then the intersection angles with the downed wood are guaranteed to be random by design; this is the assumption employed by Kaiser (1983).

With LIS, it is important to understand two potential sources of confusion and field error. The first is that the line length S should be measured horizontally, not along the slope. Measuring S along the slope introduces errors that are directly analogous to those that occur when slope distance, rather than horizontal distance, is used to evaluate borderline trees in plot or horizontal point sampling. The second is that $L_{horizontal,i}$ (along with w_i) is also defined as a horizontal distance. When a piece is not lying horizontally, it is critical to record sufficient information to recover any information necessary for volume estimation, as well as information to allow recovery of $L_{horizontal,i}$. For example, suppose the total (nonhorizontal) length of a piece, L_i, is recorded, as well as the angle of inclination relative to the horizontal, α, in degrees. Then, one can recover $L_{horizontal,i}$ by simple trigonometry as

$$L_{horizontal,i} = L_i \cos \alpha$$

For pieces lying at a very shallow angle, this kind of correction is often ignored. Although neglecting to account for slope does induce a small bias in any estimates, for angles less than 11.5° the error is less than 2%, and for angles less than 17.7° the error is less than 5%. Nonetheless, where terrain is steep, or where partially fallen trees that are not yet in contact with the ground are treated as downed wood for inventory purposes, accounting for angle of inclination can be important. For example, Monleon (2009) found that ignoring slope could lead to a bias of 4–8% in downed wood estimates for the state of Oregon in the US national forest inventory.

Finally, for pieces that lie close to the boundary of the stand or stratum, a portion of the inclusion zone may also fall outside the stand or stratum and slopover bias will be present. Fortunately, when the sample line is centered on the sample point, the reflection method outlined by Gregoire and Monkevich (1994) completely eliminates the bias and is very easy to implement. Whenever a sample line crosses over the boundary, the portion of the sample line that falls outside the stand is simply folded back onto the original line. Any pieces that occur on the folded portion are double-tallied.

12.2.2. Estimating Downed Wood Parameters

Following the same logic as used for sampling standing trees with variable probability (Chapter 9, especially eq. (9.2)), the expansion factors for other downed wood characteristics (XF_i), such as surface area or volume, are obtained by multiplying the value of the characteristic, X_i, by the piece factor:

$$XF_i = X_i \cdot PF_i \tag{12.6}$$

This expansion factor XF_i represents the contribution of an individual tallied piece, to the total estimate of attribute X per unit area that we would obtain from a single sample line. If we have tallied a total of m pieces of downed wood on n sample lines, we can estimate the amount of X per unit area as

$$\hat{X} = \frac{1}{n} \sum_{i=1}^{m} \mathrm{XF}_i \tag{12.7}$$

Examining eq. (12.4), we see that the piece factor, PF_i, is inversely proportional to $L_{\mathrm{horizontal},i}$, thus the horizontal length factor, LF, is constant in LIS. The LF is SI units is

$$\mathrm{LF}_i = L_{\mathrm{horizontal},i} \left(\frac{10{,}000\pi}{2SL_{\mathrm{horizontal},i}} \right) = \frac{10{,}000\pi}{2S} \tag{12.8}$$

and in imperial units is

$$\mathrm{LF}_i = L_{\mathrm{horizontal},i} \left(\frac{43{,}560\pi}{2SL_{\mathrm{horizontal},i}} \right) = \frac{43{,}560\pi}{2S} \tag{12.9}$$

From these relationships, we see that because LIS samples pieces of dead wood with probability proportional to their horizontally projected length, a particular LIS design is associated with a LF that depends only on the length of the sample line and the units of measure. For example, if we are working in SI units and our sample lines are 100 m long, the LF associated with our design is just $\mathrm{LF} = 10{,}000\pi / (2 \times 100) = 157.1\,\mathrm{m}/\mathrm{ha}$. In that case, every tallied piece on a single line represents 157.1 m/ha of horizontally projected length. Similarly, suppose we are working in imperial units, and our design involves sample lines that are 4 chains, or 264 ft, long. Then, the LF for our design is simply $\mathrm{LF} = 43{,}560\pi / (2 \times 264) = 259.2\,\mathrm{ft}/\mathrm{acre}$ and every tallied piece represents 259.2 ft/acre of horizontally projected length. Moreover, after a little algebra, it is easy to show that a shortcut way to calculate PF_i is just

$$\mathrm{PF}_i = \frac{\mathrm{LF}}{L_{\mathrm{horizontal},i}} \tag{12.10}$$

By itself, the LF is not of much practical interest because the total horizontally projected length of downed wood is not in itself very interesting either. However, the utility of the LF becomes much more obvious when we consider estimating attributes that can be expressed as integrals over the length of each piece. These include, among others, surface area, volume, weight, and nutrient content.

As a simple example, suppose we are interested in volume per unit area, and despite the potential for significant bias in the volumes of individual pieces (Fraver et al., 2007), we are satisfied with measuring the cross-sectional area of each piece at the basal and upper ends (A_b and A_u) and using Smalian's formula (Section 6.1.3). Thus, we have

$$V_i = \frac{A_b + A_u}{2} L_i$$

as the estimated volume for each piece, and

$$VF_i = V_i \times PF_i$$

$$= \frac{A_b + A_u}{2} L_i \times \frac{LF}{L_{horizontal,i}}$$

$$= \frac{A_b + A_u}{2} \left(\frac{L_{horizontal,i}}{\cos \alpha} \right) \times \frac{LF}{L_{horizontal,i}} \qquad (12.11)$$

$$= \frac{A_b + A_u}{2 \cos \alpha} LF$$

Close examination of the formula for VF_i shows that it is not necessary to measure piece length to estimate volume per unit area.

We may simplify the measurements further and entirely eliminate the bias associated with Smalian's formula (or any other assumed formula for piece volume) by recognizing that the point where the sample line crosses a piece of downed wood is distributed uniformly at random along the length of the piece. Denoting that point as x, the cross-sectional area A_x represents a single random sample from the distribution of cross-sectional areas along the length of the piece, and thus $A_x L_i$ is an unbiased, crude Monte Carlo estimate of the volume of the piece (i.e., one based on the use of sampling to estimate average cross section). Substituting this expression for piece volume, we have

$$VF_i = V_i \times PF_i$$

$$= A_x L_i \times \frac{LF}{L_{horizontal,i}} \qquad (12.12)$$

$$= \frac{A_x}{\cos \alpha} LF$$

Thus, an unbiased estimate of volume per unit area can be obtained with only a single cross-sectional area measurement, along with a measurement of the slope of the piece. Eliminating the bias of an assumed volume equation does come at the price of a slight increase in variance because the estimates for the volumes of the individual pieces contain a random error. However, most thorough studies, such as that of Monleon (2009), have found that the increase in variance is quite small and the combination of simpler field techniques and the elimination of a key source of bias makes it more than worthwhile. It may be valuable to measure large-end diameters to characterize the value of different pieces of downed wood for wildlife habitat (Bate et al., 2009); however, the suggestion that $A_b \times L_i$ should be used as the estimator for piece volume in the calculation of VF_i is erroneous and leads to a gross overestimate of volume as A_b is almost always larger than the average cross-sectional area along the axis of the piece.

Extending the Monte Carlo approach to other downed wood attributes is straightforward (Valentine et al., 2001). For an arbitrary attribute X, the value of XF_i is just the amount of X per unit piece length, divided by $\cos \alpha$ to adjust for piece slope, multiplied by the LF. For example, the surface area covered by a piece of downed wood, as measured on the slope, is simply its diameter (measured horizontally) integrated over piece length, multiplied by a unit conversion k ($k=1/12$ when diameter is in inches and length is in feet; $k=1/100$ when diameter is in centimeters and length is in meters). Thus, by substituting horizontally measured diameter at the crossing point, $D_{\text{horizontal},x}$, in place of cross-sectional area, we obtain a surface area factor SAF_i for the piece as

$$\text{SAF}_i = k \times D_{\text{horizontal},x} L_i \times \frac{\text{LF}}{L_{\text{horizontal},i}}$$

$$= k \times \frac{D_{\text{horizontal},x}}{\cos \alpha} \text{LF} \tag{12.13}$$

Likewise, if one determines the wood density ρ_x at the crossing point (density must be the average over the cross-sectional area), then the corresponding Monte Carlo factor for downed wood mass MF_i is

$$\text{MF}_i = \rho_x \frac{A_x}{\cos \alpha} \text{LF}$$

Note that the quantity $\rho_x A_x$ is just mass per unit length; if a section of known length, centered on x, can be removed from a sampled piece of downed wood, it may be possible to determine it directly without separate measurement of density and area. Alternatively, density may be estimated from tables or equations using species, decay class, and possibly size as predictors (Section 6.7.2). Larjavaara and Muller-Landau (2011) provide some additional discussion on the Monte Carlo approach.

12.2.3. Choosing a Line Length and Design

The most important design attributes for any survey using LIS are the line length per sample point, and the total number of sample points (or equivalently, the total line length in the survey). Broadly speaking, the longer the individual sample lines are, the less variable the estimates, just as larger plot sizes produce less variable estimates when sampling standing trees. Specifically, the variance of estimates from individual sample lines is approximately inversely proportional to line length. Thus, fewer lines are needed to achieve comparable confidence limits if the individual lines are long, while more are needed if the individual lines are short. Across a broad range of designs, therefore, the standard error in downed wood estimates from LIS is driven primarily by the total line length in the survey (Marshall et al., 2000). However, both statistical and nonstatistical criteria should be considered in the design to ensure that measurements can be done efficiently and accurately in the field.

One common design error is to seriously underestimate the total line length that will be required to obtain reliable estimates of downed wood volume, mass, or carbon content for a stand or stratum. The simulation studies by Pickford and Hazard (1978) and Hazard and Pickford (1986) suggested that an aggregate line length of 1–4 km would be needed for reliable estimates of volume or any closely related quantities in an inventory stratum. Their general conclusions seem to be borne out in field studies in a range of forest conditions in temperate or boreal forests in North America (e.g., Jordan et al., 2004; Kenning, 2007; Sikkink and Keane, 2008). However, required line lengths may be even greater when downed wood is especially sparse. For example, Böhl and Brändli (2007) describe the experience of the Swiss National Forest Inventory, which uses 30 m of LIS transects in a Y-shaped cluster. No downed wood was found on 40% of sample plots, and 10% of the sample plots contributed 50% of the total fallen volume. Evans and Ducey (2010) used statistics from Böhl and Brändli (2007) to calculate that 95% confidence limits of ±20% on downed wood volume for an inventory stratum would require slightly over 600 sample points, or a total of nearly 20 km of sample line.

Because downed wood is rarely the sole or even primary focus of a forest inventory, an LIS design must be compatible with the designs used for other forest attributes and may also need to support inferences at multiple scales. In the simplest case, where the number of sample points in an inventory is driven by the desired precision for some other attribute (say, timber volume), some knowledge of the total length of LIS line needed and the number of sample points to be installed can help guide the LIS design. For example, if one supposes that a total of 2 km of sample line will be needed in a typical hardwood forest in the northeastern United States, and a timber inventory is planned with $n = 40$ horizontal point samples, then the length needed per sample line can be computed as 2000 m/40 points, or 50 m/point. A simple design in which a sample line is run 25 m from the sample point on a randomly chosen bearing, and another line is run 25 m from the sample point on the backsight of the first, will be sufficient. However, in some cases it may be desirable to use individual sample lines to draw some inferences about the spatial patterns or ecological associations of downed wood. In these situations, short sample lines may not be enough; individual sample lines in excess of 100 m long may be preferable (Harmon and Sexton, 1996; Woldendorp et al., 2004). Conversely, ease of installation, quick work when no wood is actually present, and minimization of nonsampling errors all argue for many short lines rather than a few long ones. Ringvall and Ståhl (1999a) studied field errors in LIS and found that field personnel tended to stray from the correct sampling line, avoiding pieces that should have been tallied. Such errors are easier to minimize when sample lines are short.

Finally, there is the question of whether a single straight line should be preferred or whether the sample line should be laid out in some other geometric configuration. Although the straight-line design is simple to implement, a wide variety of other approaches have been used. Most often, these have been motivated by the hope that they will ameliorate any bias when pieces of downed wood are not randomly oriented (though as discussed above, this assumption is unnecessary if the sample lines themselves are oriented randomly). For example, the US FIA protocol uses a Y-shaped cluster of three sample lines that takes advantage of the orientation and position of the fixed-area subplots for standing trees; the placement of three sample lines at 120° angles to each other provides some protection against directional bias (Woodall and Williams, 2005; Woodall et al., 2008). Other examples include equilateral triangles (Delisle et al., 1988) and an L-shaped layout used by the British Columbia Ministry of Forests (Marshall et al., 2000). Usually, estimation in these designs is conducted by treating the cluster as equivalent to a single line of the same total length; when pieces cross more than one line in the cluster, each crossing is treated as an independent event. Gregoire and Valentine (2003) and Affleck et al. (2005a) are critical of this approach, and give alternative estimators to be used when clusters of sample lines are employed. Affleck et al. (2005b) show that the traditional reflection method does not correct slopover bias in such designs, though the proposed remedy seems challenging to implement in the field. Alternative perspectives have been offered by Barabesi (2007) and Van Deusen and Gove (2011). Gregoire and Valentine (2003) argued that the statistical advantages of such clusters have not been established objectively. Woldendorp et al. (2004) found no increase in precision when samples of equal length were laid out in clusters to capture orientation variability. However, the use of better estimating equations, such as those proposed by Van Deusen and Gove (2011), could theoretically change the picture. The efficiency of LIS designs that are not simple lines, and the best equations to use with data from those designs, remains a contested area in the biometrics literature.

12.2.4. Adaptation for Fine Fuels

Much of the interest in downed woody material focuses on its role in wildland fire, and fine fuels (typically those <7.5 cm in diameter) are critically important for predicting flame height and rate of spread under typical conditions. At least superficially, the sampling issues for fine fuels are similar to those associated with coarser downed wood, and the most important parameter (dry weight or biomass) can be approached in an identical fashion. However, the smaller diameters, irregular and patchy distribution, and need to streamline fieldwork led Brown (1971, 1974) to develop a simplified inventory method closely related to LIS. The key simplifications consist of

- tallying material along a sample line using slope distance, rather than horizontal distance;
- tallying material by diameter class only, rather than measuring diameter or cross-sectional area, and using an assumed average diameter within the class based on prior measurement in another population;
- assuming an average angle of inclination based on size class, rather than measuring α for each piece; and
- taking an assumed representative bulk density for each species and size class.

In the United States, fine fuels are typically tallied in 0–0.25 in., 0.25–1 in., and 1–3 in. classes, which correspond to 1-, 10-, and 100-h moisture timelags (Deeming et al., 1972).

Brown (1974) details the field procedures and computations associated with this simplified method, and also gives practical guidance on selection of line lengths. For example, in all but the heaviest slash, a sample line 6 ft long is recommended for 1- and 10-h fuels, and a line 10–12 ft long is recommended for 100-h fuels. Brown (1974) specifies that 15–20 sample lines should be located in areas smaller than 50 acres. This aggregate line length is substantially shorter than what would be needed for coarse material in similar habitats.

12.3. ANGLE GAUGE METHODS

With increasing interest in the ecology of downed wood and in carbon accounting, many foresters and ecologists also became familiar with the frustrations of fixed-area plot sampling and with the time investment required for adequate estimates from LIS. This stimulated a search for sampling alternatives, and a variety of new PPS designs were developed.

12.3.1. Transect Relascope Sampling

Ståhl (1997, 1998) developed transect relascope sampling (TRS) to address excessive line lengths required by LIS, especially in European forests where downed wood is usually sparse. In TRS, the term "relascope" refers to an angle gauge generally, not to a Spiegel-relascope specifically. Indeed, in TRS the angles used are usually in the tens of degrees—far wider than those used for horizontal point sampling of standing trees. TRS uses sample lines, typically laid out in a fashion similar to those for LIS. Pieces of downed wood are tallied not only if they cross the line, but also if from any point on the line, the two ends of the piece appear wider than the angle gauge.

The zone within which a piece appears wider than the angle gauge is shaped as the union of two overlapping circles, one on each side of the axis that connects the two ends of the piece. In point relascope sampling (PRS) (Section 12.3.2), this blob-shaped zone serves as the inclusion zone for the piece; in TRS, a piece is eligible for inclusion when the sample line crosses this blob, so the actual inclusion zone is larger. The degree of overlap depends on the angle of the gauge (Fig. 12.3). For any given angle, the linear dimension of the blob is proportional to piece length, and it can be shown that TRS samples pieces with probability proportional to their length (Ståhl, 1998). When a 90° (or $\pi/2$ rad) gauge is used, the two circles overlap completely, and the blob collapses to a single circle, with diameter equal to piece length, centered on the midpoint of the piece. With a 90° angle, piece orientation becomes irrelevant to its inclusion, so the question of piece or sample line orientation disappears (Ståhl, 1997). This eliminates much of the rationale for geometric sample line cluster designs seen in LIS, along with the statistical complications such designs create.

In its original development, TRS used lines that spanned the entire tract or stratum being inventoried. Ringvall and Ståhl (1999b) outline the field procedures to be used when short

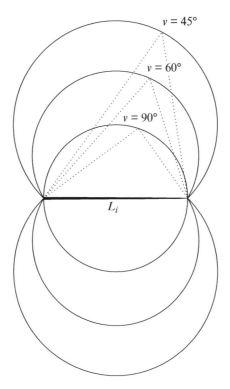

FIG. 12.3. A piece of downed wood and its corresponding TRS blobs (inclusion zones), equivalent to PRS inclusion zones, for three different gauge angles.

sample lines are employed. Pieces are only considered for inclusion in the tally if their center point falls within the region defined by boundaries perpendicular to the sample line, and passing through the ends (analogous to the way trees are counted in HLS and VLS, see Section 11.3). Because of the oblong shape of the TRS blob, it is possible that the blob for a piece will cross the extension of the sample line past its end, even though the center of the piece is within the zone. Thus, a limited amount of checking past the end of the sample line is needed for certain pieces to correctly determine their inclusion. TRS with a short sample line, and the corresponding inclusion zones for a series of pieces, is shown in Figure 12.4. Ståhl et al. (2002) detail the corrections for sloping pieces and terrain in TRS; here, we assume that the gauge is operated horizontally, sighting on the projection of the two ends of the piece into a horizontal plane. However, as Ståhl et al. (2002) point out, for moderate angles the effect of piece's slope on estimates in TRS can often be ignored. Moreover, we do not deal explicitly with slopover bias; the mirage method for LIS detailed by Gregoire and Monkevich (1994) can also be applied to TRS, so field correction is easy.

Let ν be the angle of the gauge in degrees, and let S be the length of the sample line just as in LIS. Let w_i be the width of the blob for a piece, measured perpendicular to the sample line, and let θ_i be the angle between the piece and the sample line, in degrees ($\theta_i = 0$ when the piece is parallel to the line). Then, w_i can be calculated as

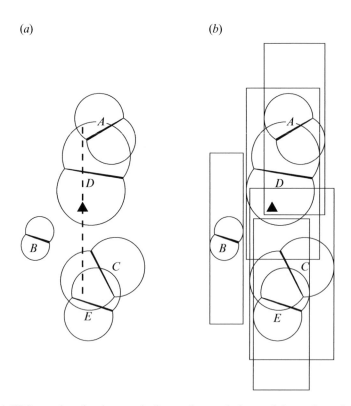

(a) (b)

FIG. 12.4. A TRS sample point, its sample line, and several pieces of downed wood (a). Pieces A, C, and D appear wider than the gauge from at least one point on the line, so those pieces are tallied. Piece E appears wider, but its center is past the end of the line, so it is not tallied. The corresponding inclusion zones are shown in (b).

$$w_i = L_{\text{horizontal},i}\left(\frac{1}{\sin v} + \cot v \cos\theta_i\right)$$

Since the area of the inclusion zone for a piece equals $S \cdot w_i$, we can calculate the piece factor PF_i as

$$PF_i = \frac{A}{Sw_i} = \frac{A}{SL_{\text{horizontal},i}\left(\dfrac{1}{\sin v} + \cot v \cos\theta_i\right)}. \qquad (12.14)$$

When working in SI units, that is, all length measurements are in meters, $A = 10,000$, and PF_i is in pieces/ha, and when working in imperial units, all length measurements are in feet, $A = 43,560$, and PF_i is in pieces/acre. Direct use of eq. (12.14) would require measuring θ_i for all pieces. This is a challenge but is not as fussy as in LIS because PF_i does not approach infinity as θ_i approaches zero as it would in LIS. Still, it may be viewed as an unnecessary measurement. If the sample lines are laid out on a random orientation, if we are willing to assume that pieces have random orientation, or if $v = 90°$, then, just as in LIS, we may integrate over the possible values of θ_i to obtain the expected value of w_i as

$$E[w_i] = L_{\text{horizontal},i}\left(\frac{1}{\sin v} + \frac{2}{\pi}\cot v\right)$$

and the piece factor becomes

$$PF_i = \frac{A}{S \cdot L_{\text{horizontal},i}\left(\dfrac{1}{\sin v} + \dfrac{2}{\pi}\cot v\right)} \qquad (12.15)$$

Since $\sin(90) = 1$ and $\cot(90) = 0$, when $v = 90°$ eq. (12.15) simplifies to

$$PF_i(v = 90) = \frac{A}{S \cdot L_{\text{horizontal},i}} \qquad (12.16)$$

Again, the LF is the constant sampling proportion, and is calculated as

$$LF = L_{\text{horizontal},i} \times PF_i = \frac{A}{S\left(\dfrac{1}{\sin v} + \dfrac{2}{\pi}\cot v\right)}, \qquad (12.17)$$

and we see that the LF is a constant independent of any piece attributes and depends only on the length of the sample line and the gauge angle. Just as in LIS, $PF_i = LF / L_{\text{horizontal},i}$ provides a convenient shortcut formula for calculating the piece factor when it is needed.

For example, suppose we are performing TRS with a gauge angle of 60°, on sample lines that are 50 m long. Then, the LF is

$$LF = \frac{10,000}{50\left(\dfrac{1}{0.866} + \dfrac{2}{\pi}0.577\right)}$$

$$= \frac{10,000}{50(1.522)}$$

$$= 131.4\,\text{m/ha}$$

Under this design, a tallied piece that is 10 m long in the horizontal plane represents 131.4/10 = 13.14 pieces/ha, and a tallied piece that is 20 m long in the horizontal plane represents 131.4/20 = 6.07 pieces/ha.

12.3.2. Point Relascope Sampling

With the exception that the object being sampled is an approximately linear piece of downed wood, rather than an approximately circular bole of a standing tree, TRS is very much like horizontal line sampling (Sections 9.3.4 and 11.3.1). Thus, a natural extension of TRS is to examine the use of a sample point in place of a sample line, just as in horizontal point sampling. The resulting technique is called point relascope sampling. The original theoretical development of PRS is presented by Gove et al. (1999); a less-technical summary, with examples in imperial units, is given by Gove et al. (2001).

In PRS, a wide-angle gauge (with an angle in the tens of degrees) is used to sight on the two ends of a piece of downed wood. However, instead of sighting on pieces while moving along a sample line, the gauge is turned around a sample point (just as an angle gauge is turned about the point in horizontal point sampling of standing trees). The blobs shown in Figure 12.4 become the inclusion zones for the individual pieces. Unlike TRS, there is no assumption about intersection angles because a point rather than a sample line is the basic sample unit. The sampling situation in PRS is illustrated in Figure 12.5. When a piece is borderline, measuring the distance from the sample point to each of the two ends of the piece, and the piece length, allows checking the inclusion of the piece via the law of sines (Gove et al., 1999, 2001).

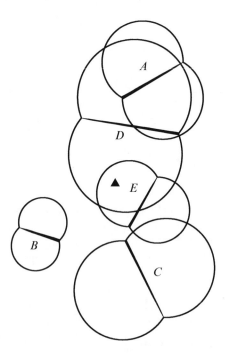

FIG. 12.5. A PRS sample point and several pieces of downed wood along with their inclusion zones. The sample point falls within the inclusion zones for pieces *D* and *E*, so those pieces are tallied.

Ståhl et al. (2002) present methods for correction for sloping pieces and terrain in PRS. As before, for simplicity, we assume here that the angle gauge is operated horizontally, sighting on the projection of the ends of pieces into a horizontal plane at eye level. Most of the standard techniques for boundary slopover correction are applicable to PRS. However, because of the irregular inclusion zone shape, the walkthrough method (Ducey et al., 2004) may be simplest.

As with TRS, let ν be the angle of the gauge in degrees. Let

$$\phi = \frac{\pi\left(1 - \nu/180\right) + \sin\nu\,\cos\nu}{2\sin^2\nu}$$

Then, the area of the blob (the inclusion zone area) associated with an individual piece can be calculated as

$$a_i = \phi L^2_{\text{horizontal},i} \tag{12.18}$$

From the form of eq. (12.11), it is easy to see that PRS is a PPS technique that samples pieces of downed wood with probability proportional to length squared. The piece factor can be calculated as

$$PF_i = \frac{A}{\phi L^2_{\text{horizontal},i}} \tag{12.19}$$

and the corresponding length-square factor (LSF) is

$$LSF = \frac{A}{\phi} \tag{12.20}$$

As above, when working in SI units $A = 10,000$ and when working in imperial units, $A = 43,560$ in eqs. (12.19) and (12.20). Whenever LSF is known in advance, a simple shortcut formula for the piece factor is $PF_i = LSF / L^2_{\text{horizontal},i}$.

As an example, suppose we are conducting PRS with a gauge angle of 60°. Then,

$$\phi = \frac{\pi\left(1 - 60/180\right) + \left(0.866\right)\left(0.5\right)}{2\left(0.866^2\right)}$$

$$= 1.685$$

and the LSF equals $10,000/1.685 = 5,935$. Under this design, a tallied piece that is $10\,\text{m}$ long in the horizontal plane represents $5935/10^2 = 59.35$ pieces/ha, and a tallied piece that is $20\,\text{m}$ long in the horizontal plane represents $5935/20^2 = 14.84$ pieces/ha.

12.3.3. Gauge Construction and Choice of Angle

For either TRS or PRS, it is extremely useful to have a suitable angle gauge. In principle, construction of a gauge can be very simple. Ståhl et al. (2001) and Gove et al. (2001) both depict easily constructed but slightly different instruments. Both feature a pair of sighting pins, held at a fixed and known distance on a rigid beam. The beam should be held perpendicular to the line of sight at a fixed distance from the observer's eye. Although a single cord or chain attached to the center of the beam would suffice to ensure a fixed distance, using two cords of equal length, attached to the beam at positions equidistant from the center and also to a fingerhold to be positioned near the eye, guarantees the beam is perpendicular. Such a gauge can be constructed out of any reasonably durable and inexpensive material. For example, Gove et al. (2001) used a piece of seasoned hardwood, with finish nails as sighting pins.

The angle associated with the gauge can be described in terms of ν, the inclusion angle in degrees. However, for purposes of construction, it is often simpler to work in terms of the width to reach ratio, calculated as

$$\text{Width} : \text{reach} = 2\tan\left(\frac{\nu}{2}\right)$$

This ratio is the ratio of the distance between the sighting pins to the distance between the observer's eye and the beam. For example, if the pins are spaced 50 cm apart and the beam is designed to be held 40 cm from the observer's eye, the width to reach ratio is 50:40 or 1.25. Note that Gove et al. (2001) present similar design values, only using the inverse of one-half of this ratio. Because neither the length per unit area nor the length-squared per unit area are of inherent interest, there is little value in ensuring that the LF or the LSF are round numbers. It is more important that the geometric relationship implied by the gauge is intuitive and that the gauge can be constructed easily and accurately. Table 12.2 gives useful figures for gauges constructed either to a round value of ν or to a rational value of the width to reach ratio across the range of angles that are likely to be useful in practice.

TABLE 12.2. Useful Constants and Relationships for Gauges to Be Used in Transect Relascope Sampling and Point Relascope Sampling

			SI		Imperial	
ν (°)	φ	Width : Reach	LF (m/ha)	LSF (m²/ha)	LF (ft/acre)	LSF (ft²/acre)
90	0.7854	2.00	10,000/S^a	12,732	43,560/S	55,462
80	0.9880	1.68	8,868/S	10,122	38,628/S	44,091
70	1.2691	1.40	7,717/S	7,880	3,3614/S	3,4324
60	1.6849	1.15	6,569/S	5,935	28,615/S	25,853
50	2.3528	0.93	5,436/S	4,250	23,679/S	18,514
45	2.8562	0.83	4,876/S	3,501	21,240/S	15,251
40	3.5528	0.73	4,321/S	2,815	18,821/S	12,261
35	4.5603	0.63	3,770/S	2,193	16,421/S	9,552
30	6.1020	0.54	3,223/S	1,639	14,040/S	7,139
90.0	0.7854	2.00	10,000/S	12,732	43,560/S	55,462
73.7	1.1520	1.50	8,148/S	8,680	35,491/S	37,812
64.0	1.4965	1.25	7,028/S	6,682	30,615/S	29,108
53.1	2.1049	1.00	5,789/S	4,751	25,216/S	20,694
41.1	3.3763	0.75	4,444/S	2,962	19,357/S	12,902
28.1	6.9244	0.50	3,013/S	1,444	13,126/S	6,291

[a]S=sample line length in meters when calculating LF in SI units, and length in feet when calculating LF in imperial units.

In many temperate and boreal forests, a gauge with a 90° angle (or 2 : 1 width to reach ratio) is likely to be most useful for TRS, while a gauge between 50° and 60° (perhaps with a 1.25 : 1 or 1 : 1 width to reach ratio) is likely to be most useful for PRS. When either TRS or PRS is implemented using a 90° angle, a gauge is not necessarily even needed: one may visualize driving a spike through the center of a candidate piece. If spinning the piece around the spike would cause the piece to sweep over the sample line or point, the piece should be tallied.

12.3.4. Estimating Downed Wood Parameters

For either TRS or PRS, the downed wood parameters most likely to be of interest (number of pieces per unit area, volume, biomass, or carbon content) are approached through a conventional factor approach, starting with the piece factor. Recall that for an arbitrary attribute X the expansion factors XF_i for individual pieces are obtained by multiplying the value of the characteristic, X_i, by the piece factor (e.g., eq. (12.6)):

$$XF_i = X_i \cdot PF_i$$

This expansion factor XF_i represents the contribution of an individual tallied piece to the total estimate of attribute X per unit area that we would obtain from a single sample (a line in the case of TRS, or a point in PRS). If we have tallied a total of m pieces of downed wood on n sample points or lines, the amount of X per unit area is estimated as (e.g., eq. (12.7))

$$\hat{X} = \frac{1}{n} \sum_{i=1}^{m} XF_i$$

Neither TRS nor PRS creates an automatic "crossing point," as in LIS, that would create a Monte Carlo procedure leading to cancellation of terms and a simplification of measurements. Thus, in their original forms, both TRS and PRS require measuring the dimensions of a piece of downed wood and computing the volume using an assumed formula (e.g., Section 6.7), and this constitutes an important source of potential bias. To overcome these limitations in the context of PRS, Gove et al. (2005) devised critical PRS. In this variant, the angle gauge is held so that one sighting pin aligns with one end of a tallied piece, and the position on the piece that aligns with the second pin is taken as a measurement point for cross-sectional area and decay class. Gove et al. (2005) provide estimators and a simulation study examining the efficiency of aligning the pin with the large end, the small end, or using both points and averaging. Either PRS or TRS could be employed in a double-sampling approach (Ringvall et al., 2001), allowing for more detailed measurements on a subset of pieces while a simple piece count serves as a first-stage sample. Gove et al. (2002) outline a multistage procedure involving PRS and randomized branch sampling for situations where branched pieces are common.

12.3.5. Practical Aspects

TRS does increase the piece count over what would be expected using LIS with similar line lengths and is less sensitive to piece orientation than LIS even when a gauge angle other than 90° is used. TRS clearly showed superior time efficiency to LIS in a series of early computer simulations (Ståhl, 1997, 1998). However, Ringvall and Ståhl (1999b) found troubling downward bias in field estimates using TRS, along with significant inter-observer variation, especially when a narrow angle gauge was used. These biases may have been associated with inappropriate use of the gauge and/or nondetection errors; the observers in their study were experienced cruisers but had not received extensive training in TRS. Ringvall and Ståhl (1999b) suggested that TRS might be most useful as the first phase in a multistage sampling effort (with later stages correcting any observer bias). They also suggested very wide angle gauges (80–90°) when considerable downed wood is present, when visibility is poor due to understory vegetation, or when approximately unbiased estimates are desired. Given the theoretical advantages of TRS with a 90° gauge, especially in comparison with LIS, further field evaluation in other systems may be warranted.

Of the newer field methods for downed wood, PRS has been the most extensively tested in North American conditions. Brissette et al. (2003) found no detectable bias in comparison with LIS in managed and recently harvested conifer stands in Maine. Jordan et al. (2004) found PRS to be typically more efficient than LIS or fixed-area sampling for volume in northern hardwood stands, although nondetection was apparently problematic using a gauge with a 1 : 2 width to reach ratio. Kenning (2007) found similar results in coastal Douglas-fir stands in British Columbia as well as in ponderosa pine and lodgepole pine stands in Colorado. (However, the perpendicular distance sampling (PDS) approach (Section 12.4) outperformed PRS under most conditions in that study.) In a study in Finland, PRS was the most efficient among several methods (including fixed plots, TRS, and adaptive cluster sampling; but not PDS) (Pesonen et al., 2009). PRS has also been used successfully in mapping fuels in the Chihuahuan desert (Poulos, 2009). However, PRS does present some challenges. Many tallied pieces are at some distance from the sample point, so PRS is most efficient with field crews of two or three (in which one observer can stand at the sample point, sight on pieces with the gauge, and record, while one or two assistants perform measurements, search for decayed or hidden pieces, and help the observer correctly locate piece ends). Gove et al. (1999) suggest PRS may not be the most suitable method when one person is working alone. Taken together, these experiences suggest that with a moderately wide angle gauge and under the right conditions, PRS can be a useful technique for downed wood inventory.

12.4. PERPENDICULAR DISTANCE SAMPLING (PDS)

One shortcoming of PPS sampling approaches such as LIS, PRS, and TRS is that sampling is proportional to a size (length or length squared) that is correlated with volume, weight, and carbon or nutrient content but is not, in itself, of significant practical interest. By contrast, the class of methods known as PDS was designed with just such a goal in mind. In the original development of PDS by Williams and Gove (2003), sampling is with probability proportional to volume: a simple count of pieces gives a direct estimate of volume per unit area. Subsequent developments of PDS have led to designs that sample with probability proportional to other attributes or that provide practical advantages over the initial version in some circumstances. As a result, this class of methods is extremely flexible and has also proven to be efficient in the field, with measurements that can often be performed effectively by a single worker.

12.4.1. PDS for Volume

In PDS, the observer sights on pieces of downed wood around a sample point, just as in horizontal point sampling. But instead of sighting at breast height on a standing tree, one sights on each piece of wood at the point where the line of sight from the sample point is perpendicular to the horizontal axis connecting the two ends of the piece. A simple geometric rule relating the size of the piece at the perpendicular point, to the limiting distance R_i, is used to determine whether the piece should be tallied or not. When PDS is done with probability proportional to volume, the rule that gives R_i is

$$R_i = \frac{A_\perp}{k_A} \tag{12.21}$$

where $A_\perp$ = the cross-sectional area of the piece (m² or ft²)

   k_A = a suitably chosen constant that plays a role similar to the gauge constant in horizontal point sampling

   (Note that Williams and Gove (2003), and later authors writing on PDS, have usually used k to denote the quantity equal to $1/k_A$ here; the notation in this chapter is used for consistency with that of the gauge constant in Chapter 9 and elsewhere in this text. Note also that k_A has units of meters in SI units, and feet in imperial units.)

For sloping or forked pieces, $A_\perp$ is taken as the sum of the cross-sectional areas measured in the vertical plane that includes the line of sight (taking the cross-sectional areas perpendicular to the piece, as normally done, and correcting based on piece slope α is also acceptable; Williams et al., 2005). Whenever the actual distance from the sample point to the perpendicular point on a piece is less than or equal to R_i, the piece is tallied. The sampling situation for a population of pieces, and their corresponding inclusion zones, is shown in Figure 12.6.

Consider the inclusion zone for an individual piece in more detail, as illustrated in Figure 12.7. The inclusion zone is symmetric about the piece's axis, but its width perpendicular to that axis equals $2R_i$, and it varies as the cross-sectional area of the piece varies with length. Actually, calculating the inclusion zone area would be exceedingly tedious; it would require measurements of many values of A along the axis just to get a close approximation. Indeed, the exact area of the inclusion zone is

$$a_i = \int_{h=0}^{L_i} 2R_i(h)\,dh$$

$$= \int_{h=0}^{L_i} \frac{2}{k_A} A(h)\,dh$$

$$= \frac{2}{k_A} \int_{h=0}^{L_i} A(h)\,dh$$

But we also know (Section 6.1.4) that

$$V_i = \int_{h=0}^{L_i} A(h)\,dh$$

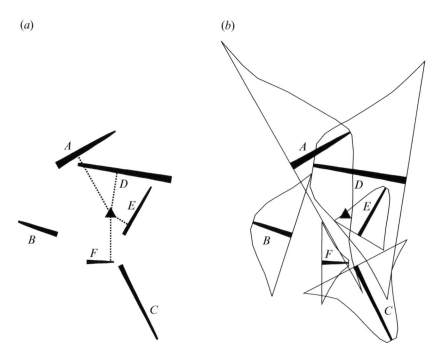

FIG. 12.6. A PDS sample point, and several pieces of downed wood (*a*). The line of sight from the sample point to the perpendicular point on each piece is indicated by a dashed line; pieces *B* and *C* have no perpendicular point. The corresponding inclusion zones are shown in (*b*); the sample point falls within the zones for pieces *A*, *D*, and *E*, so those pieces are tallied.

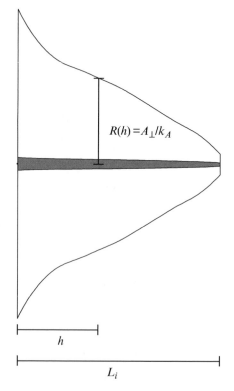

$$R(h) = A_\perp / k_A$$

FIG. 12.7. Close-up of the inclusion zone for a piece of downed wood using PDS with probability proportional to volume.

and, therefore, we have

$$a_i = \frac{2V_i}{k_A}$$

and it is immediately apparent that PDS has sampled downed wood with probability proportional to piece volume. If we were to measure V_i exactly, we would obtain a piece factor as

$$PF_i = \frac{10,000k_A}{2V_i} \quad \left(SI \; units \right) \qquad (12.22a)$$

$$PF_i = \frac{43,560k_A}{2V_i} \quad \left(imperial \; units \right) \qquad (12.22b)$$

We will see in Section 12.4.5 how to obtain a satisfactory piece factor without actually measuring the volume of the piece. However, recalling that the volume factor VF_i is just PF_i times V_i, we arrive immediately at expressions for the volume factor in PDS:

$$VF = \frac{10,000}{2}k_A \quad \left(SI \; units \right) \qquad (12.23a)$$

$$VF = \frac{43,560}{2}k_A \quad \left(Imperial \; units \right) \qquad (12.23b)$$

Now recall that in horizontal point sampling if one tallies m trees on n points using a given basal area factor (BAF), one can estimate basal area per unit area immediately as (m/n)BAF, and no measurements are needed on the individual trees (except any measurements needed to check borderline trees). In PDS with probability proportional to volume, if m pieces are tallied on n sample points using a given VF, one can estimate volume per unit area immediately as (m/n)VF, and no measurements are needed on the individual pieces (except any measurements needed to check borderline pieces). With experienced timber cruisers, a quick check of an ocular estimate of piece diameter at the perpendicular point (which gives an implied value of $A_\perp$) and an ocular estimate of distance between the perpendicular point and the sample point, against a table of limiting distances, is adequate to establish whether or not most pieces should be included.

As an example, Table 12.3 gives limiting distances for VF=200 in imperial units. A cruiser uses this table at a sample point and sweeps around starting from the north to spot pieces of downed wood:

- First, she spies two small pieces that are approximately 3" at the perpendicular point and about 10 ft away. These pieces are clearly out, so there is no need to measure them.
- Another piece is 12" at the perpendicular point, which is only 4' from the sample point. This piece is clearly in and is tallied as decay class I based on its attributes at the perpendicular point.
- The next piece encountered sweeping around the point is borderline: it appears approximately 10" at the perpendicular point and approximately 60' away. The cruiser walks over to the piece, measures $D_\perp$ (the piece is round to a close approximation,

TABLE 12.3. Limiting Distances (R_i) for PDS with Probability Proportional to Volume, Using VF = 200 ft²/acre (k_A = 0.009183 ft)

$A_\perp$ (ft²)	$D_\perp$ (in.)	R_i (ft)	$A_\perp$ (ft²)	$D_\perp$ (in.)	R_i (ft)
0.0491	3	5.35	0.5454	10	59.40
0.0873	4	9.50	0.6600	11	71.87
0.1364	5	14.85	0.7854	12	85.53
0.1963	6	21.38	0.9218	13	100.38
0.2673	7	29.10	1.0690	14	116.42
0.3491	8	38.01	1.2272	15	133.64
0.4418	9	48.11	1.3963	16	152.05

Values of $D_\perp$ assume an unforked piece with a circular cross section in the vertical plane.

with an actual diameter of 11″), and quickly registers the distance back to the sample point using an ultrasound device (the distance is 65′, less than the limiting distance of 74.8′, so the piece is tallied as a decay class III based on its attributes).

- Another piece is heavily decayed and appears to be about 5″ in diameter at a distance of 14′—close enough, perhaps, to be borderline. When the cruiser goes to check this piece, she realizes that it is quite flattened due to decay: although 5″ across horizontally, it is only about 2.5″ deep at its thickest point, so it has only about half the cross-sectional area of around 5″ piece. Even with a generous estimate of its cross-sectional area, its limiting distance could not reach 10′, so this piece is clearly out. It is not tallied.

- Finally, the cruiser spots a freshly fallen tree. The line from the pith at the base of the tree, to the terminal apex, provides the central axis for the piece, which the cruiser judges to be no more than 40 ft away. The stem forks below the perpendicular point, so there are multiple stems to contend with. The main stem is 7″ at the perpendicular point, and two 4″ branches also cross the line of sight to the perpendicular point. Consulting the chart, the cruiser adds the cross-sectional areas of these three stems and estimates the total to be approximately 0.44 ft². Since a piece with cross-sectional area of 0.44 ft² has a limiting distance of just over 50′ and the piece axis is clearly closer than that, the piece is tallied as a single piece with decay class I.

In total, three pieces have been tallied (two of decay class I and 1 of decay class III), so multiplying by VF the cruiser quickly estimates that there are 600 ft³/acre of downed wood in total based on this sample point, of which 400 ft³/acre are in decay I and 200 ft³/acre are in decay III.

Note that as a point-based method no assumptions about downed wood orientation are involved in PDS. Moreover, no volume formula is assumed. Although other methods could be used for slopover correction, the walkthough method (Ducey et al., 2004) is the simplest with PDS in the field. At first, it would seem that PDS can only be used in practice to estimate volume as obtaining the piece factor would require tedious measurement of the volume of each piece. Indeed, this was initially believed to be true, but later work showed that PDS could be used as a general-purpose inventory technique, as outlined in Section 12.4.4.

12.4.2. Distance-Limited PDS

In ordinary PDS with probability proportional to volume, R_i varies proportionally to $A_\perp$. In other words, for pieces with an approximately circular cross section, R_i varies proportionally to $D_\perp^2$, where $D_\perp$ is diameter at the perpendicular point. One implication is that for large-diameter material—often precisely the material that is of the most ecological interest from a wildlife habitat or carbon accounting standpoint—R_i can quickly become quite large. For example, in Table 12.3, pieces with $D_\perp$ larger than about $11''$ have limiting distances that would exceed visibility in many forested conditions. In practice, that could necessitate a time-consuming search for large-diameter material at some distance from the sample point to avoid the possibility of nondetection bias. Alternatively, one could select a very large VF to keep search distances short; but that could result in so few pieces being tallied that a very large number of sample points would be needed to mitigate an unacceptable level of sampling variability.

Another possibility is to modify the limiting distance and impose a fixed upper limit irrespective of the cross-sectional area at the perpendicular point. The maximum distance is set in advance based on expected visibility and other operating conditions within the stand. For example, one might use a maximum distance of one chain in stands with relatively open understories, but use 0.5 chain in stands with dense understories or where travel is made difficult by steep slopes or heavy slash. This option, called "distance-limited PDS," is described in mathematical detail by Ducey et al. (2013b), who also present the results of an early field trial.

Capping the limiting distance in PDS changes the inclusion zone shape for pieces that include large-diameter portions (strictly speaking, for ones that include a portion where the cross-sectional area is large enough that the ordinary limiting distance would exceed the cap). For those pieces, it also means PDS no longer samples with probability proportional to volume. However, the change in field procedure and in volume estimation is relatively simple. First, we must solve for the value of $A_\perp$ where the ordinary limiting distance just equals the cap on limiting distance, R_{max}, by rearranging eq. (12.21):

$$A_{max} = k_A R_{max}$$

For example, with an imperial VF of $200\,\text{ft}^3/\text{acre}$, $k_A = 0.009183$. If we impose an R_{max} of one chain, or $66'$, then $A_{max} = 0.6061\,\text{ft}^2$, corresponding to a round piece with a diameter of just over $10.5''$. If we were to impose an R_{max} of half a chain, or $33'$, then A_{max} would be cut in half, to $0.3030\,\text{ft}^2$, corresponding to a round piece with a diameter of just under $7.5''$.

In distance-limited PDS, whenever a piece is tallied with $A_\perp \le A_{max}$, it is handled just as in ordinary PDS and the volume factor for that piece equals the ordinary volume factor VF. However, when a piece is tallied with $A_\perp > A_{max}$, its volume factor must be adjusted for the fact that its limiting distance is now R_{max}. The adjustment is relatively simple, but it does require that A_{max} be measured and recorded for the piece, no matter how close to the sample point it may be:

$$VF_i = VF \frac{A_\perp}{A_{max}} \tag{12.24}$$

Unless the piece is borderline, the distance from the sample point to the perpendicular point is not required, and no piece length measurement is needed to estimate volume. An example of a limiting distance table, with associated adjusted volume factors, is shown in Table 12.4.

Now suppose our cruiser uses this table, and as before sweeps around the sample point to identify candidate pieces by sighting on their perpendicular points:

- First, she sees a piece that is 24″ in diameter some distance from the sample point. But she quickly realizes that it is over a chain away. No matter how large it may be, this piece is out.

- Next, she spots a clump of three trees that has blown down. All three stems are about 6″ in diameter at their perpendicular point, and none of the perpendicular points are over 10′ away, so all three are tallied as decay class I. These stems are all smaller than A_{max}, so they are handled just as in ordinary PDS and no additional measurements are needed.

- Finally, there is a large stem fallen right next to the sample point. It is clearly in, and based on its characteristics it is tallied as a decay class IV. Because it is larger than A_{max}, its cross-sectional area is needed to compute its volume factor. Careful measurement establishes $D_{\perp} = 16$, and its adjusted volume factor is 461 ft³/acre.

Computation of the volume estimate associated with this sample point is also easy: three ordinary pieces have been tallied with VF=200, and one large piece with VF=461, so the total downed wood volume estimate is $3 \times 200 + 461 = 1061$ ft³/acre. Of that, 600 ft³/acre is in decay I and 461 ft³/acre is in decay IV.

12.4.3. PDS for Other Attributes

In principle, it is possible to develop a PDS technique for any attribute that can be represented as an integral of some measureable quantity over the length of a piece of downed wood. Of these, two are worth particular mention: surface area coverage and downed wood length per unit area. The first is of interest because (in conjunction with the techniques in Section 12.4.4) it offers an especially attractive alternative as a general-purpose sampling

TABLE 12.4. Limiting Distances (R_i) for Distance-Limited PDS with Probability Proportional to Volume, Using VF = 200 ft²/acre (k_A = 0.009183 ft) and an R_{max} of 66′

$A_{\perp}$ (ft²)	$D_{\perp}$ (in.)	R_i (ft)	VF_i (ft³/acre)	$A_{\perp}$ (ft²)	$D_{\perp}$ (in.)	R_i (ft)	VF_i (ft³/acre)
0.0491	3	5.35	200	0.5454	10	59.40	200
0.0873	4	9.50	200	0.6600	11	66.00	218
0.1364	5	14.85	200	0.7854	12	66.00	259
0.1963	6	21.38	200	0.9218	13	66.00	304
0.2673	7	29.10	200	1.0690	14	66.00	353
0.3491	8	38.01	200	1.2272	15	66.00	405
0.4418	9	48.11	200	1.3963	16	66.00	461

Values of $D_{\perp}$ assume an unforked piece with a circular cross section in the vertical plane.

technique. The second is of interest in part because of its close relationship with distance-limited PDS.

PDS for surface area was originally devised by Williams et al. (2005). Suppose that we use piece diameter at the perpendicular point, measured horizontally, in place of cross-sectional area in eq. (12.24). Then, the equation for the limiting distance becomes

$$R_i = \frac{D_{\text{horizontal},\perp} \times u}{k_D} \tag{12.25}$$

where k_D is a suitably chosen, unitless constant, and u is a unit conversion between those of diameter and of distance ($u=0.01$ when D is in centimeters and R is in meters; $u=1/12$ when D is in inches and R is in feet). Then, the formula for the inclusion zone area becomes

$$a_i = \int_{h=0}^{L_i} 2R_i(h)\,dh$$

$$= \int_{h=0}^{L_i} \frac{2}{k_D} D(h)\,dh$$

$$= \frac{2}{k_D} \int_{h=0}^{L_i} D(h)\,dh$$

where $D(h)$ is understood to be horizontally measured diameter at a distance h from the base of the piece, and it is evident that the inclusion zone area is just $(2/k_D)$ times the horizontally measured surface area coverage of the piece. The associated surface area factor can be calculated as

$$\text{SAF} = \frac{10,000}{2} k_D \quad (\text{SI units}) \tag{12.26a}$$

$$\text{SAF} = \frac{43,560}{2} k_D \quad (\text{imperial units}) \tag{12.26b}$$

For example, suppose we are working in SI units and use $k_D = 0.04$. (Note that this value of k_D is identical to the value $k=0.04$ for horizontal point sampling of standing trees with a BAF $4\,\text{m}^2/\text{ha}$ gauge; indeed, because R_i is proportional to piece diameter, we can use the same prism or relascope to provide an initial assessment of whether a piece is in or out, by sighting on the perpendicular point.) Then, the SAF for this PDS design is

$$SAF = \frac{10,000}{2} 0.04 = 200\,\text{m}^2 \,/\,\text{ha}$$

and every tallied piece under this design represents $200\,\text{m}^2/\text{ha}$, or a ground coverage of 2%. If m pieces are tallied on n sample points, then the aggregate estimate of surface area coverage is $(m/n) \times \text{SAF}$.

Alternatively, suppose we make R_i a constant for all pieces at all positions on the piece, irrespective of piece size or characteristics. (In a sense, this is what has been done in distance-limited PDS, but only for the large material.) Then, it is easy to see that the inclusion zone for each piece is just a rectangle, with length L_i and width $2R$ (the factor of 2 remains because we can tally the piece from either side). Since the inclusion zone area equals $2RL_i$, it is easy to see that pieces have been tallied with probability proportional to length (Gove et al., 2012). Moreover, the LF under this design is

$$LF = \frac{10,000}{2R} \quad \left(\text{SI units}\right) \tag{12.27a}$$

$$LF = \frac{43,560}{2R} \left(\text{imperial units}\right) \tag{12.27b}$$

So, for example, if we were to employ fixed-distance PDS with a very short limiting distance of 6 ft, the LF would be LF=43,560/12=3630 ft/acre. Such an approach might be attractive in a survey of heavy slash or other logging residue. In particular, fixed-distance PDS has the same probability proportional to length attributes as LIS, but without any directional assumptions whatsoever.

12.4.4. Estimating Multiple Downed Wood Attributes

PDS was found quite early on to be a very efficient sampling approach, but was also thought to be useful only for a single variable at a time. For example, PDS with probability proportional to volume was thought to be useful only for estimating volume (perhaps with mass estimated later by multiplying by an assumed wood density) because obtaining the piece factor directly would require measurement of piece volume, which would destroy the efficiency of the technique (Williams and Gove, 2003). However, Ducey et al. (2008) showed that there is a very close connection between PDS and importance sampling, and that PDS can be used as an omnibus sampling method. In particular, suppose that pieces of downed wood (or portions of pieces, as in distance-limited PDS) are sampled with probability proportional to some characteristic X_i, with a constant associated factor XF. But we are interested in some other characteristic, Y_i. We assume

$$X_i = \int_{h=0}^{L_i} f(h)\,dh$$

$$Y_i = \int_{h=0}^{L_i} g(h)\,dh$$

For example, if we are sampling pieces with probability proportional to volume, then f is cross-sectional area A, because volume is the integral of cross-sectional area over length. If we are interested in surface area coverage, then g is horizontally measured diameter (after conversion between the usual diameter measurement units and those of length) because surface area coverage is the integral of diameter over length. Then, Ducey et al. (2008) prove that

$$YF_i = XF\frac{g_\perp}{f_\perp} \tag{12.28}$$

provides an unbiased factor for Y. In the example case where pieces are tallied with a constant volume factor, the quantity $g_\perp / f_\perp$ is horizontal diameter divided by cross-sectional area, with both measured at the perpendicular point; care must be taken here as diameter is often measured in different units than length.

Indeed, this approach can even be used to develop a piece factor PF_i for each piece, without actually measuring piece volume (in the case of constant VF) or surface area (in the case of constant SAF). Let $g = 1/L_i$ over the entire piece length. Then,

$$Y_i = \int_{h=0}^{L_i} \frac{1}{L_i} dh = 1$$

which is precisely the number of pieces in a piece. Therefore, when sampling with a constant volume factor,

$$PF_i = \frac{VF}{A_\perp L_i} \tag{12.29}$$

which requires measuring only the piece length and the cross-sectional area at the perpendicular point. Likewise, when sampling with a constant surface area factor,

$$PF_i = \frac{SAF}{u \times D_{\text{horizontal},\perp} L_i} \tag{12.30}$$

which requires measuring only the piece length and the horizontal diameter at the perpendicular point. Here, u is 0.01 m/cm when working in SI units, and 1/12 ft/in. in imperial units; it is needed to ensure the numerator and denominator of the fraction are in identical units.

The base factors, and derived factors for pieces, length, surface area, volume, and mass for all three common PDS designs (constant LF, constant SAF, and constant VF), are summarized in Table 12.5.

Note that for nearly all factors piece length L_i, horizontally measured diameter $D_{\text{horizontal},\perp}$, and cross-sectional area $A_\perp$ are sufficient to estimate all of the downed wood attributes except mass. Furthermore, note that the quantity $\rho_{\text{wood}} A_\perp$ is just mass per unit length; thus, if it is desired to establish the mass factor based on direct measurement, the mass of a section of known length, centered on the perpendicular point, is sufficient. If wood density is to be estimated by the use of tables or functions of decay class, then decay class at the perpendicular point is needed: it is not necessary to try to estimate an overall decay class for a heterogeneous piece. Multiplying the mass factor by the concentration per unit mass of carbon or any other constituent gives the appropriate factor; see Valentine et al. (2008) for further discussion of carbon estimation using PDS.

The basic relationship in eq. (12.28) is valid even when the sampling approach switches between different pieces or between portions of a piece, as it does in distance-limited PDS. (In distance-limited PDS, factors from the volume column of Table 12.5 are used for those pieces with $A_\perp \leq A_{\text{max}}$, and factors from the length column are used for pieces with $A_\perp > A_{\text{max}}$.) All that is required is to take the necessary measurements on each piece to allow calculating XF_i for each factor X that is of interest, on each piece. Then, as always under the factor approach, if data are collected on m pieces at n sample points, the equation

$$\hat{X} = \frac{1}{n} \sum_{i=1}^{m} XF_i$$

yields an unbiased estimate of the amount of attribute X per unit area.

TABLE 12.5. Limiting Distances, Base Factors, and Derived Factors for Other Downed Wood Attributes, When PDS Is Implemented Using Probability Proportional to Length, to Surface Area, and to Volume

	Sampling Is with Probability Proportional to		
	Length	Surface Area	Volume
Limiting distance	$R = $ constant	$R_i = \dfrac{u \times D_{horizontal,\perp}}{k_D}$	$R_i = \dfrac{A_\perp}{k_A}$
Base factor	$LF = \dfrac{\text{Unit area}}{2R}$	$SAF = \dfrac{\text{Unit area} \times k_D}{2}$	$VF = \dfrac{\text{Unit area} \times k_A}{2}$
Piece factor	$PF_i = \dfrac{LF}{L_i}$	$PF_i = \dfrac{SAF}{u \times L_i D_{horizontal,\perp}}$	$PF_i = \dfrac{VF}{L_i A_\perp}$
Length factor	LF	$LF_i = \dfrac{SAF}{u \times D_{horizontal,\perp}}$	$LF_i = \dfrac{VF}{A_\perp}$
Surface area factor	$SAF_i = LF \times D_{horizontal,\perp}$	SAF	$SAF_i = VF \dfrac{u \times D_{horizontal,\perp}}{A_\perp}$
Volume factor	$VF_i = LF \times A_\perp$	$VF_i = SAF \dfrac{A_\perp}{u \times D_{horizontal,\perp}}$	VF
Mass factor	$MF_i = LF \times \rho_{wood} A_\perp$	$MF_i = SAF \dfrac{\rho_{wood} A_\perp}{u \times D_{horizontal,\perp}}$	$MF_i = VF \times \rho_{wood}$

Unit area equals $10,000\,m^2/ha$ when working in SI units, and $43,560\,ft^2/acre$ when in imperial units. The unit conversion u equals $0.01\,m/cm$ when in SI units, and $1/12\,ft/in.$ when in imperial units.

12.4.5. Choosing a Design and Factor

When only a single downed wood attribute is of primary interest, the choice of a PDS design is quite simple. For example, if one only wishes to obtain quick estimates of volume, perhaps by species and decay class, so that these estimates can be multiplied by a representative carbon content to obtain carbon per unit area, the original form of PDS with probability proportional to volume is a natural choice. An intelligent choice of volume factor, and perhaps an upper distance limit, will make the method extremely fast in the field and should provide reliable estimates. Field tests in several North American systems suggest that estimates from distance-limited PDS compare quite favorably to those from LIS and have a much greater time efficiency (Kenning, 2007; Ducey et al., 2013b). In some cases, it will be sensible to combine PDS with another method. For example, if estimates of the number of pieces with large diameters at the basal end are needed for wildlife purposes, while the volume of all pieces is needed primarily for carbon inventory, it may be most efficient to use PDS with probability proportional to volume for volume estimation and a quick count of large-ends of pieces on a fixed radius plot to establish those densities (Williams and Gove, 2003).

Where there is an interest in multiple downed wood attributes, including perhaps some that cannot be entirely anticipated at the time of the survey, then either distance-limited PDS with probability proportional to volume or PDS with probability proportional to surface area coverage can be used along with measurements on each piece to allow calculating the factors in Table 12.5. Simulation results from Gove et al. (2013) suggest that while there are some differences in efficiency among the designs, they are subtle. Therefore, the primary focus should be on choosing a design, associated factors, and maximum distances (if needed) that facilitate rapid, accurate field work. As with horizontal point sampling of standing trees, consideration of the largest pieces that are likely to be encountered, realistic visibility conditions, and probable downed wood loads can be very helpful in establishing an appropriate factor. For PDS with probability proportional to surface area, a value of k_D that corresponds to the gauge angle k used for a reasonable BAF in horizontal point sampling is likely to prove a good starting place as the same factors (maximum tree diameter, and visibility of material of that diameter) play into the selection of both. Moreover, for experienced cruisers, using the same value of k_D and k may help minimize field errors as pieces of downed wood that are "in" will have a visual angle similar to that of "in" trees in HPS.

12.5. OTHER METHODS

With the expansion of interest in downed wood, many other methods besides TRS, PRS, and PDS have been developed. Not all have been widely field tested, though some are particularly promising and may yet prove of widespread interest.

12.5.1. Diameter Relascope Sampling

PRS uses a very wide angle gauge to sight on the ends of pieces, but it is natural to ask what would happen if an ordinary HPS angle gauge were used to sight on downed wood. Although some authors have suggested ad hoc approaches, the most thoughtful and creative method is that described by Bebber and Thomas (2003), who called it diameter relascope sampling (DRS) to distinguish it from PRS. Like PRS and HPS, DRS is centered on sample points. The cruiser uses an ordinary prism or relascope to sight on the midpoints of pieces of downed wood. If a piece is "in" (i.e., wider than the gauge or the portions of the piece overlap when viewed through the prism), then it is tallied. The estimate of volume per unit area at a sample point is $\hat{V} = \text{BAF} \times \sum_i L_i$. To obtain number of pieces per unit area, one uses the midpoint diameter in place of dbh in the usual tree factor formula from HPS. The Bebber and Thomas (2003) approach implicitly uses Huber's formula for volume. Thus, there is a nonsampling error in the use of Huber's formula that should typically introduce a very slight downward bias in estimates of volume, even though the method is unbiased from a sampling perspective. Because of its similarity to HPS, all of the standard techniques for dealing with boundary slopover are applicable to DRS.

For pieces that are borderline in DRS, it is necessary to measure the length of the piece in order to find the midpoint. Then, the diameter at the midpoint should be measured (or an effective diameter should be computed if the piece is not nearly circular at that point), and finally, the distance from the midpoint to the sample point should be measured. That distance can be compared to a conventional limiting distance table for the BAF being used.

A particular advantage of DRS is that, like PDS with probability proportional to surface area, it uses angle gauges and often the same angles that are operationally familiar to experienced foresters. For example, Bebber and Thomas (2003) and Bebber et al. (2005) used a BAF $2\,m^2$/ha prism in their original field trial and subsequent field work in Canada and found it to be satisfactory. The approach should also be very efficient when midpoint diameter and piece length are well correlated, which should be true when much of the volume is in relatively intact pieces (such as whole trees rather than tops and slash). DRS does not yet appear to have been widely tested in different systems.

12.5.2. Critical Length Sampling

Another approach to sampling downed wood using an ordinary angle gauge is critical length sampling (CLS), developed by Ståhl et al. (2010). CLS is to DRS as critical height sampling (Section 11.3.3) is to ordinary HPS. Suppose that a piece of downed wood is determined to be "in" using an angle gauge, sighting on at least one point along its length from a given sample point. Then, the critical length for that piece, CL_i, is the total length of the piece included as "in" with the gauge from that point, and

$$\hat{V} = BAF\sum_i CL_i$$

gives an unbiased estimate of the volume per unit area. Note that unlike DRS, no volume formula is assumed in CLS, so the method is free of such errors. Unfortunately, the price appears to be a significant increase in the complexity of the field work. Like critical height sampling, CLS is a theoretically clever approach that could stimulate a range of further applications. However, it has seen even less practical use than critical height sampling, if such a thing is possible.

12.5.3. Line Intersect Distance Sampling

Line intersect distance sampling (LIDS) is among newest methods for downed wood inventory (Affleck, 2008). It combines the linear sampling strategy of LIS, which typically has little or no nondetection error, with the PPS advantages of PDS. In LIDS, one or more sample lines are laid out from an initial sample point. As in LIS, the line arrangement may be simple (i.e., a single line centered on the sample point) or more complex (e.g., a radial cluster of lines emanating from the point). As in LIS, only pieces that cross the sample line are candidates to be tallied. Furthermore, pieces are only tallied if they are closer to the sample point than a critical distance that is proportional to cross-sectional area at the crossing point with the sample line. As a result, pieces are selected with probability proportional to volume, but the search is along a line. This should simplify search patterns and avoid some of the potential for nondetection bias that can arise with visual methods such as PDS. Affleck (2009, 2010) provides results of field trials in the northern Rocky Mountains, in which LIDS performs quite favorably relative to LIS.

When volume is the only parameter of interest, LIDS requires actual measurements only for borderline pieces (i.e., those whose inclusion cannot be determined from an ocular estimate of diameter and distance along the sample line). A simple count of pieces by decay class, multiplied by a volume factor, yields volume by decay class. This, in turn, can be converted to estimates of biomass or carbon using the usual methods. Affleck (2008) implemented LIDS using the walkthrough method of Ducey et al. (2004) to correct for boundary slopover; the variable line length for different piece sizes could make implementation of the original reflection method for LIS challenging.

LIDS combines the minimal nondetection bias of LIS with the PPS efficiencies of PDS. As one of the newest methods, however, it has not seen broad field testing in different environments. Additional applied research could well establish it as a serious contender for the downed wood method of choice in many situations.

12.6. DESIGN CONSIDERATIONS AND SELECTION OF METHODS

From the two methods that were available when Harmon and Sexton (1996) wrote their manual for the U.S. Long Term Ecological Research program (fixed-area sampling and LIS), we have passed to an era where many new approaches for downed wood sampling are available. Some of these are still new enough that basic questions (such as what factor to use for length, surface area, or volume) have not been thoroughly studied in many forest types. What should a practitioner do?

Field practice in forestry and ecology is often conservative, justifiably preferring simple, tried-and-true methods. That may be one reason why, of the countries surveyed by Woodall et al. (2009), nearly two-thirds of those that did inventory downed wood used fixed-area sampling in their national inventory design. When permanent, fixed-area plots are used to track growth and yield of live trees over time (Chapter 14), a compatible fixed-area design for downed wood can facilitate tracking individual trees through mortality, fall, and decomposition. Moreover, there is a long history of using fixed-area plots for a wide variety of ecological investigations, and that is especially true of downed wood (e.g., see Stokland et al., 2012). Nonetheless, as discussed in Section 12.1, fixed-area sampling of downed wood presents some design challenges. Moreover, it can impose a heavy time cost on field crews, along with substantial opportunities for nondetection and other field errors. Thus, when individual plots are not themselves the focus of primary interest, but the search is for a more general inventory approach to be used over large areas, fixed-area plots may not be the design choice of first resort.

By contrast, LIS is a proven approach in a variety of contexts and ecosystems. The fact that field workers must travel the sample line to perform measurements should minimize the kinds of nondetection errors that are problematic with other methods. In short, if a forester misses a piece they should have tallied in LIS, they will likely trip on it. However, Ringvall and Ståhl (1999a) did find potential for field errors even in LIS. Moreover, the kinds of confusion even over the basic theoretical underpinnings of LIS identified by Gregoire and Valentine (2003), and the current disagreement in the literature over best estimating equations and field practices when LIS is employed in complex designs, suggest that LIS may need some care in its design and implementation. Moreover, LIS is not necessarily fast. For example, suppose LIS is added to a routine timber inventory, for which there is a target of completing 20 plots in a field day. The LIS design involves 100 m of sample line at each sample point. This line needs to be walked twice: once while outbound to establish the line and take measurements on downed wood, and once inbound returning to the original sample point. In this situation, LIS adds 4 km of walking distance per day for the crew member on downed wood duty, in addition to the time actually spent measuring pieces.

For those who are not satisfied with either fixed-area plots or LIS, and who can afford the luxury of some experimentation, the array of other options provides many alternatives that could be worth exploring. TRS with a 90° gauge, PRS with a gauge of 50° or wider, distance-limited PDS with probability proportional to volume, or PDS with probability proportional to surface area all seem promising, especially when visual search is not made too challenging by a dense understory or crushing slash loads. For those hoping to stick with familiar tools and sighting angles, PDS with probability proportional to surface area, or the DRS approach, could prove quite successful. Finally, the LIDS approach, which combines the theoretical efficiency of PDS with the low nondetection error of LIS, is extremely promising and deserves wider attention. Finally, there may be, as yet, some undiscovered technique that outperforms any of the currently known ones; it may build on one of the techniques outlined in this chapter or it may derive from some entirely new approach. Given the relatively recent upsurge in interest in downed wood, the area of downed wood inventory should still be considered a developing one, and there is ample room for innovation.

13

INTEGRATING REMOTE SENSING IN FOREST INVENTORY

One of the most obvious changes in forest inventory practice in recent years has been the widespread availability and integration of remotely sensed data. Remotely sensed data have a long history in forest inventory through the use of aerial photography, and many current applications of remotely sensed data can trace their roots directly to those early practices. However, the diversification of kinds and sources of remotely sensed data from airborne and spaceborne platforms, the increasing availability of such data, and rapid progress in the power, portability, and affordability of computing devices that can handle large geospatial datasets have the potential to transform forest inventory practice over the next several decades.

A text on forest mensuration cannot hope to cover all aspects of remote sensing. However, all practitioners should be broadly acquainted with the kinds of remotely sensed data that are available, and the ways such data can be used to improve forest inventory from the stand to the national and global scales. In this chapter, we briefly summarize the types of remotely sensed data that are currently available, and their strengths and weaknesses for forest and ecological inventory applications. Then, we examine the roles such data can play in inventory design, estimation, and mapping.

13.1. TYPES OF REMOTELY SENSED DATA

In the broadest definition, remotely sensed data include any data acquired without making physical contact with the objects of interest. However, in modern usage, remote sensing usually refers to the acquisition of geospatial data using sensors mounted on an aircraft or satellite (Jensen, 2004; Campbell and Wynne, 2011). The vehicle to which a sensor is

Forest Mensuration, Fifth Edition. John A. Kershaw, Jr., Mark J. Ducey, Thomas W. Beers and Bertram Husch.
© 2017 John Wiley & Sons, Ltd. Published 2017 by John Wiley & Sons, Ltd.

mounted is called the *platform*. When similar sensors are mounted to a ground-based vehicle or a fixed point, the process may be called *terrestrial remote sensing*. The combination of a platform, a sensor, and a mode of operation is referred to as a *sensor system*.

Different types of remote sensing are distinguished not only by the type of platform employed, but also by the nature and characteristics of the sensors used. Nearly all sensors used in remote sensing operate by detection of electromagnetic radiation (also known as light) from different portions of the spectrum. The electromagnetic spectrum is composed of radiation of different wavelengths, as depicted in Plate 3. Some, but not all, remote sensing methods use light from the visible portion of the spectrum. Indeed, many sensors of particular interest in forestry and forest ecology use light from outside the visible range. Sensor systems can also be divided into *active* and *passive sensors*. Active sensors record radiation that has been emitted by the sensor system itself and reflected back from the objects being measured. By contrast, passive sensors depend on radiation sources in the environment, such as the sun.

Sensor systems may also be distinguished by their *resolution*. Resolution has three main components: spatial, spectral, and radiometric. Spatial resolution is the type of resolution most commonly considered. Intuitively, spatial resolution reflects the level of detail that can be captured in an image by a particular sensor system. Plates 4–6 illustrate the effect of different resolutions for optical remote sensing data from a forested region of eastern Oregon. The finer the detail that can be represented clearly, the higher the spatial resolution of the sensor system. Spatial resolution is affected not only by the sensor and the platform, but also by the mode of operation. For example, a photograph taken from an aircraft flying 1,000 m above ground level should be able to resolve much finer details than a similar photograph taken with the same camera from the same aircraft, flying at 10,000 m above ground level. Spatial resolution is related to, but not identical to, the *pixel size*. Many modern digital sensors use a rectangular array of pixels ("picture elements") to capture two-dimensional images. In principle, all else being equal, a sensor with a large number of pixels should be able to resolve finer details than a sensor with fewer pixels. However, hardware design, methods of operation, physical behaviors of light (such as diffraction), and postprocessing of the digital signal may act singly or in combination to reduce resolution, such that sensor pixel count may not be the limiting factor for resolution, and features several pixels wide cannot be resolved with clarity.

Most sensors can only detect radiation across a portion of the electromagnetic spectrum. The ability of a sensor to distinguish radiation of different wavelengths is called its *spectral resolution*. At one extreme lies traditional black and white photographic film, which depends on grains of silver halide. Although the photosensitive halides react to a wide range of wavelengths (especially when other chemicals are added to the emulsion to make it *panchromatic* or sensitive to the full visible spectrum), black and white film itself cannot record the different intensities of different wavelengths of light at any one point in an image, and hence it has essentially no spectral resolution. At the other end of the extreme are hyperspectral sensors (Section 13.1.6), which may resolve different wavelengths of light to only a few nanometers. In between are many more conventional optical remote sensing systems, which resolve wavelengths only to a few discrete bands.

The final type of resolution is *radiometric resolution*, which is related to the property of *dynamic range*. Dynamic range is the ratio between the strongest and weakest signals a sensor system can record, and in remote sensing it is usually expressed in base 2 terms (or bits). For example, if the strongest signal a system can record is 1024 (or 2^{10}) times as strong as the weakest, the system has a dynamic range of 10 bits. If the strength of these signals is captured in a 10-bit binary number such that the weakest measurable signal has value 0000000001 (i.e., 1), and the strongest has value 1111111111 (i.e., 1023), then the system will also have 10 bits of radiometric resolution. However, if in postprocessing the signals are recorded as digital numbers ranging from 0 to 255 (i.e., binary 8-bit integers) from weakest to strongest, the resulting imagery will have only 8 bits of radiometric resolution even though the dynamic range remains the same.

In theory, the ideal remote sensing approach for forestry and forest ecology would be operated from a satellite, would provide frequent or on-demand coverage anywhere in the world (with seamless, wall-to-wall coverage also available over large regions), and would use a combination of active and passive sensors (so that it could be operated at night, and so that in the daytime the imagery would be unaffected by nuisances such as terrain and solar angle). It would have very fine spatial resolution, high spectral resolution across a broad swath of the electromagnetic spectrum, and high dynamic range along with radiometric resolution and fidelity. It could be used not only for two-dimensional mapping, but also for three-dimensional measurements such as ground elevation and tree height. Moreover, data could be acquired even through dust, smoke, or clouds. The data would be freely available and could be stored and processed using simple techniques using inexpensive computers or even handheld devices in the field. Of course, no such data exist or are likely to in the future. Real data sources are associated with tradeoffs between coverage, availability, resolution, and other characteristics. All of these factors, plus cost (which includes both the cost of acquiring data and of processing it into useful form), interact with the intended use of the data to determine the "best" kind of remotely sensed data to use in any particular application.

13.1.1. Aerial Analog Photography

Experimentation with the potential uses of aerial photography began within two decades of the first heavier-than-air flight by the Wright brothers in 1903, though initial work was focused primarily on the opportunities and challenges for mapping that arose during the First World War (e.g., see MacLeod, 1919; Whitlock et al., 1919). During the 1920s, aerial photography became an operational part of the forest inventory and mapping toolkit, but was not universally used, and methods for stock mapping and cover typing tended to be qualitative rather than quantitative (e.g., see Robbins, 1929; Parsons, 1930). The use of aerial photography expanded dramatically in the years immediately following the Second World War, with growing sophistication in the use of stereophotography for quantitative measurements. This expansion was stimulated in part by the availability of surplus aircraft and trained pilots, and by the movement of technology that had been developed for military purposes to civilian use. About this time, the first textbooks on aerial photography in forestry also appeared (e.g., Spurr, 1948a). Aerial photography based on analog film technology rapidly became an indispensable part of the forest inventory toolkit and remained so until the revolution in digital imaging in the early twenty-first century.

Early uses of aerial photography were very much limited by available technology, both in terms of aircraft and photography. They emphasized oblique photographs (those taken at an angle looking downwards, but not vertically), often using handheld cameras operated by the pilot or by an observer working in an open cockpit. However, by the 1940s technology had shifted to specialized cameras with high-quality optics that provided minimal geometric

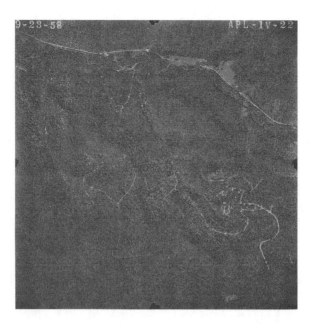

FIG. 13.1. An analog aerial photograph of a portion of the Allegheny National Forest in Pennsylvania, taken in September 1958 at a nominal 1 : 20,000 scale with a camera using 9-in.^{2} panchromatic film. Forest roads, different silvicultural treatments, and alternative land uses such as agriculture are clearly discernable. At high magnification, the size and shape of individual tree crowns is measurable, and using an adjacent, overlapping photo, tree heights can be measured using stereophotogrammetry.

distortion, mounted to provide stable, truly vertical photographs that could be used for accurate stereophotogrammetry. Often, such sensor systems were used by government agencies for wide-area mapping tasks, but yielded imagery that was also useful for a range of forestry purposes. For example, the U.S. Department of Agriculture, through its Agricultural Stabilization and Conservation Service, provided imagery using a Fairchild aerial camera that gave a 9-inch (23 cm) square negative. An example of this type of photograph is shown in Figure 13.1. The basic technology of image capture evolved as well, from the glass-plate cameras used by early aviators, to panchromatic films of increasing speed and resolution, to color films with emulsions sensitive primarily to red, green, and blue, and finally to films that included emulsions sensitive to the near-infrared (NIR). Chlorophyll is a strong reflector in the NIR, so these latter films are especially useful for clearly distinguishing vegetated and nonvegetated areas, and in some cases for identifying loss of photosynthetic capacity due to stress or disease. The combination of color and texture can allow accurate forest typing even within regions with complex species composition (e.g., Riemann Hershey and Befort, 1995). However, even with panchromatic images, the spatial resolution of analog photographs taken for moderate-scale mapping purposes is often fine enough that individual conifer trees can be identified to genus or species by their branching habit.

Analog aerial photographs were often specified in terms of their scale on the film (and on any contact prints made without enlargement). The scale of a photograph can be calculated as $RF = f/H$, where RF is the scale expressed as a relative fraction, H is the height of the aircraft, and f is the focal length of the lens expressed in the same units as H.

For example, if a camera used a 210 mm lens and was flown on an aircraft operating at 2,100 m above ground level, RF = 0.21/2100 or 1 : 10,000. At that scale, 1 cm on the film (or on a contact print) would represent 100 m on the ground. If the negative were a square 23 cm on a side, the total area on the ground captured in a single image would be a square 2300 m on a side, or 529 ha. At such a scale, assuming high-quality optics and good atmospheric conditions, it should be possible to resolve objects on the ground 1 m or slightly less in size. Most aerial photography used for operational forestry purposes would be taken at a scale between 1 : 10,000 and 1 : 20,000.

Despite the growing sophistication and quality of aerial photography, its primary uses changed little following the description by Spurr (1948b). These included first and foremost the mapping of forested areas for planning purposes, and for stratification by composition, density, and height. Secondary uses included attempts to measure tree height and crown closure directly from photographs, and to attempt to recover individual tree attributes by regression of diameter or volume on visible crown dimensions. The latter techniques were more sophisticated but ultimately proved more challenging. Because tree form, cull, and other subcanopy attributes cannot be observed directly on a vertical photograph, some level of ground measurement has always been needed to make full use of aerial photography in timber inventory. For example, volume tables prepared to allow the estimation of individual tree volumes from photogrammetric measurements in Finland were associated with errors of plus or minus approximately 30% (Nyyssönen, 1955), necessitating field plots to corroborate and correct inventory estimates from photos. Nonetheless, aerial photogrammetry has saved considerable field time and expense, especially when inventories have been conducted over large areas or those with difficult access. Techniques for photogrammetric measurement and image interpretation are reviewed by Paine and Kiser (2012).

Spurr's (1948b) intuition that aerial photography could best be used in concert with ground measurements as part of an integrated, well-planned campaign, ultimately proved true. The general approach he envisioned came to represent a main avenue of forest inventory during the twentieth century and continues to inform practice in the twenty-first; but the technology is changing dramatically.

13.1.2. Moderate-Resolution Optical Data

Spaceborne remote sensing developed rapidly following the launch of the Sputnik satellite by the Soviet Union in 1957. Early applications were overwhelmingly military, as exemplified by the U.S. CORONA program, which began operation in 1960. However, civilian applications soon followed. The LANDSAT program, which began in 1972, is the longest continuously operating program for spaceborne Earth observation. Data collection by LANDSAT has involved a series of different satellite platforms and sensors, with slightly different imaging characteristics. Until recently, most LANDSAT data were comprised of data from eight discrete spectral bands. The current sensor system, LANDSAT 8, collects data in 11 discrete bands (Table 13.1). In principle, LANDSAT images the entire earth every 16 days. However, in cloudy regions, the frequency of useful data for forest inventory can be much lower.

Several other programs provide moderate-resolution optical data from satellite platforms. Notable among these are the French SPOT (Satellite Pour l'Observation de la Terre) program, which provides imagery in blue (450–525 nm), green (530–590 nm), red

TABLE 13.1. Spectral Bands of the LANDSAT 7 and LANDSAT 8 Sensor Systems

	LANDSAT 7 Bands			LANDSAT 8 Bands		
Number	Bandwidth (nm)	Spatial Resolution (m)	Number	Bandwidth (nm)	Spatial Resolution (m)	Purpose
1	450–515	30	1	433–453	30	Coastal mapping; haze and aerosol detection
2	525–605	30	2	450–515	30	Visible blue
3	630–690	30	3	525–600	30	Visible green
4	775–900	30	4	630–680	30	Visible red
5	1550–1750	30	5	845–885	30	Near infrared; vegetation mapping
7	2090–2350	30	6	1,560–1,660	30	Shortwave infrared; soil moisture and geology
8	520–900	30	7	2,100–2,300	30	Shortwave infrared; soil moisture and geology
			8	500–680	15	Panchromatic; enhanced resolution
			9	1,360–1,390	30	Mapping thin, high clouds
			10	10,600–11,200	100	Thermal infrared; heat detection and irrigation management
			11	11,500–12,500	100	Thermal infrared; heat detection and irrigation management

Source: Adapted from Irons et al. (2012).

(625–695 nm), and NIR (760–890 nm) bands at 6 m spatial resolution, and in a panchromatic (450–745 nm) band at 1.5 m resolution (Astrium, 2013). The RESOURCESAT-2 platform of the Indian Space Research Organization provides data in two visible (520–590 and 620–680 nm) bands and one NIR (770–860 nm) band with 5.8 m resolution, with additional infrared bands at coarser resolution (ISRO, 2011). Moderate-resolution optical data are also provided by several private-sector sources.

Of a somewhat different character are the data from the NASA Moderate-resolution Imaging Spectroradiometer (MODIS) program, which provides data in 36 discrete spectral bands at spatial resolutions from 250 to 1000 m. A pair of satellites images the entire earth surface every 1–2 days (NASA, n.d.) The combination of frequent imaging with sensors that can readily detect fire and smoke has made MODIS an important tool for fire detection and mapping. MODIS data are too coarse for most tactical and strategic inventory purposes, but can be helpful in supporting regional-scale inventories.

A key advantage of satellite-based moderate resolution data is its wide area coverage. For some programs, another advantage is the temporal depth of available imagery. For example, the entire record of LANDSAT imagery spanning several decades is currently made freely available by the U.S. Geological Survey, Earth Resources Observation and Science (EROS) Center. This free data availability has greatly enhanced the utility of LANDSAT data for a range of applications (Wulder et al., 2012). Moderate-resolution optical data have been widely used for mapping land cover and its change (Hansen and Loveland, 2012), including deforestation monitoring (e.g., Skole and Tucker, 1993). Some studies have demonstrated an ability to predict stand structural attributes, such as biomass (e.g., Steininger, 2000; Zheng et al., 2007) and its change, including effects of partial harvest (Healey et al., 2006) and insects and drought (Vogelmann et al., 2009). However, moderate-resolution optical data tend to be most successful in forests with simple structure and species composition, and even in those conditions, the inaccuracy of direct predictions can prove frustrating (Lu, 2006). Moreover, care must be taken in interpreting the results of moderate-resolution analyses since the resolution of the data can have a strong influence on estimates of quantities as basic as forest area (e.g., Moody and Woodcock, 1995; Zheng et al., 2009) as well as assessments of fragmentation (e.g., Staples et al., 2012).

13.1.3. High-Resolution Optical Data

One characteristic of most of the moderate-resolution data described in Section 13.1.2 is that individual tree crowns cannot be resolved in the imagery. By contrast, with high-resolution data, as with analog aerial photographs (Section 13.1.1), individual tree crowns can be identified. With the passing of the Cold War era, governmental restrictions on access to high-resolution spaceborne data, and reductions in the cost of developing and launching satellite systems, led to rapid developments in the availability of high-resolution imagery from spaceborne platforms (Aplin et al., 1997). At the same time, technological advances in digital photography led to a shift away from film-based systems in the airborne arena.

Much of the widely available high-resolution spaceborne data is currently provided by commercial vendors. The first commercial satellite to provide such imagery was IKONOS, which included a panchromatic band with 1 m spatial resolution, along with blue, green, red, and NIR bands at 4 m spatial resolution (Gerlach, 2000). The 1999 launch of IKONOS was swiftly followed by that of QuickBird in 2001. Subsequent systems have provided even higher resolution (WorldView-1 in 2007, GeoEye in 2008, WorldView-2 in 2009).

The WorldView-3 platform, which was launched in 2014, provides commercial data with a spatial resolution below 0.5 m in the panchromatic band and less than 1.5 m in the multi-spectral bands.

The acquisition of digital imagery from airborne platforms has also become routine, and some of these data are freely available. For example, the USDA National Agricultural Imagery Program collects airborne digital imagery during the peak growing season on an annual basis over much of the conterminous United States. Imagery includes blue, green, red, and NIR bands at 1 m resolution (with 0.5 m resolution available in some areas). These data may be freely downloaded as orthocorrected imagery (USDA, 2013). In most of the developed world, contractors are available who can fly digital imagery to a variety of spec-ifications. A recent development is the widespread availability of unmanned aerial vehicles (UAVs), many of which are built on open-source software, and some of which can be oper-ated by nonspecialist personnel (Anderson and Gaston, 2013). Such craft may be fixed- or rotary-wing, and depending on payload capacity can carry one or more camera systems. It remains to be seen whether this emerging technology can offer foresters decentralized, on-demand imaging, and, if so, whether the investment in hardware, software, and training will be justifiable for the individual forester or small organization.

Most of the techniques of manual measurement, mapping, and photointerpretation applicable to traditional photography are also useful with high-resolution digital data (Paine and Kiser, 2012). However, the digital format of the data also offers new possibil-ities for automated processing. Lu (2006) reviews some early work on the use of high-res-olution data for biomass mapping, and Falkowski et al. (2009) outline some of the challenges and opportunities for integrating high-resolution optical data with large-area forest inventories. The identification of individual trees in high-resolution data, and the extraction of crown diameters and other characteristics from such data, is an area of very active research in the remote sensing community (Section 13.3).

13.1.4. LiDAR

LiDAR is an active remote sensing technique. The term LiDAR was coined by Ring (1963) by collapsing the words "light" and "radar," but most subsequent authors interpret it as an acronym for *L*ight *D*etection *A*nd *R*anging (akin to the original development of the term radar from "*Ra*dio *D*etection *A*nd *R*anging"). Unlike optical remote sensing, which relies on reflected light from the environment (usually the sun) to detect and describe landscape features, LiDAR uses a collimated beam of light, usually from a laser, to illuminate targets and describe the distance to those targets based on the characteristics of the return reflec-tion. When the laser is oriented downward from an aircraft or satellite, LiDAR is effec-tively measuring the range from the sensor to the vegetation and ground surfaces below it; hence, another synonym for LiDAR is laser altimetry. One natural application of LiDAR is to map canopy height as the difference between the highest laser returns in an area and those associated with the ground. LiDAR was originally explored in the 1980s for forest inventory applications (Lim et al., 2003), but its use has expanded dramatically since the turn of the century with advances in sensor technology, cost, and processing capabilities (van Leeuwen and Nieuwenhuis, 2010). Many organizations now use LiDAR in some aspect of forest inventory planning or analysis, and this growth is expected to continue.

LiDAR systems have a wide variety of specifications, and these affect their potential uses for forestry applications (Dubayah and Drake, 2000). Most LiDAR systems rely on a

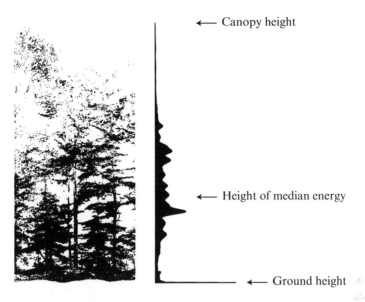

FIG. 13.2. A vertical slice through a northern hardwood–spruce–fir stand in New Hampshire, and a corresponding idealized airborne LiDAR waveform.

combination of the motion of the platform (aircraft or satellite) as well as rotation of a mirror or prism (scanning) to collect range measurements at a large number of points. The speed of the platform and/or scanner, and the distance from the LiDAR to the target, influences the density of pulses or probes. The divergence or spread of the laser beam itself also influences the characteristics of the data. Imagining the beam as akin to that from a flashlight, the illuminated area or footprint may be very small (centimeters for some airborne systems and millimeters for terrestrial ones), up to tens of meters (for some spaceborne systems). The combination of probe density and footprint, along with postprocessing, determines the ultimate spatial resolution of a LiDAR system. Finally, LiDAR systems differ in the handling of raw data from the reflected beam. For time-of-flight LiDAR (which includes nearly all of the systems used in operational forestry), the raw data consist of a distribution of reflected energy over time since emission of the signal (Fig. 13.2). Single-return LiDAR systems compute a single distance for each laser probe, usually based on the strongest, first, or last return. Again, considering the laser footprint as like a flashlight beam, a single-return LiDAR gives the distance to the brightest, nearest, or farthest object illuminated by the beam. Using the origin and orientation of each probe, one can convert the distances associated with multiple probes into a set of [x,y,z] coordinates or "point cloud." Multiple-return LiDAR can return multiple distances for each probe using different peaks in the signal. Finally, full-waveform LiDAR samples the returns from each probe to provide a representation of the entire return signal. Full-waveform LiDAR provides much more detailed information about the physical structure of objects contacted by each probe, and many in the LiDAR research community have emphasized its advantages especially with large-footprint systems (e.g., Dubayah and Drake, 2000; Pirotti, 2011). However, full-waveform data are typically much more expensive to collect and process. Most LiDAR systems used for operational mapping applications are single-return systems.

The availability of spaceborne LiDAR data is limited. Between 2003 and 2009, the Geosciences Laser Altimetry System (GLAS), a large-footprint, full-waveform LiDAR, was operated from the NASA IceSAT satellite. Although GLAS was originally designed to measure the Arctic and Antarctic ice sheets, researchers soon discovered the potential of this sensor for large-area mapping of forest structure and disturbance. For example, Lefsky (2010) combined a MODIS-derived forest cover map with GLAS altimetry data to develop a global forest canopy height map. The spatial resolution of GLAS-derived products is limited by the large interval between successive probes (approximately 175 m) and by the frequency of cloud cover in some regions. As of the writing of this textbook, there are no current, publicly available spaceborne LiDAR data suitable for forestry applications. However, successor missions to IceSAT are planned.

The vast majority of operational forest assessments using LiDAR employ airborne platforms. Although some airborne LiDAR employs full-waveform sensors, such as the NASA SLICER and LVIS sensors (Blair et al., 1999), most commercial LiDAR is flown using single-return systems. An example of a point cloud from airborne LiDAR data is shown in Plate 7. The characteristics of airborne LiDAR data are heavily influenced by scanner design, by probe density (dependent not only on the scanner but on aircraft altitude and velocity), and by seasonality (with scans during leaf-on season providing better resolution of the upper canopy, but scans during leaf-off providing better identification of the ground surface). Wulder et al. (2008) outline potential uses for airborne LiDAR data in forest management, and van Leeuwen and Nieuwenhuis (2010) review airborne LiDAR approaches for estimating canopy height, stand-level volume and biomass, and individual tree parameters. The most common approach is the "area-based approach" (Næsset, 2002), in which empirical relationships are developed between field data on the stand parameters of interest and a variety of statistics that summarize features of the LiDAR returns. van Leeuwen et al. (2011) suggest that because airborne LiDAR data may also be useful for mapping variables such as crown size and shape, site quality, and stand density, such data could also be used to help map and predict wood quality. Bergseng et al. (2015) provide a recent comparison of alternative approaches for estimating forest structural parameters from airborne LiDAR data; many of these approaches are reviewed in Sections 13.2, 13.3, and 13.4. When airborne LiDAR data are collected at a high probe density (several strikes per square meter) and under favorable conditions (e.g., leaf-off season), the resulting elevation maps and ability to detect ground features can also provide data that are extremely useful for management. For example, Johnson and Ouimet (2014) describe the ability of airborne LiDAR data to map cultural and archaeological features beneath the forest canopy. Figure 13.3 shows a high-resolution elevation map of the ground surface beneath the canopy of a northern hardwood forest in New Hampshire, USA; features of an abandoned farmstead, including foundations, a well, and disused roads, can be discerned easily.

Finally, LiDAR units can be deployed terrestrially, either mounted on a tripod or on a vehicle (Fig. 13.4). In this configuration, LiDAR can provide extremely detailed three-dimensional information over relatively small areas, typically with millions of points in each point cloud. It is sometimes possible to coregister point clouds from multiple scans, resulting in greater areal coverage and the ability to detect trees and other objects that would have been hidden in a single scan. An example of a single terrestrial LiDAR scan is depicted in Plate 8. Terrestrial LiDAR is data-rich, but the extraction of relevant information from such data remains an active area of research. One avenue of work concerns the automated extraction of tree diameter, position, taper, and volume from terrestrial LiDAR

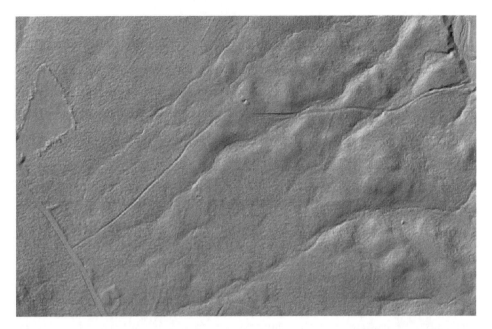

FIG. 13.3. A 1-m resolution digital elevation model for a portion of the White Mountain National Forest in New Hampshire generated from airborne LiDAR data. In addition to landforms, cultural features (including stone walls, disused roads, and an open well) can be discerned (Data courtesy R.A. Colter, U.S.D.A. Forest Service).

FIG. 13.4. A terrestrial LiDAR unit collecting a dense point cloud within a tropical rainforest.

data (Henning and Radtke, 2006b). Some authors have highlighted the potential of terrestrial LiDAR as a sort of relascope (Jupp et al., 2007), but the problem of hidden trees (and trees that are not detected because of imperfect postprocessing algorithms) is pervasive. Recent work suggests that distance sampling techniques may offer a partial solution (Ducey and Astrup, 2013; Astrup et al., 2014). Other research has focused on the use of terrestrial LiDAR for measuring canopy parameters at the tree or stand scale (Hilker et al., 2010; Huang and Pretzsch, 2010). Terrestrial LiDAR appears well-suited for this task, though results have been shown to be sensitive to scanner characteristics such as beam width (Ducey et al., 2013a). Although terrestrial LiDAR is currently too expensive for widespread application, the continuing improvement in scanner speed and robustness, and reductions in size and expense, suggest that terrestrial LiDAR may begin to see more operational use in the near future. At least one private company currently offers terrestrial LiDAR services for forest inventory on a commercial basis.

13.1.5. Synthetic Aperture Radar

Synthetic aperture radar (SAR) began to emerge as a viable remote sensing method for land cover mapping and the estimation of forest characteristics in the mid-1990s (Kasischke et al., 1997; Dobson, 2000). SAR is an active remote sensing technique, relying on measurements of the backscatter of electromagnetic radiation with long wavelengths, typically from the microwave portion of the spectrum (Fig. 13.5). SAR has a significant advantage over optical remote sensing and LiDAR using lasers in the optical or NIR range, in that radar can penetrate cloud cover as well as dust and smoke. This makes SAR especially valuable for ongoing monitoring in the humid tropics and temperate rainforests, where cloud-free conditions can be rare. Depending on the band or wavelength used, SAR can also penetrate forest canopies to reveal subcanopy structures as well as ground surface and moisture conditions. Kasischke et al. (1997), Dobson (2000), and Balzter (2001) provide basic reviews on SAR technology and applications.

In recent decades, a variety of SAR systems have been deployed on spaceborne or airborne platforms, though most were designed for primary purposes other than forest monitoring. The imaging characteristics of SAR data depend on system attributes as well as the

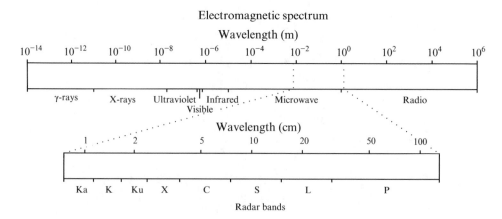

FIG. 13.5. The RADAR portion of the electromagnetic spectrum.

characteristics of the vegetation or other target material. Wavelength is a primary consideration as radiation in the radar spectrum interacts most strongly with material that is similar in dimensions to the wavelength or slightly larger. For example, X-band radiation with a wavelength of only a few centimeters readily penetrates water droplets associated with cloud cover and atmospheric dust, but interacts strongly with foliage and other canopy elements, providing information about canopy structure but little penetration to describe tree stems or ground surface attributes. By contrast, L- or P-band SAR, with wavelengths in the tens of centimeters, largely passes through the canopy and interacts strongly with coarse branches, tree boles, and the ground. Radar backscatter from vegetation is also strongly influenced by polarization of the transmitted and received signals, and by the angle at which the radiation strikes vegetative surfaces. Holding the system attributes constant, the size, spatial distribution, and angular orientation of vegetative surfaces or other scattering elements play a dominant role in determining the intensity and polarization of backscatter. The moisture content of the scattering elements also plays a strong role as radiation at these wavelengths interacts strongly with water. In general, increasing moisture content decreases the ability of radar to penetrate vegetative canopies, while flat water surfaces are highly reflective. Thus, forested wetlands can often be discriminated in SAR imagery because of the strong backscatter from the water surface, even though the water would be concealed by the overhanging canopy in conventional optical imagery.

The widespread use of SAR data has historically faced several obstacles. SAR data are inherently complicated: while the phase, amplitude, and polarization of SAR returns may contain a great deal of information, that information is not as intuitive as optical remote sensing with its close similarity to more familiar digital imaging technologies. The geometric and radiometric complexities of SAR data require specialized software for data assimilation, visualization, and analysis. Finally, interactions with topography and with heterogeneous scattering targets (resulting in "speckle" in SAR imagery) can make interpretation of SAR data challenging (Kasischke et al., 1997). Great advances have been made within the research community, but the use and interpretation of SAR remain somewhat specialized. Within that framework, SAR has been shown to be valuable for large-scale assessments that hint of the possibilities for more operational work. For example, Saatchi et al. (2007) used SAR to map fuel conditions in the Yellowstone ecosystem, while Cartus et al. (2012) developed a high-resolution aboveground forest biomass map for the northeastern United States. Dobson (2000) wrote, "Radar is often depicted as a promising technology not quite ready for prime time." Some observers might claim the statement remains true today. However, SAR does seem poised to move into an operational role in coming decades.

13.1.6. Hyperspectral Data

Many sources of optical remote sensing data are multispectral. A good example of a multispectral data source is LANDSAT, which provides images from 8 to 11 relatively narrow spectral bands, separated by gaps within which no data are collected. By contrast, hyperspectral imagery involves the collection of data from a large number of bands with little or no gap between, so that the full spectrum of reflected light is measured across a given range. Examples of hyperspectral imagery sources include the Hyperion sensor mounted on the NASA EO-1 satellite, and the NASA AVIRIS sensor, which has been flown on several airborne platforms. AVIRIS records data across 224 spectral channels, continuously spanning the range from 400 to 2500 nm (Green et al., 1998).

In principle, hyperspectral data contain a great deal of information about the physical and chemical composition of forest canopies, and this has been exploited in research settings. For example, hyperspectral data have been used to map canopy N content and forest productivity (Martin and Aber, 1997; Smith et al., 2002). Hyperspectral data also have been used to map species distribution in complex temperate and tropical forests (Martin et al., 1998; Clark et al., 2005). The rich spectral information has also been used to improve tasks that might otherwise have been challenging using conventional multispectral data, such as the identification of individual tree crowns in high-resolution imagery (Bunting and Lucas, 2006). Treitz and Howarth (1999) provide a general overview of the potential of hyperspectral imagery.

Despite its evident potential, hyperspectral remote sensing has yet to achieve widespread use in operational forestry. In part, this may reflect the expense and relatively sophisticated technology associated with sensors, the large amount of data storage and processing involved, and the relatively sophisticated statistical techniques needed to make full use of the information contained in the data. However, the same could easily have been said of LiDAR only a decade ago. Indeed, airborne research platforms have now been developed that combine simultaneous LiDAR and hyperspectral data collection (Asner et al., 2007; Cook et al., 2013). If such platforms were to move into operational forestry settings, the benefits could be significant.

13.2. REMOTE SENSING FOR STRATIFICATION

One of the earliest uses of aerial photography in forestry was mapping of forested areas, and basic stand mapping (e.g., Robbins, 1929). Mapping is very closely related to the process of stratification in forest inventory design (Section 10.5). Recall that in stratification the area or tract to be inventoried is divided into two or more strata of known area, and sample plots or other measurement units are then located independently within each stratum. The estimates for the individual strata are then combined to produce a more accurate overall estimate for the tract. When tracts are small, the individual strata are often single stands; when tracts are large, the individual strata may be comprised of many stands grouped into likely stand types. A key issue in stratification is that the areas of the strata are presumed to be known exactly, and this is greatly facilitated by the mapping capability provided by aerial photography (or more recently by other types of remotely sensed data). However, it is important to remember that the goals of mapping strata in the design phase of an inventory are not necessarily identical with those of mapping observed stand or stand type boundaries for an end user. This can create confusion about what to do when the on-the-ground situation does not conform to what was expected during stratification, an issue to which we will return in Section 13.2.4.

13.2.1. Photo Interpretation and Stand Mapping

The advantages of aerial photography for delineating and mapping forest stands were recognized early in the development of modern forest inventory methods. Given imagery of adequate spatial resolution, one could map stand boundaries and compute stand areas much more quickly, and in many cases more accurately, than could be done using ground effort alone. Accurate stand areas required that the imagery be orthorectified or that specialized equipment be used with overlapping stereo pairs of images.

TABLE 13.2. Stand Typing System for Aerial Photographs

Composition	
H	Hardwood species comprise over 80% of canopy cover
M	Mixedwood; neither hardwood nor softwood species exceed 80% canopy cover
S	Softwood species comprise over 80% of canopy cover
Tree height	
1	Dominant trees <10 m in height
2	Dominant trees between 10 and 20 m in height
3	Dominant trees between 20 and 30 m in height
4	Dominant trees exceed 30 m in height
Canopy cover	
A	Canopy cover >75%. Fully stocked, not recently harvested
B	Canopy cover 50–75%. Harvested at some point in the past, but likely with considerable standing merchantable volume (as appropriate to height class)
C	Canopy cover 25–50%. Recently harvested, standing volume greatly reduced
D	Some trees but with canopy cover <25%. Any merchantable volume restricted to scattered individual trees (such as seed trees)
E	Open/regenerating

Source: Adapted and modified from Spurr (1948a).
A given stand type is designated by the combination of a letter indicating composition, a number indicating size, and another letter indicating density (e.g., H4A is a tall, dense hardwood stand).

A stand-typing system to be used with aerial photographs must respect the limitations of the photographs themselves. For example, while trained experts may be able to identify tree species on high-resolution photographs with greater or lesser accuracy, some species determinations may not be possible except on the ground. Likewise, while it might be desirable to stratify stands based on the average diameter of the trees, it is not possible to measure dbh directly on photographs. Nonetheless, some variables related to composition, site quality, and structure can be measured or interpreted with some reliability on conventional aerial imagery, especially when stereo pairs are used. These include broad species composition, tree heights (Section 13.3.2), crown diameters (Section 13.3.1), crown cover, and density of individual trees (at least for canopy dominants visible in the image). It is also typically possible to ascertain topographic position, which is often correlated with species composition and site quality. Stand typing systems based on these variables have been widely used and adapted to different needs and contexts, and have often persisted into present use even when aerial photographs have been supplemented or replaced by other sources of data (such as airborne LiDAR). An example of such a system is described in Table 13.2.

Stand typing systems, even those as crude as in Table 13.2, can convey a great deal of information that is useful for forest management. Thus, accurate stand type maps can be invaluable tools for a variety of purposes, and for many of those purposes the accuracy of the map should be as high as possible. However, when stratification for the primary purpose of improving efficiency of an inventory is a goal, the burden on a stand typing system is much lower. Recall that the statistical purpose of stratification is to divide the tract of interest into units that are of known area and internally similar, so that field effort can be allocated efficiently and the resulting estimates of tract total will have low standard error. It is not necessary for stratification that all stands or other elements within a stratum be identical or that they all match the stand type with which a stratum may have been labeled.

It is only necessary that stands or elements that are similar tend to fall into the same stratum. Indeed, where some stand types (or other relevant categories) cannot be distinguished reliably in the imagery, it may be advantageous to group those types into a single stratum that will have high variance and then allocate significant field resources to that stratum while assigning somewhat lower effort to strata that are likely to be much more homogenous.

13.2.2. Pixel-Based Classification

The advent of digital imagery and the availability of faster and less expensive computing power made it possible (at least in principle) to replace some of the time-consuming, subjective delineation of stands and stand types with automated image processing. Until recently, the vast majority of remote sensing analyses have operated on a pixel-by-pixel basis. For example, the pixels in a LANDSAT scene might be binned into a number of categories based on their spectral characteristics, such that the pixels in each category are relatively similar to one another. Then, contiguous patches of pixels within the same category would be delineated and designated as forest stands (or patches of other land cover types).

The task of grouping pixels into different categories is called *classification*. Classification can be done according to either of two major approaches: supervised or unsupervised. In *supervised classification*, field observations of forest cover and characteristics are matched to the corresponding pixels in the image. A series of variables are calculated for each pixel. For example, the reflectances in each band of a multispectral image or ratios of those such as the Normalized Difference Vegetation Index (NDVI; Tucker, 1979) might be used. These variables comprise the *feature space* for the classification. The variables in the feature space are then used to predict the category to which each pixel belongs, using the field observations as *training data*. Once a predictive model is developed, it is applied to the pixels in the remainder of the image, and each pixel is assigned the category predicted to be most likely for that pixel.

In *unsupervised classification*, pixels are clustered together and assigned categories based solely on their similarity in the feature space; no training data are required. However, once the categories have been developed, they may be compared to field data to determine whether the categories match age classes, species groups, or other useful attributes. Unsupervised classification is not always successful for mapping a desired set of stand types across the landscape (because the categories developed in the classification may not break cleanly across the desired types). However, it is wholly appropriate for stratification when there is no particular need for the strata to match a set of predetermined types.

The number of statistical techniques available for supervised and unsupervised classification is enormous.

Duda et al. (2001) is the standard text for classification methods. When it is important that a classification yield an accurate map or spatial representation of forest types, an accuracy assessment should also be conducted. Techniques for accuracy assessment for pixel-based classifications are described in Congalton and Green (2009).

13.2.3. Object-Oriented Classification

Pixel-based approaches to classification are most widely used when individual pixels are larger than the objects of interest (such as trees). For example, the reflectances in a 30 m LANDSAT pixel from a forested area reflect the average reflectance over multiple dominant tree crowns. The individual pixels within a single stand are likely to have very similar characteristics. However, as spatial resolution becomes finer, more and more detail is apparent. While useful for human perception, such detail is not necessarily advantageous for pixel classification. For example, a 1 m IKONOS pixel from the sunlit side of an individual tree crown is likely to have very different spectral characteristics from a pixel on the shaded side of the same tree crown. Moreover, the differences between sunlit and shaded pixels in the same crown may be greater than the differences between those on another tree, even if that tree is of a very different species. Correct classification and interpretation of the individual pixels is nearly impossible without context.

The field of *object-oriented classification* has arisen in remote sensing largely in response to these challenges (Blaschke, 2010). In an object-oriented classification, pixels are first grouped into objects based not only on their spectral characteristics but also on their spatial arrangement. Then, either supervised or unsupervised classification can be applied to the resulting objects. Because the objects are comprised of multiple pixels, they can be classified not only using their average spectral characteristics, but also using their shape, size, and internal variability as elements of the feature space. For example, knowing that riparian stands tend to have long, sinuous shapes can help separate riparian from upland formations in an object-based classification. Although object-based classification is most commonly used when pixels are smaller than the objects of interest (e.g., to identify individual trees in high-resolution optical imagery), object-based classification can also be applied to coarser-resolution data (e.g., to map stands using moderate-resolution optical data). In the latter case, object-oriented approaches mimic that taken by a human interpreter, first delineating a relatively homogenous or otherwise recognizable area, then assigning it to a stand type or other category based on its attributes.

Object-oriented approaches are relatively new, and the software used to identify individual pixels can be complex. For example, one popular commercial package requires the user to select parameters that govern how objects are defined based on size, color (or spectral characteristics), and shape, with shape controlled in turn by compactness and smoothness of the boundary (Baatz and Schäpe, 2000). The results of the classification can be highly sensitive to the parameter choices. Moreover, while object-oriented classification often produces maps that are intuitively appealing, it is not always the case that such maps have higher accuracy. For example, Campbell et al. (2015) found higher accuracy with pixel-based approaches both for land cover mapping and change detection in a forested portion of eastern Oregon using LANDSAT. Moreover, accuracy assessment itself can be more challenging for object-based approaches as the objects can be quite extensive. An individual field plot may be adequate to evaluate the contents of a single moderate-resolution pixel, but assessing an entire stand will often require subsampling. Identification of appropriate sampling approaches and sample sizes to support accuracy assessment of object-based forest classifications is an active area of research (e.g., MacLean et al., 2013).

13.2.4. Effects of Misclassification on Estimation

Whether classification and mapping of stands or stand types is done by manual interpretation or automated digital processing, some amount of error is inevitable. This raises the question of what to do when a plot location, or an entire stand, is found to differ in the field from what was expected based on the initial stratification. For example, suppose that stands have been mapped following the types in Table 13.2, and the initial stand map has been used to generate stratum areas and to assign plot numbers and locations in a stratified inventory. Upon arriving at a plot location that is supposed to be in the S3A stratum (dominated by softwoods), the field crew determines that the plot actually contains a significant amount of hardwood basal area and cover, and the correct stand type is M3A (a mixed-wood type). What should be done with this information?

The answer to this question lies in the distinction between the act of stratification, on the one hand, and the production of a forest type map for end users, on the other. The numbers and locations of plots in the inventory are based on the stratum assignments. The actual contents of a field plot or other sample do not impact the sample selection process retroactively. Thus, it is an incorrect procedure to reassign the stratum of a field sample when the contents of the sample do not agree with what was expected to be typical for that stratum. In the example given above, the stratum labeled as S3A may be comprised mostly of stands that actually are S3A, but it will inevitably include small areas of other stand types, and those areas contribute to the overall mean and variability within the stratum. Excluding "atypical" plots when calculating the sample mean and variance for the stratum (e.g., eqs. (10.22) and (10.25)) will bias those estimates for the stratum. Likewise, including any "incorrect" plots from the S3A stratum when calculating means and variances for the M3A stand type can bias the results for M3A, especially if the two strata were sampled with unequal intensity. Unless great care is taken, even the areas of the different stand types may be estimated with bias (note that the areas of the strata, by contrast, are already known exactly: these were mapped in advance). The practice of reassigning strata when field samples do not conform to the labels associated with those strata is incorrect and should be strictly avoided. For an extended discussion of this issue, see Iles (2003, chapter 8).

Should a management plan or other document include a stand type map that contains the "uncorrected" types? That depends on the purpose of the map. If the map is intended to show the strata used in the design and reporting of the inventory, there is no error to correct: the stratum associated with the sample plot in this example is the S3A stratum because that is how it was selected. If the purpose of the map is to provide operational guidance, however, it makes little sense to present the strata as known cover types, when the correct cover type is known for some areas. On such a map, however, it may be useful to distinguish between those locations whose cover type is known (because they have been visited in the field, and the type corrected as needed) and those where it is suspected but not confirmed (because those locations did not comprise part of the inventory sample). In the example here, it may be reasonable to map the nonsampled stands in the S3A stratum as S3A, but we suspect that some of these stands may actually belong in other types (we simply cannot state which ones). A properly conducted accuracy assessment (Congalton and Green, 2009) can help describe the level of confidence end users should place in such maps.

13.3. INDIVIDUAL TREE MEASUREMENTS

Early in the development of aerial photography, it was recognized that if imagery had sufficient spatial resolution, one could identify and measure attributes of individual trees. In principle, such measurements could be used to substitute for ground-based plot measurements, though in practice a number of difficulties are usually encountered with such an approach. However, individual tree measurements can be used to inform cover type mapping for stratification (Section 13.2) or to develop sampling covariates (Section 13.4). Many, though not all, of the techniques used with analog aerial photographs can also be used with digital imagery of sufficiently high resolution, while airborne LiDAR has proven invaluable for remote measurement of tree and canopy height.

13.3.1. Crown Widths

One of the most straightforward tree measurements on vertical aerial photographs is that of crown width. Provided the scale of the photograph is known, a simple measurement of crown width on the negative or print can be translated directly into field units. Spurr (1948b) suggests that even on typical 1 : 15,000 scale photographs a trained expert should be able to measure crown width to the nearest meter. On higher-resolution imagery, crown area and shape could be measured using a planimeter (Section 4.4.3). Using more modern equipment and techniques, crown width, area, or shape can easily be digitized on-screen using imagery captured by an UAV.

A desire to automate the extraction of crown width or area has accompanied the availability of digital imagery and a variety of algorithms have been proposed. Such algorithms typically involve two major tasks: first, recognizing individual trees within the imagery, and second, delineating and measuring their crowns. In comparison with manual interpretation, automated methods are typically faster and less subjective, and can be used to construct "wall to wall" maps from imagery, but manual methods are typically more accurate. Examples of individual tree extraction and crown measurement from high-resolution optical imagery can be found in Leckie et al. (2005) and Palace et al. (2008). For example, the algorithm of Palace et al. (2008) finds bright spots (assumed to be the sunlit upper portion of individual tree crowns), then searches outward to find the dark edge where the shaded portion of the crown falls away into the understory. Individual tree crowns may also be identified and measured in high-resolution aerial LiDAR data, where the 3D information contained in the point cloud simplifies many of the challenges but does not make it foolproof (Popescu et al., 2003).

It stands to reason that if dbh and crown width are highly correlated (Chapter 5), it should be possible to reverse the usual predictive allometric relationship and estimate dbh once crown width has been measured. Moreover, it should also be possible to predict tree attributes (such as volume or biomass) that would usually be predicted from dbh using crown width as a proxy. Indeed, this strategy has been employed widely for a range of purposes (e.g., Minor, 1951; Bonnor, 1964b). However, it is often important to develop predictive equations that employ crown width or area measured on the imagery, rather than in the field, because these may differ in systematic ways. It is also advisable to develop equations that predict the variable of interest (such as volume or biomass) as a function of crown width or area, treated as the independent variable, rather than inverting the more typical allometric equations that predict crown width or area as a dependent variable using dbh as the independent variable because these two approaches involve very different statistical assumptions about the error distribution and which variable(s) are known versus which are unknown and predicted.

13.3.2. Tree Heights

Measurement of tree height from aerial imagery is somewhat more complicated than for crown width, but some reasonably accurate approaches are available. One of the simplest relies on measuring the length of the shadow of sunlit trees. If the location and time of image capture are available, it should be possible to calculate the solar angle at that moment. Then, computing tree height from shadow length becomes a nearly trivial exercise in trigonometry. However, the method is challenging in closed canopy forests, and complicated corrections are needed on sloping terrain or the measured heights may be wildly inaccurate. Thus, obtaining tree height from shadow length is typically restricted to trees in open stands or near edges, situated on level terrain (Spurr, 1948b).

Another technique for measuring tree heights from aerial imagery relies on the parallax difference between images in a stereo pair. This approach was pioneered as early as the 1930s (Andrews, 1936). On analog photographs, this required a parallax wedge or other specialized equipment. However, Spurr (1948b) suggests that tree height could be measured to within 1.5 m on typical imagery using this technique. In principle, related approaches could be used to reconstruct tree heights from densely overlapping photographs captured using UAVs; however, some technical challenges remain for obtaining accurate measurements (Dandois and Ellis, 2010).

Airborne LiDAR currently stands as the best available data for remote mapping of tree or stand height over large areas (van Leeuwen and Nieuwenhuis, 2010). Provided the density of probes is high enough to capture both the crown and ground elevation, inferring tree height is straightforward. However, except at very high densities, airborne LiDAR data typically cannot identify or resolve the crowns of suppressed trees or those in the understory.

13.3.3. Estimating Stand Characteristics

If reliable predictions of tree attributes such as volume or biomass can be made from characteristics such as crown width and total height, it seems as though one could conduct an inventory using imagery alone, without ever putting boots on the ground. In principle, it is possible to count and measure all the trees in an area using imagery alone, or to subsample the imagery using strips or circular plots. Some authors have suggested counting trees using techniques analogous to horizontal point sampling, tallying them with probability approximately proportional to crown area (Bitterlich, 1984; McTague, 1988; Gering and May, 1997).

Unfortunately, such a direct attack on the problem of forest inventory seems predisposed to failure. A principal challenge is the difficulty of detecting all trees. For example, Solberg et al. (2006), using airborne LiDAR with a moderately high pulse density (5 pulses/ m^2), were able to identify only about 2/3 of live trees in Norway spruce stands. Over 90% of dominant trees could be identified and measured, but this fraction dropped to 2/3 for codominants and 1/3 for trees in lower crown classes. Challenges associated with nondetection can be especially severe in stands with complex structure. Moreover, many important tree attributes (such as grade, presence of cull, and sometimes species) cannot be measured reliably or at all from above. Thus, even when conditions are favorable, some ground effort seems necessary to adjust or correct inventory estimates derived from the measurement of individual trees in remotely sensed imagery.

13.4. REMOTE SENSING FOR COVARIATES

Even if remotely sensed data cannot provide direct estimates of the quantities of interest in an inventory, or if measurements of tree or stand attributes they provide are imperfect and potentially biased, those same data can provide a rich set of covariates to aid in sampling design and estimation. Sampling with covariates was introduced in Section 3.9 and more fully elaborated in Section 10.8. Here, we explore how remotely sensed data can be used to develop covariates, and how they can be used in ratio and regression sampling. We also look at some emerging approaches that extend the covariate concept in useful ways.

13.4.1. Tree and Stand Attributes and Sampling Covariates

Broadly speaking, any variable that is relatively easily measured and is correlated with the variables of interest in an inventory could potentially be useful as a covariate. Once remotely sensed data have been acquired for an inventory area, there may be many ways to extract useful covariates from the same raw data. Given modern computing power, even large sets of imagery can often be processed in an automated fashion to produce spatial data layers that (one hopes) will be correlated with forest characteristics such as canopy cover, species composition, timber volume, and biomass.

Conceptually, any variable that can be calculated from remotely sensed data can be considered as a surface over two-dimensional space: the location of a particular point (e.g., its latitude and longitude) define the x- and y-axes, while the value of the remote sensing–derived variable provides the z-axis. Many types of remotely sensed data are grids or rasters of individual pixels; in this case, the surface may not be strictly smooth or continuous, but if we can associate any given x and y coordinate with the value of the pixel that contains the coordinate, the concept still applies. Call this surface the *covariate surface*. Now, imagine moving around the same set of x- and y-coordinates, taking a conventional field-based sample (such as a fixed-area plot or horizontal point sample) centered on the same coordinate. The results of the field-based sample, if we were to perform measurements at all possible coordinates, would define what Williams (2001a) calls the *sampling surface*. In ordinary field work, we cannot know the entire sampling surface; we only know the value of the sampling surface at a relatively small number of coordinates. Physically traveling to a location, and conducting a field sample, can be relatively expensive; by contrast, remotely sensed data can potentially provide a covariate surface over much, or even all, of the area of interest. By combining the two sources of data, we may be able to calculate estimates of means or totals for the inventory area at much higher accuracy or lower cost than using field work alone. As an important benefit, we may also be able to map variables of interest (albeit usually with imperfect accuracy) at a much higher resolution than would be possible with conventional data. From a statistical perspective, it matters little what the original source of the remotely sensed data might be or what procedures were used to transform the raw data into a covariate surface; what matters most is the correlation between the covariate surface and the sampling surface.

Conceptually, there are four main ways to construct covariates from remotely sensed data; some approaches lend themselves better to specific types of raw data than to others. The four broad types of covariates are indices, direct measurements, physical models, and object aggregation.

Indices. Any variable that can be calculated from remotely sensed data and is not directly interpretable as a simple physical variable but is likely to be correlated with variables that are of interest can be considered an index. In optical remote sensing, the classic example is the NDVI (Tucker, 1979). When multispectral optical remote sensing data include reflectance in the NIR and red portions of the spectrum, NDVI can be calculated as

$$NDVI = \frac{NIR - red}{NIR + red}$$

NDVI is widely used as a measure of the overall "greenness" or health of vegetation, and is often used as a covariate to predict leaf area, canopy cover, and biomass. In LiDAR remote sensing, the height of median energy (or HOME) is defined as the height of the median aboveground return within a given area. HOME is often a strong predictor of the vertical structure of forests and hence is typically among the LiDAR-derived variables used to predict timber volume, biomass, and canopy structure. In a similar fashion, reflectances and polarizations of different bands in SAR data can be considered as indices.

Direct Measurements. A remotely sensed variable that is calculated with few assumptions from the data recorded by the sensor, and has a simple, clear interpretation in terms of a variable that could be measured in principle using some other method, can be considered as a direct measurement. For example, if pixels in high-resolution optical data can be classified into canopy and ground pixels, then the proportion of ground pixels within an area can be taken as a measure of canopy cover, and the proportion calculated within a suitable moving window around a given x- and y-coordinate may provide a useful covariate surface. In principle, it would be possible to perform this same measurement using one of the field methods described in Section 8.6.1, though it would be prohibitively expensive to do so over a large area. Similarly, the height difference between the highest LiDAR return and the ground returns in a suitable moving window can provide an estimate of forest canopy height. In principle, it would be possible to measure this same variable from the ground. Note, however, that to be useful as a sampling covariate the remotely sensed variables do not need to correspond perfectly to their ground-based counterparts or even to provide unbiased estimates of the same physical variables. What is required is that they be well correlated with the variables that are of interest in the inventory.

Physical Models. Some useful covariates can be calculated from remotely sensed data using a physical model that entails additional assumptions or parameters beyond those provided by the data themselves. One common example is the computation of the quantity and vertical distribution of foliage from airborne or spaceborne LiDAR data (e.g., Coops et al., 2007). This typically involves using the Beer–Lambert law, which relates the attenuation of light beams to the optical density of the material through which they pass, along with assumptions or auxiliary measurements of the distribution of leaf angles within the canopy. The reconstruction of vertical foliage distribution usually follows the approach developed by MacArthur and Horn (1969), though the close connection of that approach to the statistical techniques of survival analysis suggests other alternatives may be promising (Maynard et al., 2013). Although there is a strong physical basis for the calculation, the extraction of the quantity and distribution of foliage does depend on some assumptions about foliage randomness and angular distribution, which may not be exactly correct; thus, the variables calculated by this method would be classified as the results of a physical model rather than direct measurements. Again, these variables need not be strictly unbiased estimates of the physical quantities they represent to be useful as covariates; the utility of a covariate depends on its correlation with the other variables that are the targets of the inventory.

Object Aggregation. When high-resolution optical or LiDAR data are available, it may be possible to identify and locate individual trees, and to compute estimates of basal area, volume, or biomass using allometric relationships with crown dimensions and sampling methods akin to fixed-area plots or horizontal point sampling (Section 13.3.3). Because a substantial fraction of trees may not be detected, the estimates of stand variables are likely to be biased downward. Imperfect or off-site allometric equations, and the inability to assess form or defect in the remotely sensed imagery, could also contribute to bias. However, if the fraction of trees that escapes detection is consistent, or if large trees that are detected relatively frequently also contribute disproportionately to the inventory variables of interest, then the estimates of those variables calculated from the remotely sensed data are likely to be highly correlated with those obtained by ground sampling. For example, one might use the algorithms of Palace et al. (2008) to identify, locate, and estimate the crown dimensions of trees over a large inventory area, then use a number of ground plots paired with comparable samples from the imagery to calibrate and adjust the remote sensing-based estimates. The success of such an approach would likely depend on accurate geopositioning of the ground plots relative to the imagery.

13.4.2. Applications to Ratio and Regression Sampling

General sampling approaches that take advantage of covariates were described in Section 10.8. Many different approaches are applicable with covariates developed from remotely sensed data. In practical work, the majority of applications involve regression sampling or double sampling using either a regression or ratio estimator. In these situations, the remotely sensed data provide a covariate (x) while the ground data provide reference values for the inventory variable of interest (y); note that we are not using x and y to refer to spatial coordinates here. Using the classification in Section 10.8, we can outline several common situations:

1. Regression sampling (true mean of covariate, X, known by direct calculation from the remotely sensed data over the entire inventory area)
   - Regression estimation: $\bar{y}_R = \bar{y} + b\left(\mu_x - \bar{x}\right)$, where $\bar{y}$ is estimated from the ground sample data, $\bar{x}$ is estimated from the matching locations in the remotely sensed data, b is estimated by regression, and μ_x is the known population mean computed from the remotely sensed data for the entire inventory area.
   - Ratio estimation: $\bar{y}_R = \hat{R}\mu_x$, where $\hat{R}$ is an estimated ratio obtained from the ground sample data and the matching locations in the remotely sensed data by either $\hat{R} = \bar{y}/\bar{x}$ (ratio of means) or $\hat{R} = \sum\left(\dfrac{y}{x}\right)/n$ (mean of ratios).
2. Double sampling (true mean of covariate, X, is unknown because remotely sensed data do not cover the entire inventory area, or because exhaustive computation of

the covariate over the entire area would be too expensive, so the mean must be estimated from a sample).

- Regression estimation: $\bar{y}_{Rd} = \bar{y}_S + b\left(\bar{x}_L - \bar{x}_S\right)$, where $\bar{y}_S$, b, and $\bar{x}_S$ are determined from a small sample comprised of the ground data and the matching locations in the remotely sensed data, and $\bar{x}_L$ is determined from a larger sample (the entire area covered by the remotely sensed data, or the area that can be processed for the covariate).
- Ratio estimation: $\bar{y}_{Rd} = \hat{R}\bar{x}_L$, where $\hat{R}$ is either a ratio of means or a mean of ratios obtained from the small sample and $\bar{x}_L$ is obtained from the large sample.

Double-sampling scenarios often arise when airborne platforms provide data for strips across the inventory area; in this case, the remotely sensed strips constitute the large sample while plots or other ground samples located within those strips define the locations of the small samples. For recent examples of double-sampling strategies using a combination of field plots and LiDAR, see Gregoire et al. (2011) and Ståhl et al. (2011).

The simple classification outlined above is appropriate when the ground samples are located independently of one another. However, it is often advantageous (both in terms of the efficiency of field work, and for the ability to draw inferences about stands or other management units as well as individual sample points) to conduct a nested ground sample. First, a set of stands or other polygons is selected from a list or map (possibly with equal probability, but more commonly with probability proportional to area or to predicted volume). Then, multiple ground sample points are located within each selected polygon. Strictly speaking, this represents a form of *multistage sampling*, and the computation of estimates from a multistage sample can easily take advantage of remotely sensed covariates (though the estimating equations used differ from those presented in Section 10.8 to reflect the hierarchical nature of the ground sampling). The number of potential design options (including stratification at one or more levels of the sampling hierarchy, which could also be facilitated by the remotely sensed data) is enormous. An exhaustive list of the possible approaches is outside the scope of this book; general sampling texts such as those by Cochran (1977) and Thompson (2012), as well as advanced sampling texts specifically targeting forestry and environmental applications such as Mandallaz (2008) and Gregoire and Valentine (2008), are useful in analyzing such designs.

Sometimes, the data available for regression sampling or double sampling do not satisfy the usual sampling assumptions. For example, the flight lines of an airborne platform are rarely assigned at random and will often include some departures from the original plan even if they were intended to provide data across a set of systematically spaced strips. Even when remotely sensed data are available for the entire inventory area, it is sometimes necessary from a practical perspective to confine ground sampling to operationally accessible portions of the area. Such situations require a shift from a conventional inventory approach in which the random location of samples forms the basis for inference (often called *design-based* or *model-assisted* inference), to one where the observed data are considered to be a random departure from a specified underlying model (often called *model-based* or *model-dependent* inference). In model-dependent inference, the adequacy of the underlying model is critical to the credibility of the inventory estimates and to any associated estimates of uncertainty (such as standard errors). Model-assisted and model-dependent inference often employ similar formulae for calculating estimates, but not always. The theoretical and practical distinctions between model-assisted and model-dependent inference remain contentious in the forest biometrics literature, though most resource managers and many researchers remain unaware of the distinction or controversy. Recent discussions relevant to forestry include those by Gregoire (1998) and Magnussen (2015).

13.4.3. Imputation and Mapping

Often, mapping the variables associated with a forest inventory is of equal (or even greater) importance than estimating means and totals. In principle, one can apply the linear models used in ratio or regression estimation (Sections 10.8 and 13.4.2) to predict inventory variables for those pixels or polygons where no ground data were collected. However, such models can produce unrealistic values, such as negative or implausibly high estimates of volume or biomass, especially when pixels or polygons have covariate values outside the range of those used to develop the models. Where this does not occur, maps of regression predictions typically show less variability than is seen in the original sample data because only a proportion of the variance of the variable of interest can be predicted from the remotely sensed data. Finally, modern forest inventories are often concerned with a large number of variables, and developing separate regression or ratio models for each variable can be tedious. Moreover, when separate prediction equations are used for different variables, sometimes the results can be incongruous. For example, one equation might predict a reasonable basal area, while another predicts a reasonable number of trees per hectare; but the two equations in combination may not give predictions that imply a reasonable quadratic mean diameter.

An increasingly popular approach to overcome these limitations of the ratio or regression estimation approach is to "fill in" pixels or polygons using the ground data from similar pixels or polygons, a process known as *imputation*. The use of imputation techniques in forest inventory was pioneered by the National Forest Inventory in Finland, where a technique known as *k-nearest neighbor* (*k*-NN) imputation has been used for nearly three decades (Tomppo and Katila, 1991). In *k*-NN imputation for a raster image, the characteristics of a target pixel (such as its reflectance values in different spectral bands, in the case of optical imagery) are compared to those of the pixels for which ground data are available. Those that are judged to be most similar, or nearest to the target pixel using a measure of distance between sets of characteristics, are used to inform the imputation. The parameter *k* determines the number of nearest neighbors used; for example, if $k=7$, then the seven most similar pixels associated with the ground data are used in imputing a value to the target pixel. In the simplest form of imputation, the values of inventory variables associated with the *k*-NN pixels are averaged and assigned to the target pixel. More commonly, a weighted average is used, so that the most similar neighbors contribute more heavily to the average than those that are less similar. Alternatively, but less commonly, one of the *k* neighbors is chosen at random and its value is assigned to the target pixel directly.

The simplicity of the *k*-NN approach, and the availability of efficient software (e.g., Crookston and Finley, 2008), has led to its growing use by other national forest inventories (McRoberts and Tomppo, 2007) and in an increasing number of operational inventories as well. Although originally pioneered using optical remote sensing, the technique is also suitable for LiDAR (Hudak et al., 2008). More sophisticated imputation approaches, including the gradient nearest-neighbor technique (Ohmann and Gregory, 2002) and Random Forest imputation (Breiman, 2001; Eskelson et al., 2009), have also been explored. Comparison of these alternative techniques, and the assessment of their strengths and weaknesses, is an active area of current research (e.g., Fuchs et al., 2009; Temesgen and Ver Hoef, 2015).

13.4.4. Areal Importance Sampling

Another potential use of remote sensing–derived covariates is to drive the sampling probabilities themselves. For example, in a typical stratified inventory, the location of ground
plots within a stratum would usually be chosen uniformly at random, or systematically.
Williams (2001b) proposed allocating the sample locations within each stratum with probability proportional to the predicted value of a target variable, such as volume or number of
trees per acre. When space is considered as continuous, and the number of possible sample
points is infinite, this approach is properly considered as an areal form of *importance
sampling*. When raster-based remote sensing provides the covariate, this approach can be
considered a form of list sampling (Section 10.9): the list of pixels serves as the population,
and sampling from the list is done with probability proportional to the covariate value. (If
pixels are large, then selecting a random sample point within the pixel may be a useful final
step.) If the covariate is well correlated with the variable of interest, then substantial reductions in the number of ground samples required may be possible. Williams (2001b) provides
equations for estimating the stratum means and standard errors.

A challenge for this approach is how to choose the sampling probabilities when multiple
variables are of interest in a forest inventory: sampling with probability based on a covariate that is highly correlated with one variable may lead to probabilities that are poorly
correlated with others. This does not create bias in the resulting estimates, but it can lead to
dramatically larger standard errors for the nontarget variables. Perhaps for this reason, areal
importance sampling has not seen widespread use in practical work to date. However, it
might prove valuable for applications where a single variable (such as volume or biomass)
is of primary importance for an inventory.

14

MEASUREMENT OF TREE
AND STAND GROWTH

The individual tree and plot measurements and sample designs discussed in the previous chapters serve to provide resource managers with a snapshot, or static assessment, of the tree, plot, stand, or forest. Because trees are living organisms, they grow, die, and new trees are born. As a result, stands and forests are dynamic systems that are constantly changing. The size, abundance, and composition of stands and forests are influenced by many factors including endogenous factors such as individual tree life cycles and inter- and intraspecific competition, and exogenous factors such as fire, insects, and environmental and geological disturbances (Oliver and Larson, 1996). Human interventions also influence stand and forest structure. Because of these dynamics, tree and plot measurements and the resulting stand- and forest-level estimates are valid for a limited period of time. The length of time over which inventory estimates are applicable to a given forest varies widely depending upon species composition and growth rates, stage of stand development, and frequency of exogenous disturbances. Inventory estimates for a mature boreal forest, free of exogenous disturbances, might be representative for a decade or more, while estimates from a fast-growing plantation of pine in the southern United States or eucalyptus in the subtropics might only be applicable for a single growing season.

The term *growth* is generally used to describe changes in size or content over time, while the term *increment* is used to describe differences or rates of change (Weiskittel et al., 2011). Increases in tree dimensions or stand attributes should be qualified by the period of time during which the increments occurred. The period may be a day, a month, a year, a decade, and so on. When the period is a year, the increase, termed *current annual increment* (CAI), is the difference between the dimensions measured at the beginning and at the end of the year's growth. Since it is difficult to measure some characteristics, such as volume, for a single year, the average annual growth for a period of years, termed *periodic annual*

Forest Mensuration, Fifth Edition. John A. Kershaw, Jr., Mark J. Ducey, Thomas W. Beers and Bertram Husch.
© 2017 John Wiley & Sons, Ltd. Published 2017 by John Wiley & Sons, Ltd.

increment (PAI), is often used in place of CAI. This is found by obtaining the difference between the dimensions measured at the beginning and at the end of the period, say 5 or 10 years, and dividing by the number of years in the period. If the difference is not divided by the number of years, it is termed *periodic increment*. The average annual increase to any age, termed *mean annual increment* (MAI), is found by dividing the cumulative size by the age.

Forest managers have developed a number of tools and approaches to deal with changes in tree size and forest composition and structure. Periodic forest inventories, networks of permanent sample points, and growth and yield models are the more common tools used by resource managers to predict changes, update inventories, and make long-term forecasts to be used as inputs into management planning software. In this chapter, the mensurational aspects of measuring or estimating tree, stand, and forest growth are presented. An overview of growth and yield models is presented and the role of these models in forest mensuration explored. For a more detailed discussion of growth and yield models and their development, see Weiskittel et al. (2011).

14.1. INDIVIDUAL TREE GROWTH

Tree growth consists of elongation and thickening of roots, stems, and branches. Growth causes trees to change in volume and weight (size), and form (shape). As discussed in Section 6.1.9, carbon and nutrient contents tend to be tissue specific and change little with tree age (Table 6.4), so while total amounts increase with increasing tree size, concentrations are fairly stable (Swamy et al., 2003, 2004).

Linear growth of all parts of a tree results from activities of the primary meristem; stem thickening (i.e., diameter growth) results from activities of the secondary meristem, or cambium, which produces new wood and bark between the old wood and bark. Tree growth is influenced by the genetic capabilities of a species interacting with the environment. Environmental influences include *climatic factors* (air temperature, precipitation, wind, and insolation); *soil factors* (physical and chemical characteristics, moisture, and microorganisms); *topographic characteristics* (slope, elevation, and aspect); and *competition* (influence of other trees, lesser vegetation, and animals). The sum of all these environmental factors is expressed as site quality (Section 8.8).

Total and merchantable height growth, diameter growth at breast height, and diameter growth at points up the stem are elements of tree growth traditionally measured by mensurationists; from these elements, volume or weight growth of sections of the stem, or the entire stem, may be determined. Changes in roots, branches and crown dimensions, such as crown length, crown width, and height to the base of the crown, are also measured in many situations. Carbon and nutrient contents are generally calculated from the new values of volume or weight.

14.1.1. Tree Growth Curves

The pattern of an individual organism's growth can be summarized by plotting size (e.g., volume, weight, diameter, or height for a tree) over its age, the curve so defined is commonly called the *growth curve* (Fig. 14.1*a*). Such curves, characteristically S- or sigmoid-shaped, show the cumulative size at any age. Thus, they are more descriptively termed *cumulative growth curves*. A true growth curve, which shows increment at any age, results from plotting increment over age (Fig. 14.1*b*). The S-shaped form of the cumulative growth

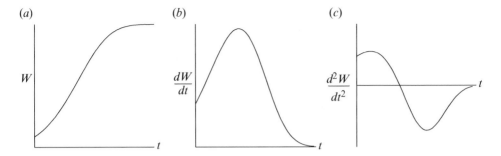

FIG. 14.1. The curve of (*a*) cumulative growth, (*b*) growth rate, and (*c*) growth acceleration (*W*=size, *t*=age).

curve is evident for individual cells, tissues, and organs, and for individual plants and animals over their full life span. Also, the pattern of growth for short growing periods, such as a growing season, tends to follow an S-shaped curve.

Although the exact form of the cumulative growth curve will change when the tree dimension (height, diameter, basal area, volume, or weight) plotted over age is changed, the cumulative growth curve has characteristics that hold for all dimensions of a tree. With this in mind, an insight into tree growth can be obtained by studying Figure 14.1*a* and the derived curves in Figure 14.1*b* and *c*. During youth, the growth rate increases rapidly to a maximum at the point of inflection in the cumulative growth curve, and the acceleration first increases and then drops to zero at the point of inflection in the cumulative growth curve. During maturity and senescence, the growth rate decreases with related changes in acceleration.

There are several ways to mathematically express the curves shown in Figure 14.1. One of the more common formulations in forestry and ecology literature is the Chapman–Richards generalization (Richards, 1959; Chapman, 1961) of the von Bertalanffy (1957) growth curve (Pienaar and Turnbull, 1973):

$$W_t = W_{max}\left(1-e^{-k \cdot t}\right)^m \tag{14.1}$$

where W_t = the size of the organism at age t
         W_{max} = the asymptotic maximum size
              k and m are species-specific growth parameters
         t = age

Equation (14.1) corresponds to the curve shown in Figure 14.1*a* and is, therefore, a cumulative growth curve. The equation for Figure 14.1*b* (increment equation) is the derivative of eq. (14.1) with respect to *t*:

$$\frac{dW_t}{dt} = W_{max} \cdot k \cdot m \cdot e^{-k \cdot t}\left(1-e^{-k \cdot t}\right)^{c-1} \tag{14.2}$$

and the equation of Figure 14.1*c* is the second derivative of eq. (14.1) with respect to *t*:

$$\frac{d^2W}{dt^2} = \frac{dW}{dt}\left(\frac{k\left(m \cdot e^{-k \cdot t}-1\right)}{1-e^{-k \cdot t}}\right) \tag{14.3}$$

The parameters, W_{max}, k, and m, can be obtained for a given species and attribute using nonlinear regression analysis (Pienaar and Turnbull, 1973; Zeide, 2004; Weiskittel et al., 2011).

Curves of CAI, PAI, and MAI may also be derived from a cumulative growth curve by computing increments from sizes read from the cumulative growth curve at chosen ages and by plotting the increments over age. Figure 14.2 shows curves of PAI and MAI derived from a cumulative height growth curve. From Figure 14.2, we can see that MAI culminates (is maximal) when it equals PAI (this is also true when MAI equals CAI). A formal proof of this could be given using eqs. (14.2) and (14.3), but the reason is obvious: MAI will lag behind PAI if it is smaller than PAI. When PAI drops below MAI, MAI must decrease; MAI, therefore, reaches its maximum when equal to PAI.

When developing or working with growth curves, one should realize that each species, perhaps each tree, matures at its own rate. This physiological time varies from one tree species to another, and from one stage of development to another. The span of ages required to fit a growth curve that accurately depicts the true development of a species, therefore, varies. For aspen species, this time span may be 40–60 years, while for long-lived species like Douglas-fir, it may be 150 years or more. Because of the long timeframes required, the data are generally collected from a chronosequence of trees of different ages from different sites. The data may be gathered by remeasuring the same trees at specific time intervals or by felling trees and reconstructing their development using stem analysis techniques. Diameter growth also can be estimated from increment cores. These techniques are described in detail below.

14.1.2. Growth Percent

Growth percentage is a means of expressing increment of a tree parameter in relation to total size of the parameter at the initiation of growth. Although growth percentage is most frequently used for volume and basal area growth, it is applicable to any parameter.

In terms of simple interest, growth percent p is

$$p = \frac{s_n - s_o}{ns_o}(100) \qquad (14.4)$$

where s_o = size of parameter at beginning of growth period

s_n = size of parameter at end of growth period

n = number of units of time in growth period

In this equation, average growth per unit of time is expressed as a percentage of the initial size s_o. To illustrate, if the present volume of a tree is 400 board feet and the volume 10 years ago was 300 board feet,

$$p = \frac{400 - 300}{10(300)}(100) = 3.3\%$$

(a)

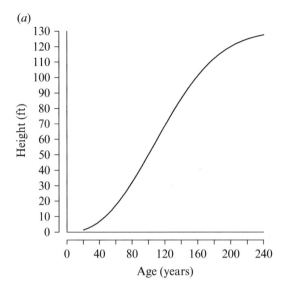

(b)

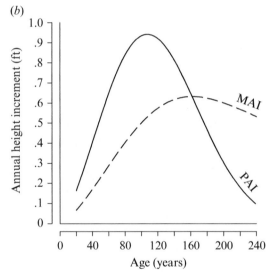

FIG. 14.2. Height growth curves: (a) the cumulative height growth curve and (b) the curves of periodic annual height growth (PAI) and mean annual height growth (MAI); PAI and MAI curves derived from cumulative height growth curve.

In terms of compound interest, growth percent p is

$$p = \left(\sqrt[n]{\frac{S_n}{S_o}} - 1 \right) 100 \tag{14.5a}$$

In this form, p may be computed by logarithms; however, a more convenient form of the equation is

$$\left(1+p\right)^n = \frac{s_n}{s_o} \tag{14.5b}$$

The compound interest rate for the previously mentioned tree is then

$$\left(1+p\right)^{10} = \frac{400}{300} = 1.333$$
$$p = 2.9\%$$

The compound interest rate is based on the premise that the increment for each unit of time is accumulated, resulting in an increasing value of s_o. Thus, as the period increases, the simple and compound interest rates will diverge more and more. For short periods, however, they will be almost the same.

To avoid the use of compound interest, Pressler (1860) based a simple rate of interest on the average for the period, $(s_n + s_o)/2$, which has the effect of reducing the rate to near the compound interest rate. Pressler's growth percent, p_p, is

$$p_p = \left(\frac{s_n - s_o}{s_n + s_o}\right)\frac{200}{n} \tag{14.6}$$

For the previous example,

$$p_p = \left(\frac{400 - 300}{400 + 300}\right)\frac{200}{10} = 2.86\%$$

Growth percentage is essentially the same parameter as relative growth rate (e.g., Ford, 1975; Cannell et al., 1984), which is often advocated as an expression of growth independent of tree size; however, it is essential to remember that growth percentages are ratios between increment and initial size. Thus, percentages change as the amount of increment, and the base on which it is accrued, changes. As trees grow, the base of the percentage constantly increases, and the growth percentage declines even though the absolute increment may be constant or even increasing slightly (Fig. 14.3). In early life, the growth percentage for a tree is at its highest because the base of the ratio is small; the percentage falls as the size of the tree increases. Although young trees may grow at compound rates for limited periods, growth percent is generally an unsafe tool for predicting tree or stand growth because of the uncertainty in extrapolating growth curves.

14.2. DIRECT MEASUREMENT OF TREE GROWTH

Changes in diameter, total and merchantable heights, height to crown base, form, and other physical attributes of individual trees can be obtained by remeasuring the same individuals at periodic intervals. Individual tree attributes are measured using the tools and techniques described in Chapter 5. Extra care and quality control of measurements are required when trees are designated to be remeasured relative to trees that are only going to be measured for a single inventory (Curtis and Marshall, 2005). Measurement error tolerances that are acceptable for a single inventory may be too large for repeated measurements, resulting in increment estimates that are too large, too small, or, in some cases, even negative.

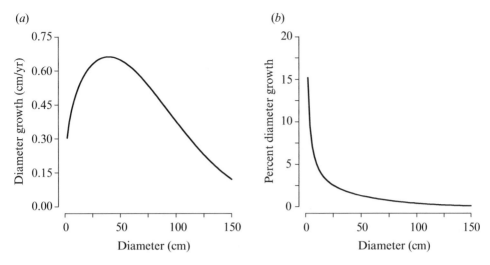

FIG. 14.3. Diameter growth curves: (*a*) absolute annual growth and (*b*) percent annual growth.

Careful field procedures and field checking new measurements against previous measurements can minimize these types of errors.

14.2.1. Diameter Growth Measurement

Diameter growth of individual trees can be obtained by measuring tree diameters at the beginning and at the end of specified periods and calculating differences. Since annual diameter increment is small for many species, when instruments such as calipers and diameter tapes are used, measurements are commonly taken at intervals of several years. Diameter growth at breast height is the most common diameter growth measure obtained in most long-term studies, though diameter growth measures at any height can be obtained. It is generally recommended to mark the point on the stem where diameter is measured. Weather-proof loggers' paint or a bark scribe can be used to mark the point where diameter was measured. When a bark scribe is used, diameter is generally measured directly above the scribe, though this practice should be clearly documented. Care should be exercised when using bark scribes on thin-barked trees to avoid damaging the cambium. Tags nailed or stapled to the tree also can be used. In some species, especially fast-growing species, nails can cause stem deformities, thus altering stem diameter. In some cases, nails or pins are placed near the base of the tree and a standard staff is placed on top of the nail to determine the height at which diameter is measured. The method of determining the point on the stem to measure diameter needs to be clearly documented so that variations in measurement heights do not influence measurements of diameter and estimates of diameter growth.

Precise measurement of minute increments in diameter may be required in research. Such changes, which may be for periods as short as an hour, cannot be detected with calipers or diameter tapes. Consequently, dendrometer bands, dial-gauge micrometers, recording dendrographs, and transducers are used to measure minute changes. Dendrometer bands, as described by Liming (1957), Bower and Blocker (1966), and Yocom (1970), consist of aluminum or zinc bands with Vernier scales. These bands are placed around tree stems and held taut with springs. Changes in diameter as small as 0.001 in. can be read.

Another type of band dendrometer is the "Dial Dendro" (Relaskop–Technik). This requires fixing a support bracket to the stem and then a steel strap is hooked onto the level or a dial disk. The instrument can measure changes as small as 0.1 mm. Reineke (1932) first described a dial-gauge micrometer. This instrument utilized a stationary reference point, a hook screwed into the xylem. The micrometer measures the distance from this fixed point to a metal contact glued to the bark. Changes in diameter as small as 0.001 in. could be read. Daubenmire (1945) modified this instrument by inserting three screws into the xylem as fixed reference points. Dendrometer bands are readily available from a number of manufacturers and are the most common method currently used for such precise measurements in forestry research.

Fritts and Fritts (1955) devised a precision dendrograph. The instrument consists of a pen on an arm bearing on a fixed point on the tree stem. The pen records diameter changes to 0.001 in. on a chart mounted on a drum of an eight-day clock. Phipps and Gilbert (1960) designed an electrically operated dendrograph similar in principle to the dial-gauge micrometer. A potentiometer was fixed to a tree by screws anchored in the xylem. A movable shaft was fixed to the outer bark; any displacement of the shaft was measured by a change in electrical resistance and recorded on a continuous strip chart. Imprens and Schalck (1965) used a variable differential transformer rather than a potentiometer in a similar instrument. Kinerson (1973) designed a transducer that uses a linear motion potentiometer fixed to an invar plate. When the device is fixed to a tree stem, changes in the stem diameter move the potentiometer shaft. Of these tools, point dendrographs are the only ones currently available commercially.

14.2.2. Height Growth Measurement

For many coniferous species, height increments can be measured by measuring distances between branch whorls; however, this is generally only practical on small trees or felled trees. Height increment on hardwood species can be similarly measured by determining the location of bud scale scars. Height growth of individual trees most often is obtained by measuring the total (or merchantable) heights of standing trees at the initiation and at the cessation of specified growing periods and calculating differences. Tree heights up to 75 ft (23 m) can be accurately and precisely determined with telescopic measuring poles that can be purchased from forestry and engineering supply companies. However, for measuring heights over 30 ft, poles are slow and cumbersome; then, it is more convenient to use a transit or some precise tripod-mounted instrument that gives height indirectly.

Repeated measurements of trees with mechanical hand-held hypsometers (Fig. 5.14) generally do not give sufficiently precise increment measurements. The newer electronic hypsometers (Fig. 5.15) produce height estimates that are extremely precise and repeatable (Skovsgaard et al., 1998); however, these devices tend to underestimate total height in many situations. If the error from these tools is consistent across the range of heights, then precise estimates of increment can be obtained from the differences in estimated heights from two periods; however, if the error varies with height, then the increment estimates will not be reliable.

As with diameter measurements, measurement tolerances of heights that are acceptable for single inventories are often too large to produce acceptable growth measurements. The best practice for repeated measurements is field checking current measurements against past measurements. Some measurement protocols actually record the distance and azimuth from which height measurements were made and require field crews to use the same point for remeasurement. While this does have potential to reduce interobserver bias, it should be noted that optimal distances for measuring tree heights vary with tree height (Fig. 5.17), and tree tops are not always going to be visible from the same point over time. It is better practice to have observers measure heights from locations where tree tops are clearly visible.

Damage from animals, harvest operations, neighboring trees, storm events, and fire can all influence subsequent diameter and height measurements. In the case of diameter measurement, it may be necessary to move the measurement point on the stem. In this case, detailed field notes can aid in subsequent data entry and error checking. Top damage to trees can result in negative height growth estimates. Again, detailed field notes are invaluable to subsequent data users.

It may be possible to estimate height increments from repeated LiDAR (Section 5.3.4) flights (Meyer et al., 2013). While the work by Meyer et al. (2013) focused on estimation of changes in mean canopy height, application to individual trees is theoretically possible. The technology and postprocessing capabilities are rapidly improving, and LiDAR technology has the potential to eliminate or minimize many of the field measurements limitations described above.

14.2.3. Crown Growth Measurements

Like height and diameter, crown measurements can be made over time using the same tools described in Chapter 5. Obtaining accurate measurements so that reliable estimates of crown recession, crown expansion, and leaf area changes can be challenging. Crown dimensions can be very difficult to delimit in the field, thus these measurements are subject to much variation between observers and over time. Height to the base of live crown, crown length, and crown ratio are easier to measure on coniferous trees that have more determinant growth forms than on hardwood trees with more indeterminant growth forms. On coniferous trees, branches are generally clustered in distinct whorls, thus identification of crown base is easier, though it is important to have a clear definition of crown base (Section 5.5.1). On hardwood trees, foliage is often dispersed more widely and determining the base of live foliage is subject to more interpretation by observers. Again, clear definitions of crown base and field checking new measurements against previous measurements will minimize measurement errors and produce more consistent and useful results.

14.2.4. Belowground Growth Measurements

Belowground growth, like belowground biomass, is very difficult to directly measure, and most often is estimated from allometric equations (Section 14.3.3) using component ratios or other diameter-based equations (Heath et al., 2009). The recent interest in carbon accounting has resulted in an increased interest in root biomass dynamics; however, studies of root growth and development are not new and date back to Theophrastos of Lesbos (300 BCE), who observed that roots start growing before shoots in spring. Hartig (1863) is one of the earliest systematic studies of root morphology, while Resa (1877) is one of the earliest systematic studies of seasonal root growth dynamics. In both of these studies, roots were periodically exposed and measurements made, a practice that continued well into the

(*a*)   (*b*)

(*c*)

June 2005   July 2005   August 2005

FIG. 14.4. Root growth measurements: (*a*) soil coring, (*b*) minirhizotron tube and camera, and (*c*) sequence of minirhizotron images.

twentieth century (e.g., Stevens, 1931). Excavation of roots results in soil mixing and root damage, both of which alter root growth, thus these methods only provided very imprecise measures of root growth (Lyr and Hoffmann, 1967). An alternative to periodically excavating roots is to install a dendrometer band on roots. There are a few commercially available dendrometer bands specifically made for measuring root growth (e.g., see http://www.phytogram.com/RootDendrometer.htm; last accessed August 31, 2016).

Changes in belowground biomass can be estimated from repeated sampling using soil cores (Fig. 14.4*a*); however, biomass estimates based on soil cores often are highly variable. As a result, very large samples are required in order to obtain reliable growth estimates. Additionally, soil cores are generally only effective at sampling smaller roots. Repo et al. (2005) describe a nondestructive/noninvasive method of determining root growth based on analysis of electrical impedance. While this technique shows promise, the method was only tested on willow species grown in a hydroponic situation.

Transparent viewing planes have been widely used for observing and measuring root growth (Glinski et al., 1993). The most common form of transparent viewing plane is the *rhizotron*, a trench or pit dug in the soil and lined with clear acrylic or glass. Roots growing along the transparent plane can be measured over time, and growth, production, and mortality can be estimated. Rhizotron studies normally require several years of observation to yield reliable results (Glinski et al., 1993) because the disturbance created by installing the rhizotrons greatly disrupts the root systems of the trees being observed. *Minirhizotrons* (Fig. 14.4*b*), which consist of clear acrylic tubes inserted into the soil, are similar to rhizotrons, but are less disruptive to the soil and root systems. A camera is inserted in the tubes (Fig. 14.4*b*) and photos are obtained at preset intervals over time. Root growth can then be determined from the photo sequences (Fig. 14.4*c*). Minirhizotrons are primarily used for measuring fine root production in trees. All transparent viewing plane techniques provide an estimate of root length density (Glinski et al., 1993). Glinski et al. (1993) provide a comprehensive review of several techniques used to transform estimates of root length density into biomass per unit volume of soil. Bernier and Robitaille (2004) propose an extension of the line intersect method (Section 12.2) that they call the plane intersect method for estimating biomass per unit volume of soil using minirhizotron data.

14.3. RECONSTRUCTING TREE GROWTH

Direct measurement of tree growth requires repeated observations on the same set of trees over time. Not only does this require several years to obtain the data, it is also a very expensive exercise. Timeframes for many studies and analyses require data more quickly than most remeasurement programs can provide. Remeasurement data are valuable assets to forest researchers and managers, and the establishment and maintenance of permanent sample plot (PSP) networks are strongly encouraged; however, many alternative methods for gathering tree growth data are available.

14.3.1. Stem Analysis

As discussed in Section 5.1, many tree species around the world have an annual record of growth stored as alternating rings of lighter earlywood and darker latewood. A count of these rings provides a measure of tree age, and, by measuring the width of these rings, stem diameter growth can be obtained. Thus, a record of the past growth of a tree may be obtained by a stem analysis (Duff and Nolan, 1953, 1957, 1958). Such a study shows how a tree grew in height and diameter and how it changed in form as it increased in size. In making a stem analysis, one counts and measures growth rings on stem cross sections at different heights above the ground. Measurements may be taken on a standing tree by using an increment borer (Sections 5.1 and 14.3.2), if the tree is not too big or the wood too hard. It is more convenient and more accurate, however, to obtain the measurements from cut cross sections.

Stem analysis has been used by foresters for centuries (Büsgen and Munch, 1929). One the most complete and comprehensive presentations of stem analysis is the work by Canadian physiologist Dr. G. H. Duff and his students (Duff and Nolan, 1953, 1957, 1958; Forward and Nolan, 1961a, 1961b) on red pine. The procedure for making a stem analysis on cut cross sections is simple and follows the procedures used by Dr. Duff:

1. Fell the tree and cut the stem into sections of desired lengths.
2. Determine and record species, dbh, total height, years to attain stump height, and total age.
3. Measure and record the height of the stump, length of each section, and length of tip.
4. Measure and record average diameter at top of each section.

5. If only one radius is measured, the average radius should be used. The average radius should be located on each cross section and marked with a line along it with a soft pencil. In many cases, it is desirable to measure more than one radius because of stem eccentricity. For example, the longest and shortest radii or two radii at right angles might be measured. Each radius to be measured should be located and marked with a line using a soft pencil.

6. Along each radius, count the annual rings from the cambium inward, marking the beginning of each ring at the desired interval (e.g., every year, every 5th year, every 10th year, etc.). Record the total number of rings at each cross section.

7. From the center of each cross section, measure outward toward the cambium along each radius, recording the distance from the center to each interval. The fractional part of a decade, or other desired period, will be measured and recorded first. The radii on each disk can be averaged using the arithmetic mean or geometric mean (Section 3.3).

Table 14.1 shows an example of how the measurements should be recorded, and Table 14.2 shows how the height measurements should be summarized.

In making the stem analysis (Fig. 14.5), the first step is to draw a curve of *height* above ground of section tops over years to attain height at section tops (i.e., *age*) from data in Table 14.2. (This curve appears on the left side of the graph in Fig. 14.5.) Next, diameters for each section (i.e., double the radial measurements) are plotted for the appropriate height from data in Table 14.1 (e.g., the seven radial measurements for section 2 are doubled and plotted at 2.37 m). Finally, diameters within a year (or age) column are connected to form the taper curves for specific years (or ages); the terminal position of each taper curve is estimated from the curve of height over age. For a more detailed description of stem analysis methods and data analysis and interpretation, see Duff and Nolan (1953, 1957, 1958) and Forward and Nolan (1961a, 1961b).

Ideally, trees selected for stem analysis would be a probabilistic sample from the population of trees being studied; however, stem analysis is costly and often is conducted on a very small proportion of the population. As a result, purposeful selection of trees is often undertaken to insure that the range of species and size classes are represented. Care should be exercised when extrapolating growth rates or applying models built from such data to the larger population. If the selected trees are not a true sample of the population, then the growth estimates are not true estimates of population growth and may be biased. Use of ratio estimators based on basal area growth at breast height could be used to adjust stem analysis growth estimates and reduce this bias.

14.3.2. Estimating Diameter Growth from Increment Cores

Stem analysis provides a comprehensive whole-tree growth analysis; however, often only diameter growth at breast height is needed. Increment cores are an efficient and nondestructive way to obtain quick estimates of diameter growth. Increment borings should be taken from trees on sample plots or sample points. For example, if we used 1/5–acre plots on a cruise, we might establish 1/20–acre plots within selected 1/5–acre plots (say one in four) and bore trees on the 1/20–acre plots. If we used horizontal point sampling (using BAF = 10), we might use a 40-factor gauge at selected points to choose trees to bore. Boring

TABLE 14.1. Measurements for Stem Analysis of a 39-Year-Old Western Hemlock

Species:	Western hemlock	dbh:	27.8 cm	Total height:	18.5 m
Years to attain base height:	2			Total age:	39
Date:	November 15, 1990		Measured by:	JAK	

Section no.	Length (m)	Top dib (cm)	No. of rings at top	Average radial distance from pith to ring corresponding to nth year (mm)[a]							
				1955	1960	1965	1970	1975	1980	1985	1990
Base	0.37	25.4	37	3	12	29	46	57	73	105	127
1	1.00	25.3	35	0	8	26	45	59	76	107	126
2	1.00	24.6	31	0	1	14	38	54	72	104	123
3	1.00	23.4	29	0	0	7	31	49	68	98	117
4	1.00	22.9	26	0	0	1	23	46	66	97	115
5	1.00	22.3	24	0	0	0	13	35	60	91	111
6	1.00	20.6	23	0	0	0	8	30	51	83	103
7	1.00	19.6	21	0	0	0	2	20	44	77	98
8	1.00	18.0	19	0	0	0	0	13	38	69	90
9	1.00	16.2	18	0	0	0	0	6	29	58	81
10	1.00	15.1	16	0	0	0	0	2	22	51	76
11	1.00	12.0	14	0	0	0	0	0	11	37	60
12	1.00	9.7	12	0	0	0	0	0	2	26	49
13	1.00	7.5	9	0	0	0	0	0	0	14	37
14	1.00	5.4	7	0	0	0	0	0	0	5	27
15	1.00	3.6	6	0	0	0	0	0	0	1	18
16	1.00	2.2	4	0	0	0	0	0	0	0	11
Tip	2.16	0.0	0	0	0	0	0	0	0	0	0

[a] Average radius = the geometric mean of the longest and shortest radii; values are doubled to give average diameter when plotting taper curves.

467

TABLE 14.2. Height Summary for Stem Analysis of a 39-Year-Old Western Hemlock

Section No.	Length (m)	Height Above Ground at Top of Section (m)	Ring Count, Top of Section	Years to Grow Section	Years to Attain Height at Top of Section
Base	0.37	0.37	37	2	2
1	1.00	1.37	35	2	4
2	1.00	2.37	31	4	8
3	1.00	3.37	29	2	10
4	1.00	4.37	26	3	13
5	1.00	5.37	24	2	15
6	1.00	6.37	23	1	16
7	1.00	7.37	21	2	18
8	1.00	8.37	19	2	20
9	1.00	9.37	18	1	21
10	1.00	10.37	16	2	23
11	1.00	11.37	14	2	25
12	1.00	12.37	12	2	27
13	1.00	13.37	9	3	30
14	1.00	14.37	7	2	32
15	1.00	15.37	6	1	33
16	1.00	16.37	4	2	35
Tip	2.16	18.53	0	4	39

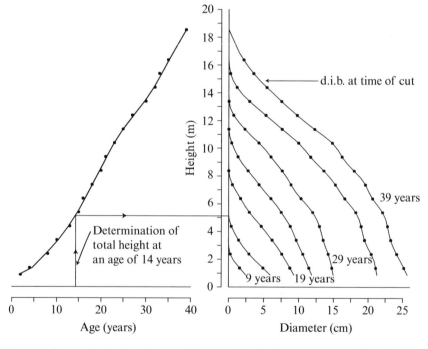

FIG. 14.5. Stem analysis for a 39-year-old western hemlock; taper curves at 5-year intervals.

one, two, or any predetermined number of trees per plot, rather than sampling, and then using the mean growth of the bored trees to estimate average growth rates will most likely result in an overestimation of growth because in open stands the sample will represent a larger proportion of the trees than in dense stands, and trees in open stands generally grow faster than trees in dense stands.

For each species or species group studied, a reliable estimate of average diameter increment by diameter classes can be obtained from a representative sample of about 100 increment measurements. Because trees are not perfectly round, it is better to extract multiple cores from each tree and average the increment measures. Cores extracted from the long and short axes of the tree or taken at right angles from each other generally provide more reliable estimates of diameter increment than single cores. Bakker (2005) provides additional methods for adjusting diameter increment estimates based on pith location. If single cores are extracted, there should be clear instructions for determining the azimuth of the coring (e.g., obtain the core facing the sample location point), otherwise significant selection biases or large interobserver variation can result. An example form for recording the field data is shown in Table 14.3. This table indicates the recommended degree of accuracy for measurements and calculation. Generally, only columns 1, 2, 3, 4, and 6 are completed in the field. Table 14.3 only records the past 10-year growth record; if more detailed growth is required, including all rings or all rings in sets of 5- or 10-year increments, then a form like Table 14.1 may be required.

The constant K, the average ratio of diameter outside bark to diameter inside bark, is the inverse of the ratio used to calculate bark volume (Section 6.1.6). Thus, the equation of the straight line expressing diameter outside bark, D, as a function of diameter inside bark, d, is $D = Kd$. K varies by species and, to some extent, by locality. Average values of K are calculated as a ratio of means (Section 10.8):

$$K = \frac{\Sigma D}{\Sigma d} \tag{14.7}$$

Thus, for the trees listed in Table 14.3, $K = 1.10$.

The calculations for columns 5, 7, and 10 in Table 14.3 are self-explanatory. To calculate column 8, that is, past diameter at breast height outside bark, D_p, past diameter at breast height inside bark, d_p, is multiplied by K (eq. (14.7)). To calculate column 9, that is, periodic diameter increment outside bark, Δd, the past 10 years' radial wood growth, L, is doubled and multiplied by K because, if in a given period, $2L$ in. of wood are laid on a given past diameter inside bark, d_p, in terms of present diameter inside bark, d, past diameter inside bark will be

$$d_p = d - 2L$$

and

$$2L = d - d_p$$

If in a given period ΔD in. of wood and bark are laid on a given past diameter outside bark, D_p, in terms of present diameter outside bark, D, past diameter outside bark will be

$$D_p = D - \Delta D \tag{14.8}$$

TABLE 14.3. Determination of Diameter Increment from Increment Cores

[Column 1]	[Column 2]	[Column 3]	[Column 4]	[Column 5]	[Column 6]	[Column 7]	[Column 8]	[Column 9]	[Column 10]
Tree No. (inches)	Species	Present dbh o.b. (D)	Double Bark Thickness $(2b)$	Present dbh i.b. (d)	Past 10 Year's radial Wood Growth (L)	Past dbh i.b. (d_p)	Past dbh o.b. (D_p)	Periodic dbh o.b. Increment (ΔD)	Periodic Annual dbh o.b. Increment $(\Delta D/10)$
1	Hard maple	16.2	1.4	14.8	1.32	12.2	13.4	2.90	0.290
2	Hard maple	12.6	1.1	11.5	0.75	10.0	11.0	1.65	0.165
3	Hard maple	10.4	1.0	9.4	0.80	7.8	8.6	1.76	0.176
4	Hard maple	12.2	1.1	11.1	1.08	8.9	9.8	2.38	0.238
⋮	⋮							⋮	⋮
100	Hard maple		1.6	16.2	1.04	14.1	15.5	2.29	
Totals		1285.0	117.0	1168.0	98.70	970.6	1068.0	217.14	21.714
Averages		12.85	1.17	11.68	0.987	9.71	10.68	2.171	0.2171

and

$$\begin{aligned}
\Delta D &= D - Dp \\
&= K(d) - K(d_p) \\
&= K(d - d_p) \\
&= K(2L)
\end{aligned} \qquad (14.9)$$

When the data are presented as shown in Table 14.3, it is convenient to plot periodic growth or periodic annual diameter growth over present diameter outside bark, or over past diameter outside bark, and thus obtain average diameter growth by diameter classes. When a straight line can represent the relationship, it can easily be fitted by the method of least squares; however, if the trend is curvilinear, Mawson (1982) demonstrated that the following nonlinear model could be used to predict diameter growth Δd from diameter at breast height:

$$\Delta D = b_0 e^{b_1/D} \qquad (14.10)$$

where b_0 and b_1 = regression constants
$\qquad e$ = base of natural logarithms
$\qquad D$ = dbh

Equation (14.10) can be transformed into a linear equation using the natural logarithm and fitted using linear regression techniques:

$$\ln(\Delta D) = \ln(a) + b\left(\frac{1}{D}\right) \qquad (14.11)$$

Other useful equations for modeling diameter growth are discussed in Vanclay (1994) and Weiskittel et al. (2011).

Whether one plots diameter growth over present diameter outside bark or over past diameter outside bark, the final growth determinations will generally be about the same. The second alternative, which assumes that trees of a given diameter will have the same average diameter growth that trees of that diameter had in the past, is preferred to the first alternative, which assumes that future average diameter growth will equal past average diameter growth.

Although most increment borings are made at breast height, diameter increment determined at stump height may be converted to diameter increment at breast height because the relationship between diameter at breast height, D, and diameter at stump height, D_s, for most species or species groups, may be expressed by the simple linear regression equation:

$$D = a + bD_s$$

Thus, diameter increment at breast height, ΔD, may be obtained from diameter increment at stump height, ΔD_s, as follows:

$$D + \Delta D = a + b(D_s + \Delta D_s)$$

and

$$\Delta D = a + b\left(D_s + \Delta D_s\right) - D$$

By substituting $a + bD_s$ for D:

$$\Delta D = a + b\left(D_s + \Delta D_s\right) - \left(a + bD_s\right)$$

and simplifying

$$\Delta D = b\Delta D_s$$

Therefore, if the diameter growth at stump height for an individual tree is multiplied by the slope coefficient b of D over D_s for the appropriate species or species group, one obtains diameter increment at breast height.

14.3.3. Allometric Relationships

Allometric relationships, as described in Section 5.6, are widely used to estimate tree attributes that are either difficult to measure with acceptable precision and/or costly to measure directly. Height, height to crown base, crown ratio, leaf area, volume, and below-ground components are frequently estimated using allometric relationships. Theoretically, growth of various tree attributes can be estimated from differences in predictions from allometric equations:

$$\Delta X_t = \hat{X}_t - \hat{X}_0$$
$$= f\left(\mathrm{dbh}_t, \ldots\right) - f\left(\mathrm{dbh}_0, \ldots\right)$$

where ΔX_t = the estimated change in attribute X over time t

$\hat{X}_t$ = the predicted value of attribute X at time t

$\hat{X}_0$ = the predicted value of attribute X at time 0

$f()$ = the allometric equation

dbh_t = the diameter at time t

dbh_0 = the initial diameter

… indicates additional covariates in the allometric equation

This approach is widely used, especially for belowground growth estimates (e.g., Ravindranath and Ostwald, 2008). For well-studied variables, such as volume, where data are available across a range of age, size, stand, and site conditions, and where the allometric equations are based on more than one tree attribute (e.g., dbh and height), the approach is generally considered robust. However, using differences between predictions from an allometric equation as estimates of growth should be approached with caution, especially if these estimates are subsequently used to fit or calibrate growth and yield models. Potential problems with using differences between two predictions as an estimate of growth include

1. The algebraic equation forms implied by taking the differences between two predictions from allometric equations are very different from the equation forms presented in Section 14.1.1. This issue has been recognized in the growth and yield literature for decades and led to the development of compatible growth and yield equations (Buckman, 1962; Clutter, 1963).

2. In most allometric equations, dbh is the only covariate (or is at least the main covariate). dbh growth is very sensitive to stand density, stand structure, and competition from other trees, whereas other tree attributes may differ in their sensitivity to these factors (Oliver and Larson, 1996; Smith et al., 1997). The pattern of growth observed in dbh will be reflected in the patterns of growth that result from the differences in predictions.

3. Allometric equations are derived from regression analysis and each prediction has an unknown residual error component. While the expected value (average) of the residual errors is zero, the expected value of the differences between two residual error components is not likely to be zero (as is often implicitly assumed). The nonzero differences between error components can result in substantial biases in growth predictions. The use of mixed effects models with tree-level random effects may help reduce this potential bias.

4. Allometric equations are often derived from data collected across a range of sizes based on sample trees selected across a range of stand ages (i.e., a chronosequence). The allometric curve derived from trees of different sizes due to age may be substantially different from the curves derived from trees of different sizes within an age class. For example, Kershaw et al. (2008) show that the shape of height–dbh curves for many hardwood species in Indiana varied by dominant height of the stands.

5. If the allometric equations include covariates that are measures of stand structure and/or tree competitive status, such as those developed by Sharma and Zhang (2004), Sharma and Parton (2009), and Rijal et al. (2012), changes in these covariates over time, especially changes resulting from silvicultural treatments, can result in erroneous predictions and thus erroneous growth estimates from the differences.

Lack of data across a full range of site, stand, and tree conditions are one of the greatest limitations to the use of most allometric equations for growth and yield estimation. Simplistic models based on traditional measurements are another contributing factor (Ketterings et al., 2001). Fehrmann and Kleinn (2006) proposed using diameter measurements at a specified relative height rather than at breast height as a means of developing robust allometric equations across a range of site qualities. They found that diameter measured at 10% of total height provided a much more stable equation than those developed using the traditional dbh measurement and the allometric coefficients corresponded better to those derived theoretically from process models. It is likely these models would be more robust over time as well. Another potential solution that has not been widely developed would be use an approach similar to compatible growth and yield equations (c.f., Buckman, 1962; Clutter, 1963) for individual tree components.

14.4. STAND AND FOREST GROWTH

Tree growth involves the increase in size of various attributes (e.g., dbh, height, volume). Stands and forests are collections of trees commonly referred to as *populations*. Population growth includes not only the increase in size of individuals, but also the births and deaths of individual trees. As a result, stand dynamics and forest dynamics deal not only with change in sizes of individuals, but also changes in species composition and stand structure. In this section, we deal with the measurement issues associated with growth and development of forests. For more information on stand dynamics, see Oliver and Larson (1996) or Smith et al. (1997).

14.4.1. Components of Stand Growth

The structure of a stand—that is, the distribution of trees by species and size classes—changes from year to year because of the birth (regeneration) of new trees and the increase in size, death, and cutting of individual trees that compose a stand. These changes can be expressed in terms of various stand parameters: volume, weight, basal area, average stand diameter, height, and so on (see Chapter 8 for a complete description of stand parameters). The net effect of these changes (stand growth) may be positive, indicating increases, or negative, indicating decreases. Many problems of stand growth are best understood by considering a stand to be a population of trees and by studying the changes in the structure of the population. For example, consider the even-aged stand in Figure 14.6 for which two successive 100% inventories were collected. If the periodic diameter growth of all trees were 2 in., the periodic growth of this stand would be characterized by a displacement of the diameter distribution 2 in. to the right. Basal area or volume growth can be calculated as the differences between the first and second inventories; however, depending upon how *ingrowth*, *mortality*, and *cut* are included, very different expressions of growth can be produced.

The importance of ingrowth, mortality, and cut in any expression of stand growth is illustrated in Figure 14.6. Thus, before stand growth is considered, these important terms must be clearly defined. The terms are equally appropriate for any stand-level parameter (e.g., stand basal area, volume, weight, or carbon content), though for number of individuals, some of the definitions have slightly different meanings.

Ingrowth is the number or amount (volume, biomass, etc.) of trees periodically growing into a measurable size. There will normally be ingrowth between any two successive inventories, particularly when measurements are made above a minimum diameter threshold, such as 6 or 8 in. The amount of ingrowth will be variable from one period to another and dependent upon stand structure, species composition, and any disturbances (natural or human induced). For example, uneven-aged stands may have a more continuous rate of ingrowth, whereas even-aged stands may have high rates of ingrowth when young and little or no ingrowth as they age.

Mortality is the number or amount of trees periodically dying from natural causes such as old age, competition, insects, diseases, wind, and ice. Mortality may be insignificant to catastrophic and may occur at any time during a growth period. As with ingrowth, mortality varies with stand structure, species composition, and stage of development.

Cut is the number or amount of trees periodically felled or salvaged, whether removed from the forest or not. A cut may be light, medium, or heavy and may occur at any time during the period.

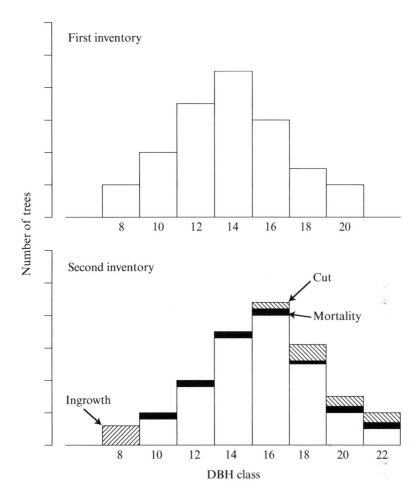

FIG. 14.6. Schematic representation of the changes in stand structure of an even-aged stand due to growth over a 10-year period (Beers, 1962).

14.4.2. Types of Stand Growth

With the above definitions of ingrowth, mortality, and cut in mind, the generally accepted stand growth terms (Beers, 1962) can be defined by the following equations:

$$G_g = V_2 + M + C - I - V_1 \tag{14.12}$$

$$G_{g+i} = V_2 + M + C - V_1 \tag{14.13}$$

$$G_n = V_2 + C - I - V_1 \tag{14.14}$$

$$G_{n+i} = V_2 + C - V_1 \tag{14.15}$$

$$G_d = V_2 - V_1 \tag{14.16}$$

where G_g = gross growth

G_{g+i} = gross growth including ingrowth

G_n = net growth

G_{n+i} = net growth including ingrowth

G_d = net increase

V_1 = value of the stand attribute (e.g., basal area or volume) at beginning of growth period

V_2 = value of the stand attribute at end of growth period

M = mortality

C = cut

I = ingrowth

In the above equations, mortality and cut may be defined in two different ways:

1. M and C represent the amount of M and C trees at the time of their death or cutting.
2. M and C represent the amount of M and C trees at the time of the first inventory—that is, the initial volume of M and C trees.

The above growth terms are best considered in the context of repeated measurements of PSPs or entire woodlands. Then, it will be clear that the method of inventory generally dictates the most applicable definition of mortality and cut. For example, in inventory systems where the trees are not numbered, it is necessary to measure mortality trees at the time of the second inventory, which amounts to measuring them at the time of death, and to measure cut trees at the time of cutting. Under these conditions, definition 1 would apply and computations of growth that included M and C would then include growth put on by trees that died or were cut during the period between inventories.

If the inventory system utilizes numbered trees, as in continuous forest inventory (CFI) procedures, one can use the initial amounts of cut and mortality trees and avoid measuring cut trees at the time of cutting. Under these conditions, definition 2 would apply, and computations of growth that included M and C would not include growth put on by trees that died or were cut during the period between inventories. Of course, if numbered trees are used, cut trees may be measured at the time of cutting and mortality trees measured at the second inventory, but this practice, which is seldom followed, requires extra care and record keeping.

When gross growth (eq. (14.12)) is computed using definition 2 for M and C, it includes only the growth on the trees that survived the period and is often called *survivor growth*. When gross growth is computed using definition 1 for M and C, it includes growth on trees that later died or were cut, and is often called *accretion*. (Note that growth on trees that later died or were cut may be an important component if cutting or mortality has been heavy, or if the interval between inventories is long.) Marquis and Beers (1969) recommended that "the terms *survivor growth* and *accretion* be used where appropriate, and that *gross growth* be considered a general term applicable only when either of these more precise terms is not appropriate." Gross growth has no meaning when numbers of trees is the stand attribute of interest since an individual tree only increases in size, not number.

Equations (14.12)–(14.16) apply when tree values are first totaled and the resulting sums manipulated. If PSPs are used, such as in the CFI system, and observations are made at the individual tree level, the following equations may be used to calculate the various types of growth:

$$G_g = V_{s2} - V_{s1} \tag{14.17}$$

$$G_{g+i} = G_g + I \tag{14.18}$$

$$G_n = G_g - M \tag{14.19}$$

$$G_{n+i} = G_g + I - M \tag{14.20}$$

$$G_d = G_g + I - M - C \tag{14.21}$$

where G_g, G_{g+i}, G_n, G_{n+i}, and G_d are the same variables as defined previously

V_{s1} = initial attribute value of *survivor trees* (live trees measured at both inventories)

V_{s2} = final attribute value of survivor trees

M = initial attribute value of mortality trees

C = initial attribute value of cut trees

I = attribute value of ingrowth trees

Since M and C represent the initial amounts of mortality and cut trees, gross growth, G_g, is correctly termed survivor growth. Table 14.4 illustrates the differences between eqs. (14.12)–(14.16) and eqs. (14.17)–(14.21).

If the volume totals from Table 14.4 are used, the net growth including ingrowth is obtained from eq. (14.15):

$$G_{n+i} = V_2 + C - V_1 = 749.3 + 241.4 - 744.4 = 246.3$$

Neither mortality volume nor ingrowth volume enters into the calculation. But if the growth per tree is first calculated, then the net growth including ingrowth is obtained from eq. (14.20):

$$G_{n+i} = G_g + I - M = 273.8 + 34.6 - 62.1 = 246.3$$

Or, in an approach that typifies the use of PSPs (such as in CFI), G_{n+i} may be obtained by totaling the last column in Table 14.4. (Also note that G_g may be obtained by totaling the "survivor growth" column in Table 14.4.) Clearly, the other growth terms may be computed by the alternate equations, and the same results will be obtained by consistent use of either eqs. (14.13)–(14.16) or eqs. (14.17)–(14.21).

Note that Table 14.4 uses the term *sound volume*. This is done to avoid the use of the terms *net* and *gross* when referring to tree or stand soundness; that is, amount of defect. This follows the recommendation of Meyer (1953), who suggested that tree or stand volume before defect deduction be termed *total* tree volume (rather than gross), and that tree or stand volume after deduction be termed *sound* tree volume (rather than net).

TABLE 14.4. Growth Data From a 1/5-acre Fixed-Area Permanent Sample Plot—Growth Period: 10 Years

Tree Number	Tree Status[a]	Sound Volume (Board Feet) of						
		First Inventory	Second Inventory	Survivor Growth	Mortality	Cut	Ingrowth	Net Change
1	20	62.1	—	—	62.1	—	—	-62.1
2	24	81.3	—	—	—	81.3	—	—
3	24	66.8	—	—	—	66.8	—	—
4	22	42.4	62.3	19.9	—	—	—	19.9
5	22	63.3	122.5	59.2	—	—	—	59.2
6	22	106.0	163.8	57.8	—	—	—	57.8
7	24	93.3	—	—	—	93.3	—	—
8	22	82.0	119.8	37.8	—	—	—	37.8
9	22	147.2	246.3	99.1	—	—	—	99.1
10	12	—	34.6	—	—	—	34.6	34.6
Plot totals		744.4	749.3	273.8	62.1	241.4	34.6	246.3
Symbol		V_1	V_2	G_g	M	C	I	G_{n+i}

Source: Adapted from Beers (1962).

[a]Tree status as used here defines the class of tree from a growth-contribution standpoint. Status at each inventory is coded as follows: 0=not present, 1=pulpwood size, 2=sawlog size, 3=cull, 4=cut. By combining the tree classes at successive inventories, then 20=sawlog mortality, 24=sawlog cut, 22=sawlog survivor tree, 12=sawlog ingrowth from pulpwood size, and so on.

By employing this terminology, one can have gross and net stand growth in terms of total or sound stand volume.

14.5. MEASUREMENT OF STAND AND FOREST GROWTH AND YIELD

Stand and forest growth and yield information can be obtained from a variety of sources. Field methods include stand reconstruction, temporary sample plots, and PSPs. A variety of growth and yield models also can be used to estimate stand and forest yield information. Growth and yield data are used to develop growth models, to make harvest decision, as inputs into long-term strategic forest management planning systems, and for predicting or monitoring responses to silviculture treatments and forest management strategies (to name a few). This section focuses on field methods for measuring or estimating stand and forest growth, and Section 14.7 focuses on growth and yield models.

14.5.1. Stand Reconstruction

Stand reconstruction is a stand analysis technique that combines historical information, observations, measurements, and destructive sampling to recreate the disturbance history of a stand (Oliver and Stephens, 1977). The technique, first formalized by Stephens (1955), is widely used in stand development and ecology research (Henry and Swan, 1974; Oliver and Stephens, 1977; Oliver, 1982; Oliver and Larson, 1996), but has only had limited use in growth and yield analyses (Bealle Statland, 1996). The primary factors contributing to a lack of widespread use in growth and yield studies include (1) the ad hoc nature of the selection process of individual trees used in many ecological and stand dynamics research projects; (2) the intense field data collection efforts limit the spatial extent and number of sample locations that can be examined; and (3) the resulting data often are not precise enough to be used in growth and yield analyses and modeling projects. However, stand reconstruction can yield very useful information for growth and yield analyses, especially in areas where long-term, permanent networks of sample plots are not available (Hann, 1983; Bealle Statland, 1996).

The long-term (>50 years), historical stand reconstructions presented in many ecological and stand dynamics studies (e.g., Stephens, 1955; Henry and Swan, 1974; Oliver and Stephens, 1977; Lorimer, 1985) yield valuable data about disturbance frequencies, disturbance intensities, changes in species composition, and changes in species dominance. For example, Plate 9a shows a reconstruction from a mixed conifer–hardwood stand in central New Brunswick, Canada. From this figure, we can see that the eastern hemlock are older remnants from a previous stand that originated at least 250 years ago. Around 110 years ago, there was a major disturbance, apparently killing everything except the eastern hemlock (most likely a fire); the major disturbance is deduced from the change in growth rates observed in the hemlock and the appearance of the mix of species that dominate the stand today. Another, smaller-scale disturbance is observed to have occurred about 50 years ago (most likely a small canopy gap following blowdown of trees that died during selfthinning). Again changes in the growth trajectories and the appearance of new trees, mostly balsam fir, were used to deduce this disturbance event. Plate 9b shows a stand reconstruction based on height growth for a mixed conifer stand from the interior forests of British Columbia. This stand originated from fire about 60 years prior to reconstruction.

Aspen quickly regenerated on the site, and, with its fast early growth rates, dominated the site for the first 40 years. Douglas-fir, spruce, and several other species also regenerated following the disturbance, but with far smaller abundances. The slow height growth rates prevented site dominance until about age 40 (20 years before present), when Douglas-fir began to overtop the aspen. The appearance of ingrowth trees around this time also indicates that the aspen were self-thinning and the stand had entered the understory reinitiation stage (Oliver and Larson, 1996). While these data are more qualitative than quantitative and do not provide exact estimates of density and species composition over time, they are useful for understanding forest dynamics and the processes driving the development of stand structures we observe today. This information can be used to develop simulation scenarios and to benchmark long-term projections from forest growth and yield models.

The procedures and processes for conducting a stand reconstruction depend upon the intensity and detail required by the study. A number of sources describe various approaches and techniques used in stand reconstruction and should be consulted for additional details, including Henry and Swan (1974), Oliver and Stephens (1977), Oliver (1982), Lorimer (1985), and Bealle Statland (1996). The key steps important for forest mensuration and growth and yield are discussed here:

Stand Selection. Stand reconstructions generally focus on specific stand types (Oliver, 1982; Hann, 1983; Bealle Statland, 1996). Because of the intensity of field data collection, many stand reconstruction studies focus on a single or very small number of stands. For growth and yield studies, stand selection should represent a chronosequence across the type(s) of interest (Bealle Statland, 1996). By selecting stands as a chronosequence, individual reconstructions can be linked with each other and other temporary inventory data to refine growth and yield models. Systematically sampling from a list of candidate stands sorted by age is an efficient means of insuring stands are selected across a chronosequence. Bealle Statland (1996) discusses additional selection criteria based on structure variability.

Plot Selection. Subjective selection of sample plots is widely used in stand reconstruction research (e.g., Stephens, 1955; Henry and Swan, 1974; Oliver and Stephens, 1977). This method is not recommended because it limits the inferences that may be derived from the collected data and does not provide unbiased estimates of current and past stand conditions. Systematic sampling is reported in some studies (e.g., Stubblefield, 1978; Cobb, 1988; Bealle Statland, 1996). Bealle Statland (1996) describes a sampling approach in which a grid of sample points was located across the stand. At each sample point, stand structure was measured so that stand-level estimates could be obtained, and a subsample of points were selected for destructive sampling. This type of approach lends itself well to double-sampling with ratio or regression estimation to scale plot-level measurements to the stand level.

Standing Tree Measurements. At the very least, species should be identified, and diameter at breast height (dbh) and total height measured. Additional tree measurements include height to the base of live crown and crown width. The presence of visible defects, such as stem crooks, forks, and fire scars, is useful for determining the number and type of disturbances, and which trees were present during the disturbance events.

Destructive Tree Measurements. Destructive measurements can vary from extraction of single increment cores (Section 14.3.2) to full stem analysis (Section 14.3.1). The data shown in Plate 9*a* are from single increment cores, while the data in Plate 9*b* are heights reconstructed from full stem analysis. While single increment cores can be very informative, the patterns of stand development observed in diameter increment may be very different from the patterns observed in height growth (Oliver and Stephens, 1977; Bealle Statland, 1996). In addition, increment cores extracted at breast height do not depict the early development patterns of trees and stands. Increment cores taken near the base of the tree will be more informative and will reflect the stand development sequences more fully. Increment cores extracted through fire scars and other visible defects can be used to more precisely date disturbances.

Other Measurements. Many additional measurements may be made during stand reconstructions. The presence of charcoal in the soil is often used to determine whether fire has been a factor in stand development (Stephens, 1955; Oliver and Stephens, 1977; Lorimer, 1985; Bealle Statland, 1996). Soil features such as plow lines can reveal past land use changes (Stephens, 1955). In more intensive stand reconstructions, increment cores from standing dead trees and down dead trees can be extracted and cross-dated to determine the year of death. The presence of old rotted logs, their felling direction, and species are often determined (Oliver, 1982). In some studies, detailed stand maps are drawn depicting standing and down trees (Stephens, 1955; Bealle Statland, 1996). Again, the level of detail is dependent upon the study objectives.

14.5.2. Estimation of Stand Growth and Yield from Temporary Sample Plots

A forest inventory gives estimates of forest attributes at a given point in time. Differences between successive independent inventories can be used to estimate net increase (eq. (14.13) or (14.18)). These differences are generally estimated at the stand level since successive temporary inventories often do not use the same sample locations and sample trees (Bickford, 1954; Winner and Stott, 1954). As a result, these estimates have a high degree of variation since they contain sampling errors from both inventories as well as variation in growth. Yield curves developed from temporary plots can be used to infer net increase as well (Clutter et al., 1983). Estimates of net increase can be improved by remeasuring the same sample locations over time (Bickford, 1954); however, PSPs, where the individual trees are permanently identified and measured over time, are the most effective approach for estimating growth and yield (Winner and Stott, 1954).

14.5.3. Estimation of Stand Growth and Yield from Fixed-Area Permanent Sample Plots

Fixed-area plots are commonly used because of their simplicity of data compilation and calculations of growth. With a single-sized fixed-area PSP, there are four types of trees commonly encountered:

1. Survivor trees—live trees above the minimum dbh threshold at time 1, and still present and alive at time 2 (trees 4, 5, 6, 8, and 9 in Table 14.4 are survivor trees).
2. Mortality trees—live trees above the minimum dbh threshold at time 1, but dead at time 2 (tree 1 in Table 14.4 is a mortality tree).

3. Cut trees—live trees above the minimum dbh threshold at time 1, but cut between time 1 and time 2 (trees 2, 3, and 7 in Table 14.4 are cut trees).
4. Ingrowth trees—live trees below the minimum dbh threshold at time 1, but grow enough to cross the dbh threshold and become new measured trees (tree 10 in Table 14.4 is an ingrowth tree).

As discussed in Section 14.4.2 and shown in Table 14.4, the various expressions of stand growth are the sums of these different classes of trees.

Diameter and height increment of individual trees are calculated as differences in successive measures, and may be expressed on an annual basis by dividing by the measurement interval. For example, dbh increment is calculated using

$$\Delta dbh_i = \frac{dbh_{i+t} - dbh_i}{t} \tag{14.22}$$

where Δdbh_i = diameter increment between time i and $i + t$
      dbh_i = diameter at time i (initial diameter)
      dbh_{i+t} = diameter at time $i + t$ (diameter at remeasurement)
      t = time

Time is generally expressed in years, but for some faster-growing species, especially in tropical regions, time may be expressed in shorter units. Similarly, basal area increment is calculated by using the cross-sectional area formula (eq (5.3) or (5.4)) and substituting into eq. (14.22):

$$\begin{aligned} \Delta BA &= \frac{BA_{i+t} - BA_i}{t} \\ &= \frac{c \cdot dbh_{i+t}^2 - c \cdot dbh_i^2}{t} \\ &= c \frac{dbh_{i+t}^2 - dbh_i^2}{t} \end{aligned}$$

where c is the appropriate cross-sectional area conversion factor (0.005454 in imperial units and 0.00007854 in SI units). Increment values for volume, biomass, and other tree attributes are usually obtained by applying appropriate allometric equations (Section 14.3.3).

When a single-sized fixed-area PSP is used (e.g., the data shown in Table 14.4), all trees have the same probability of being sampled and this probability does not change over time; therefore, all trees have the same per unit area expansion factor over time (Section 9.1). In this situation, growth can be scaled in the same way the individual sample tree attributes are scaled to per unit area using the factor approach (Section 9.1). For example, the average annual "volume-growth" factor for a single tree could be expressed as

$$VF_i = TF \cdot V_i$$
$$VF_{i+t} = TF \cdot V_{i+t}$$
$$VGF = \frac{VF_{i+t} - VF_i}{t}$$
$$= \frac{TF \cdot V_{i+t} - TF \cdot V_i}{t} = TF \frac{V_{i+t} - V_i}{t}$$

For example, tree 4 in Table 14.4 had 42.4 board feet at time 1 and 62.3 board feet at time 2. Assuming these measures were over a 10-year period, then the average annual growth for this tree was

$$\Delta V = \frac{62.3 - 42.4}{10} = \frac{19.9}{10} = 1.99 \text{ board feet / year}$$

And the volume growth factor would be

$$VGF = 5\frac{63.2 - 42.4}{10} = 5 \cdot 1.99 = 9.95 \text{ board feet / acre / year}$$

The growth factor described above would be most correctly termed the survivor growth factor. Factors for mortality, ingrowth, and cut could be determined using the same process.

The total growth per unit area is then calculated as the sum of the growth factors or by applying the tree factor to scale plot totals to per unit area totals. Gross growth per acre per year for the plot shown Table 14.4 would then be

$$VG_g / \text{acre} / \text{year} = \sum_{i=1}^{10} VGF_i = \sum_{i=1}^{10} (TF \cdot \Delta V_i) = TF\sum_{i=1}^{10} \Delta V_i$$
$$= 5(0 + 0 + 0 + 1.99 + 5.92 + 5.78 + 0 + 0 + 3.78 + 9.91)$$
$$= 5 \cdot 27.38$$
$$= 136.9 \text{ board feet / acre / year}$$

Similarly, mortality per acre per year would be

$$M / \text{acre} / \text{year} = \sum_{i=1}^{10} MF_i = TF\sum_{i=1}^{10} M_i$$
$$= 5(6.21 + 0 + 0 + 0 + 0 + 0 + 0 + 0 + 0 + 0)$$
$$= 5 \cdot 6.21$$
$$= 30.16 \text{ board feet / acre / year}$$

All the other components of stand growth presented in Section 14.4.1 can be calculated on an annual per unit area basis using the same process.

Some PSP systems use nested plots (e.g., FIA, 2013). A larger plot is used to sample trees above some minimum dbh threshold, and a smaller plot (or set of smaller plots) used to sample trees below this threshold. If the two systems are measured independently and trees are not tracked as they move across the dbh threshold (i.e., switch from the smaller to the larger plot), then the procedures described above can be used. However, there will be two sets of ingrowth trees, and it will also be necessary to account for trees that grow out of the smaller plot. On the other hand, if the small trees are tagged and followed from the small plot onto the large plot, then the procedures must be modified to account for the change in tree factor. These methods are covered in detail for variable probability permanent sample designs in the next section.

14.5.4. Estimation of Stand Growth and Yield from Permanent Variable Probability Sample Points

Most PSP systems utilize fixed-area plots rather than variable probability sampling. This preference is largely a legacy of Dr. Cal Stott and other researchers from the USDA Forest Service, North Central Forest Experiment Station, and their "Forest Control by Continuous Inventory" series.* Iles (2003, chapter 15) identifies three justifications for this preference:

1. The logic and calculations are relatively simple.
2. Equal emphasis placed on large and small trees.
3. A large number of trees is available for studying spatial interactions.

As shown in Section 14.5.3, the calculations for determining stand growth are relatively straightforward when using a single fixed-area plot size. Since all trees have the same tree factor, stand-level growth is easily calculated as the sum of individual tree differences multiplied by the tree factor. Additionally, all trees would have the same weight when fitting regression equations for predicting individual tree growth. With variable probability sampling, different sized trees not only have different tree factors, but these factors change over time as trees grow. This obviously complicates calculations and is the primary argument supporting the first justification. The other two justifications are less exact and really depend upon objectives of the PSP system. The second justification is the exact reason variable probability sampling is favored over fixed-area plots in normal inventory operations, and can lead to an excessive number of small trees being measured on fixed-area PSPs. This is also one of the main reasons for employing nested PSPs with a smaller fixed-area plot for sampling small trees; however, this solution also results in variable sampling probabilities and similar accounting issues as trees "grow" from the smaller plot onto the larger plot. Argument 3 is a reasonable concern for research installations, especially ones where different silvicultural treatments will be applied; however, when the primary purpose of the PSP system is monitoring or simple growth estimations, variable probability designs can be quite effective (Iles, 2003).

The high costs and relatively low efficiencies of fixed-area PSPs (Bickford, 1954) should favor variable probability PSPs; however, a lack of understanding of (or mystification by) the methods of calculating growth has led many organizations to favor more costly fixed-area plots. Some of the confusion originates from the tendency to (mis)conceptualize fixed area samples as measures and variable probability samples as estimates. Visualizing a fixed area sample as an area of land rather than a sample point that intersects a set of

---

*The "Forest Control by Continuous Inventory" series, commonly referred to as the "CFI Notes" is a series of 153 short notes (often one page) on methods, applications, and results of continuous forest inventory (CFI) mainly focused on the upper Midwest of the United States. These notes were published between 1954 and 1967.

fixed, equal-sized inclusion zones reinforces the confusion. If the focus shifts to the sample point and what is happening at the sample point, rather than trying to think about what is happening over some unit of land, the methods for calculating growth using variable probability sampling are more easily understood. Iles (2003), building upon the logic of Bitterlich (1984), categorizes the methods for handling growth calculations into three classes: (1) subtraction methods, (2) constant inclusion-zone-size methods, and (3) compatible methods. These methods are illustrated using the data shown in Table 14.5.

Iles (2003, chapter 15) advocates conceptualizing and calculating growth on variable probability plots using the basal area ratio approach (Section 11.2.1) rather than the factor approach (Section 9.3.1). Because this view point helps dispel the myth that a fixed-area plot sample is a measure of some land area while a horizontal point sample is only an estimate, basal area ratios will be presented; however, it is more important to understand that the per unit area estimates are merely scalars of the sample point estimators whether fixed area or horizontal point samples are used. For example, volume per unit area values at a sample point are obtained by summing the volume factors over all "in" trees. Similarly, volumes per unit area are obtained by multiplying the average volume:basal area ratio by basal area per unit area, which in turn is the basal area factor multiplied by the tree count. When VBAR is determined for all sample trees (as is often the case for permanent point samples), rather than a subsample (as with big BAF sampling), the two methods are algebraically equivalent:

$$
\mathrm{Vol} = \sum_{i=1}^{n} \mathrm{VF}_i = \sum_{i=1}^{n} \mathrm{TF}_i \cdot \mathrm{Vol}_i = \sum_{i=1}^{n} \left(\frac{\mathrm{BAF}}{\mathrm{BA}_i} \right) \mathrm{Vol}_i
$$

$$
= \mathrm{BAF} \sum_{i=1}^{n} \left(\frac{\mathrm{Vol}_i}{\mathrm{BA}_i} \right) = \mathrm{BAF} \sum_{i=1}^{n} \mathrm{VBAR}_i = \mathrm{BAF} \cdot n \cdot \left(\frac{\sum_{i=1}^{n} \mathrm{VBAR}_i}{n} \right) = \mathrm{BAF} \cdot n \cdot \overline{\mathrm{VBAR}}.
$$

Like fixed-area PSPs, variable probability PSPs have survivor trees, mortality trees, cut trees, and ingrowth trees; however, ingrowth can be viewed as a more complicated category of trees, with three types of ingrowth trees recognized (Beers and Miller, 1964; Martin, 1982):

4a. Ingrowth trees—live trees below the minimum dbh threshold and included on the sample point at time 1, but grow enough to cross the dbh threshold and become new measured trees (in this situation, ingrowth has the same definition on fixed-area and variable probability sample points).

4b. Ongrowth trees—live trees below the minimum dbh threshold but not included on the sample point at time 1, but grow enough to cross the dbh threshold and onto the sample point, thus becoming new measured trees (similar to the definition for ingrowth on fixed-area samples except for the plot size change).

4c. Nongrowth trees—live trees above the minimum dbh threshold but far enough away from the sample point to be not included on the sample point at time 1, but grow enough to grow onto the sample point, thus becoming new measured trees (Iles (2003, chapter 15) points out that the term nongrowth is a misnomer since these trees are indeed growing, otherwise they would not have appeared on the sample point at time 2).

TABLE 14.5. An Example Permanent Horizontal Point Sample and Two Measurement Periods (BAF = 20F (ft²/acre/tree tallied), 5 Years Between Measurements, dbh Measured in Inches, Height in Feet, and Volume in Cubic Feet)

Tree Number	Distance (ft)	Status[a]	Measurement 1					Measurement 2				
			dbh (in.)	Height (ft)	Volume (ft³)	Volume Factor (ft³/acre)	VBAR (ft³/ft²)	dbh (in.)	Height (ft)	Volume (ft³)	Volume Factor (ft³/acre)	VBAR (ft³/ft²)
1	10.9	22	13.0	54.9	23.1	501.1	25.1	13.4	55.2	24.7	503.5	25.2
2	5.1	23	4.9	36.7	2.3	345.5	17.3					
3	3.6	22	9.3	50.4	10.9	463.7	23.2	9.7	51.1	12.0	469.5	23.5
4	33.6	22	17.3	56.9	42.3	518.0	25.9	17.7	57.0	44.3	518.9	25.9
5	11.0	22	14.3	55.7	28.3	508.2	25.4	15.3	56.2	32.7	512.3	25.6
6	21.0	22	15.1	56.1	31.8	511.5	25.6	15.5	56.3	33.6	513.0	25.6
7	26.3	22	16.7	56.8	39.3	516.6	25.8	17.0	56.8	40.8	517.3	25.9
8	24.3	22	12.8	54.7	22.3	499.9	25.0	12.9	54.8	22.7	500.5	25.0
9	1.6	23	4.8	36.2	2.1	341.1	17.1					
10	5.0	22	4.0	31.8	1.3	302.4	15.1	4.0	31.8	1.3	302.4	15.1
11	18.7	22	10.5	52.3	14.4	479.6	24.0	11.3	53.3	17.0	488.0	24.4
12	7.8	22	7.2	45.5	6.0	422.2	21.1	7.5	46.3	6.6	429.5	21.5
13	6.6	23	4.3	33.5	1.6	317.8	15.9					
14	9.6	22	7.1	45.2	5.8	419.6	21.0	7.2	45.5	6.0	422.2	21.1
15	3.3	22	8.4	48.6	8.6	448.4	22.4	9.6	55.3	25.5	504.7	25.2
16	28.5	01						15.9	56.5	35.5	514.3	25.7
17	17.6	01						9.7	51.1	12.0	469.5	23.5

[a]Status: 01 = not present time 1, ingrowth (new tree) time 2; 22 = present and live, time 1, present and live, time 2; and 23 = present and live, time 1, dead time 2

The "nongrowth" trees are typically the source of concern and/or confusion regarding variable probability PSPs. Their appearances on PSPs cause sudden, large shifts in basal area ratios (Fig. 14.7a); however, sudden shifts are exactly what PSPs are supposed to avoid (Bickford, 1954; Winner and Stott, 1954; Grosenbaugh, 1960; Martin, 1982; Iles, 2003) and are one of the major factors leading some agencies to not use variable probability PSPs. Grosenbaugh (1960) argued that the desire for continuity should be weighed against the costs of data collection, the ultimate use of the data, and the need for accuracy; however, Grosenbaugh's views have been largely ignored, dismissed, or forgotten (Iles, 2003). The various methods of calculating growth on variable probability PSPs primarily have been developed to deal with the issues around these so-called "nongrowth" trees. The discussion below focuses on volume; however, any stand attribute can be used.

Subtraction Method. With the subtraction method, sums of either VBAR or VF can be used, and growth is simply the differences of these sums over time (Table 14.6). In this example, the totals for the VBAR approach differ from the totals for the VF approach by a factor of 20, the BAF used to collect the data shown in Table 14.5. As can be seen in Table 14.6, all components of stand growth (Section 14.4.1) are represented*; therefore, all types of stand growth (Section 14.4.2) can be calculated. For example, using eq. (14.16), net change can be calculated from either the volume factor approach:

$$G_d = V_2 - V_1 = 6665.6 - 6595.8 = 69.8 \, \text{ft}^3 \, / \, \text{acre}$$

or using the VBAR approach:

$$G_d = V_2 - V_1 = 333.3 - 329.8 = 3.5 \, \text{ft}^3 \, / \, \text{ft}^2$$

and G_d/acre is obtained by multiplying by the BAF: 3.5·20=69.8 ft³/acre. Likewise, G_d could be obtained using eq. (14.21):

$$G_d = G_g + I - M - C = 90.4 + 983.8 - 1004.4 = 69.8 \, \text{ft}^3 \, / \, \text{acre}$$

for the volume factor approach and

$$G_d = G_g + I - M - C = 4.5 + 49.2 - 50.2 - 0 = 3.5 \, \text{ft}^3 \, / \, \text{ft}^2$$

for the VBAR approach. Again, G_d/acre is obtained by multiplying the G_d for VBAR by the BAF (20, in this case). The components of stand growth and the types of stand growth can be calculated for any stand parameter; however, gross growth or survivor growth of basal area does not really have a meaning with horizontal point samples in the same way it has no meaning for density (stems/unit area) with fixed-area plots.

The subtraction method simply incorporates the sudden jumps in VF or VBAR when a tree grows "on" to a sample point. For trees not included on a sample point, the VF and VBAR are both 0; however, when a tree grows enough to be included on a point (either by growing across the minimum threshold diameter or by "growing" its inclusion zone enough to overlap the sample point), the whole VF or VBAR at time 2 is added in. For the data shown in Table 14.5, 4 in. was the minimum measured dbh. For a fixed-area plot, all trees within the plot boundary will be considered ingrowth trees once they reach 4.0 in. dbh.

---

*Cut is not shown in Table 14.6 because there were no trees harvested on this plot between times 1 and 2.

(*a*)

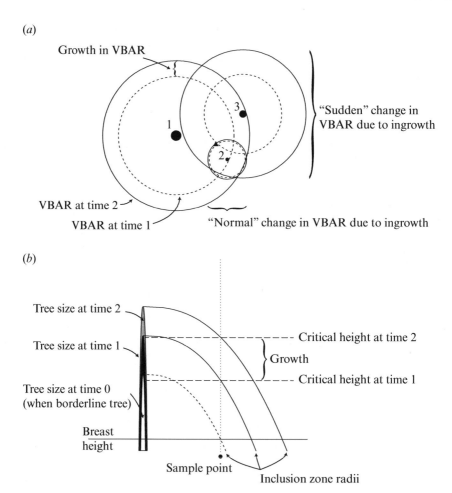

FIG. 14.7. Geometric representation of growth on permanent variable probability point samples: (*a*) using the subtraction method (circles depict increases in volume:basal area ratio not inclusion zone) and (*b*) critical height sampling for growth (adapted from Iles (2003)).

On a horizontal point sample, a 4.0 in. tree has a rather small inclusion zone (for BAF = 20 F, the inclusion zone radius is 7.78 ft). As trees get farther away from the sample point, they must be larger and larger to become ingrowth (such as trees 16 and 17 in Table 14.5)—a tree 20 ft away must be 10.3 in. dbh before it is on the sample point. For the trees in Table 14.5, a 4.0 in. dbh tree has a VBAR of approximately 14 while a 10 in. dbh tree has a VBAR of 24. This is illustrated geometrically in Figure 14.7*a*. Tree 1 was included on the sample point at both measures, so growth for this tree is the difference in VBARs at the two times. Tree 2 was less than the minimum dbh at time 1, grew over the minimum dbh threshold and on to the sample point; while the VBAR goes from 0 at time 1 to 14.8 at time 2, this is just considered normal ingrowth (technically "ongrowth" using Martin's (1982) definitions) and such changes in VBAR total would be observed on fixed-area sample points as well as variable probability sample points. Tree 3 was greater than the minimum dbh threshold, but far enough away from the sample point to not be on the sample point at

TABLE 14.6. The Subtraction Method for Calculating Growth on Variable Probability Sample Points

Tree Number	Status	Volume Factor Approach						VBAR Approach					
		1st meas	2nd meas	Survivor Growth	Ingrowth	Mortality	Net Change	1st meas	2nd meas	Survivor Growth	Ingrowth	Mortality	Net Change
1	22	501.1	503.5	2.4	—	—	2.4	25.1	25.2	0.1	—	—	0.1
2	23	345.5	—	—	—	345.5	−345.5	17.3	—	—	—	17.3	−17.3
3	22	463.7	469.5	5.8	—	—	5.8	23.2	23.5	0.3	—	—	0.3
4	22	518.0	518.9	0.9	—	—	0.9	25.9	25.9	0.0	—	—	0.0
5	22	508.2	512.3	4.1	—	—	4.1	25.4	25.6	0.2	—	—	0.2
6	22	511.5	513.0	1.5	—	—	1.5	25.6	25.6	0.1	—	—	0.1
7	22	516.6	517.3	0.8	—	—	0.8	25.8	25.9	—	—	—	0.0
8	22	499.9	500.5	0.7	—	—	0.7	25.0	25.0	—	—	—	0.0
9	23	341.1	—	—	—	341.1	−341.1	17.1	—	—	—	17.1	−17.1
10	22	302.4	302.4	—	—	—	0.0	15.1	15.1	—	—	—	0.0
11	22	479.6	488.0	8.3	—	—	8.3	24.0	24.4	0.4	—	—	0.4
12	22	422.2	429.5	7.3	—	—	7.3	21.1	21.5	0.4	—	—	0.4
13	23	317.8	—	—	—	317.8	−317.8	15.9	—	—	—	15.9	−15.9
14	22	419.6	422.2	2.5	—	—	2.5	21.0	21.1	0.1	—	—	0.1
15	22	448.4	504.7	56.2	—	—	56.2	22.4	25.2	2.8	—	—	2.8
16	01	0.0	514.3	—	514.3	—	514.3	0.0	25.7	—	25.7	—	25.7
17	01	0.0	469.5	0.0	469.5	0.0	469.5	0.0	23.5	0.0	23.5	0.0	23.5
Total		6595.8	6665.6	90.4	983.8	1004.4	69.8	329.8	333.3	4.5	49.2	50.2	3.5

Based on the tree data shown in Table 14.5.

time 1; however, by time 2, it was large enough and the entire VBAR was added to the total at time 2, resulting in a sudden change of 25.7 in VBAR.

A primary advantage of the subtraction method is its simplicity. It automatically produces consistent estimates that can account for all the components of growth. However, for some purposes (such as modeling growth) the sudden changes in VBAR associated with ingrowth can be problematic, especially when it is desirable to model growth at the individual sample point level. When inferences are drawn over large numbers of sample points, the sudden changes in VBAR on a small number of plots tend to average out, so estimates of growth and its components are more stable.

Constant Inclusion-Zone-Size Method. In this method, the size of the inclusion zone is fixed at either the beginning or end of the measurement interval. The most common approach is to fix the inclusion zone at the beginning of the measurement interval; however, as Grosenbaugh (1960) points out, if ingrowth is to be included, fixing the inclusion zone at the end of the interval is more practical (Beers and Miller (1964), Furnival (1979), and Martin (1982) discuss alternative methods for estimating ingrowth rates). Intuitively, the factor approach (Section 9.3.1) is a logical way to explain the constant inclusion-zone-size method (Beers and Miller, 1964). The tree factor is set based on measurements at either time 1 or time 2 and volume or any other tree attribute is measured, or estimated from measures, at times 1 and 2. Growth is then estimated as the difference. Fixing the tree factor at time one, for volume we get

$$\Delta V = TF_1 \cdot V_2 - TF_1 \cdot V_1 = TF_1 \left(V_2 - V_1 \right) = \frac{BAF}{c \cdot dbh_1^2} \left(V_2 - V_1 \right) \qquad (14.23a)$$

and fixing the tree factor at time two we get

$$\Delta V = TF_2 \cdot V_2 - TF_2 \cdot V_1 = TF_2 \left(V_2 - V_1 \right) = \frac{BAF}{c \cdot dbh_2^2} \left(V_2 - V_1 \right) \qquad (14.23b)$$

For the basal area ratio approach, the basal area is fixed at either time one:

$$\Delta V = BAF \left(VBAR_2^* - VBAR_1 \right) = BAF \left(\frac{V_2}{c \cdot dbh_1^2} - \frac{V_1}{c \cdot dbh_1^2} \right) \qquad (14.24a)$$

or at time two:

$$\Delta V = BAF \left(VBAR_2 - VBAR_2^* \right) = BAF \left(\frac{V_2}{c \cdot dbh_2^2} - \frac{V_1}{c \cdot dbh_2^2} \right) \qquad (14.24b)$$

where ΔV = volume growth (for annual growth eq. (14.23) would need to be divided by years $(t_2 - t_1)$

TF_1 = tree factor (eqs. (9.9) or (9.12)) at time 1

V_i = volume[†] at time i

[†]While the emphasis in the examples is on volume, any tree or stand attribute can be used.

dbh_i = diameter at breast height at time i

c = conversion factor for diameter in inches or cm and area in ft² (0.005454) or m² (0.00007854)

$VBAR_1$ = the volume to basal area ratio at time 1

$VBAR_2^*$ = the modified volume to basal area ratio at time 2 (volume at time 2/basal area at time 1)

$VBAR_1^*$ = the modified volume to basal area ratio at time 1 (volume at time 1/basal area at time 2)

$VBAR_2$ = the volume to basal area ratio at time 2

Table 14.7 shows the VBAR growth estimates[†] for the data shown in Table 14.5, based on fixing the inclusion-zone-sizes at both measurement times (for ingrowth, size was the inclusion zone at time 2—when the tree was first measured, and for mortality size was the inclusion zone at time 1—the last time the tree was measured). Because different inclusion zone sizes are used, the two estimates of survivor growth and net change are not the same. Comparing Table 14.7 with Table 14.6, it also can be seen that neither estimate corresponds to the estimate obtained by subtraction. Furthermore, based on eq. (14.16):

$$V_2 = V_1 + G_d = V_1 - G_g + I - M - C$$

For the estimates based on fixing the inclusion zone size at time one, we get

$$\sum VBAR_2 = 329.8 + 29.8 + 49.2 - 50.2 = 358.6$$

and for time 2 we get

$$\sum VBAR_2 = 329.8 + 24.6 + 49.2 - 50.2 = 354.4$$

neither of which equal 333.3, the observed $\sum VBAR_2$ at time 2. This arises because the growth estimates obtained by fixing inclusion zone size are not compatible (c.f., Buckman, 1962; Clutter, 1963) with yield estimates obtained by allowing inclusion zone size to change. Given the differences in calculation methods, compatibility should not be expected (Iles, 2003); however, the compatibility issue arises frequently in the literature (e.g., Beers and Miller, 1964; Myers and Beers, 1968; Martin, 1982). Several adjustments have been proposed to force compatibility. For example, Martin (1982), fixing inclusion zone size at time 1, proposed adjusting ingrowth using

$$I_a = I + \sum VBAR_{2(\text{survivors})} - \sum VBAR_{2(\text{survivors})}^* \tag{14.25}$$

and $\sum VBAR_2$ becomes

$$\sum VBAR_2 = \sum VBAR_1 + I_a - M - C$$

So from Table 14.7, we get $\sum VBAR_{2(\text{survivors})}^* = 309.4$ and $\sum VBAR_{2(\text{survivors})} = 284.1$, therefore, the adjusted ingrowth is

$$I_a = 49.2 + 284.1 - 309.4 = 23.9$$

[†]Per acre values are obtained by multiplying the values in Table 14.7 by 20, the BAF for the data shown in Table 14.5.

TABLE 14.7. The Fixed-Inclusion-Zone Method for Calculating Growth on Variable Probability Sample Points

Tree	Status	Plot Size Fixed at Time 1						Plot Size Fixed at Time 2					
		$VBAR_1$	$VBAR_2^*$	Survivor Growth	Mortality	Ingrowth	Net Change	$VBAR_1^*$	$VBAR_2$	Survivor Growth	Mortality	Ingrowth	Net Change
1	22	25.1	26.8	1.7			1.7	23.6	25.2	1.6			1.6
2	23	17.3	0.0		17.3		−17.3	17.3	0.0		17.3		−17.3
3	22	23.2	25.5	2.4			2.4	21.3	23.5	2.2			2.2
4	22	25.9	27.2	1.3			1.3	24.7	25.9	1.2			1.2
5	22	25.4	29.3	3.9			3.9	22.2	25.6	3.4			3.4
6	22	25.6	27.0	1.4			1.4	24.3	25.6	1.4			1.4
7	22	25.8	26.8	1.0			1.0	24.9	25.9	0.9			0.9
8	22	25.0	25.4	0.4			0.4	24.6	25.0	0.4			0.4
9	23	17.1	0.0		17.1		−17.1	17.1	0.0		17.1		−17.1
10	22	15.1	15.1	0.0			0.0	15.1	15.1	0.0			0.0
11	22	24.0	28.3	4.3			4.3	20.7	24.4	3.7			3.7
12	22	21.1	23.3	2.2			2.2	19.5	21.5	2.0			2.0
13	23	15.9	0.0		15.9		−15.9	15.9	0.0		15.9		−15.9
14	22	21.0	21.7	0.7			0.7	20.4	21.1	0.7			0.7
15	22	22.4	33.0	10.5			10.5	17.2	25.2	8.1			8.1
16	01	0.0	25.7			25.7	25.7	0.0	25.7			25.7	25.7
17	01	0.0	23.5			23.5	23.5	0.0	23.5			23.5	23.5
Total		329.8	358.6	29.8	50.2	49.2	28.8	308.7	333.3	25.6	50.2	49.2	24.6

Based on the tree data shown in Table 14.5.

and

$$\Sigma \text{VBAR}_2 = 329.8 + 25.6 + 23.9 - 50.2 = 333.3$$

Comparing this to the total for VBAR_2 in Table 14.6, we see that the two values are identical. A similar adjustment to ingrowth can be developed to insure compatibility for estimates based on fixing inclusion zone at time 2. In this case, adjusted ingrowth would be

$$I_a = I + \Sigma \text{VBAR}^*_{1(\text{survivors})} - \Sigma \text{VBAR}_{1(\text{survivors})}$$

While these types of adjusted estimates insure compatibility and reduce sudden changes in VBAR due to ingrowth, they simply shift some of the change from ingrowth into survivor growth. There is little theoretical justification for these adjusted estimates and they are not design-based estimators. The next section will discuss compatible approaches that do not have these limitations.

Compatible Methods. As discussed above, the subtraction method results in sudden jumps in VBAR as trees much greater than the minimum dbh threshold grow onto the sample point. Fixing the inclusion zone size at either end of the measurement period results in more realistic estimates of survivor growth, but does not prevent the sudden jumps if ingrowth is included and results is estimates where the growth sums are not compatible with the resulting yields. In Section 11.3.3, critical height sampling is described. One of the more practical applications of critical height sampling is growth monitoring. The critical height, Hc, is defined as the height on each tree where the tree's stem diameter would be exactly borderline given the projected horizontal angle (Fig. 11.9) and the associated estimator for cubic volume per unit area is given by (Kitamura, 1962)

$$V = \text{BAF} \cdot \sum_{j=1}^{m} \text{Hc}_j$$

As shown above, and in Section 11.2.1, volume per unit area also is obtained by multiplying the average VBAR by basal area per unit area or by multiplying BAF by the sum of VBARs across all trees:

$$V = (n \cdot \text{BAF}) \cdot \overline{\text{VBAR}} = \text{BAF} \Sigma \text{VBAR}$$

Comparing the two equations, it can be seen that ΣHc plays the same role as ΣVBAR. Critical height sampling is essentially sampling for VBAR, and $\overline{\text{Hc}} \approx \overline{\text{VBAR}}$ in the long run (Iles, 2003). This makes critical height sampling particularly useful for monitoring growth on permanent horizontal point samples for several reasons:

1. No matter how large the tree is, when it becomes exactly borderline with the sample point, the critical height is breast height (1.3 m in SI units, 4.5 ft in imperial units) (Fig. 17.7*b*); therefore, every borderline tree contributes an Hc=breast height addition. As a result, trees grow slowly onto the plot rather than the sudden jumps in VBAR that results from the other methods.

2. Volume estimates from critical height measures do not require volume tables or volume equations. This avoids incorporation of any errors or biases from these

tables or equations into the estimates of volume per unit area and volume growth (McTague and Bailey, 1985).

3. While precisely measuring critical height on every tree would yield the best results, critical height measures can be quite time consuming and costly. Taper equations or even simple tree shapes such as a paraboloid or cone can be used (Van Deusen, 1987). Estimates based on taper functions can be further refined by determining VBAR of a subsample of trees and using ratio estimation (Iles, 2003, chapter 15).

Table 14.8 shows growth estimates based on critical heights derived for the data shown in Table 14.5. In this case, critical heights were estimated using Honer et al.'s (1983) volume percentages (Section 6.4.1). A quick check shows that the growth estimates are compatible with the yield estimates (324.1 + 44.8 − 59.2 + 19.6 = 329.4). Comparing Tables 14.6 and 14.8, it can be seen that, for this example, the sums of VBARs and sums of Hcs are in close agreement, this is due to total volume being estimated using Honer's (1967) total volume equations for calculating VBAR and Honer et al.'s (1983) volume percentages to estimate critical height. This will not always be the case, especially if Hc is measured in the field or if the volume equation is not derived from the taper function. Critical height estimation reduced the impact of ingrowth on the overall growth estimation and shifted more estimated growth into the survivor growth category, similar to the estimates obtained by forcing compatibility using the constant plot size method, but based on design-based considerations rather than forced compatibility.

While the growth calculations for variable probability samples are slightly more complicated than for fixed area samples, the above examples show that growth can be estimated from variable probability plots, and that the calculations are not overly complex. Concern over the sudden jumps caused by ingrowth trees and the subsequent splitting of ingrowth into ingrowth, ongrowth, and nongrowth have only served to further the confusion and distrust of growth estimates from variable probability samples. Using critical height sampling minimizes this issue, but requires identification and measurement of critical heights or the estimation of critical heights from taper equations. However, these issues are no different and no more complex than those that arise when using nested fixed-area plots. Given that many PSP programs are often abandoned because of high costs, permanent variable probability samples are a cost-effective alternative for obtaining long-term growth and yield data needed by growth and yield modelers and forest managers.

14.6. CONSIDERATIONS FOR THE DESIGN AND MAINTENANCE OF PERMANENT SAMPLE PLOT SYSTEMS

PSP systems represent a significant commitment of time and financial resources in order for results to be beneficial (Bickford, 1954; Winner and Stott, 1954; Curtis and Marshall, 2005). Because of these commitments, careful planning and execution is required. Despite over a century of experience working with PSPs and thousands of PSP networks, common errors in design and maintenance persist. Curtis and Marshall (2005) compiled a detailed list of problems frequently encountered:

1. Many PSP networks are plagued by inadequate documentation. Field procedures for measuring tree or stand attributes are not fully documented in written protocols or the protocols themselves are not preserved. This makes consistent

TABLE 14.8. Growth Estimates Derived From Critical Height Estimates

Tree	Status	Critical Distance (ft)	Critical Diameter (in.)	Measurement 1 dbh (in.)	Measurement 1 Total Height (ft)	Measurement 1 Critical Height (ft)	Measurement 2 dbh (in.)	Measurement 2 Total Height (ft)	Measurement 2 Critical Height (ft)	Growth Survivor	Growth Mortality	Growth Ingrowth	Growth Net Change
1	22	10.9	5.6	13.0	54.9	34.6	13.4	55.2	39.5	4.9			4.9
2	23	5.1	2.6	4.9	36.7	20.8					20.8		−20.8
3	22	3.6	1.8	9.3	50.4	36.8	9.7	51.1	41.7	4.9			4.9
4	22	33.6	17.3	17.3	56.9	7.3	17.7	57.0	6.1	−1.2			−1.2
5	22	11.0	5.7	14.3	55.7	36.6	15.3	56.2	42.5	5.9			5.9
6	22	21.0	10.8	15.1	56.1	22.0	15.5	56.3	25.7	3.8			3.8
7	22	26.3	13.5	16.7	56.8	16.7	17.0	56.8	19.7	3.0			3.0
8	22	24.3	12.5	12.8	54.7	7.3	12.9	54.8	6.8	−0.5			−0.5
9	23	1.6	0.8	4.8	36.2	26.6					26.6		−26.6
10	22	5.0	2.6	4.0	31.8	15.3	4.0	31.8	17.1	1.8			1.8
11	22	18.7	9.6	10.5	52.3	9.3	11.3	53.3	15.0	5.7			5.7
12	22	7.8	4.0	7.2	45.5	24.5	7.5	46.3	28.8	4.3			4.3
13	23	6.6	3.4	4.3	33.5	11.8					11.8		−11.8
14	22	9.6	5.0	7.1	45.2	18.9	7.2	45.5	21.7	2.8			2.8
15	22	3.3	1.7	8.4	48.6	35.5	9.6	55.3	45.1	9.6			9.6
16	01	28.5	14.7				15.9	56.5	10.5			10.5	10.5
17	01	17.6	9.0				9.7	51.1	9.1			9.1	9.1
Total						324.1			329.4	44.8	59.2	19.6	5.3

Source: Based on the data shown in Table 14.5 and Honer et al.'s (1983) volume percentages.
Estimates are point estimates, per acre estimates are obtained by multiplying by the BAF (20 in this example).

measurement across different time periods challenging. Likewise, records of silvicultural treatments or disturbances that impact individual plots are often incomplete, so drawing inferences about the reasons for change can be difficult.

2. PSPs are occasionally too small to yield reasonable estimates of growth and stand characteristics, especially at the scale of the individual plot. In PSPs used for inventory, inappropriate procedures for dealing with slopover near property or stratum boundaries may be used, while in experimental work inadequate buffers between adjacent treatments can be problematic.

3. The subjective assignment of treatments to plots in experiments is a source of bias. For example, in a thinning trial, if thinning treatments are not assigned randomly but instead the plots that "need thinning" are thinned, it becomes challenging to separate the effects of the thinning from the initial condition of the plots on subsequent growth.

4. Height measurements are neglected, or when they are performed, have been too few, inappropriately distributed among the sample trees, or inaccurate.

5. Minimum diameters for measurement are often too large and associated with merchantability limits at the time of measurement. This compromises the ability to understand the biological and silvicultural basis for growth and leads to truncated diameter distributions and growth results that can be difficult to compare with those of other studies or inventories.

6. Definitions and procedures for estimating tree and stand age are absent, ambiguous, or inconsistent, and sampling trees for age is often subjective or inadequate.

7. Evidence of initial stand conditions or other factors that could influence initial growth or treatment responses are not recorded.

8. Plot locations and sizes may be unreliable because of poor records and sloppy initial layout.

9. Changes in experimental treatments, in plot sizes, or in measurement procedures (such as minimum measured diameter) are often made for reasons of efficiency or expediency, but with inadequate documentation and little consideration of how those changes will affect analysis.

10. Measurement standards, data coding, and error checking are incompletely documented, inconsistent, and sometimes incompatible across measurement periods or within subunits of a large organization. This prevents or complicates shared data analysis and can require considerable investment in data cleaning and conversion to produce any meaningful analysis, even one that entails a substantial loss of information relative to what could otherwise have been provided.

Many of these problems or errors can be avoided by carefully planning and documenting PSP programs and research trials. Clear objectives for the program are required before any relevant planning can be carried out. Understanding how the data are going to be ultimately analyzed, interpreted, and applied is an important design consideration. In the next few sections, we briefly review the important considerations for designing and implementing PSP programs. For more detailed descriptions of considerations, we recommend reviewing the CFI Notes compiled by Cal Stott and others for the North Central Research Station in the 1950s and 1960s (e.g., Winner and Stott, 1954), the overview by Curtis and Marshall (2005), and the national forest inventory book by Kangas and Maltamo (2006). In addition to these resources, most national forest inventory programs maintain extensive websites with detailed establishment and measurement guidelines. Although the goals of any inventory or experimental effort will likely differ from those of a national-level inventory, so that field procedures and eventual analysis will also differ, such websites can offer valuable guidance on field procedures.

14.6.1. Sampling Unit Type

Fixed-area samples dominate PSP programs throughout the world. This is partially explained because many programs began collecting data before variable probability sampling was well developed. Advocacy by Cal Stott and others through the Continuous Forest Inventory Notes (e.g., Winner and Stott, 1954) influenced the development of many public and private PSP programs throughout North America in the 1950s and 1960s. Their emphasis on continuity in forest inventory favored fixed-area samples over variable probability samples. Concerns over the sudden jumps caused by larger ingrowth trees, and the further confusion resulting from the classification of ingrowth into ingrowth, ongrowth, and nongrowth added complexity to the calculations of growth and resulted in many researchers and managers favoring the more simple analyses provided by fixed area data over the potentially more cost-effective variable probability data.

  While the dominance of the use of fixed-area samples in PSP programs is largely the result of historic precedence, the choice should not be considered entirely arbitrary, and there are several factors to consider:

1. How are the data going to ultimately be used? If aggregate, per unit area growth estimates at the stand or forest level are the primary objective, then variable probability samples are likely to be more efficient than fixed-area plots. For a given number of sampling units, typically fewer trees will be measured on variable probability samples than on fixed-area samples. Alternatively, for a fixed expenditure, more variable probability samples can be established and measured. Averaging over several sampling units will reduce the impacts of large ingrowth trees entering the PSP program. On the other hand, if individual tree growth is the primary objective, then fixed-area samples are often a better choice.

2. Are treatments being applied? Single variable-probability samples are rarely adequate to characterize treatments and to monitor responses following treatments. If treatments are applied at the sampling plot level, fixed-area samples are inherently better at depicting the treatment than variable probability samples. If treatments are applied at larger scales (e.g., a 1 or 2 ha or acre treatment area, or even at the stand level), either multiple variable probability samples or multiple fixed-area samples can be used to characterize the treatment across the treatment area.

3. If information on individual tree competition is desired, fixed-area plots are recommended since variable probability plots often have too few trees to compute competition indices, many smaller trees are not frequently sampled (not because of error, but because of sampling probabilities), and "out" trees may influence the competitive status of "in" trees, resulting in competition values that are not reflective of true competitive statuses. (This latter challenge is also true of very small fixed-area plots. Thus, experiments and other research programs often benefit from the use of larger fixed-area plots than would be most efficient simply for estimating volume or growth at stand or larger scales.)

14.6.2. Sampling Unit Size and Shape

Size and shape of PSPs depend upon common practice, forest conditions, and goals of the program. Curtis and Marshall (2005) recommend using square or rectangular plots because it is generally easier to determine borderline trees with square or rectangular plots, especially if plots are large. In plantations, care needs to be taken to ensure that plot locations are not biased by plantation rows. The influence of plantation rows can be minimized by orienting the plot sides at an angle to the row orientation. In research installations, aligning the plot boundaries exactly midway between plantation rows is an acceptable and recommended practice (Curtis and Marshall, 2005). In many cases, circular plots are utilized (e.g., Winner and Stott, 1954; FIA, 2013). With large circular plots, borderline trees can be difficult to check. Using an electronic distance measurement device (Section 4.2.4) can minimize the field effort required to check borderline trees. Circular plots only require one location to be established in the field, while square or rectangular plots require four (5 if plot center is also located). The corners must be carefully surveyed in, permanently marked, and it is important to assure that the plots are square by checking the diagonals. Circular plots minimize the edge per unit plot area and are generally not influenced by tree spatial pattern.

Size of PSPs depends upon (1) program objectives, (2) forest conditions (even-aged vs. uneven-aged; single species versus mixed species), and (3) types of treatments (if any) to be applied. If the primary goal is aggregate per unit area estimates, then a greater number of smaller plots with fewer trees per plot will be more efficient. If the goal is individual tree growth and its response to competition and/or silvicultural treatment, then larger plots are required. Smaller plots often have large unknown edge effects (Curtis and Marshall, 2005) and may not necessarily reflect the stand conditions following silvicultural treatment (e.g., small plots in a thinning study may have either all trees removed or no trees removed). In addition, only rough estimates of stand-level development and competition measures can be obtained from small plots (Garcia, 1998, 2006; Magnussen, 1999; Curtis and Marshall, 2005). Curtis and Marshall (2005) provide a comprehensive review of the literature regarding recommended plot sizes. They conclude that the best way to determine plot size is to choose the desired number of measurement trees per plot at the end of the study and derive plot size based on anticipated final stand conditions. In research installations where treatments create a large range of stand conditions, this may result in excessive measurements in untreated or lightly treated areas, and a compromise in numbers of plot trees across the range of stand conditions may need to be considered. Two or more plot sizes may initially appear to be the ideal compromise; however, caution should be exercised since plot size and treatment will be confounded and may influence results. It is recommended that a single fixed-area plot size be selected (or a single BAF in the case of variable probability samples) and that this size be fixed over the entire duration and spatial extent of the study. Field testing measurement protocols and plot sizes to determine level of effort and total resource commitments should be carried out prior to installation.

14.6.3. Sampling Unit Layout

If the objective of the study is monitoring growth and yield over an extensive forest area, sample locations should be determined using an appropriate statistical design (see Chapter 10). Systematic sampling is widely used and is most likely the most efficient design for large area estimation. Random sampling, stratified sampling, and two-stage sampling can also be used. In all cases, sample locations should be marked on a map or photo, or entered into a GPS, prior to crews going into the field.

In some cases, either because of budgetary constraints or a large amount of diversity within the forest, it may be desirable to purposefully determine sample locations in order to maximize the diversity of conditions observed. While these plots can yield useful data for modeling stand-level or individual tree growth and development, estimates from these plots cannot be scaled up to provide estimates of forest level changes. Care should also be exercised when inferring changes in large regional-level processes over time from these measures.

In research installations, plots will be established in purposefully selected stand conditions. If treatments are performed at the plot level, a buffer around each plot should be established to prevent edge effects from adjacent treatments influencing the observations. Buffers need to be large enough to mitigate effects of neighboring treatments and should be treated identically to the measurement portion of the treatment unit. At a minimum, buffers should be at least as large as the expected maximum crown width at the end of the study (Curtis and Marshall, 2005); however, such a narrow width will not be adequate for fertilization studies (since tree roots may extend far from the tree bole) or for studies where regeneration is of interest (because sunlight penetrates canopy gaps at an angle, so treatments may influence forest floor conditions a tree height or more away from the direct area of impact). Treatments should be assigned to units using one of the commonly accepted experimental designs (see Green, 1979; Hicks and Turner, 1999; Zar, 2009 for descriptions of experimental designs). Failure to properly design experiments frequently results in data being confounded with other nontreatment factors and limits inferences one is able to make from the results. If treatments are applied at a larger scale (e.g., at the stand), a number of sample locations may be required to capture treatment variability. In either case, plots should be established prior to treatment and all trees measured prior to treatment and immediately following treatment to be able to clearly document before and after conditions.

14.6.4. Measurement Protocols

Detailed measurement protocols are required for best results (Curtis and Marshall, 2005). Field guides and establishment guides should detail exactly how plots are established and the order of tree measurement and numbering. Field sheets or hand-held data recorders should be organized such that they follow the measurement order. Any codes for species, crown class, health, defects, and so on should be documented. If hand-held data recorders are being used, they should be programmed so that only valid codes can be used. When plots are being remeasured, the previous field data should be available so that current measurements can be checked against previous measurements. Any discrepancies and/or errors should be corrected in the field and notes made regarding these corrections. In this section, the major considerations are highlighted; for more detailed descriptions of field measurement protocols, see Curtis and Marshall (2005) or national forest inventory field guides such as FIA (2013).

Plot Establishment. If circular plots are being used, the plot center should be permanently marked with a stake or other permanent monument. With square or rectangular plots, the four corners need to be marked and it is useful to mark plot center. Wooden stakes can be used, but must be replaced periodically. Metal or plastic stakes are longer lasting. Plastic stakes will not damage harvesting equipment. Plastic-coated surveyor's magnets can be buried and relocated with an appropriate detector in situations where visible aboveground markers would be inappropriate. All monuments should be geolocated with a GPS and triangulated to at least three witness trees. At establishment, plot areas should be systematically searched for "in" trees. With circular plots, it is useful to start at a cardinal direction (say north) and work clockwise around sequentially numbering and tagging trees. With square plots, it is useful to work in a zig-zag fashion parallel to one of the plot sides in strips of about 2 m (6 ft), again sequentially numbering and tagging trees. Tree numbers can be painted on trees; however, metal tags are preferred. Tags can be placed either near the ground line or near breast height (1.3 m in SI units, 4.5 ft in imperial units). Tags can be nailed into the tree, leaving room for tree growth or placed on a cable tie and stapled to the tree. Stapling reduces potential damage due to fastening tags to trees. Tags should face toward plot center. A paint line at the height of dbh measurement is recommended. It is recommended that tree numbers start at 1 for each plot and continue sequentially. When plots are remeasured, any ingrowth trees are numbered starting at the last number used during the previous measurement. It is not recommended to reuse tree numbers because of the potential for confusion later on when data are being compiled.

In some situations, excessive visual impact on a stand is socially unacceptable (e.g., near heavily used recreational trails), and in others, it can lead to eventual bias. For example, where PSPs are used to inventory large areas of managed forest, the obvious presence of a PSP can influence marking and treatment decisions by field personnel. Over time, the tendency of PSPs to be treated differently than the surrounding forest causes the PSP network to become less representative of broader forest conditions. In these cases, hidden monuments (such as surveyor's magnets) must be used for plot centers and corners. Instead of tags or painted numbers to identify individual trees, all trees on the plot should be mapped. This can be an expensive undertaking using traditional tools, but the use of electronic distance measurement along with an accurate compass can dramatically reduce the cost for plots of moderate size (up to perhaps 0.1 ha). Maintaining a consistent height for dbh measurements without a visible reference (such as a painted or scribed line) becomes especially problematic. Where no markings at all are allowable, procedures for determining breast height must be clear and ironclad (such as the use of a reference pole) and even so it should be expected that individual tree growth estimates will have considerably greater error than would otherwise be possible.

Plot Remeasurement. When plots are being remeasured, all previous trees should be located and their statuses noted (live, dead, cut, missing, etc.). Any ingrowth trees should be added in the same sequence as plot establishment, as described above, and numbered sequentially starting with the last tree number used during the previous measurement. The tree numbers for trees that have died or been harvested should not be reassigned to new trees. Ideally, if plots are measured during the growing season, the timing of remeasurement should be near the date of plot establishment so that whole years of growth are captured. When this is not logistically possible, measurement dates should be incorporated into the data record so that average annual growth can be calculated correctly.

Tree Measurement. Precision and tolerance of tree measurements on PSPs need to be greater than those used in a normal forest inventory. If measurements are not made precisely enough, tree growth might not be detected in subsequent measurements. For example, if tree diameter is measured to the nearest 1.0 cm and average annual diameter growth is 0.1 cm, for a 5-year remeasurement cycle, many trees would still have the same diameter measurement as at the previous measurement. Tolerances likewise need to be controlled to insure that growth is detected and that any measurement tolerances do not result in negative growth results. Checking current measurements against previous measures helps insure that measurement tolerances are maintained and any errors corrected immediately. With the rapid development of more technologically advanced tree measurement tools, especially with hypsometers, it is important that crews know how to use these tools and that the measurements are consistent with the previous tools.

Common tree measurements obtained on PSPs include

1. *Species.* Species should be identified in the field and an appropriate species code used to designate the species used. Character-based codes are recommended over numeric codes because it is easier to recover character codes should measurement protocols become lost. Two-letter codes based on common names are frequently used; however, it is recommended to use a more complex code such as those developed by the USDA, Natural Resources Conservation Service (http://plants. usda.gov/java/; accessed September 4, 2016), to avoid ambiguity between different common names.

2. *Age.* Tree age is a useful parameter in even-aged stands and for determining site index (Section 8.8.3); however, for uneven-aged stands, age is not always a useful or meaningful parameter. If stand age or site index age is desired, it should be collected at plot establishment and updated based on years between measurements. Some protocols require that trees just off the plot be cored for age determination so as to avoid any potential damage to plot trees. While this has some intuitive appeal, it also introduces the potential for observer selection bias. It is better to have a probabilistic selection criteria associated with the plot trees or some other criterion such as tallest tree, most dominant species, largest dbh, and so on. If trees are being aged for site index determination, then the trees must be dominant trees that have been free to grow throughout their lives; visually determining free to grow trees can be very difficult in the field, especially in uneven-aged stands. In mixed-species stands, it may be useful to determine ages of all species present.

3. *dbh.* Diameter at breast height (1.3 m in SI units; 4.5 ft in imperial units) should be measured to the nearest 1 mm or nearest 0.1 in. Diameters should be measured with a diameter tape at the height indicated on the tree by either the tag, paint line, or bark scribe. Using a diameter tape minimizes interobserver error and eliminates errors associated with caliper orientation on eccentric trees. Ideally, all trees breast height or taller would be measured (Curtis and Marshall, 2005) for dbh and assigned a unique tree number; however, for many practical reasons, often only trees above a minimum threshold dbh are measured. Differences in minimum threshold diameter is one of the most common sources of incompatibility across different PSP programs. If minimum threshold diameters must be set, they should be as small as possible to ensure maximum compatibility with other PSP programs and to more fully capture vegetation dynamics. Trees smaller than the minimum threshold diameter are often tallied by species, dbh class and/or height class on a smaller concentric plot or plots. In some programs, such as the US Forest Service Forest

Inventory and Analysis program, trees on these smaller plots are tagged and measured just like the larger trees on the main plot (FIA, 2013). When trees grow past the minimum dbh threshold, they grow out of the small plot and on to the larger plot, resulting in a change in sampling probability, and growth estimates have to be treated using methods similar to those discussed for variable probability sampling (Section 14.5.4).

4. *Height.* Height is essential to most growth and yield analyses (Curtis and Marshall, 2005). Height improves estimates of bole volume and biomass, is required for site index estimation, and is useful for characterizing stand development. Heights should be measured to the nearest 0.1 m or 0.25 ft. Most electronic hypsometers report values to the nearest 0.1 units. Previous field measurements are vital in the field. Current measurements can be checked against previous measurements and any errors corrected in the field preventing heights from growing and shrinking over time. Reductions in height should only be acceptable when there is visible top damage to the tree. Some PSP programs record the distance and azimuth from which heights were measured; while this may reduce inter-measurer biases, it may not always result in better height measurements because tree tips may not be as clearly visible at the second measurement time. Because of the time required to obtain accurate height measurements, height is frequently either subsampled or not measured at all on many PSP programs. If height trees are a subsample of all plot trees, it is best to ensure that the selection is probabilistic rather than subjective to reduce observer biases.

5. *Crown.* Height to crown base, crown length, crown width, and crown ratio are all important variables for understanding and predicting individual tree growth. Height to crown base should be measured with the same precision as total height. With measurements of total height and height to crown base, crown length and crown ratio can be calculated. In many PSP programs, crown ratio is often visually estimated to the nearest 10%, and while this reduces the field work required, the time required to measure height to crown base is minimal if total height is being measured as well. There are many definitions of crown base (Section 5.5.1); clear definitions of crown base need to be developed in order to obtain consistent measurements. Crown width can be measured as described in Section 5.5.2. The azimuth of the crown width measurement is often prescribed (e.g., E-W or N-S, or toward and away from the plot center on circular plots). When it is left up to the field crew, the azimuth used at plot establishment should be recorded and used in subsequent measurements.

6. *Visual Defects and Damage.* Many PSP programs will include visual assessment of tree quality. Large knots, cracks, holes, bark damage, sweep, broken tops, and insect damage are examples of common defects or damage recorded. Curtis and Marshall (2005) and FIA (2013) provide a very comprehensive list of defects and damages. Defects and damages are generally numerically coded and trees often are allowed to have multiple codes.

7. *Regeneration.* Seedlings and small trees shorter than breast height are usually assessed on smaller subplots. Concentric or nested subplots are frequently used (e.g., FIA, 2013). Because of the high site disturbance associated with establishment and measuring PSPs, others advocate that regeneration subplots be placed

along the periphery of the main plot (Curtis and Marshall, 2005). In either case, it is good practice to locate these subplots, measure them first, and mark their locations to minimize disturbance during remeasurement. Seedlings and small trees are usually tallied by species and sometimes height classes. In research installations where the focus of the study is on regeneration, individual seedlings or small trees may be tagged and followed over time; however, these would be rare in PSP programs.

8. *Other Vegetation.* Herbaceous vegetation and smaller woody shrubs are often assessed on the same subplots used for assessing regeneration. As with regeneration, a summative tally rather than tagging of individuals is generally used. In some programs, only the presence of species is recorded and in others a visual assessment of cover is made. Because of differences in size and the fact that many stems can arise from the same root system, cover rather than count gives a better measure of dominance for many herbaceous species. Visual estimation of cover is a subjective measure and subject to large amounts of inter-measurer variation. The use of visual guides for estimating cover can minimize this variation.

Which of these measures are made, the intensity of the PSP system, and the frequency of measurement depends on the goals and objectives of the PSP program. At the very least, species and dbh should be measured on all tagged trees and status assessed (live, dead, cut). Total height should be measured on all trees or on an appropriately designed subsample. For monitoring programs, 5- or 10-year measurement cycles are adequate for most temperature forest species. For research installations, more frequent intervals may be required. Adequate planning and training is required to obtain the best, long-term results, and the development of a field guide describing plot establishment, measurement protocols, and documenting all codes used is invaluable to the longevity of the program.

14.7. GROWTH AND YIELD MODELS

Growth and yield models play an important role in forest mensuration. Forest inventories are expensive, as a result, forest inventories are carried out periodically, often over intervals of 10 or more years. Growth and yield models provide a mechanism for projecting inventory data and providing inventory updates for the time between inventories. In addition, forest planning horizons often span 80–100 years, depending on species and rotation lengths, and growth and yield models provide the projections needed to forecast flows of volume and other resources over these long planning horizons.

Many different types of growth and yield models have been developed. Each of these model types have particular assumptions, data requirements, computational requirements, and projection outputs. The major growth and yield model types are reviewed here and their role in forest mensuration discussed. For a more detailed treatment on the development of growth and yield models and their data requirements, see Weiskittel et al. (2011).

The type of model selected will depend upon application and availability. The model selected should provide predictions that are sufficiently detailed and accurate for the intended use. Overly complicated models are not necessarily better than models utilizing a simpler approach.

Vanclay (1994) sets out the following guidelines for selecting an appropriate model:

1. Does the approach make sense?
2. Will the model work for my application and input data?
3. What range of data was used to develop the model?
4. Do model assumptions and inferences apply in my situation?
5. What confidence can I place in model predictions?
6. Be skeptical and demand proof!

Buchman and Shifley (1983) provide a more detailed checklist.

14.7.1. Stand Table Projection

Stand table projection is a simple growth modeling approach, generally applied to individual stands for short projection periods. Stand table projection, in its simplest form, is completely data driven and can be developed from summaries of measurements of diameter increment, mortality, and ingrowth. More complicated stand table projections are possible that involve fitting regression equations and making predictions of changes from the resulting equations (e.g., Borders and Patterson, 1990; Lindsay et al., 1996).

To apply the method of stand table projection, the following data are required:

1. Diameter growth information
2. Present stand table
3. Local volume table
4. Information to calculate ingrowth
5. Estimates of mortality

Diameter growth information is most commonly obtained from increment cores (Section 14.3.2). However, excellent diameter growth information may be obtained from repeated measurements of permanent plots. In any case, there are three basic ways that diameter growth information may he applied to the present stand table, in conjunction with a local volume table, to obtain a growth estimate.

1. *Assume that all trees in each diameter class are located at the class midpoint, and that all trees will grow at the average rate.* Table 14.9 illustrates this approach.
   - Column 2 is obtained as explained in Section 14.3.2.
   - Column 3 = column 1 + column 2.
   - Column 4 gives the local volume table values for the diameters given in column 3. The values may be read from a curve of volume over tree diameter (column 6 over column 1) or calculated by an appropriate volume equation.
   - Column 5 is obtained from inventory data.
   - Column 6 gives the local volume table values for the diameters given in column 1.
   - Column 7 = column 4 × column 5.
   - Column 8 = column 6 × column 5.
   - Column 9 = column 7 − column 8.

TABLE 14.9. Calculation of 10-Year Predicted Volume Growth per Acre, Assuming that All Trees in Each Diameter Class Are Located at the Class Midpoint, and that All Trees Grow at the Average Rate

[Column 1] Present dbh Class (in.)	[Column 2] 10-year dbh Increment (in.)	[Column 3] Future dbh (in.)	[Column 4] Future Volume per Tree (ft³)	[Column 5] Present Stand Table (Number)	[Column 6] Present Volume per Tree (ft³)	[Column 7] Future Stock Table (ft³)	[Column 8] Present Stock Table (ft³)	[Column 9] Volume Production (ft³)
6	2.02	8.02		41.73				
8	1.88	9.88		28.73				
10	1.74	11.74	17.0	21.73	12.5	369.4	271.6	97.8
12	1.60	13.60	24.2	17.33	18.4	419.4	318.9	100.5
14	1.46	15.46	31.9	12.87	25.6	410.6	329.5	81.1
16	1.32	17.32	40.7	9.47	34.2	385.4	323.9	61.6
18	1.18	19.18	50.1	8.27	44.1	414.3	364.7	49.6
20	1.04	21.04	62.3	5.00	55.6	311.5	278.0	33.5
22	0.90	22.90	75.3	3.47	68.5	261.3	237.7	23.6
24	0.76	24.76	89.8	2.87	83.5	257.7	239.6	18.1
26					100.1			
Total				151.47		2829.6	2363.9	465.7

Note that the sum of column 9 equals the periodic gross growth of the initial volume. However, if the periodic diameter growth of the 8-in. diameter class had been over 2.00 in., all trees in this class would have grown to measurable size, that is, to the 10-in. class, and would have been *ingrowth*. This, of course, would have increased volume production. Thus, results may be inconsistent if an attempt is made to include ingrowth. But when no attempt is made to determine ingrowth, the method gives good estimates of gross growth of the initial volume.

2. *Assume trees in each diameter class are evenly distributed through the class, and each tree will grow at the average rate.* Table 14.10 illustrates this approach. In this case, a future stand table is predicted by first calculating the movement ratio M for each diameter class:

$$M = \frac{I}{C} \qquad (14.26)$$

where I = periodic diameter increment
          C = diameter class interval in same units as I

Thus, the movement ratio for the 12-in. diameter class (Table 14.10) is

$$M = \frac{1.60}{2} = 0.80$$

The two digits to the right of the decimal point indicate the proportion of the trees in the class that will move one class more than indicated by the digit to the left of the decimal point. Therefore, for the 12-in. class, $0.80 \times 17.33 = 13.86$ trees move one class, and $0.20 \times 17.33 = 3.47$ trees move zero classes.

In Table 14.10, the movements for all classes are shown in columns 6, 7, and 8, and the future stand table in column 9. The arrows show how the trees are moved into the future stand. Columns 4 and 5 are the same as columns 5 and 6 in Table 14.9. Columns 10, 11, and 12 are determined as follows.

- Column 10 = column 9 × column 5
- Column 11 = column 4 × column 5
- Column 12 = column 10 − column 11

Note that the sum of column 12 in Table 14.10 equals the periodic gross growth since neither ingrowth nor mortality have been considered in this analysis.

3. *Recognize the actual position of trees in each diameter class and apply the diameter growth for individual trees in the class.* In this approach movement, percentages are calculated by applying actual individual increments to individual tree diameters. A graphic solution (Wahlenberg, 1941) may be used, but a simple tabular solution is equally satisfactory and lends itself to electronic data processing. For example, in Table 14.11 tree movement percentages are computed for the 8-in. diameter class from the data used to compute column 2 in Table 14.10.

TABLE 14.10. Calculations of 10-Year Predicted Volume Growth per Acre, Assuming Trees in Each Diameter Class Are Evenly Distributed through the Class, and Each Tree Grows at the Average Rate

[Column 1] dbh Class (in.)	[Column 2] 10-year dbh Increment (in.)	[Column 3] Movement Ratio (M)	[Column 4] Present Stand Table (Number)	[Column 5] Volume per Tree (ft³)	[Column 6] Number of Trees Moving 2 Classes	[Column 7] 1 Class	[Column 8] 0 Classes	[Column 9] Future Stand Table (number)	[Column 10] Future Stock Table (ft³)	[Column 11] Present Stock Table (ft³)	[Column 12] Volume Production (ft³)
6	2.02	1.01	41.73		0.42	41.31	0.00	0.00			
8	1.88	0.94	28.73			27.01	1.72	43.04			
10	1.74	0.87	21.73	12.5		18.91	2.82	30.25	378.1	271.6	106.5
12	1.6	0.80	17.33	18.4		13.86	3.47	22.37	411.6	318.9	92.8
14	1.46	0.73	12.87	25.6		9.40	3.47	17.34	443.9	329.5	114.4
16	1.32	0.66	9.47	34.2		6.25	3.22	12.61	431.4	323.9	107.6
18	1.18	0.59	8.27	44.1		4.88	3.39	9.64	425.2	364.7	60.5
20	1.04	0.52	5.00	55.6		2.60	2.40	7.28	404.7	278.0	126.7
22	0.9	0.45	3.47	68.5		1.56	1.91	4.51	308.8	237.7	71.1
24	0.76	0.38	2.87	83.5		1.09	1.78	3.34	279.0	239.6	39.3
26				100.1				1.09	109.2		109.2
Total			151.47					151.47	3191.9	2363.9	828.0

507

TABLE 14.11. Determination of Tree Movement Ratios From Raw Data for 8-in. dbh Class

	Raw data				Summary		
dbh Class (in.)	Present dbh (in.)	10-year dbh Increment (in.)	Future dbh (in.)	Classes Moved (Number)	Classes Moved (Number)	Trees Moving (Number)	Trees Moving (%)
8	7.1	1.5	8.6	0	0	3	30
	7.3	1.6	8.9	0	1	5	50
	7.5	1.5	8.9	0	2	2	20
	7.7	1.8	9.3	1	Total	10	100
	7.9	2.5	10.4	1			
	8.1	1.6	9.7	1			
	8.3	1.8	10.1	1			
	8.5	2.6	11.1	2			
	8.7	1.7	10.4	1			
	8.9	2.2	11.1	2			

These percentages are applied to the present stand table to obtain number of trees moving. Except for this calculation, the future stand table and the future stock table are predicted as in Table 14.10.

If the method depicted in Table 14.10 is used to predict growth, reasonable estimates of *ingrowth* may he determined by including, in the initial stand table, several diameter classes below the merchantable limit. For example, if we consider the 6- and 8-in. dbh classes as below merchantable, then ingrowth could be estimated from the number of trees growing into the 10-in. dbh class ($I = 0.42 + 27.01 = 27.43$ trees/acre) and the volume ingrowth would be $27.43*12.5 = 342.8 \, ft^3$/acre. In this case, the total column 12 would represent gross growth plus ingrowth. If the method depicted in Table 14.9 is used, inconsistent estimates of ingrowth may he obtained by this procedure. In either case, estimates of ingrowth are unreliable for long prediction periods or for rapidly growing stands.

Mortality, which was not considered in the preceding examples (Tables 14.9 and 14.10), may be accounted for in one of the following ways:

1. By deducting predicted number of trees dying from each diameter class of the present stand table prior to projecting the present stand table.
2. By deducting predicted number of trees dying from each diameter class of the future stand table after projecting the present stand table, but before computing the future stock table.

In healthy, middle-aged stands following the stem exclusion stage or in stands under intensive management, mortality will not be large and can be accurately predicted. In young stands and in old stands, mortality will often be great and, because of its erratic nature, cannot always be accurately predicted. In either case, allowances are made only for normal mortality resulting from old age, competition, typical insects and diseases, ordinary wind, and so on. No allowance is made for catastrophic mortality resulting from fire, epidemics, great storms, and so forth.

Good information on mortality may be obtained from PSPs. From such information for any given stand, one can determine functions relating mortality to age, diameter, stand density, and species, and thus apply the mortality information to other stands. Often when we desire to make predictions using stand-table projection, we will lack suitable PSP data. Then, mortality estimates must be obtained from a stand inspection, which is normally made during the cruise. In the inspection, which is quite subjective, one estimates on plots or points the number of trees, by species and diameter classes, that died during some past period, say 10 years, or that will die during some future period. When the mortality information is summarized, it is expressed as percentages of the trees in each diameter class of the stand table.

If accurate diameter growth information is used, any stand-table projection method will give an excellent estimate of gross growth of the initial basal area. Of course, basal area growth is an important component of volume growth; so is height growth. Therefore, determination of gross growth of initial volume also depends on the stability, during the prediction period, of the height–diameter relationship for which the local volume table was constructed. It also assumes no change in form. It has been demonstrated that the future height–diameter relationship will not necessarily be the same as the present height–diameter relationship (Chapman and Meyer, 1949; Kershaw et al., 2008). For large areas and for uneven-aged stands, this change may he slight, but for small areas and for even-aged stands the change may be substantial, even for periods of 10–20 years. With the exception of abnormal conditions, form changes may be safely ignored for short periods.

Stand-table projection will give good results for uneven-aged stands of immature timber that are understocked. Then, mortality will be small and predictable. Ingrowth may be accurately predicted, and the height–diameter relationships will change only slightly. In even-aged stands, young dense stands, and overmature stands, stand-table projection will often give inaccurate results because of the change in the height–diameter relationships, and because of the high and unpredictable mortality. Regardless of stand conditions, the application of stand-table projection to long projection periods becomes unreliable because growth rates, mortality, and ingrowth will all change as stand structure changes. Accounting for such changes requires a more complicated approach.

14.7.2. Yield Tables and Equations

Stand-level models make predictions about changes in per unit area values. Numbers of trees, basal area, and volume are typical stand parameters modeled at the stand level. Yield tables are one of the most common examples of a stand-level model. A yield table is a tabular presentation of volume per unit area and other stand characteristics of even-aged stands by age classes, site classes, species, and density.

Most early yield tables used in America were prepared by graphical procedures described by Bruce (1926) and Reineke (1927) and were improved on by Osborne and Schumacher (1935). Departure from the graphical approach was evident in the classic study by MacKinney et al. (1937), who used least-squares regression techniques applied to a logarithmic transformation of the Pearl–Reed logistic curve. The multiple regression applications by Schumacher (1939), Smith and Ker (1959), among others, further demonstrated the superiority of sound statistical techniques over the purely graphical approach for the preparation of yield tables. Although the tabular form of yield relationships has endured years of use in forest management activities, the formula form is the choice for developing yield curves now. The construction techniques employed for most of the older yield tables have been supplanted by modern, mathematically sophisticated approaches using computers and statistical analyses.

Even-aged yield tables are prepared from yield studies of the relationship between a dependent variable, such as volume, basal area, or number of trees, and independent variables describing stand conditions such as age, site quality, and stand density. Site quality is most often measured in terms of site index, although discrete site quality classes have been used. Density has been most commonly measured in terms of basal area, although stand density index often is more convenient to use. Yield tables are valuable in such forest management activities as regulating cut, determining rotation length, making growth estimates, and forest valuation.

Almost by definition yield tables are not applicable to uneven-aged stands since there is no one representative average age. A similar type of table has been produced for uneven-aged stands showing the volumes produced in growth for given periods with a certain level of growing stock on land of different site qualities (Meyer, 1934). Yield records for uneven-aged stands over long periods are required to prepare this kind of table.

There are several types of yield tables for even-aged stands, depending on the independent variables used to predict growth:

Normal Yield Tables. An example of a normal yield table is shown in Table 14.12. These tables predict stand growth and yield based on stand age and site index. Normal yield tables originated before analytical methods for handling more than two independent variables were available. Since normal yield tables use only two independent variables, historically they were conveniently constructed by graphical means. Density is held constant by attempting to select sample plots of the same density. The density required has been called *full* or *normal stocking*. Full or normal stocking is supposed to describe the density of a stand that completely occupies a given site and makes full use of its growth potential. Since it is difficult to describe full stocking quantitatively, qualitative and somewhat subjective guides often have been used. For example, the guides might include complete canopy closure, no openings in the stand, and regular spacing of the trees. Such specifications leave much to the judgment of the individual in choosing so-called fully stocked stands for samples.

The values of the stand parameters for successive ages shown in normal yield tables are generally prepared from averages of a number of temporary sample plots established in stands considered fully stocked. Each stand in which sample plots are measured might have varying patterns of development. In past years, some stands may have been overstocked or understocked in terms of the definition of normality. When data from these samples are compiled, average relationships are developed that represent the development of a theoretically fully stocked stand over its entire life. It is quite unlikely that any existing stand will show the same pattern as is represented in a normal yield table. In reality, very few stands are encountered that can be called fully stocked. Both understocked and overstocked stands can be encountered, with understocking the usual case.

The volume of an existing stand can be estimated from a normal yield table, by measuring its age, site index, and stocking percentage relative to normality. Relative stocking can be measured by comparing the volume, basal area, or number of trees per unit area estimated for a stand with the values predicted from the yield table for a stand of the same age and site index. There is little point in using volume to measure normality when volume is

TABLE 14.12. Example of a Normal Yield Table

Yield per Acre for Fully Stocked Spruce–Aspen Stands; Good Site – Index 80[a]

	Tolerant Softwoods (Mainly Spruce)							Intolerant Hardwoods (Mainly Aspen)			
					Volume Inside Bark						
						Merchantable	Scribner			Volume to Basal Area	Entire Stand
Spruce Total Age (Years)	Average Dominant Height (ft)	Number of Trees	Average dbh (in.)	Basal Area (ft²)	Entire Stem (ft³)	Stem[b] (12 in.+) (ft³)	Rule[c] (12 in. +) (bd. ft.)	Basal Area (ft²)	Composition[d] (%)	Factor[e] (Units)	Basal Area (ft²)
30	23	1091	1.6	16	155	0	0	88	83	16.7	104
40	34	1379	2.5	49	620	5	0	88	64	19.9	137
50	45	1344	3.3	79	1280	20	0	85	52	22.6	164
60	58	1091	4.1	100	2010	55	150	82	45	25.0	182
70	70	845	5.0	114	2740	105	600	79	41	27.1	193
80	80	643	6.0	125	3445	310	1,540	76	38	28.7	201
90	87	484	7.1	133	4060	835	3,490	71	35	30.1	204
100	93	376	8.3	140	4565	1645	7,020	66	32	31.3	206
110	98	307	9.3	145	4995	2520	11,180	60	29	32.3	205
120	102	264	10.2	150	5370	3325	15,590	53	26	33.2	203
130	106	238	10.9	153	5680	3995	18,980	45	23	34.0	198
140	109	220	11.4	156	5945	4460	21,320	36	19	34.7	192
150	112	208	11.8	158	6165	4810	23,060	26	14	35.3	184

Source: Macleod and Blyth (1955).

[a] In this study, site index is defined as the height attained by the average dominant spruce at 80 years of age.

[b] 1-ft stump, 4-in. top inside bark.

[c] 1-ft stump, 6in. top inside bark.

[d] Hardwood composition by basal area.

[e] ×basal area=total cubic volume.

generally the parameter to be estimated. Basal area has been found to be the most satisfactory basis for expressing relative stocking, when age (or stand diameter) and height are held constant. It is easily and quickly determined, and is closely related to volume:

$$RS = \frac{BA_{obs}}{BA_{Norm}}$$
(14.27)

where   RS = relative basal area stocking
                 BA_{obs} = observed basal area
                 BA_{Norm} = basal area of a normal stand of the same age and site index

The stocking or normality percentage times the yield table volume estimates the volume of the existing stand:

$$\hat{V} = RS\left(V_{Norm}\right)$$
(14.28)

where $\hat{V}$ = predicted volume
                 RS = relative basal area stocking
                 V_{Norm} = volume of a normal stand of the same age and site index

For example, if we had a stand applicable to the yield table shown in Table 14.12, which was 90 years old and had a total basal area of 160 ft²/acre, then the relative stocking $RS = 160/204 = 0.78$ and the estimated volume $\hat{V} = 0.78(4060) = 3184$ ft³ / acre. This naturally assumes that relative stocking in basal area equals relative stocking in volume. This may be a tenable assumption for cubic volume but can lead to serious error for board–foot volume estimation. Normality percentages for the same stand calculated from basal area, cubic volume, and board feet can give widely differing results.

Because of the several subjective decisions necessary in their preparation and use, normal yield tables have been challenged: indeed, the entire normality concept has been seriously questioned. To overcome the fully stocked assumption of normal yield tables, two other types of tables have been developed and used—the empirical yield table and the variable density yield table.

Empirical Yield Tables. These tables are similar to normal yield tables but are based on sample plots of average rather than full stocking. The judgment necessary for selecting fully stocked stands is eliminated, simplifying the collection of field data.

Variable Density Yield Tables. When stand density is used as an independent variable, variable density yield tables result. These tables show yields for various levels of stocking. This approach also has the advantage of not requiring samples to be fully stocked. Sample plots of any density can be used since the density is measured as an independent variable.

Yield studies incorporating stand density as an independent variable were carried out in the 1930s and 1940s as the use of statistical techniques increased. MacKinney et al. (1937) used age, stand density, site index, and a stand composition index as independent variables. This resulted in a more general type of variable density yield table. In a later study, MacKinney and Chaiken (1939), in one of the first applications of multiple regression to

yield estimation, used age, site index, and stand density as independent variables employing the following model for loblolly pine:

$$\log(Y) = b_0 + b_1 \frac{1}{A} + b_2 S + b_3 \log(\text{SDI}) + b_4 C \qquad (14.29)$$

where Y = volume/acre, ft^3

          A = age in years

          S = site index

       SDI = stand density index

         C = composition index (basal area of loblolly pine divided by total basal area
            per acre)

Smith and Ker (1959) used site index, age, maximum height, average stand diameter, basal
area per acre, and number of trees per acre in preparing yield equations by multiple
regression.

    Variable density yield tables are easily created from any yield function that includes
density as a variable. To illustrate this, we will use the yield equation developed by Sullivan
and Clutter (1972) for natural loblolly pine stands:

$$\ln(V) = 2.8837 + 0.014441S - 21.326A^{-1} + 0.95064\ln(B)$$

where V = total cubic–foot volume, inside bark, per acre

          S = site index in ft

          A = stand age (years)

          B = basal area/acre in ft^2

    By substituting in values for B and A, the variable density yield table shown in
Table 14.13 was created.

    The fact that regression techniques provide a "yield formula" as well as yield table was
recognized as a distinct advantage especially in early computer applications. Through all
this development, however, a major problem was ignored or overlooked. Because yield
functions (that predict stand volume at a specified age) and growth functions (that predict
volume growth over shorter periods) were often derived independently, summation of a
succession of periodic *growth estimates* added to an initial volume would not necessarily
lead to the final stand volume indicated by the *yield function estimate*.

    The application of calculus to growth and yield studies led to the resolution of this
inconsistency between growth summation and terminal yield. The independent, essentially
simultaneous, works of Buckman (1962) and Clutter (1963) began a new era in yield
studies. A brief description of their work is appropriate.

    Working with even-aged red pine, Buckman (1962) emphasized that growth and yield
are not independent phenomena and should not be treated as such. Furthermore, he
employed methods of calculus that had been neglected in virtually all previous yield studies
of this type. Beginning with a *basal area growth equation* of the form:

$$Y = b_0 + b_1 B + b_2 B^2 + b_3 A + b_4 A^2 + b_5 S \qquad (14.30)$$

TABLE 14.13. Variable Density Yield Table for Loblolly Pine

	Basal Area (ft²/acre)				
	80	100	120	140	160
Stand Age (Years)	(Total ft³, Inside Bark, per Acre)				
30	2771	3426	4075	4718	5356
40	3310	4093	4867	5635	6398
50	3683	4553	5415	6269	7118
60	3954	4888	5814	6731	7642
70	4160	5143	6116	7082	8040
80	4322	5343	6354	7357	8352

Source: Sullivan and Clutter (1972).

where Y = periodic net annual basal area increment (i.e., dB/dA, change in basal area with respect to age)

B = basal area in ft²/acre

A = age in years

S = site index

Yield tables were prepared by iterative solution and accumulation; that is, the least-squares fit of eq. (14.30) is solved for a particular site, age, and stand density. Basal area growth is then added to the stand density, 1 year is added to age, and the equation is again solved. Summation of the n successive annual growth estimates to the initial basal area provides a yield estimate *n* years hence.

Clutter (1963) working with even-aged loblolly pine clearly indicated the relationship between growth and yield models by the following definition:

Such models are here defined as compatible when the yield model can be obtained by summation of the predicted growth through the appropriate growth periods or, more precisely, when the algebraic form of the yield model can be derived by mathematical integration of the growth model.

In the research reported by Clutter, a yield model was first prepared of the form

$$\ln(V) = a + b_1 S + b_2 \ln(B) + b_3 A^{-1} \tag{14.31}$$

where $\ln(V)$ = logarithm to the base *e* of volume

A = stand age in years

S = site index in ft

B = basal area/acre in ft²

and differentiated with respect to age, obtaining

$$\frac{dV}{dA} = b_2 VB^{-1}(dB/dA) - b_3 VA^{-2} \tag{14.32}$$

where dV/dA = rate of change of volume with respect to age or instantaneous rate of
          volume growth

   dB/dA = rate of change of basal area with respect to age or instantaneous rate of
          basal area growth

Since the rate of basal area growth is not ordinarily available, regression analysis was used
to obtain dB/dA as a function of age, site, and basal area. The model finally adopted was

$$\frac{dB}{dA} = -B(\ln B)A^{-1} + c_0 A^{-1}B + c_1 BSA^{-1} \tag{14.33}$$

Substituting this relation for dB/dA in eq. (14.32) led to the equation:

$$\frac{dV}{dA} = -b_2 V(\ln B)A^{-1} + b_2 c_0 VA^{-1} + b_2 c_1 VSA^{-1} - b_3 VA^{-2} \tag{14.34}$$

Using the form of this equation as a model, and based on data gathered on PSPs, a least
squares regression equation was obtained, thus relating volume growth with present basal
area, age, site index, and volume (estimated using eq. (14.31)). Subsequent integration of
this regression equation led to the final yield function, from which volume yield at some
future time could be predicted from given initial age, basal area and volume, projected age,
and site index. Numerous other yield functions have been developed for even-aged stands.
For a review of these equations, see Clutter et al. (1983), Vanclay (1994), and Weiskittel
et al. (2011).

The majority of yield functions historically have been developed for even-aged stands.
Functions for uneven-aged stands are still relatively rare. An example of an uneven-aged
growth and yield function is the one proposed by Moser and Hall (1969) for uneven-aged
mixed northern hardwood stands. In their model, volume growth was expressed as the first
derivative with respect to time:

$$\frac{dV}{dt} = (b_1 VB^{-1})\frac{dB}{dt} \tag{14.35}$$

where dV/dt = instantaneous rate of change of stand volume

   V = stand volume

   B = stand basal areas

   dB/dt = instantaneous rate of change of basal area

   b_1 = a constant

Note that VB^{-1} is just the stand-average VBAR (Section 11.2.1); Equation (14.35)
implies that VBAR remains constant over time, which may be approximately true for an
uneven-aged stand with a near-constant proportional representation of trees in different
height classes. For dB/dt, they used the Von Bertalanffy's generalized growth rate equation
(Richards, 1959):

$$\frac{dB}{dt} = nB^m - kB \tag{14.36}$$

where dB/dt = instantaneous rate of change of basal area

B = stand basal area

n, m, and k are constants

By substituting eq. (14.36) into eq. (14.35) and integrating Moser and Hall produced the following yield function:

$$V = V_0 / \left(B_0^{b_1} \right) \left[n/k - \left(n/k - B_0^{1-m} \right) e^{-(1-m)kt} \right]^{b_1/(1-m)} \tag{14.37}$$

where V = future volume

V_0 = initial volume

B_0 = initial basal area

b_1, n, m, k are constants

t = elapsed time

e = natural logarithmic exponent

14.7.3. Diameter Distribution Models

Diameter distribution models represent a hybrid between stand-level models and tree-level models. In these models, future stand-level parameters are predicted. Then, based on these predictions, the parameters of a distribution model such as the Weibull distribution (Section 8.3.3) are obtained. The resulting distribution is then used to predict the number of individual trees across various diameter classes.

There are two general approaches to diameter distribution models. In the first method, the future parameters of the distribution model are directly predicted from the current parameters and other information about the stand (density, basal area, volume, etc.). This approach is referred to as the *parameter prediction approach*. In the second approach, the future values of the stand parameters are directly predicted, and the parameters of the diameter distribution model are recovered using estimates of the moments of the distribution derived from the stand parameters. This approach is referred to as the *parameter recovery approach*. The parameter recovery approach generally yields better results than the parameter prediction approach (Reynolds et al., 1988). Hyink and Moser (1983) present a generalized framework for developing diameter distribution models. Reynolds et al. (1988) discuss various approaches to developing models and discuss model selection procedures.

14.7.4. Individual Tree Growth and Yield Models

Both distance-dependent and distance-independent individual tree models have been developed for a variety of species and stand conditions. Individual tree models have been developed for single species even-aged stands (plantations and natural stands) as well as for mixed species stands (even-aged and uneven-aged).

Distance-dependent models require information about the spatial location (x–y coordinates) of individual trees as well as tree characteristics (e.g., species, dbh, height, etc.) as inputs. These models use measures of point density (Section 8.9.7) to estimate the level of competition for each tree, and model growth as a function of tree size and competition. Examples of distance-dependent models include the model FOREST developed by Ek and Monserud (1974), and the Tree and Stand Simulator (TASS) used in British Columbia, originally developed for managed stands of Douglas-fir (Mitchell, 1975). TASS is one of the best examples of a distance dependent model being used for forest management decision-making.*

Distance-independent models only require information about tree characteristics as inputs. These models predict tree growth based on initial tree characteristics and general expressions of competition (e.g., stand density index, total basal area, basal area of larger trees, relative height, etc.). Distance-independent models are more common than distance-dependent models primarily because detailed information about tree locations is not readily available.

Examples of distance-independent models include Prognosis (Stage, 1973; Wykoff et al., 1982) originally developed for mixed-species conifer stands in Idaho and Montana, ORGANON (Hester et al., 1989) originally developed for mixed-species conifer forests in southwestern Oregon, and TWIGS (Miner et al., 1988) developed for mixed-species stands in the central United States. The PROGNOSIS model is now referred to as the Forest Vegetation Simulator and a number of variants for many regions of the United States have been developed.† Examples of other individual tree models and a detailed discussion of the process of building individual tree models can be found in Vanclay (1994) and Weiskittel et al. (2011).

14.7.5. Other Types of Models

The above discussion of models has been limited to those models that have been widely applied in forest management. There are numerous other approaches to modeling forest growth and development. For a thorough review of these, see Vanclay (1994) and Weiskittel et al. (2011).

Two modeling approaches that merit brief mention here are process models and forest succession models. *Process models* attempt to model growth using physiological processes such as light absorption, nutrient and water uptake, photosynthesis and respiration (e.g., Mäkelä, 1992). Such models have shown some success in modeling changes in stem form, crown length, and knot size as well as carbon storage in relation to stand density and other treatments (e.g., Valentine et al., 2012b; Niinimäki et al., 2013). *Forest succession models* attempt to model changes in species composition over long periods of time (e.g., Botkin, 1993). These models are useful for understanding growth and stand dynamics, but have not been widely applied for predicting forest yields. These models have been widely applied to assess potential impacts of global climate change and increases in atmospheric carbon dioxide.

Process models have great potential in future forest management. Several efforts have attempted to apply process models to forest management problems (e.g., Battaglia and

---

*Current information on TASS and its applications for forest management can be obtained from the BC Ministry of Forests, Lands, and Natural Resource Operations (https://www.for.gov.bc.ca/hts/growth/; accessed September 4, 2016).
† Current information on FVS variants and copies of the software can be obtained from the USDA Forest Service, Forest Management Service Center (http://www.fs.fed.us/fmsc/fvs/; accessed September 4, 2016).

Sands, 1998; Johnsen et al., 2001; Monserud, 2003). Other developments have included linking process models with empirical growth and yield models (e.g., Somers and Nepal, 1994; Baldwin et al., 2001) and collaborating process models with stand growth and yield data (Sievänen and Burk, 1994). Examples of applying forest successional models to forest management can be found in Taylor et al. (2009) and Ashraf et al. (2012).

14.7.6. Feedbacks between Growth and Yield Models and Forest Mensuration

The outputs from growth and yield models traditionally are used in a variety of forest management activities. Detailed forest management planning uses yield predictions to model the flows of timber and other resources. Models can be used to assess tree and stand responses to silviculture treatments and aid in the design and selection of appropriate treatments given management objectives. When either temporary or PSP inventory results will be used as input into growth and yield models, it is critical to ensure that the field protocols include collecting the data required for the models that are likely to be used. For example, when PSP data for an inventory stratum that contains monospecific plantations will be used as input to a growth and yield model that requires site index as a descriptor of stand productivity, it would be wise to include collection of total height and age of dominant trees in the data collection protocol. Some variables that have proven valuable as predictors of stand growth in models, such as crown length, are rarely collected in traditional operational inventories focused primarily on evaluating current economic yields. There is a tradeoff between inventory expense and utility. Failure to anticipate future needs in designing field protocols can impose substantial constraints on the types of growth and yield models (and other decision support tools) that can be used to inform management.

Forest inventories provide valuable information to forest managers; however, these inventories represent the forest at a particular point in time. Forest inventories are expensive; therefore, inventories are generally conducted periodically, often at 10-year intervals in slow-growing temperate and boreal forests. Nonetheless, the results of an inventory may be outdated late in the interval. Growth and yield models have been used to predict changes in tree and stand values for periods between successive inventories. This problem is particularly challenging for national inventories and other large-scale inventories that use a panel design, where a portion of the plots are measured every year, and it may be desirable to "update" the older plots to provide current estimates of forest conditions. A variety of modeling approaches, from relatively simple growth and yield models to imputation techniques, can be used in such situations (McRoberts, 2001).

Predictions from growth and yield models could be used as selection criteria for PSP remeasurements. For example, Bokalo et al. (1996) proposed a sampling with partial replacement scheme that incorporated predictions from growth and yield models. Growth and yield models could be used in the prediction phase of a 3P sample design as well.* These approaches have not been widely used, but might be suggested as a way to reduce the costs of maintaining PSP programs. The use of such methods could have resulted in fewer PSP programs being abandoned in recent decades and could facilitate their adoption in coming decades as the value of long-term data on forest growth and dynamics is increasingly recognized for a variety of management and policy purposes.

---

*Though not articulated in this context, Grosenbaugh (1965) suggested this application.

APPENDIX

Forest Mensuration, Fifth Edition. John A. Kershaw, Jr., Mark J. Ducey, Thomas W. Beers and Bertram Husch.
© 2017 John Wiley & Sons, Ltd. Published 2017 by John Wiley & Sons, Ltd.

TABLE A.1. Some Conversion Factors for Common Units of Measure

A. Length

	mm	cm	m	km	in.	ft	yd	link	rod	chain	mile
1 millimeter	1	0.1	0.001	10^{-6}	0.03937	0.0032808	0.001094	0.004971	0.0001988	4.971×10^{-5}	6.214×10^{-7}
1 centimeter	10	1	0.01	0.00001	0.3937	0.0328084	0.01094	0.04971	0.001988	0.0004971	6.214×10^{-6}
1 meter	1,000	100	1	0.001	39.37	3.28084	1.094	4.971	0.1988	0.04971	0.0006214
1 kilometer	10^6	10^5	1000	1	39,370	3280.84	1094	4971	198.8	49.71	0.6214
1 inch	25.4	2.54	0.0254	2.54×10^{-5}	1	0.08333	0.02778	0.12626	0.005050	0.001263	1.578×10^{-5}
1 foot	304.8	30.48	0.3048	0.0003048	12	1	0.3333	1.51515	0.06061	0.01515	0.0001894
1 yard	914.4	91.44	0.9144	0.0009144	36	3	1	4.54545	0.18182	0.04545	0.0005682
1 link	201.168	20.1168	0.20117	0.0002012	7.92	0.66	0.22	1	0.04	0.01	0.000125
1 rod	5029.2	502.92	5.0292	0.0050292	198	16.5	5.5	25	1	0.25	0.003125
1 chain	20116.8	2011.68	20.1168	0.0201168	792	66	22	100	4	1	0.0125
1 mile[a]	1,609,344	160934.4	1609.344	1.609344	63,360	5280	1760	8000	320	80	1

[a] 1 international nautical mile = 1852 m.

B. Area

	cm²	m²	ha	km²	in.²	ft²	yd²	chain²	acre	mile²
1 centimeter²	1	0.0001	10^{-8}	10^{-10}	0.155	0.001076	1.196×10^{-4}	2.471×10^{-7}	2.471×10^{-8}	3.86×10^{-11}
1 meter²	10^4	1	0.0001	10^{-6}	1550.	10.76	1.196	0.002471	2.471×10^{-4}	3.86×10^{-7}
1 hectare[a]	10^8	10^4	1	0.01	1.55×10^7	1.076×10^5	11959.9	24.7105	2.47105	0.003861
1 kilometer²	10^{10}	10^6	100	1	1.55×10^9	1.076×10^7	1,195,990	2471.05	247.105	0.3861
1 inch²	6.4516	6.452×10^{-4}	6.452×10^{-8}	6.452×10^{-10}	1	0.006944	7.716×10^{-4}	1.594×10^{-6}	1.594×10^{-7}	2.491×10^{-10}
1 foot²	929.0304	0.092903	9.2903×10^{-6}	9.2903×10^{-8}	144	1	0.1111	0.000230	2.296×10^{-5}	3.587×10^{-8}
1 yard²	8361.2736	0.83613	8.361×10^{-5}	8.361×10^{-7}	1,296	9	1	0.00207	2.066×10^{-4}	3.228×10^{-7}
1 chain²[b]	4.0469×10^{6}	404.69	0.04047	4.047×10^{-4}	627,264	4,356	484	1	0.1	1.5625×10^{-4}
1 acre[c]	4.0469×10^{7}	4046.9	0.4047	0.004047	6,272,640	43,560	4,840	10	1	0.0015625
1 mile²	2.59×10^{10}	2588988.11	258.999	2.58999	4.014×10^9	27,878,400	3,097,600	6400	640	1

[a] 1 hectare = 100 acres.

[b] 1 chain² = 10,000 link².

[c] 640 acres = 1 section; 36 sections = 1 township (6 miles × 6 miles).

Basal area per unit land area conversion factors:

1 ft²/acre = 0.2296 m²/ha

1 m²/ha = 4.3560 ft²/acre

521

C. Volume

	cm³	liter[a]	m³	in.³	ft³	yd³	oz[b]	pint[b]	qt[b]	gal[b]	qt[c]
1 centimeter³	1	9.9997×10^{-4}	10^{-6}	0.061024	3.531×10^{-5}	1.308×10^{-6}	0.03381	0.002113	0.001057	2.642×10^{-4}	9.0808×10^{-4}
1 liter	1000.027	1	0.001	61.0254	0.03531	0.001308	33.81	2.113	1.0567	0.26417	0.9081
1 meter³	10^6	999.973	1	61,024	35.31	1.308	33,813	2113.36	1056.68	264.17	908.076
1 inch³	16.387	0.016387	1.6387×10^{-5}	1	0.000579	2.143×10^{-5}	0.5541	0.03463	0.01732	0.004329	0.01488
1 foot³	28316.8	28.31608	0.028317	1728	1	0.037037	957.5	59.84	29.922	7.4805	25.714
1 yard³	764,555	764.5342	0.764554	46,656	27	1	25852.6	1615.79	807.895	201.97	694.28
1 ounce[b]	29.5737	0.029573	2.9573×10^{-5}	1.8047	0.001044	3.868×10^{-5}	1	0.0625	0.03125	0.0078125	27.4997
1 pint[b]	473.179	0.473166	0.0004732	28.875	0.01671	0.000619	16	1	0.5	0.125	1.7187
1 quart[b]	946.359	0.946333	9.4635×10^{-4}	57.75	0.03342	0.001238	32	2	1	0.25	0.8594
1 gallon[b]	3785.43	3.785328	0.0037854	231	0.13368	0.004951	128	8	4	1	3.4375
1 quart[c]	1101.23	1.1012	0.001101	67.2	0.03889	0.00144	0.03636	0.5818	1.16365	0.2909125	1

[a] Volume of 1 kg water at 4°C and 760 mm pressure.
[b] U.S. fluid measure. British imperial gallon = 277.24 in.³ = 1.2009 U.S. gallons.
[c] U.S. dry measure. 32 quarts = 1 bushel.

Volume per unit land area conversion factors:

1 ft³/acre = 0.06997 m³/ha

1 m³/ha = 14.29 ft³/acre

D. Mass

	grain[a]	ounce[a]	pound[a]	ton[a,b]	gram	kilogram	metric ton
1 grain[a]	1	0.00228571	0.00014286	7.1428×10^{-8}	0.0647989	6.4799×10^{-5}	6.47989×10^{-8}
1 ounce[a]	437.5	1	0.0625	3.125×10^{-5}	28.3495201	0.0283495	2.83495×10^{-5}
1 pound[a]	7000	16	1	0.0005	453.59232	0.4535923	0.00045359
1 ton[a,b]	1.4×10^{7}	32,000	2000	1	907184.6	907.18464	0.9071846
1 gram	15.43236	0.035274	0.0022046	1.10231×10^{-6}	1	0.001	10^{-6}
1 kilogram	15432.36	35.273966	2.2046228	0.001102311	1000	1	0.001
1 metric ton	1.543236×10^{7}	35273.966	2204.6228	1.102311	10^{6}	1000	1

[a] Avoirdupois system.
[b] U.S. ton; 1 imperial ton = 2240 pounds.
Mass per unit land conversion factors:

1 U.S. ton per acre = 2.2417 metric tons per hectare; 1 lb/acre = 1.120851 kg/ha

1 metric ton per hectare = 0.44609 U.S. tons per acre; 1 kg/ha = 0.892179 lb/acre

TABLE A.2. Areas of Some Plane Figures

Figure	Diagram	Formula
Rectangle		$A = 1 \times w$
Parallelogram		$A = b \times h$
Triangle		$A = \dfrac{b \times h}{2}$ or $A = \sqrt{S(S-a)(S-b)(S-c)}$ where $S = \dfrac{1}{2}(a+b+c)$
Trapezoid		$A = \dfrac{1}{2}(a+c) \times h$
Circle		$A = \pi \times r^2$ or $A = \dfrac{\pi \times D^2}{4}$ where $r = \dfrac{D}{2}$ Circumference $= 2\pi \times r = \pi \times D$
Circular sector		$A = \dfrac{\theta \times r^2}{2}$ where θ is in radians $A = \dfrac{\theta \times \pi \times r^2}{360}$ where θ is in degrees
Circular segment		$A = \dfrac{1}{2} r^2 (\theta - \sin\theta)$ where θ is in radians
Ellipse		$A = \pi \times a \times b$ Perimeter $= 2\pi \sqrt{\dfrac{a^2+b^2}{2}}$ (approximately)
Parabola		$A = \dfrac{2}{3} l \times d$ Length of arc $= l\left[1 + \dfrac{2}{3}\left(\dfrac{2d}{l}\right)^2 - \dfrac{2}{5}\left(\dfrac{2d}{l}\right)^4 + \cdots\right]$

TABLE A.3. Volume and Surface Areas of Some Solids

The following symbols are used in the formulas:

V = volume $\qquad\qquad\qquad$ L = slant height

A_b = area of base $\qquad\qquad\quad$ P_b = perimeter of lower base

A_m = area of midsection parallel to A_b and A_u $\quad$ P_u = perimeter of upper section

A_u = area of upper section $\qquad\quad$ P_r = perimeter of right section

A_r = area of right section $\qquad\quad$ S_l = lateral surface area

h = altitude or height $\qquad\qquad$ S_t = total surface area

$\qquad\qquad\qquad\qquad\qquad\qquad$ d = distance from base to intermediate section

Solid	Diagram	Formula
Prismoid		$V = \dfrac{h}{6}(A_b + 4A_m + A_u)$ $S_l = \dfrac{L}{2}(P_b + P_u)$ $S_t = S_l + A_b + A_u$
Prism or cylinder		$V = A_b h$ $S_l = P_r L$ $S_t = S_l + 2A_b$
Pyramid or cone		$V = \dfrac{h}{3} A_b$ $S_l = \dfrac{L}{2} P_b$ $S_t = S_l + A_b$
Frustum of cone or pyramid		(treat as prismoid or) $V = \dfrac{h}{3}\left(A_b + \sqrt{A_b \times A_u} + A_u\right)$ $S_l = \dfrac{L}{2}(P_b + P_u)$ $S_t = S_l + A_b + A_u$
Sphere		$V = \dfrac{4}{3}\pi r^3$ $S_t = 4\pi r^2$

(Continued)

TABLE A.3. (Continued)

Solid	Diagram	Formula

Paraboloid

$$V = \frac{h}{2} A_b$$

$$S_l = \frac{2\pi r}{12h^2}\left[(r2 + 4h^2)^{\frac{3}{2}} - r^3\right]$$

$$S_t = S_l + A_b$$

Frustum of paraboloid

$$V = \frac{d}{2}(A_b + A_u)$$

$$S_l = S_{l(acb)} - S_{l(gck)}$$

$$S_t = S_l + A_b + A_u$$

Neiloid

$$V = \frac{h}{4} A_b$$

$$S_l = 2\pi \int_0^h X\left[\frac{3}{2}\sqrt{1 + \frac{9X}{4}}\right] dX$$

$$S_t = S_l + A_b$$

Frustum of neiloid

$$V = \frac{d}{6}(A_b + 4A_m + A_u)$$

$$S_l = S_{l(acb)} - S_{l(gck)}$$

$$S_t = S_l + A_b + A_u$$

TABLE A.4. Critical Values of Student's *t*-Distribution

| Degrees of Freedom | Two-Tailed Probability of Obtaining a Large Value | | | | | | | | | |
	0.5	0.4	0.3	0.2	0.1	0.05	0.02	0.01	0.001
1	1.0000	1.3764	1.9626	3.0777	6.3137	12.7062	31.8210	63.6559	636.5776
2	0.8165	1.0607	1.3862	1.8856	2.9200	4.3027	6.9645	9.9250	31.5998
3	0.7649	0.9785	1.2498	1.6377	2.3534	3.1824	4.5407	5.8408	12.9244
4	0.7407	0.9410	1.1896	1.5332	2.1318	2.7765	3.7469	4.6041	8.6101
5	0.7267	0.9195	1.1558	1.4759	2.0150	2.5706	3.3649	4.0321	6.8685
6	0.7176	0.9057	1.1342	1.4398	1.9432	2.4469	3.1427	3.7074	5.9587
7	0.7111	0.8960	1.1192	1.4149	1.8946	2.3646	2.9979	3.4995	5.4081
8	0.7064	0.8889	1.1081	1.3968	1.8595	2.3060	2.8965	3.3554	5.0414
9	0.7027	0.8834	1.0997	1.3830	1.8331	2.2622	2.8214	3.2498	4.7809
10	0.6998	0.8791	1.0931	1.3722	1.8125	2.2281	2.7638	3.1693	4.5868
11	0.6974	0.8755	1.0877	1.3634	1.7959	2.2010	2.7181	3.1058	4.4369
12	0.6955	0.8726	1.0832	1.3562	1.7823	2.1788	2.6810	3.0545	4.3178
13	0.6938	0.8702	1.0795	1.3502	1.7709	2.1604	2.6503	3.0123	4.2209
14	0.6924	0.8681	1.0763	1.3450	1.7613	2.1448	2.6245	2.9768	4.1403
15	0.6912	0.8662	1.0735	1.3406	1.7531	2.1315	2.6025	2.9467	4.0728
16	0.6901	0.8647	1.0711	1.3368	1.7459	2.1199	2.5835	2.9208	4.0149
17	0.6892	0.8633	1.0690	1.3334	1.7396	2.1098	2.5669	2.8982	3.9651
18	0.6884	0.8620	1.0672	1.3304	1.7341	2.1009	2.5524	2.8784	3.9217
19	0.6876	0.8610	1.0655	1.3277	1.7291	2.0930	2.5395	2.8609	3.8833
20	0.6870	0.8600	1.0640	1.3253	1.7247	2.0860	2.5280	2.8453	3.8496
25	0.6844	0.8562	1.0584	1.3163	1.7081	2.0595	2.4851	2.7874	3.7251
30	0.6828	0.8538	1.0547	1.3104	1.6973	2.0423	2.4573	2.7500	3.6460
35	0.6816	0.8520	1.0520	1.3062	1.6896	2.0301	2.4377	2.7238	3.5911
40	0.6807	0.8507	1.0500	1.3031	1.6839	2.0211	2.4233	2.7045	3.5510
45	0.6800	0.8497	1.0485	1.3007	1.6794	2.0141	2.4121	2.6896	3.5203
50	0.6794	0.8489	1.0473	1.2987	1.6759	2.0086	2.4033	2.6778	3.4960
55	0.6790	0.8482	1.0463	1.2971	1.6730	2.0040	2.3961	2.6682	3.4765

(*Continued*)

TABLE A.4. (Continued)

Degrees of Freedom	Two-Tailed Probability of Obtaining a Large Value								
	0.5	0.4	0.3	0.2	0.1	0.05	0.02	0.01	0.001
60	0.6786	0.8477	1.0455	1.2958	1.6706	2.0003	2.3901	2.6603	3.4602
70	0.6780	0.8468	1.0442	1.2938	1.6669	1.9944	2.3808	2.6479	3.4350
80	0.6776	0.8461	1.0432	1.2922	1.6641	1.9901	2.3739	2.6387	3.4164
90	0.6772	0.8456	1.0424	1.2910	1.6620	1.9867	2.3685	2.6316	3.4019
100	0.6770	0.8452	1.0418	1.2901	1.6602	1.9840	2.3642	2.6259	3.3905
150	0.6761	0.8440	1.0400	1.2872	1.6551	1.9759	2.3515	2.6090	3.3565
200	0.6757	0.8434	1.0391	1.2858	1.6525	1.9719	2.3451	2.6006	3.3398
∞	0.6745	0.8416	1.0364	1.2816	1.6449	1.9600	2.3263	2.5758	3.2905

Source: The table was generated using the Splus Statistical Software Package (Insightful Corporation, Seattle, WA).

TABLE A.5.A. Plot Radius (ft) and Tree Factor (trees/acre) for Horizontal Point Sampling in Imperial Units by Basal Area Factor (BAF, ft²/acre) and dbh (in.)

dbh (in.)	Plot Radius (ft) BAF						Tree Factor (trees/acre) BAF					
	10	20	40	60	80	100	10	20	40	60	80	100
2	5.50	3.89	2.75	2.25	1.94	1.74	458.38	916.76	1833.52	2750.28	3667.03	4583.79
3	8.25	5.83	4.13	3.37	2.92	2.61	203.72	407.45	814.90	1222.34	1629.79	2037.24
4	11.00	7.78	5.50	4.49	3.89	3.48	114.59	229.19	458.38	687.57	916.76	1145.95
5	13.75	9.72	6.88	5.61	4.86	4.35	73.34	146.68	293.36	440.04	586.73	733.41
6	16.50	11.67	8.25	6.74	5.83	5.22	50.93	101.86	203.72	305.59	407.45	509.31
7	19.25	13.61	9.63	7.86	6.81	6.09	37.42	74.84	149.67	224.51	299.51	374.19
8	22.00	15.56	11.00	8.98	7.78	6.96	28.65	57.30	114.59	171.89	229.19	286.49
9	24.75	17.50	12.38	10.10	8.75	7.83	22.64	45.27	90.54	135.82	181.09	226.36
10	27.50	19.45	13.75	11.23	9.72	8.70	18.34	36.67	73.34	110.01	146.68	183.35
11	30.25	21.39	15.13	12.35	10.69	9.57	15.15	30.31	60.61	90.92	121.22	151.53
12	33.00	23.33	16.50	13.47	11.67	10.44	12.73	25.47	50.93	76.40	101.86	127.33
13	35.75	25.28	17.88	14.59	12.64	11.31	10.85	21.70	43.40	65.10	86.79	108.49
14	38.50	27.22	19.25	15.72	13.61	12.17	9.35	18.71	37.42	56.13	74.84	93.55
15	41.25	29.17	20.63	16.84	14.58	13.04	8.15	16.30	32.60	48.89	65.19	81.49
16	44.00	31.11	22.00	17.96	15.56	13.91	7.16	14.32	28.65	42.97	57.30	71.62
17	46.75	33.06	23.38	19.09	16.53	14.78	6.34	12.69	25.38	38.07	50.75	63.44
18	49.50	35.00	24.75	20.21	17.50	15.65	5.66	11.32	22.64	33.95	45.27	56.59
19	52.25	36.95	26.13	21.33	18.47	16.52	5.08	10.16	20.32	30.47	40.63	50.79
20	55.00	38.89	27.50	22.45	19.45	17.39	4.58	9.17	18.34	27.50	36.67	45.84
21	57.75	40.84	28.88	23.58	20.42	18.26	4.16	8.32	16.63	24.95	33.26	41.58
22	60.50	42.78	30.25	24.70	21.39	19.13	3.79	7.58	15.15	22.73	30.31	37.88
23	63.25	44.72	31.63	25.82	22.36	20.00	3.47	6.93	13.86	20.80	27.73	34.66
24	66.00	46.67	33.00	26.94	23.33	20.87	3.18	6.37	12.73	19.10	25.47	31.83
25	68.75	48.61	34.38	28.07	24.31	21.74	2.93	5.87	11.73	17.60	23.47	29.34
26	71.50	50.56	35.75	29.19	25.28	22.61	2.71	5.42	10.85	16.27	21.70	27.12

(*Continued*)

TABLE A.5.A. (Continued)

dbh (in.)	Plot Radius (ft)						Tree Factor (trees/acre)					
	BAF						BAF					
	10	20	40	60	80	100	10	20	40	60	80	100
27	74.25	52.50	37.13	30.31	26.25	23.48	2.52	5.03	10.06	15.09	20.12	25.15
28	77.00	54.45	38.50	31.44	27.22	24.35	2.34	4.68	9.35	14.03	18.71	23.39
29	79.75	56.39	39.88	32.56	28.20	25.55	2.18	4.36	8.72	13.08	17.44	21.80
30	82.50	58.34	41.25	33.68	29.17	26.09	2.04	4.07	8.15	12.22	16.30	20.37
35	96.25	68.06	48.13	39.29	34.03	30.44	1.50	2.99	5.99	8.98	11.97	14.97
40	110.00	77.78	55.00	44.91	38.89	34.79	1.15	2.29	4.58	6.88	9.17	11.46
45	123.75	87.50	61.88	50.52	43.75	39.13	0.91	1.81	3.62	5.43	7.24	9.05
50	137.50	97.23	68.75	56.13	48.61	43.48	0.73	1.47	2.93	4.40	5.87	7.33
60	165.00	116.67	82.50	67.36	58.34	52.18	0.51	1.02	2.04	3.06	4.07	5.09

TABLE A.5.B. Plot Radius (m) and Tree Factor (trees/ha) for Horizontal Point Sampling in SI Units by Basal Area Factor (BAF, $m^2 ha^{-1}$) and dbh (cm)

dbh (cm)	Plot Radius (m) BAF						Tree Factor (trees/ha) BAF					
	2	3	4	5	6	8	2	3	4	5	6	8
2	0.71	0.58	0.50	0.45	0.41	0.35	6366.18	9549.27	12732.37	15915.46	19098.55	25464.73
4	1.41	1.15	1.00	0.89	0.82	0.71	1591.55	2387.32	3183.09	3978.86	4774.64	6366.18
6	2.12	1.73	1.50	1.34	1.22	1.06	707.35	1061.03	1414.71	1768.38	2122.06	2829.41
8	2.83	2.31	2.00	1.79	1.63	1.41	397.89	596.83	795.77	994.72	1193.66	1591.55
10	3.54	2.89	2.50	2.24	2.04	1.77	254.65	381.97	509.29	636.62	763.94	1018.59
12	4.24	3.46	3.00	2.68	2.45	2.12	176.84	265.26	353.68	442.10	530.52	707.35
14	4.95	4.04	3.50	3.13	2.86	2.47	129.92	194.88	259.84	324.81	389.77	519.69
16	5.66	4.62	4.00	3.58	3.27	2.83	99.47	149.21	198.94	248.68	298.41	397.89
18	6.36	5.20	4.50	4.02	3.67	3.18	78.59	117.89	157.19	196.49	235.78	314.38
20	7.07	5.77	5.00	4.47	4.08	3.54	63.66	95.49	127.32	159.15	190.99	254.65
22	7.78	6.35	5.50	4.92	4.49	3.89	52.61	78.92	105.23	131.53	157.84	210.45
24	8.49	6.93	6.00	5.37	4.90	4.24	44.21	66.31	88.42	110.52	132.63	176.84
26	9.19	7.51	6.50	5.81	5.31	4.60	37.67	56.50	75.34	94.17	113.01	150.68
28	9.90	8.08	7.00	6.26	5.72	4.95	32.48	48.72	64.96	81.20	97.44	129.92
30	10.61	8.66	7.50	6.71	6.12	5.30	28.29	42.44	56.59	70.74	84.88	113.18
32	11.31	9.24	8.00	7.16	6.53	5.66	24.87	37.30	49.74	62.17	74.60	99.47
34	12.02	9.81	8.50	7.60	6.94	6.01	22.03	33.04	44.06	55.07	66.08	88.11
36	12.73	10.39	9.00	8.05	7.35	6.36	19.65	29.47	39.30	49.12	58.95	78.59
38	13.44	10.97	9.50	8.50	7.76	6.72	17.63	26.45	35.27	44.09	52.90	70.54
40	14.14	11.55	10.00	8.94	8.16	7.07	15.92	23.87	31.83	39.79	47.75	63.66
42	14.85	12.12	10.50	9.39	8.57	7.42	14.44	21.65	28.87	36.09	43.31	57.74
44	15.53	12.70	11.00	9.84	8.98	7.78	13.15	19.73	26.31	32.88	39.46	52.61
46	16.26	13.28	11.50	10.29	9.39	8.13	12.03	18.05	24.07	30.09	36.10	48.14
48	16.97	13.86	12.00	10.73	9.80	8.49	11.05	16.58	22.10	27.63	33.16	44.21
50	17.68	14.43	12.50	11.18	10.21	8.84	10.19	15.28	20.37	25.46	30.56	40.74
52	18.38	15.01	13.00	11.63	10.61	9.91	9.42	14.13	18.83	23.54	28.25	37.67

(Continued)

TABLE A.5.B. (Continued)

dbh (cm)	Plot Radius (m) BAF						Tree Factor (trees/ha) BAF					
	2	3	4	5	6	8	2	3	4	5	6	8
54	19.09	15.59	13.50	12.07	11.02	9.55	8.73	13.10	17.47	21.83	26.20	34.93
56	19.80	16.17	14.00	12.52	11.43	9.90	8.12	12.18	16.24	20.30	24.36	32.48
58	20.51	16.74	14.50	12.97	11.84	10.25	7.57	11.35	15.14	18.92	22.71	30.28
60	21.21	17.32	15.00	13.42	12.25	10.61	7.07	10.61	14.15	17.68	21.22	28.29
70	24.75	20.21	17.50	15.65	14.29	12.37	5.20	7.80	10.39	12.99	15.59	20.79
80	28.28	23.09	20.00	17.89	16.33	14.14	3.98	5.97	7.96	9.95	11.94	15.92
90	31.82	25.98	22.50	20.12	18.37	15.91	3.14	4.72	6.29	7.86	9.43	12.58
100	35.36	28.87	25.00	22.36	20.41	17.68	2.55	3.82	5.09	6.37	7.64	10.19

TABLE A.6.A. Equations for Obtaining Common Factors and Constants for Various Forms of PPS Sampling in Imperial Units

Factor	Type of PPS Sampling			
	Horizontal		Vertical	
	Point	Line	Point	Line
Gauge constant	$k = \dfrac{D_i}{12R_i} = 2\sin\left(\dfrac{\theta}{2}\right)$		$q = \dfrac{H_i}{R_i} = \tan(\varphi)$	
Inclusion zone radius or half-width (ft)	$R_i = \dfrac{D_i}{12k} = \dfrac{\left(33\sqrt{10}\right)D_i}{12\sqrt{BAF}}$	$R_i = \dfrac{D_i}{12k} = \dfrac{330 \cdot D_i}{DF}$	$R_i = \dfrac{H_i}{q} = \dfrac{\left(66\sqrt{10}\right)H_i}{\sqrt{\pi SHF}}$	$R_i = \dfrac{H_i}{q} = \dfrac{330 H_i}{HF}$
Inclusion zone area (ft²)	$\begin{aligned} Area_i &= \pi R_i^2 \\ &= \pi\left(\dfrac{D_i}{12k}\right)^2 \\ &= \pi\left(\dfrac{10{,}890 D_i^2}{144 BAF}\right) \end{aligned}$	$\begin{aligned} Area_i &= 2R_i L \\ &= \dfrac{D_i L}{6k} \\ &= \dfrac{660 D_i L}{DF} \end{aligned}$	$\begin{aligned} Area_i &= \pi R_i^2 \\ &= \pi H_i^2 \cot^2(\varphi) \\ &= \pi\left(\dfrac{H_i}{q}\right)^2 \\ &= \dfrac{43{,}560 H_i^2}{SHF} \end{aligned}$	$\begin{aligned} Area_i &= 2R_i L \\ &= 2H_i L \cot(\varphi) \\ &= \dfrac{2H_i L}{q} \\ &= \dfrac{660 H_i L}{HF} \end{aligned}$
Horizontal distance multiplier (ft/in. D)	$\begin{aligned} HDM &= \dfrac{1}{12k} \\ &= \dfrac{33\sqrt{10}}{12\sqrt{BAF}} \end{aligned}$	$\begin{aligned} HDM &= \dfrac{1}{12k} \\ &= \dfrac{330}{DF} \end{aligned}$	$\begin{aligned} HDM &= \dfrac{1}{q} \\ &= \dfrac{66\sqrt{10}}{\sqrt{\pi SHF}} \end{aligned}$	$\begin{aligned} HDM &= \dfrac{1}{q} \\ &= \dfrac{300}{HF} \end{aligned}$
Tree factor (trees/acre/tree tallied)	$\begin{aligned} TF_i &= \dfrac{43{,}560}{Area_i} \\ &= \dfrac{10{,}890 k^2}{0.005454 D_i^2} \\ &= \dfrac{BAF}{BA_i} \end{aligned}$	$\begin{aligned} TF_i &= \dfrac{43{,}560}{Area_i} \\ &= \dfrac{261{,}360 k}{D_i L} \\ &= \dfrac{66 DF}{D_i L} \end{aligned}$	$\begin{aligned} TF_i &= \dfrac{43{,}560}{Area_i} \\ &= \dfrac{43{,}560}{\pi H_i^2 \cot^2(\varphi)} \\ &= \dfrac{43{,}560 q^2}{\pi H_i^2} \\ &= \dfrac{SHF}{H_i^2} \end{aligned}$	$\begin{aligned} TF_i &= \dfrac{43{,}560}{Area_i} \\ &= \dfrac{21{,}780}{H_i L \cot(\varphi)} \\ &= \dfrac{21{,}780 q}{H_i L} \\ &= \dfrac{66 HF}{H_i L} \end{aligned}$

(Continued)

TABLE A.6.A. (Continued)

Factor	Horizontal		Vertical	
	Point	Line	Point	Line
Basal area factor (ft²/acre/tree tallied)	$BAF_i = BA_i \cdot TF_i$ $= BA_i \left(\dfrac{10,890k^2}{BA_i} \right)$ $= 10,890k^2$	$BAF_i = BA_i \cdot TF_i$ $= \dfrac{1425.457\pi k D_i}{L}$ $= \dfrac{0.36 DF \cdot D_i}{L}$	$BAF_i = BA_i \cdot TF_i$ $= BA_i \left(\dfrac{SHF}{H_i^2} \right)$	$BAF_i = BA_i \cdot TF_i$ $= BA_i \left(\dfrac{66HF}{H_i L} \right)$
Diameter factor (in./acre/tree tallied)	$DF_i = D_i \cdot TF_i$ $= \dfrac{BAF}{0.005454 D_i}$	$DF = D_i \cdot TF_i$ $= \dfrac{261,360k}{L}$	$DF_i = D_i \cdot TF_i$ $= D_i \left(\dfrac{SHF}{H_i^2} \right)$	$DF_i = D_i \cdot TF_i$ $= D_i \left(\dfrac{66HF}{H_i L} \right)$
Squared height factor (ft²/acre/tree tallied)	$SHF_i = H_i^2 \cdot TF_i$ $= H_i^2 \left(\dfrac{BAF}{0.005454 D_i^2} \right)$	$SHF_i = H_i^2 \cdot TF_i$ $= H_i^2 \left(\dfrac{66DF}{D_i L} \right)$	$SHF_i = H_i^2 \cdot TF_i$ $= \dfrac{43,560 \, q^2}{\pi}$	$SHF_i = H_i^2 \cdot TF_i$ $= \dfrac{21,780 \, q H_i}{L}$
Height factor (ft/acre/tree tallied)	$HF_i = H_i \cdot TF_i$ $= H_i \left(\dfrac{BAF}{0.005454 D_i^2} \right)$	$HF_i = H_i \cdot TF_i$ $= H_i \left(\dfrac{66DF}{D_i L} \right)$	$HF_i = H_i \cdot TF_i$ $= \dfrac{43,560 q^2}{\pi H_i}$	$HF_i = H_i \cdot TF_i$ $= \dfrac{21,780q}{L}$
Volume factor (units/acre/tree tallied)	$VF_i = V_i \cdot TF_i$ $= (bD_i^2 H_i) \left(\dfrac{BAF}{BA_i} \right)$ $= \dfrac{bH_i BAF}{0.005454}$	$VF_i = V_i \cdot TF_i$ $= (bD_i^2 H_i) \left(\dfrac{66DF}{D_i L} \right)$ $= \dfrac{66b D_i H_i DF}{L}$	$VF_i = V_i \cdot TF_i$ $= (bD_i^2 H_i) \left(\dfrac{SHF}{H_i^2} \right)$ $= \dfrac{bD_i^2 SHF}{H_i}$	$VF_i = V_i \cdot TF_i$ $= (bD_i^2 H_i) \left(\dfrac{66HF}{H_i L} \right)$ $= \dfrac{66b D_i^2 HF}{L}$
Weight or biomass factor (lb/acre/tree tallied)	$WF_i = WT_i \cdot TF_i$	$WF_i = WT_i \cdot TF_i$	$WF_i = WT_i \cdot TF_i$	$WF_i = WT_i \cdot TF_i$

Circumference factor (in./acre/tree tallied)	$CF_i = C_i \cdot TF_i$ $= \pi D_i \left(\dfrac{BAF}{BA_i} \right)$ $= \dfrac{576BAF}{D_i}$	$CF_i = C_i \cdot TF_i$ $= \dfrac{66\pi DF}{L}$	$CF_i = C_i \cdot TF_i$ $= \pi D_i \left(\dfrac{SHF}{H_i^2} \right)$	$CF_i = C_i \cdot TF_i$ $= \pi D_i \left(\dfrac{66HF}{H_i L} \right)$
Surface area factor (ft²/acre/tree tallied)	$BSAF_i = BSA_i \cdot TF_i$ $= bC_i H_i \left(\dfrac{BAF}{BA_i} \right)$ $= \dfrac{576bH_i BAF}{D_i}$	$BSAF_i = BSA_i \cdot TF_i$ $= bC_i H_i \left(\dfrac{66DF}{D_i L} \right)$ $= \dfrac{66\pi bH_i DF}{L}$	$BSAF_i = BSA_i \cdot TF_i$ $= bC_i H_i \left(\dfrac{SHF}{H_i^2} \right)$ $= \dfrac{b\pi D_i SHF}{H_i}$	$BSAF_i = BSA_i \cdot TF_i$ $= bC_i H_i \left(\dfrac{66HF}{H_i L} \right)$ $= \dfrac{66b\pi D_i HF}{L}$
All other factors (units/acre/tree tallied)	$XF_i = X_i \cdot TF_i$	$XF_i = X_i \cdot TF_i$	$XF_i = X_i \cdot TF_i$	$XF_i = X_i \cdot TF_i$

D_i = diameter (in.); BA_i = basal area (ft²); H_i = height (ft); C_i = circumference (in.); V_i = volume (ft³ or board feet); WT_i = weight or biomass (lb); BSA = bole surface area (ft²); and L = line length (m).

535

TABLE A.6.B. Equations for Obtaining Common Factors and Constants for Various Forms of PPS Sampling in SI Units

Factor	Type of PPS Sampling			
	Horizontal		Vertical	
	Point	Line	Point	Line
Gauge constant	$k = \dfrac{D_i}{100R_i} = 2\sin\left(\dfrac{\theta}{2}\right)$		$q = \dfrac{H_i}{R_i} = \tan(\varphi)$	
Inclusion zone radius or half-width (m)	$R_i = \dfrac{D_i}{100k} = \dfrac{D_i}{2\sqrt{BAF}}$	$R_i = \dfrac{D_i}{100k} = \dfrac{250 \cdot D_i}{DF}$	$R_i = \dfrac{H_i}{q} = \dfrac{100H_i}{\sqrt{\pi SHF}}$	$R_i = \dfrac{H_i}{q} = \dfrac{250H_i}{HF}$
Inclusion zone area (m²)	$\begin{aligned} Area_i &= \pi R_i^2 \\ &= \pi\left(\dfrac{D_i}{100k}\right)^2 \\ &= \pi\left(\dfrac{D_i^2}{4BAF}\right) \end{aligned}$	$\begin{aligned} Area_i &= 2R_iL \\ &= \dfrac{D_iL}{50k} \\ &= \dfrac{500D_iL}{DF} \end{aligned}$	$\begin{aligned} Area_i &= \pi R_i^2 \\ &= \pi H_i^2 \cot^2(\varphi) \\ &= \pi\left(\dfrac{H_i}{q}\right)^2 \\ &= \dfrac{10{,}000H_i^2}{SHF} \end{aligned}$	$\begin{aligned} Area_i &= 2R_iL \\ &= 2H_iL\cot(\varphi) \\ &= \dfrac{2H_iL}{q} \\ &= \dfrac{500H_iL}{HF} \end{aligned}$
Horizontal distance multiplier (m/cm D)	$\begin{aligned} HDM &= \dfrac{1}{100k} \\ &= \dfrac{1}{2\sqrt{BAF}} \end{aligned}$	$\begin{aligned} HDM &= \dfrac{1}{100k} \\ &= \dfrac{250}{DF} \end{aligned}$	$\begin{aligned} HDM &= \dfrac{1}{q} \\ &= \dfrac{100}{\sqrt{\pi SHF}} \end{aligned}$	$\begin{aligned} HDM &= \dfrac{1}{q} \\ &= \dfrac{250}{HF} \end{aligned}$
Tree factor (trees/ha/tree tallied)	$\begin{aligned} TF_i &= \dfrac{10{,}000}{Area_i} \\ &= \dfrac{2{,}500k^2}{0.00007854D_i^2} \\ &= \dfrac{BAF}{BA_i} \end{aligned}$	$\begin{aligned} TF_i &= \dfrac{10{,}000}{Area_i} \\ &= \dfrac{500{,}000k}{D_iL} \\ &= \dfrac{20DF}{D_iL} \end{aligned}$	$\begin{aligned} TF_i &= \dfrac{10{,}000}{Area_i} \\ &= \dfrac{10{,}000}{\pi H_i^2 \cot^2(\varphi)} \\ &= \dfrac{10{,}000q^2}{\pi H_i^2} \\ &= \dfrac{SHF}{H_i^2} \end{aligned}$	$\begin{aligned} TF_i &= \dfrac{10{,}000}{Area_i} \\ &= \dfrac{5{,}000}{H_iL\cot(\varphi)} \\ &= \dfrac{5{,}000q}{H_iL} \\ &= \dfrac{20HF}{H_iL} \end{aligned}$

Basal area factor (m²/ha/tree tallied)

$$BAF = BA_i \cdot TF_i$$
$$= BA_i\left(\frac{2,500k^2}{BA_i}\right)$$
$$= 2,500k^2$$

$$BAF_i = BA_i \cdot TF_i$$
$$= \frac{12.5\pi k D_i}{L}$$
$$= \frac{0.00157 DF \cdot D_i}{L}$$

$$BAF_i = BA_i \cdot TF_i$$
$$= BA_i\left(\frac{SHF}{H_i^2}\right)$$

$$BAF_i = BA_i \cdot TF_i$$
$$= BA_i\left(\frac{20HF}{H_i L}\right)$$

Diameter factor (cm/ha/tree tallied)

$$DF_i = D_i \cdot TF_i$$
$$= \frac{BAF}{0.00007854 D_i}$$

$$DF = D_i \cdot TF_i$$
$$= \frac{500,000k}{L}$$

$$DF_i = D_i \cdot TF_i$$
$$= D_i\left(\frac{SHF}{H_i^2}\right)$$

$$DF_i = D_i \cdot TF_i$$
$$= D_i\left(\frac{20HF}{H_i L}\right)$$

Squared height factor (m²/ha/tree tallied)

$$SHF_i = H_i^2 \cdot TF_i$$
$$= H_i^2\left(\frac{BAF}{0.00007854 D_i^2}\right)$$

$$SHF_i = H_i^2 \cdot TF_i$$
$$= H_i^2\left(\frac{20DF}{D_i L}\right)$$

$$SHF_i = H_i^2 \cdot TF_i$$
$$= \frac{10,000q^2}{\pi}$$

$$SHF_i = H_i^2 \cdot TF_i$$
$$= \frac{5,000qH_i}{L}$$

Height factor (m/ha/tree tallied)

$$HF_i = H_i \cdot TF_i$$
$$= H_i\left(\frac{BAF}{0.00007854 D_i^2}\right)$$

$$HF_i = H_i \cdot TF_i$$
$$= H_i\left(\frac{20DF}{D_i L}\right)$$

$$HF_i = H_i \cdot TF_i$$
$$= \frac{10,000q^2}{\pi H_i}$$

$$HF_i = H_i \cdot TF_i$$
$$= \frac{5,000q}{L}$$

Volume factor (ft³/acre/tree tallied)

$$VF_i = V_i \cdot TF_i$$
$$= (bD_i^2 H_i)\left(\frac{BAF}{BA_i}\right)$$
$$= \frac{bH_i BAF}{0.00007854}$$

$$VF_i = V_i \cdot TF_i$$
$$= (bD_i^2 H_i)\left(\frac{20DF}{D_i L}\right)$$
$$= \frac{20bD_i H_i DF}{L}$$

$$VF_i = V_i \cdot TF_i$$
$$= (bD_i^2 H_i)\left(\frac{SHF}{H_i^2}\right)$$
$$= \frac{bD_i^2 SHF}{H_i}$$

$$VF_i = V_i \cdot TF_i$$
$$= (bD_i^2 H_i)\left(\frac{20HF}{H_i L}\right)$$
$$= \frac{20bD_i^2 HF}{L}$$

(Continued)

TABLE A.6.B. (Continued)

Factor	Type of PPS Sampling			
	Horizontal		Vertical	
	Point	Line	Point	Line
Weight or Biomass factor (lb/acre/tree tallied)	$WF_i = WT_i \cdot TF_i$	$WF_i = WT_i \cdot TF_i$	$WF_i = WT_i \cdot TF_i$	$WF_i = WT_i \cdot TF_i$
Circumference factor (in./acre/tree tallied)	$CF_i = C_i \cdot TF_i$ $= \pi D_i \left(\dfrac{BAF}{BA_i}\right)$ $= \dfrac{40{,}000 BAF}{D_i}$	$CF = C_i \cdot TF_i$ $= \dfrac{20\pi DF}{L}$	$CF_i = C_i \cdot TF_i$ $= \pi D_i \left(\dfrac{SHF}{H_i^2}\right)$	$CF_i = C_i \cdot TF_i$ $= \pi D_i \left(\dfrac{20HF}{H_i L}\right)$
Surface area factor (ft²/acre/tree tallied)	$BSAF_i = BSA_i \cdot TF_i$ $= bC_i H_i \left(\dfrac{BAF}{BA_i}\right)$ $= \dfrac{40{,}000 b H_i BAF}{D_i}$	$BSAF_i = BSA_i \cdot TF_i$ $= bC_i H_i \left(\dfrac{20DF}{D_i L}\right)$ $= \dfrac{20\pi b H_i DF}{L}$	$BSAF_i = BSA_i \cdot TF_i$ $= bC_i H_i \left(\dfrac{SHF}{H_i^2}\right)$ $= \dfrac{b\pi D_i SHF}{H_i}$	$BSAF_i = BSA_i \cdot TF_i$ $= bC_i H_i \left(\dfrac{20HF}{H_i L}\right)$ $= \dfrac{20b\pi D_i HF}{L}$
All other factors (units/acre/tree tallied)	$XF_i = X_i \cdot TF_i$	$XF_i = X_i \cdot TF_i$	$XF_i = X_i \cdot TF_i$	$XF_i = X_i \cdot TF_i$

D_i = diameter (cm); BA_i = basal area (m²); H_i = height (m); C_i = circumference (cm); V_i = volume (m³); WT_i = weight or biomass (kg); BSA = bole surface area (m²); and L = line length (m).

TABLE A.7. Plot Summary Data for the 100 m² Circular Plot Simple Random Sample Shown in Figure 10.1 (Values Are Per Plot)

Plot	X Coord.	Y Coord.	Num. Trees	Basal Area (m²)	Total Volume (m³)	Plot	X Coord.	Y Coord.	Num. Trees	Basal Area (m²)	Total Volume (m³)	Plot	X Coord.	Y Coord.	Num. Trees	Basal Area (m²)	Total Volume (m³)
1	3.6	29.0	12	0.20	1.170	51	169.8	241.2	17	0.35	2.042	101	339.1	202.3	10	0.24	1.536
2	9.1	293.0	23	0.41	2.375	52	172.5	220.1	7	0.22	1.264	102	339.4	344.4	13	0.35	2.299
3	9.9	49.8	9	0.16	0.998	53	173.5	327.4	10	0.30	1.991	103	340.9	285.0	13	0.30	1.959
4	17.2	427.3	15	0.30	1.884	54	175.9	348.9	5	0.19	1.550	104	343.1	137.3	5	0.15	0.906
5	18.3	69.9	14	0.31	1.780	55	183.0	18.3	6	0.14	0.785	105	348.8	253.4	8	0.19	1.060
6	19.5	176.2	10	0.21	1.200	56	188.1	183.2	8	0.20	1.164	106	353.2	99.2	15	0.41	2.887
7	24.1	464.6	9	0.26	1.854	57	188.4	65.0	15	0.24	1.366	107	359.4	125.5	7	0.21	1.502
8	24.9	34.2	6	0.13	0.788	58	189.1	488.4	9	0.22	1.737	108	363.5	368.6	12	0.18	1.120
9	30.7	440.3	9	0.13	0.787	59	192.0	410.5	7	0.11	0.657	109	369.4	256.6	2	0.08	0.736
10	34.9	296.2	21	0.46	2.641	60	194.8	128.3	12	0.31	1.979	110	369.5	282.8	4	0.08	0.553
11	37.0	485.7	5	0.17	1.191	61	197.3	375.5	7	0.19	1.351	111	371.5	495.3	7	0.13	0.744
12	40.3	10.8	9	0.25	1.388	62	213.2	303.0	9	0.35	2.435	112	372.0	411.1	3	0.07	0.438
13	44.9	98.3	11	0.21	1.266	63	222.4	71.4	7	0.24	1.714	113	376.3	449.3	11	0.19	1.049
14	48.9	484.5	17	0.40	2.680	64	227.0	13.1	7	0.20	1.112	114	377.8	200.2	7	0.37	2.529
15	50.6	464.3	13	0.29	1.747	65	229.4	157.1	8	0.19	1.363	115	378.8	346.8	9	0.20	1.251
16	51.7	441.1	3	0.11	0.815	66	231.9	445.7	12	0.28	1.997	116	383.4	460.9	16	0.30	1.905
17	53.7	189.1	17	0.29	1.629	67	232.9	175.3	11	0.20	1.131	117	387.0	278.0	5	0.15	1.023
18	57.0	53.3	10	0.20	1.155	68	235.2	0.3	2	0.06	0.417	118	388.7	318.8	7	0.21	1.479
19	59.8	101.9	22	0.33	1.820	69	235.7	304.4	5	0.20	1.499	119	388.9	300.2	12	0.31	2.050
20	60.3	35.5	6	0.11	0.846	70	239.2	229.4	9	0.16	1.044	120	391.0	265.2	12	0.39	3.161
21	62.7	143.0	8	0.15	0.884	71	244.2	12.9	4	0.09	0.504	121	391.8	71.5	10	0.20	1.160
22	63.0	88.9	11	0.25	1.447	72	247.8	129.4	18	0.36	2.174	122	391.9	495.0	10	0.25	1.730
23	63.8	360.7	5	0.13	0.889	73	248.4	374.1	7	0.22	1.335	123	407.4	224.6	4	0.14	0.965
24	64.9	384.2	4	0.15	1.063	74	249.7	248.5	10	0.20	1.186	124	407.5	443.0	16	0.36	2.399
25	74.7	447.3	4	0.07	0.395	75	254.9	212.9	6	0.20	1.489	125	407.6	171.4	9	0.26	1.583
26	75.2	78.9	13	0.25	1.430	76	256.2	164.2	13	0.49	3.625	126	407.7	138.0	9	0.23	1.428

(*Continued*)

Plot	X Coord.	Y Coord.	Num. Trees	Basal Area (m²)	Total Volume (m³)	Plot	X Coord.	Y Coord.	Num. Trees	Basal Area (m²)	Total Volume (m³)	Plot	X Coord.	Y Coord.	Num. Trees	Basal Area (m²)	Total Volume (m³)
27	77.4	15.3	13	0.24	1.462	77	262.8	192.8	3	0.08	0.561	127	411.9	368.5	11	0.23	1.445
28	77.4	376.7	7	0.30	2.280	78	279.2	125.2	9	0.24	1.545	128	425.9	243.1	5	0.17	1.186
29	83.6	488.0	3	0.07	0.419	79	283.4	10.3	1	0.04	0.273	129	427.7	362.2	6	0.17	0.939
30	90.5	102.4	14	0.28	1.846	80	284.5	34.3	8	0.39	2.841	130	430.2	25.4	9	0.27	1.846
31	90.9	227.2	14	0.34	2.043	81	287.2	343.2	7	0.17	1.060	131	433.8	131.8	11	0.31	2.112
32	91.6	418.0	7	0.19	1.131	82	288.6	497.1	6	0.11	0.618	132	434.0	149.5	10	0.22	1.544
33	96.9	247.8	12	0.33	1.955	83	290.5	412.6	6	0.24	2.018	133	439.8	80.8	6	0.12	0.636
34	101.9	474.6	4	0.27	2.013	84	294.2	275.8	8	0.15	0.839	134	441.8	322.3	8	0.19	1.039
35	110.6	311.3	13	0.33	1.982	85	297.5	472.5	7	0.10	0.600	135	443.5	8.4	6	0.19	1.089
36	113.9	394.9	13	0.35	2.063	86	300.1	37.8	14	0.30	1.798	136	451.2	344.1	11	0.22	1.477
37	116.6	291.1	24	0.44	2.462	87	304.4	444.6	12	0.28	1.822	137	464.7	279.1	11	0.24	1.538
38	116.8	498.3	4	0.20	1.341	88	304.8	344.8	12	0.16	0.968	138	468.5	103.3	3	0.09	0.481
39	119.2	374.2	17	0.32	1.894	89	307.0	210.5	6	0.16	1.167	139	470.1	488.9	3	0.12	0.899
40	119.4	350.9	11	0.23	1.501	90	310.6	330.2	10	0.23	1.271	140	473.5	238.1	11	0.32	2.366
41	132.4	153.1	12	0.28	1.595	91	315.4	88.5	4	0.40	3.179	141	476.8	46.1	8	0.38	3.291
42	140.4	435.9	6	0.45	3.143	92	316.5	25.9	6	0.24	1.920	142	477.0	212.8	2	0.10	0.573
43	142.9	5.1	6	0.07	0.356	93	316.7	62.5	5	0.30	2.272	143	483.0	483.5	5	0.11	0.638
44	147.6	20.6	9	0.18	1.152	94	322.2	192.5	6	0.16	0.951	144	485.2	122.1	5	0.09	0.494
45	152.3	408.2	4	0.16	0.944	95	323.8	103.2	10	0.14	0.783	145	487.0	149.2	17	0.32	2.160
46	154.9	314.9	7	0.15	1.039	96	326.8	170.3	17	0.25	1.521	146	488.5	401.3	12	0.17	0.855
47	157.2	178.3	22	0.42	2.306	97	328.0	50.1	8	0.21	1.218	147	490.8	374.9	5	0.15	1.131
48	159.9	255.3	20	0.37	2.145	98	328.8	307.4	8	0.24	1.917	148	492.7	237.9	4	0.08	0.555
49	162.2	464.8	9	0.27	1.839	99	335.4	438.3	16	0.43	3.042	149	497.8	168.0	21	0.50	3.427
50	167.9	138.6	2	0.06	0.364	100	336.5	268.1	10	0.17	0.958	150	499.5	392.0	6	0.17	1.208

TABLE A.8. Plot Summary Data for the 100 m² Circular Plot Systematic Sample Shown in Figure 10.4c (Values Are Per Plot)

Plot	X Coord.	Y Coord.	Num. Trees	Basal Area (m²)	Total Volume (m³)
1	11.3	24.1	10	0.16	0.967
2	11.3	74.1	10	0.27	1.606
3	11.3	124.1	15	0.24	1.388
4	11.3	174.1	14	0.29	1.665
5	11.3	224.1	16	0.46	2.828
6	11.3	274.1	19	0.40	2.264
7	11.3	324.1	9	0.28	2.220
8	11.3	374.1	10	0.30	2.075
9	11.3	424.1	13	0.24	1.454
10	11.3	474.1	12	0.24	1.463
11	44.6	24.1	15	0.31	1.793
12	44.6	74.1	12	0.25	1.469
13	44.6	124.1	10	0.18	1.110
14	44.6	174.1	18	0.29	1.612
15	44.6	224.1	16	0.22	1.265
16	44.6	274.1	11	0.38	2.198
17	44.6	324.1	10	0.18	1.075
18	44.6	374.1	10	0.19	1.326
19	44.6	424.1	11	0.21	1.466
20	44.6	474.1	7	0.14	0.864
21	77.9	24.1	8	0.21	1.355
22	77.9	74.1	8	0.15	0.903
23	77.9	124.1	7	0.15	0.963
24	77.9	174.1	17	0.31	1.735
25	77.9	224.1	12	0.30	1.753
26	77.9	274.1	7	0.26	1.550

Plot	X Coord.	Y Coord.	Num. Trees	Basal Area (m²)	Total Volume (m³)
51	177.8	24.1	10	0.24	1.350
52	177.8	74.1	9	0.12	0.749
53	177.8	124.1	7	0.40	3.363
54	177.8	174.1	9	0.28	1.722
55	177.8	224.1	18	0.33	1.876
56	177.8	274.1	10	0.12	0.647
57	177.8	324.1	7	0.26	2.189
58	177.8	374.1	5	0.11	0.627
59	177.8	424.1	15	0.40	2.543
60	177.8	474.1	12	0.24	1.451
61	211.1	24.1	13	0.23	1.359
62	211.1	74.1	7	0.13	0.825
63	211.1	124.1	8	0.17	1.009
64	211.1	174.1	11	0.29	2.257
65	211.1	224.1	10	0.35	2.677
66	211.1	274.1	5	0.11	0.836
67	211.1	324.1	8	0.11	0.654
68	211.1	374.1	11	0.15	0.997
69	211.1	424.1	16	0.45	3.093
70	211.1	474.1	2	0.04	0.223
71	244.4	24.1	4	0.09	0.596
72	244.4	74.1	17	0.33	1.933
73	244.4	124.1	7	0.18	1.104
74	244.4	174.1	15	0.37	2.523
75	244.4	224.1	11	0.24	1.625
76	244.4	274.1	7	0.13	0.908

Plot	X Coord.	Y Coord.	Num. Trees	Basal Area (m²)	Total Volume (m³)
101	344.3	24.1	2	0.04	0.231
102	344.3	74.1	16	0.33	2.242
103	344.3	124.1	6	0.20	1.151
104	344.3	174.1	5	0.13	0.990
105	344.3	224.1	12	0.27	1.975
106	344.3	274.1	9	0.17	0.891
107	344.3	324.1	10	0.44	3.396
108	344.3	374.1	11	0.19	1.106
109	344.3	424.1	12	0.38	2.661
110	344.3	474.1	11	0.28	1.972
111	377.6	24.1	3	0.07	0.489
112	377.6	74.1	15	0.25	1.642
113	377.6	124.1	7	0.13	0.749
114	377.6	174.1	7	0.37	2.529
115	377.6	224.1	11	0.30	1.913
116	377.6	274.1	11	0.30	2.094
117	377.6	324.1	10	0.24	1.637
118	377.6	374.1	8	0.16	0.875
119	377.6	424.1	8	0.16	0.891
120	377.6	474.1	5	0.11	0.641
121	410.9	24.1	11	0.22	1.347
122	410.9	74.1	9	0.28	2.002
123	410.9	124.1	5	0.08	0.440
124	410.9	174.1	6	0.16	0.911
125	410.9	224.1	7	0.24	1.703
126	410.9	274.1	20	0.57	3.670

(*Continued*)

TABLE A.8. (Continued)

Plot	X Coord.	Y Coord.	Num. Trees	Basal Area (m²)	Total Volume (m³)	Plot	X Coord.	Y Coord.	Num. Trees	Basal Area (m²)	Total Volume (m³)	Plot	X Coord.	Y Coord.	Num. Trees	Basal Area (m²)	Total Volume (m³)
27	77.9	324.1	22	0.45	2.662	77	244.4	324.1	9	0.26	1.771	127	410.9	324.1	12	0.33	2.452
28	77.9	374.1	7	0.33	2.469	78	244.4	374.1	11	0.43	3.274	128	410.9	374.1	12	0.20	1.202
29	77.9	424.1	8	0.19	1.104	79	244.4	424.1	9	0.33	2.492	129	410.9	424.1	3	0.09	0.572
30	77.9	474.1	6	0.17	1.235	80	244.4	474.1	5	0.11	0.768	130	410.9	474.1	5	0.30	2.497
31	111.2	24.1	6	0.35	2.421	81	277.7	24.1	6	0.12	0.876	131	444.2	24.1	15	0.38	2.345
32	111.2	74.1	2	0.09	0.566	82	277.7	74.1	8	0.11	0.668	132	444.2	74.1	11	0.10	0.524
33	111.2	124.1	5	0.17	1.055	83	277.7	124.1	11	0.30	1.941	133	444.2	124.1	13	0.21	1.130
34	111.2	174.1	6	0.21	1.260	84	277.7	174.1	11	0.24	1.543	134	444.2	174.1	5	0.20	1.295
35	111.2	224.1	14	0.31	2.032	85	277.7	224.1	11	0.22	1.309	135	444.2	224.1	7	0.17	0.986
36	111.2	274.1	23	0.49	3.010	86	277.7	274.1	7	0.21	1.653	136	444.2	274.1	12	0.14	0.731
37	111.2	324.1	8	0.14	0.902	87	277.7	324.1	13	0.37	2.421	137	444.2	324.1	7	0.19	1.201
38	111.2	374.1	11	0.32	1.888	88	277.7	374.1	8	0.27	2.061	138	444.2	374.1	6	0.17	1.317
39	111.2	424.1	4	0.12	0.763	89	277.7	424.1	13	0.32	2.153	139	444.2	424.1	8	0.14	0.732
40	111.2	474.1	4	0.12	0.675	90	277.7	474.1	5	0.11	0.589	140	444.2	474.1	1	0.04	0.235
41	144.5	24.1	9	0.19	1.143	91	311	24.1	12	0.47	3.221	141	477.5	24.1	3	0.04	0.255
42	144.5	74.1	14	0.39	2.652	92	311	74.1	8	0.17	1.233	142	477.5	74.1	4	0.13	0.864
43	144.5	124.1	4	0.05	0.285	93	311	124.1	6	0.22	1.722	143	477.5	124.1	5	0.08	0.404
44	144.5	174.1	8	0.18	1.038	94	311	174.1	13	0.22	1.207	144	477.5	174.1	7	0.26	1.749
45	144.5	224.1	17	0.38	2.191	95	311	224.1	9	0.33	2.160	145	477.5	224.1	6	0.21	1.403
46	144.5	274.1	13	0.30	1.722	96	311	274.1	4	0.09	0.659	146	477.5	274.1	10	0.25	1.298
47	144.5	324.1	8	0.19	1.103	97	311	324.1	13	0.34	2.183	147	477.5	324.1	12	0.24	1.722
48	144.5	374.1	11	0.17	0.928	98	311	374.1	7	0.17	1.220	148	477.5	374.1	1	0.05	0.374
49	144.5	424.1	7	0.43	3.229	99	311	424.1	15	0.39	2.767	149	477.5	424.1	9	0.18	1.121
50	144.5	474.1	5	0.12	0.824	100	311	474.1	14	0.19	0.995	150	477.5	474.1	3	0.07	0.421

TABLE A.9. Plot Summary Data for the 100 m² Circular Plot Stratified Random Sample Shown in Figure 10.9 (Values Are Per Plot)

Plot	X Coord.	Y Coord.	Strata Type	Num. Trees	Basal Area (m²)	Total Volume (m³)	Plot	X Coord.	Y Coord.	Strata Type	Num. Trees	Basal Area (m²)	Total Volume (m³)	Plot	X Coord.	Y Coord.	Strata Type	Num. Trees	Basal Area (m²)	Total Volume (m³)
1	2.8	140.3	BS	20	0.33	1.806	51	144.2	488.9	IH	19	0.44	2.837	101	345.9	17.0	IH	2	0.11	0.707
2	9.8	49.5	BS	9	0.16	0.998	52	146.0	66.2	IH	6	0.17	0.944	102	347.0	57.2	IH	4	0.06	0.331
3	10.8	245.5	BS	19	0.32	1.903	53	146.3	25.3	IH	10	0.19	1.156	103	356.6	430.8	IH	6	0.19	1.069
4	17.1	258.3	BS	10	0.19	1.108	54	155.5	52.5	IH	6	0.16	1.044	104	358.8	137.6	IH	6	0.16	0.866
5	26.3	246.6	BS	5	0.16	1.005	55	160.3	463.7	IH	11	0.77	6.526	105	359.0	183.7	IH	10	0.24	1.606
6	29.9	30.7	BS	18	0.37	2.149	56	161.4	315.5	IH	5	0.18	1.324	106	360.5	24.0	IH	8	0.22	1.407
7	31.8	129.7	BS	12	0.29	1.695	57	166.9	136.8	IH	2	0.06	0.368	107	361.8	380.3	IH	9	0.29	2.207
8	34.0	86.6	BS	14	0.30	1.690	58	173.2	50.8	IH	6	0.13	0.721	108	362.4	124.1	IH	10	0.22	1.542
9	40.3	264.5	BS	10	0.28	1.698	59	175.6	321.9	IH	6	0.11	0.590	109	363.0	450.8	IH	8	0.15	0.903
10	45.1	59.1	BS	12	0.31	1.798	60	179.2	300.2	IH	12	0.16	0.967	110	363.1	270.3	IH	2	0.11	0.908
11	57.9	78.5	BS	14	0.22	1.260	61	182.7	422.8	IH	10	0.27	1.498	111	372.1	425.2	IH	6	0.15	0.889
12	59.7	263.3	BS	15	0.30	1.665	62	194.7	107.1	IH	5	0.21	1.571	112	379.9	138.7	IH	6	0.09	0.564
13	59.9	319.2	BS	11	0.23	1.314	63	206.0	20.6	IH	8	0.16	1.155	113	380.5	236.5	IH	23	0.34	1.957
14	60.2	186.4	BS	18	0.38	2.154	64	213.6	262.9	IH	8	0.16	0.907	114	380.7	124.8	IH	12	0.20	1.254
15	65.2	225.8	BS	14	0.26	1.588	65	214.2	204.6	IH	14	0.20	1.155	115	380.8	448.5	IH	6	0.15	0.903
16	71.8	281.4	BS	21	0.39	2.228	66	220.9	350.7	IH	15	0.27	1.665	116	381.4	315.2	IH	5	0.24	1.618
17	85.4	335.0	BS	15	0.28	1.635	67	221.0	127.2	IH	8	0.22	1.363	117	386.6	153.3	IH	5	0.17	1.195
18	89.3	211.6	BS	13	0.42	2.682	68	226.5	89.9	IH	9	0.32	2.137	118	394.2	481.3	IH	11	0.28	2.166
19	107.5	243.1	BS	14	0.24	1.324	69	226.9	456.8	IH	2	0.06	0.454	119	398.8	225.7	IH	17	0.28	1.590
20	121.4	165.7	BS	12	0.27	1.570	70	227.4	368.0	IH	10	0.30	2.305	120	400.7	494.2	IH	9	0.21	1.548
21	123.2	272.2	BS	27	0.38	2.054	71	231.0	140.6	IH	7	0.19	1.453	121	404.2	428.1	IH	3	0.06	0.371
22	148.1	249.7	BS	15	0.43	2.745	72	236.2	315.5	IH	16	0.23	1.300	122	405.1	379.2	IH	8	0.33	2.500
23	157.4	216.0	BS	13	0.28	1.633	73	247.2	252.7	IH	11	0.27	1.773	123	406.0	295.4	IH	2	0.03	0.178
24	157.9	231.3	BS	12	0.26	1.552	74	247.4	415.6	IH	2	0.06	0.382	124	410.1	38.2	IH	15	0.27	1.702
25	170.2	211.5	BS	14	0.32	1.846	75	248.5	23.2	IH	10	0.31	2.276	125	410.6	86.9	IH	10	0.26	1.879

(Continued)

TABLE A.9. (Continued)

Plot	X Coord.	Y Coord.	Strata Type	Num. Trees	Basal Area (m²)	Total Volume (m³)
26	115.0	35.0	EH	8	0.58	4.296
27	145.0	435.0	EH	5	0.47	3.409
28	285.0	235.0	EH	7	0.46	3.351
29	4.7	455.1	IH	10	0.28	1.868
30	6.1	496.2	IH	2	0.06	0.414
31	6.7	316.6	IH	20	0.82	5.919
32	11.1	481.8	IH	7	0.09	0.535
33	26.7	360.4	IH	8	0.16	1.050
34	39.3	9.7	IH	9	0.23	1.285
35	40.5	484.4	IH	9	0.23	1.605
36	47.8	360.6	IH	10	0.16	0.812
37	58.0	15.9	IH	0	0.00	0.000
38	58.5	375.4	IH	6	0.16	1.151
39	92.2	390.0	IH	14	0.33	2.042
40	103.2	432.2	IH	2	0.09	0.698
41	104.1	97.9	IH	5	0.08	0.431
42	105.3	6.8	IH	5	0.17	1.304
43	109.8	50.5	IH	4	0.27	1.798
44	114.1	79.1	IH	10	0.25	1.641
45	130.2	339.0	IH	9	0.22	1.611
46	131.8	409.4	IH	12	0.24	1.524
47	134.6	445.5	IH	5	0.30	2.240
48	139.7	85.8	IH	9	0.17	1.198
49	139.9	471.4	IH	8	0.19	1.339
50	141.0	321.7	IH	8	0.20	1.109
76	249.6	389.5	IH	9	0.20	1.302
77	253.0	189.8	IH	11	0.23	1.327
78	257.2	130.4	IH	11	0.18	1.241
79	259.0	418.2	IH	9	0.24	1.464
80	260.2	109.0	IH	7	0.19	1.393
81	267.0	397.3	IH	5	0.13	0.879
82	268.3	350.0	IH	12	0.21	1.200
83	270.0	233.3	IH	6	0.18	1.088
84	270.2	94.7	IH	7	0.24	1.744
85	274.8	196.7	IH	22	0.26	1.343
86	277.4	66.6	IH	9	0.26	1.894
87	282.2	470.9	IH	1	0.04	0.219
88	288.9	321.7	IH	8	0.16	0.856
89	291.2	485.2	IH	1	0.03	0.190
90	293.5	118.4	IH	5	0.11	0.779
91	300.5	261.2	IH	15	0.36	2.518
92	308.9	117.4	IH	9	0.19	1.182
93	315.4	360.6	IH	14	0.22	1.477
94	316.7	301.7	IH	10	0.22	1.291
95	323.7	320.4	IH	6	0.19	1.408
96	324.1	181.1	IH	7	0.15	0.862
97	326.3	374.4	IH	14	0.41	2.639
98	332.1	246.5	IH	12	0.18	0.995
99	337.6	381.2	IH	11	0.26	1.691
100	344.5	229.2	IH	12	0.35	2.459
126	412.6	150.4	IH	9	0.32	2.822
127	416.3	163.6	IH	7	0.12	0.656
128	420.9	26.6	IH	9	0.23	1.584
129	421.0	44.9	IH	21	0.46	2.825
130	423.9	223.8	IH	13	0.20	1.156
131	427.8	285.4	IH	4	0.11	0.601
132	428.4	270.2	IH	5	0.07	0.465
133	437.9	371.7	IH	7	0.15	0.935
134	447.8	339.2	IH	10	0.18	1.049
135	449.9	308.1	IH	13	0.25	1.790
136	452.9	197.1	IH	7	0.12	0.690
137	459.7	316.6	IH	9	0.15	1.015
138	461.9	435.9	IH	10	0.22	1.555
139	464.6	112.4	IH	10	0.28	1.908
140	466.1	195.3	IH	6	0.20	1.247
141	472.0	257.8	IH	9	0.18	1.012
142	474.2	105.8	IH	12	0.29	1.805
143	476.2	227.6	IH	11	0.31	1.975
144	480.4	290.2	IH	15	0.26	1.494
145	485.7	359.6	IH	9	0.25	1.652
146	486.5	468.0	IH	12	0.23	1.565
147	488.5	123.2	IH	3	0.09	0.614
148	490.6	220.7	IH	4	0.11	0.647
149	490.7	154.0	IH	17	0.88	6.814
150	492.8	39.9	IH	8	0.14	0.948

TABLE A.10. Plot Summary Data for the 100 m² Circular Plot Systematic Sample with Multiple Starts Shown in Figure 10.12 (Values Are Per Plot)

Plot	Sample Num.	X Coord.	Y Coord.	Num. Trees	Basal Area (m²)	Total Volume (m³)
1	1	93.4	1.4	11	0.20	1.211
2	1	93.4	101.4	14	0.22	1.220
3	1	93.4	201.4	16	0.37	2.117
4	1	93.4	301.4	7	0.16	0.973
5	1	93.4	401.4	5	0.13	1.012
6	1	193.4	1.4	9	0.27	2.064
7	1	193.4	101.4	6	0.17	1.075
8	1	193.4	201.4	11	0.28	1.630
9	1	193.4	301.4	7	0.12	0.772
10	1	193.4	401.4	7	0.23	1.618
11	1	293.4	1.4	0	0.00	0.000
12	1	293.4	101.4	5	0.25	2.030
13	1	293.4	201.4	10	0.41	3.015
14	1	293.4	301.4	5	0.11	0.696
15	1	293.4	401.4	10	0.29	2.145
16	1	393.4	1.4	6	0.17	1.034
17	1	393.4	101.4	15	0.28	1.634
18	1	393.4	201.4	10	0.16	0.938
19	1	393.4	301.4	17	0.43	2.746
20	1	393.4	401.4	4	0.11	0.662
21	1	493.4	1.4	7	0.08	0.398
22	1	493.4	101.4	3	0.11	0.826
23	1	493.4	201.4	4	0.09	0.746
24	1	493.4	301.4	4	0.06	0.332
25	1	493.4	401.4	13	0.20	1.004
26	2	86.0	90.9	10	0.22	1.245

Plot	Sample Num.	X Coord.	Y Coord.	Num. Trees	Basal Area (m²)	Total Volume (m³)
51	3	10.3	48.4	9	0.13	0.777
52	3	10.3	148.4	20	0.39	2.269
53	3	10.3	248.4	15	0.26	1.520
54	3	10.3	348.4	6	0.13	0.822
55	3	10.3	448.4	7	0.17	1.224
56	3	110.3	48.4	8	0.43	3.063
57	3	110.3	148.4	6	0.15	0.916
58	3	110.3	248.4	14	0.29	1.894
59	3	110.3	348.4	9	0.15	0.957
60	3	110.3	448.4	3	0.10	0.691
61	3	210.3	48.4	5	0.08	0.507
62	3	210.3	148.4	12	0.25	1.387
63	3	210.3	248.4	15	0.25	1.505
64	3	210.3	348.4	5	0.08	0.464
65	3	210.3	448.4	10	0.28	2.133
66	3	310.3	48.4	10	0.45	3.106
67	3	310.3	148.4	5	0.17	1.335
68	3	310.3	248.4	8	0.33	2.160
69	3	310.3	348.4	13	0.26	1.715
70	3	310.3	448.4	14	0.40	2.986
71	3	410.3	48.4	5	0.11	0.669
72	3	410.3	148.4	10	0.20	1.421
73	3	410.3	248.4	8	0.12	0.699
74	3	410.3	348.4	8	0.12	0.796
75	3	410.3	448.4	14	0.27	1.755
76	4	61.0	28.3	12	0.24	1.598

Plot	Sample Num.	X Coord.	Y Coord.	Num. Trees	Basal Area (m²)	Total Volume (m³)
101	5	33.7	81.8	6	0.10	0.556
102	5	33.7	181.8	7	0.22	1.339
103	5	33.7	281.8	17	0.29	1.689
104	5	33.7	381.8	5	0.11	0.762
105	5	33.7	481.8	7	0.22	1.279
106	5	133.7	81.8	10	0.22	1.430
107	5	133.7	181.8	11	0.23	1.352
108	5	133.7	281.8	16	0.34	2.011
109	5	133.7	381.8	8	0.20	1.389
110	5	133.7	481.8	3	0.12	1.124
111	5	233.7	81.8	15	0.46	3.283
112	5	233.7	181.8	12	0.18	0.962
113	5	233.7	281.8	11	0.29	2.215
114	5	233.7	381.8	4	0.11	0.790
115	5	233.7	481.8	5	0.17	1.145
116	5	333.7	81.8	7	0.20	1.457
117	5	333.7	181.8	4	0.18	1.436
118	5	333.7	281.8	4	0.11	0.704
119	5	333.7	381.8	6	0.17	1.117
120	5	333.7	481.8	5	0.16	1.080
121	5	433.7	81.8	3	0.08	0.460
122	5	433.7	181.8	6	0.19	1.242
123	5	433.7	281.8	11	0.52	3.880
124	5	433.7	381.8	21	0.37	2.289
125	5	433.7	481.8	13	0.30	1.779
126	6	37.4	29.8	14	0.26	1.437

(*Continued*)

TABLE A.10. (Continued)

Plot	Sample Num.	X Coord.	Y Coord.	Num. Trees	Basal Area (m²)	Total Volume (m³)	Plot	Sample Num.	X Coord.	Y Coord.	Num. Trees	Basal Area (m²)	Total Volume (m³)	Plot	Sample Num.	X Coord.	Y Coord.	Num. Trees	Basal Area (m²)	Total Volume (m³)
27	2	86.0	190.9	16	0.31	1.796	77	4	61.0	128.3	25	0.44	2.464	127	6	37.4	129.8	7	0.21	1.262
28	2	86.0	290.9	17	0.41	2.395	78	4	61.0	228.3	17	0.29	1.684	128	6	37.4	229.8	14	0.31	1.909
29	2	86.0	390.9	12	0.22	1.374	79	4	61.0	328.3	11	0.18	1.013	129	6	37.4	329.8	11	0.17	0.953
30	2	86.0	490.9	9	0.16	0.847	80	4	61.0	428.3	6	0.14	0.921	130	6	37.4	429.8	13	0.33	1.984
31	2	186.0	90.9	4	0.22	1.846	81	4	161.0	28.3	13	0.48	3.440	131	6	137.4	29.8	8	0.18	1.176
32	2	186.0	190.9	6	0.17	1.080	82	4	161.0	128.3	8	0.17	1.042	132	6	137.4	129.8	9	0.18	1.237
33	2	186.0	290.9	17	0.45	3.198	83	4	161.0	228.3	11	0.27	1.637	133	6	137.4	229.8	22	0.41	2.280
34	2	186.0	390.9	5	0.16	1.149	84	4	161.0	328.3	9	0.44	3.544	134	6	137.4	329.8	8	0.18	1.191
35	2	186.0	490.9	6	0.09	0.520	85	4	161.0	428.3	9	0.22	1.407	135	6	137.4	429.8	8	0.50	3.580
36	2	286.0	90.9	15	0.33	1.957	86	4	261.0	28.3	13	0.27	1.826	136	6	237.4	29.8	23	0.55	3.784
37	2	286.0	190.9	7	0.18	1.211	87	4	261.0	128.3	7	0.17	1.244	137	6	237.4	129.8	8	0.15	0.849
38	2	286.0	290.9	5	0.08	0.459	88	4	261.0	228.3	10	0.24	1.461	138	6	237.4	229.8	9	0.13	0.688
39	2	286.0	390.9	18	0.36	2.325	89	4	261.0	328.3	11	0.22	1.340	139	6	237.4	329.8	10	0.17	1.082
40	2	286.0	490.9	5	0.15	1.017	90	4	261.0	428.3	8	0.12	0.664	140	6	237.4	429.8	13	0.25	1.686
41	2	386.0	90.9	5	0.11	0.637	91	4	361.0	28.3	7	0.22	1.383	141	6	337.4	29.8	8	0.24	1.687
42	2	386.0	190.9	14	0.47	3.256	92	4	361.0	128.3	4	0.12	0.734	142	6	337.4	129.8	8	0.18	1.047
43	2	386.0	290.9	13	0.28	1.846	93	4	361.0	228.3	10	0.15	0.918	143	6	337.4	229.8	12	0.34	2.112
44	2	386.0	390.9	6	0.09	0.470	94	4	361.0	328.3	6	0.17	1.018	144	6	337.4	329.8	13	0.39	2.480
45	2	386.0	490.9	21	0.37	2.387	95	4	361.0	428.3	3	0.15	0.829	145	6	337.4	429.8	21	0.39	2.794
46	2	486.0	90.9	6	0.17	1.231	96	4	461.0	28.3	18	0.35	2.120	146	6	437.4	29.8	9	0.20	1.242
47	2	486.0	190.9	11	0.23	1.306	97	4	461.0	128.3	8	0.17	1.128	147	6	437.4	129.8	11	0.25	1.691
48	2	486.0	290.9	1	0.02	0.084	98	4	461.0	228.3	5	0.25	2.159	148	6	437.4	229.8	7	0.18	1.092
49	2	486.0	390.9	7	0.10	0.526	99	4	461.0	328.3	8	0.10	0.600	149	6	437.4	329.8	6	0.09	0.593
50	2	486.0	490.9	4	0.12	0.794	100	4	461.0	428.3	17	0.32	1.661	150	6	437.4	429.8	6	0.13	0.783

TABLE A.11. Plot-Level Summaries for the 100 m² Circular Plot Two-Stage Sample of Noonan Research Forest (Values Are Per Plot)

Strata	Stand	Plot	Trees	Basal Area (m²)	Total Volume (m³)
BFHW	01_02	1	0	0.000	0.000
BFHW	01_02	2	0	0.000	0.000
BFHW	01_02	3	0	0.000	0.000
BFHW	01_06	1	10	0.280	1.978
BFHW	01_06	2	11	0.291	2.128
BFHW	01_06	3	10	0.280	1.978
HDWD	02_22	1	6	0.184	1.275
HDWD	02_22	2	2	0.053	0.357
HDWD	02_22	3	7	0.129	0.647
HDWD	02_24	1	5	0.167	1.267
HDWD	02_24	2	14	0.253	1.469
HDWD	02_24	3	2	0.017	0.074
HDWD	02_14	1	0	0.000	0.000
HDWD	02_14	2	0	0.000	0.000
HDWD	02_14	3	0	0.000	0.000
HDWD	02_10	1	13	0.331	2.342
HDWD	02_10	2	8	0.228	1.544
HDWD	02_10	3	13	0.331	2.342
HDWD	02_16	1	3	0.185	1.276
HDWD	02_16	2	3	0.185	1.276
HDWD	02_16	3	12	0.378	2.554
HDWD	02_03	1	1	0.077	0.588
HDWD	02_03	2	2	0.024	0.088
HDWD	02_03	3	1	0.077	0.588
HDWD	02_19	1	2	0.106	0.767
HDWD	02_19	2	2	0.106	0.767
INHW	03_07	1	3	0.141	1.110
INHW	03_07	2	5	0.224	1.857
INHW	03_07	3	3	0.141	1.110
INHW	03_05	1	6	0.178	1.584
INHW	03_05	2	10	0.243	1.776
INHW	03_05	3	4	0.176	1.315
PINE	04_01	1	9	0.230	1.534
PINE	04_01	2	4	0.236	1.888
PINE	04_01	3	4	0.236	1.888
PINE	04_02	1	6	0.419	4.295
PINE	04_02	2	8	0.293	2.208
PINE	04_02	3	4	0.405	4.100
REGEN	05_13	1	1	0.049	0.282
REGEN	05_13	2	1	0.097	0.641
REGEN	05_13	3	0	0.000	0.000
REGEN	05_11	1	0	0.000	0.000
REGEN	05_11	2	0	0.000	0.000
REGEN	05_11	3	0	0.000	0.000
REGEN	05_12	1	0	0.000	0.000
REGEN	05_12	2	0	0.000	0.000
REGEN	05_12	3	0	0.000	0.000
REGEN	05_10	1	6	0.250	2.001
REGEN	05_10	2	4	0.099	0.609
REGEN	05_10	3	4	0.099	0.609
SPBF	06_19	1	3	0.181	1.469
SPBF	06_19	2	5	0.130	0.874
SPBF	06_15	1	8	0.336	2.535
SPBF	06_15	2	13	0.308	2.169
SPBF	06_15	3	12	0.461	3.339
SPBF	06_16	1	7	0.201	1.422
SPBF	06_16	2	11	0.288	2.022
SPBF	06_16	3	11	0.288	2.022
SPBF	06_12	1	16	0.397	2.795
SPBF	06_12	2	13	0.284	1.943
SPBF	06_12	3	13	0.284	1.943
SPBF	06_18	1	3	0.148	1.227
SPBF	06_18	2	6	0.131	1.024
SPBF	06_18	3	6	0.131	1.024
SPHW	07_29	1	13	0.347	2.314
SPHW	07_29	2	9	0.184	1.087
SPHW	07_29	3	11	0.203	1.265
SPHW	07_28	1	15	0.201	0.999
SPHW	07_28	2	0	0.000	0.000
SPHW	07_28	3	12	0.254	1.633
SPHW	07_30	1	7	0.250	2.052
SPHW	07_30	2	16	0.377	2.328
SPHW	07_30	3	4	0.131	0.891
SPHW	07_17	1	2	0.101	0.699
SPHW	07_17	2	5	0.128	0.667
SPHW	07_17	3	5	0.128	0.667
SPHW	07_08	1	8	0.335	2.352
SPHW	07_08	2	7	0.303	2.103

(Continued)

TABLE A.11. (Continued)

Strata	Stand	Plot	Trees	Basal Area (m²)	Total Volume (m³)
HDWD	02_19	3	2	0.106	0.767
HDWD	02_12	1	6	0.193	1.388
HDWD	02_12	2	9	0.263	1.906
HDWD	02_12	3	6	0.195	1.382
HDWD	02_07	1	9	0.268	1.875
HDWD	02_07	2	9	0.272	2.030
HDWD	02_07	3	8	0.140	0.737
INHW	03_15	1	0	0.000	0.000
INHW	03_15	2	8	0.189	1.167
INHW	03_15	3	0	0.000	0.000
INHW	03_10	1	3	0.200	1.529
INHW	03_10	2	3	0.200	1.529
INHW	03_10	3	3	0.200	1.529
INHW	03_08	1	0	0.000	0.000
INHW	03_08	2	2	0.016	0.048
INHW	03_08	3	0	0.000	0.000
INHW	03_03	1	0	0.000	0.000
INHW	03_03	2	8	0.101	0.534
INHW	03_03	3	8	0.101	0.534
INHW	03_06	1	21	0.565	4.222
INHW	03_06	2	10	0.248	1.861
INHW	03_06	3	13	0.330	2.123
INHW	03_01	1	3	0.116	1.034
INHW	03_01	2	3	0.116	1.034

Strata	Stand	Plot	Trees	Basal Area (m²)	Total Volume (m³)
SPBF	06_19	3	7	0.165	1.099
SPBF	06_25	1	7	0.162	1.067
SPBF	06_25	2	1	0.078	0.670
SPBF	06_25	3	2	0.097	0.717
SPBF	06_22	1	6	0.336	2.659
SPBF	06_22	2	9	0.210	1.446
SPBF	06_22	3	12	0.178	0.916
SPBF	06_01	1	8	0.169	1.076
SPBF	06_01	2	17	0.309	1.922
SPBF	06_01	3	17	0.309	1.922
SPBF	06_05	1	14	0.238	1.408
SPBF	06_05	2	4	0.056	0.262
SPBF	06_05	3	14	0.238	1.408
SPBF	06_03	1	8	0.205	1.305
SPBF	06_03	2	8	0.254	1.714
SPBF	06_03	3	8	0.254	1.714
SPBF	06_04	1	9	0.098	0.457
SPBF	06_04	2	9	0.098	0.457
SPBF	06_04	3	4	0.200	1.630
SPBF	06_11	1	10	0.304	2.090
SPBF	06_11	2	10	0.304	2.090
SPBF	06_11	3	10	0.304	2.090
SPBF	06_14	1	15	0.326	1.880
SPBF	06_14	2	16	0.406	2.570

Strata	Stand	Plot	Trees	Basal Area (m²)	Total Volume (m³)
SPHW	07_08	3	8	0.335	2.352
SPHW	07_11	1	3	0.043	0.203
SPHW	07_11	2	3	0.060	0.384
SPHW	07_11	3	6	0.122	0.714
SPHW	07_15	1	8	0.269	1.930
SPHW	07_15	2	10	0.233	1.486
SPHW	07_15	3	8	0.269	1.930
SPHW	07_02	1	11	0.305	2.081
SPHW	07_02	2	5	0.191	1.380
SPHW	07_02	3	11	0.305	2.081
SPHW	07_16	1	6	0.243	1.835
SPHW	07_16	2	6	0.243	1.835
SPHW	07_16	3	9	0.213	1.431
SPHW	07_20	1	7	0.163	1.058
SPHW	07_20	2	11	0.249	1.647
SPHW	07_20	3	11	0.249	1.647
SPHW	07_03	1	6	0.240	1.689
SPHW	07_03	2	6	0.240	1.689
SPHW	07_03	3	5	0.140	1.046
TOHW	08_02	1	8	0.200	1.425
TOHW	08_02	2	6	0.255	1.833
TOHW	08_02	3	8	0.200	1.425
TOHW	08_07	1	10	0.354	2.581
TOHW	08_07	2	8	0.411	3.051

TABLE A.12. Measured Tree Data for Point-3P Sample (Section 11.2.2)

Plot	Observed Values			Pred.	V:
	Species	dbh	HT	BFV	HT	BA·HT

Plot	Species	dbh	HT	BFV	Pred. HT	V: BA·HT
1	SM	13.1	48.4	162	61.6	3.577
1	SM	14.2	49.5	184	55.9	3.382
2	BW	10.7	45.3	82	56.4	2.897
2	SM	12.7	48.0	131	55.7	3.105
2	SM	24.1	74.3	692	57.8	2.940
2	SM	11.6	51.8	0	55.8	0.000
3	SM	13.7	58.9	192	53.6	3.186
3	SM	13.1	57.5	147	48.5	2.729
3	SM	14.3	60.1	176	62.6	2.624
4	SM	11.9	52.7	128	59.5	3.146
4	SM	21.4	70.9	497	59.7	2.805
4	TP	17.1	64.1	311	69.5	3.041
5	BW	12.4	50.7	116	60.6	2.730
5	AS	19	64.8	419	79.8	3.284
5	WA	11.6	48.6	121	60.9	3.395
6	SM	13.7	49.0	178	72.3	3.549
6	SM	14	49.3	186	52	3.531
7	SM	10.1	44.4	73	49.1	2.957
7	BC	23.3	61.1	538	49.6	2.972
7	SM	24	61.4	573	68.7	2.972
8	BC	26.9	62.2	0	48	0.000
10	BW	16.4	59.9	0	66.3	0.000
10	AS	17.3	61.6	262	52.3	2.604
10	AS	15.8	58.6	258	71.5	3.233
10	HB	18	65.9	346	67.4	2.970
10	WA	16.1	63.3	0	58.4	0.000
10	AS	16.4	63.7	279	72.7	2.985
11	SM	18	66.4	314	72.4	2.677
11	BW	20.7	70.4	463	53.6	2.815
11	SM	18.9	67.8	383	70.2	2.900
12	WA	11.3	52.2	115	52.6	3.165
12	SM	11.3	52.2	96	57.9	2.642
12	SM	23.5	73.8	709	58.2	3.192
12	SM	11.8	53.5	116	58.1	2.856
12	SM	19.7	69.0	418	64.7	2.863
12	BW	14.2	59.2	174	54.2	2.673
12	TP	10.7	50.5	95	64.9	3.012
13	CO	12.7	56.1	152	61	3.080
13	WA	27.5	69.6	844	56.9	2.938
13	SM	27.5	69.6	913	63.8	3.178
14	WA	14.5	58.4	200	67.9	2.987
14	WA	10.6	51.7	93	67.2	2.934
16	SM	14.1	59.0	189	57.8	2.956
16	TP	12	54.0	130	59.8	3.065
16	TP	10.5	50.0	91	68.1	3.030
17	TP	11.7	53.2	123	58.7	3.095
17	SM	19.3	68.4	451	62.5	3.246
17	SM	18	66.4	314	69.2	2.677
17	WA	21.5	71.4	502	48.9	2.788
18	SM	9	47.7	57	67.7	2.704
18	SM	19.3	65.5	363	60.3	2.728
19	WA	13.3	57.9	152	66.5	2.720
19	SM	13.4	58.1	184	55.9	3.232
19	SM	29.2	78.0	1124	59.5	3.100
19	HB	28	77.2	1028	55.5	3.115
20	CO	17	61.4	278	65.8	2.874
20	CO	28.3	68.1	897	56.8	3.016
20	SH	13.9	57.4	166	62.3	2.744
20	CO	19.6	63.8	405	60.8	3.031
21	SM	18.8	66.1	343	52.9	2.690
22	SM	23.6	70.3	660	55.9	3.090
23	SM	16.5	62.4	237	61.7	2.557
23	SM	19.1	65.3	355	63.9	2.732
24	AS	28.7	77.6	909	64.6	2.606
24	WA	21.1	70.9	522	50.3	3.033
25	WA	22.7	71.2	562	78	2.810
25	WA	16	59.7	0	55.3	0.000
27	SM	24.3	64.7	702	71.2	3.367
27	SM	20.6	62.9	449	68.7	3.086
27	SM	29.4	66.3	973	59.9	3.115
27	SM	21.5	63.4	492	63.3	3.078
28	WA	23.2	73.4	638	53.7	2.962
28	WA	13.7	57.6	178	58.4	3.020
28	WA	13	55.9	142	63.2	2.754
29	PH	19.5	67.8	0	68	0.000
30	BG	11.1	53.4	102	56.7	2.842
30	AB	26.7	73.7	859	58.5	2.997
30	AB	20.3	68.7	481	72.5	3.115

Species: AB = American beech; AS = American sycamore; BC = black cherry, BG = black gum; BW = black walnut; CO = chinquapin oak; HB = hackberry; PH = pignut hickory; SM = sugar maple; TP = tulip-poplar; and WA = white ash. BFV = board foot volume.

REFERENCES

Aber, J. D. 1979. A method for estimating foliage-height profiles in broad-leaved forests. J. Ecol. 67:35–40.

Adams, E. L. 1971. Effect of moisture loss on red oak sawlog weight. Research Note RN-NE-133, Northeastern Forest Experiment Station, USDA Forest Service. 4p.

Adams, E. L. 1976. The adjusting factor method for weight-scaling truckloads of mixed hardwood sawlogs. Research Paper RP-NE-344, Northeastern Forest Experiment Station, USDA Forest Service. 7p.

Affleck, D. L. R. 2008. A line intersect distance sampling strategy for downed wood inventory. Can. J. For. Res. 38:2262–2273.

Affleck, D. L. R. 2009. Field results for line intersect distance sampling of coarse woody debris. In: McWilliams, W., Moisen, G., and Czaplewski, R. (Comps.) Forest Inventory and Analysis (FIA) Symposium 2008, Proceedings RMRS-P-56CD, Rocky Mountain Research Station, USDA Forest Service. 12p.

Affleck, D. L. R. 2010. On the efficiency of line intersect distance sampling. Can. J. For. Res. 40:1086–1094.

Affleck, D. L. R., T. G. Gregoire, and H. T. Valentine. 2005a. Design unbiased estimation in line intersect sampling using segmented transects. Environ. Ecol. Stat. 12:139–154.

Affleck, D. L. R., T. G. Gregoire, and H. T. Valentine. 2005b. Edge effects in line intersect sampling with segmented transects. J. Agric. Biol. Environ. Stat. 10:460–477.

Akossou, A. Y. J., S. Arzouma, E. Y. Attakpa, N. H. Fonton, and K. Kokou. 2013. Scaling of teak (*Tectona grandis*) logs by the xylometer technique: Accuracy of volume equations and influence of the log length. Diversity 5:99–113.

Alden, H. A. 1997. Softwoods of North America. General Technical Report GTR-FPL-102, Forest Products Laboratory, USDA Forest Service, Madison, WI. 151p.

Forest Mensuration, Fifth Edition. John A. Kershaw, Jr., Mark J. Ducey, Thomas W. Beers and Bertram Husch.
© 2017 John Wiley & Sons, Ltd. Published 2017 by John Wiley & Sons, Ltd.

Alder, K. 2003. The measure of all things: The seven-year odyssey and hidden error that transformed the world. Free Press, New York. 448p.

Alemdag, I. S. 1978. Evaluation of some competition indexes for the prediction of diameter increment in planted white spruce. Information Report FMR-X 108, Forest Management Institute, Ottawa, ON, Canada. 39p.

Alexander, S. A., and J. E. Barnard. 1994. Forest health monitoring: Field methods guide. EPA/620/R-94/ 027, US Environmental Protection Agency, Environment Monitoring and Assessment Program Center. 266p.

Anderson, K., and K. J. Gaston. 2013. Lightweight unmanned aerial vehicles will revolutionize spatial ecology. Front. Ecol. Environ. 11:138–146.

Andrews, G. S. 1936. Tree-heights from air photographs by simple parallax measurements. For. Chron. 12:152–197.

Angers, V. A., Y. Bergeron, and P. Drapeau. 2012. Morphological attributes and snag classification of four North American boreal tree species: Relationships with time since death and wood density. For. Ecol. Manag. 263:138–147.

Anhold, J. A., and M. J. Jenkins. 1987. Potential mountain pine beetle (*Coleoptera: scolytidae*) attack of lodgepole pine as described by stand density index. Environ. Entomol. 16:738–742.

Anhold, J. A., M. J. Jenkins, and J. N. Long. 1996. Management of lodgepole pine stand density to reduce susceptibility to mountain pine beetle attack. West. J. Appl. For. 11:50–53.

Antony, F., L. R. Schimleck, R. F. Daniels, A. Clark III, and D. B. Hall. 2010. Modeling the longitudinal variation in wood specific gravity of planted loblolly pine (*Pinus taeda*) in the United States. Can. J. For. Res. 40:2439–2451.

Aplin, P., P. M. Atkinson, and P. J. Curran. 1997. Fine spatial resolution satellite sensors for the next decade. Int. J. Remote Sens. 18:3873–3881.

Arner, S. L., S. Woudenberg, S. Waters, J. Vissage, C. D. MacLean, and M. Thompson. 2003. National algorithms for determining stocking class, stand size class, and forest type for forest inventory and analysis plots. Supplemental Documents, Forest Inventory and Analysis, USDA Forest Service. 65p.

Aronoff, S. 1989. Geographic information systems: A management perspective. WDL Publishers, Ottawa, ON, Canada. 294p.

Arrhenius, O. 1921. Species and area. J. Ecol. 9:95–99.

Ashraf, M. I., C. P.-A. Bourque, D. A. MacLean, T. Erdle, and F.-R. Meng. 2012. Using JABOWA-3 for forest growth and yield predictions under diverse forest conditions of Nova Scotia, Canada. For. Chron. 88:708–721.

Asner, G. P., D. E. Knapp, T. Kennedy-Bowdoin, M. O. Jones, R. E. Martin, J. Boardma, and C. B. Field. 2007. Carnegie Airborne Observatory: In-flight fusion of hyperspectral imaging and waveform light detection and ranging for three-dimensional studies of ecosystems. J. Appl. Remote Sens. 1:013536.

Astrium. 2013. SPOT 6/SPOT 7 technical sheet. Available online at: http://www.satimagingcorp. com/satellite-sensors/spot-6/; Last accessed August 31, 2016.

Astrup, R., M. J. Ducey, A. Granhus, T. Ritter, and N. von Lüpke. 2014. Approaches for estimating stand-level volume using terrestrial laser scanning in a single-scan mode. Can. J. For. Res. 44:666–676.

Avery, T. E., and H. E. Burkhart. 2002. Forest measurements. McGraw-Hill, New York. 480p.

Baatz, M., and A. Schäpe. 2000. Multiresolution segmentation: An optimization approach for high quality multi-scale image segmentation. In: Strobl, J., Blaschke, T., and Griesebner, G. (Eds.) Angewandte Geographische Informationsverarbeitung, Wichmann-Verlag, Heidelberg. 12–23.

Bailey, G. R. 1970. A simplified method of sampling logging residue. For. Chron. 46:288–303.

Bailey, R. L., and T. R. Dell. 1973. Quantifying diameter distributions with the Weibull function. For. Sci. 19:97–104.

Bakker, J. D. 2005. A new, proportional method for reconstructing historical tree diameters. Can. J. For. Res. 35:2515–2520.

Baldwin, V. C., Jr., H. E. Burkhart, J. A. Westfall, and K. D. Peterson. 2001. Linking growth and yield and process models to estimate impact of environmental changes on growth of loblolly pine. For. Sci. 47:77–82.

Balzter, H. 2001. Forest mapping and monitoring with interferometric synthetic aperture radar (InSAR). Prog. Phys. Geogr. 25:159–177.

Barabesi, L. 2007. Some comments on design-based line-intersect sampling with segmented transects. Environ. Ecol. Stat. 14:483–494.

Barbati, A., P. Corona, and M. Marchetti. 2007. A forest typology for monitoring sustainable forest management: The case of European forest types. Plant Biosyst. 141:93–103.

Barbati, A., M. Marchetti, G. Chirici, and P. Corona. 2014. European forest types and forest Europe SFM indicators: Tools for monitoring progress on forest biodiversity conservation. For. Ecol. Manag. 321:145–157.

Bartlett, M. S. 1948. Determination of plant densities. Nature. 162:621.

Barton, C. V. M., and K. D. Montagu. 2004. Detection of tree roots and determination of root diameters by ground penetrating radar under optimal conditions. Tree Physiol. 24:1323–1331.

Bartoo, R. A., and R. J. Hutnik. 1962. Board foot volume tables for timber tree species in Pennsylvania. Research Paper 30, School of Forest Resources, University of Maine, Pennsylvania State University, State College, PA. 35p.

Baskerville, G. 1965. Estimation of dry weight of tree components and total standing crop in conifer stands. Ecology. 46:867–869.

Baskerville, G. 1972. Use of logarithmic regression in the estimation of plant biomass. Can. J. For. Res. 2:49–53.

Bate, L. J., T. R. Torgersen, M. J. Wisdom, and E. O. Garton. 2009. Biased estimation of forest log characteristics using intersect diameters. For. Ecol. Manag. 258:635–640.

Bates, C. G., and R. Zon. 1922. Research methods in the study of forest environment. Bulletin 1059, USDA Forest Service. 209p.

Battaglia, M., and P. J. Sands. 1998. Process-based forest productivity models and their application in forest management. For. Ecol. Manag. 102:13–32.

Battles, J. J., J. G. Dushoff, and T. J. Fahey. 1996. Line intersect sampling of forest canopy gaps. For. Sci. 42:131–138.

Bealle Statland, C. 1996. Stand reconstruction. In: Comeau, P. G. and Thomas, K. D. (Eds.) Designing mixedwood experiments. B.C. Ministry of Forestry, Research Branch, Richmond, B.C. 12–15.

Bebber, D. P., and S. C. Thomas. 2003. Prism sweeps for coarse woody debris. Can. J. For. Res. 33:1737–1743.

Bebber, D. P., W. G. Cole, S. C. Thomas, D. Balsillie, and P. Duinker. 2005. Effects of retention harvests on structure of old-growth *Pinus strobus* L. stands in Ontario. For. Ecol. Manag. 205:91–103.

Bechtold, W. A. 2003. Crown-diameter prediction models for 87 species of stand-grown trees in the eastern United States. South. J. Appl. For. 27:269–278.

Bechtold, W. A., and P. I. Patterson. 2005. The enhanced forest inventory and analysis program— National sampling design and estimation procedures. General Technical Report GTR-SRS-80, Southern Research Station, USDA Forest Service. 2p.

Beers, T. W. 1962. Components of forest growth. J. For. 60:245–248.

Beers, T. W. 1964a. Composite hardwood volume tables. Research Bulletin 787, Agriculture Experiment Station, Purdue University, West Lafayette, IN. 12p.

Beers, T. W. 1964b. Cruising for pulpwood by the ton without concern for tree diameter: Point sampling with diameter obviation. Extension Mimeo F-49, Purdue University, West Lafayette, IN. 8p.

Beers, T. W. 1966. The direct correction for boundary-line slopover in horizontal point sampling. Research Progress Report 224, Agriculture Experiment Station, Purdue University, West Lafayette, IN. 8p.

Beers, T. W. 1969. Slope correction in horizontal point sampling. J. For. 67:188–192.

Beers, T. W. 1973. Revised composite tree volume tables for Indiana hardwoods. Research Progress Report 417, Agriculture Experiment Station, Purdue University, West Lafayette, IN. 4p.

Beers, T. W. 1974a. Optimum upper-log viewing distance. Bulletin 39, Agriculture Experiment Station, Purdue University, West Lafayette, IN. 9p.

Beers, T. W. 1974b. Vertical line sampling for regeneration surveys. In: Frayer, W., Hartman, G., and Bower, D. (Eds.) Proceedings of a Workshop on Inventory Design and Analysis, Fort Collins, CO, 1974, Society of American Foresters, Washington, DC. 246–260.

Beers, T. W. 1977. Practical correction of boundary overlap. South. J. Appl. For. 1:16–18.

Beers, T. W., and C. I. Miller. 1964. Point sampling: Research results, theory, and applications. Research Bulletin 786, Agriculture Experiment Station, Purdue University, West Lafayette, IN.

Beers, T. W., and C. I. Miller. 1973. Manual of forest mensuration. T&C Enterprises, West Lafayette, IN.

Beers, T. W., and C. I. Miller. 1976. Line sampling for forest inventory. Research Bulletin 934, Agriculture Experiment Station, Purdue University, West Lafayette, IN.

Beers, T. W., P. E. Dress, and L. C. Wensel. 1966. Aspect transformation in site productivity research. J. For. 64:691–692.

Behre, C. E. 1927. Form-class taper curves and volume tables and their application. J. Agric. Res. 35:673–744.

Behre, C. E. 1935. Factors involved in the application of form-class volume tables. J. Agric. Res. 51:669–713.

Bell, J. F., and J. R. Dilworth. 1993. Log scaling and timber cruising. OSU Bookstore, Corvallis, OR. 444p.

Bell, J. F., K. Iles, and D. Marshall. 1983. Balancing the ratio of tree count-only sample points and VBAR measurements in variable plot sampling. In: Bell, J. F. and Atterbury, T. (Eds.) Proceedings: Renewable resource inventories for monitoring changes and trends. College of Forestry, Oregon State University, Corvallis, OR. 699–702.

Berger, W. H., and F. Parker. 1970. Diversity of planktonic foraminifera in deep-sea sediments. Science. 168:1345–1347.

Bergseng, E., H. O. Ørka, E. Næsset, and T. Gobakken. 2015. Assessing forest inventory information obtained from different inventory approaches and remote sensing data sources. Ann. For. Sci. 72:33–45.

Bernier, P. Y., and G. Robitaille. 2004. A plane intersect method for estimating fine root productivity of trees from minirhizotron images. Plant Soil. 265:165–173.

Besley, L. 1967. Importance, variation and measurement of density and moisture. In: Buckingham, F. (Ed.) Wood Measurement Conference Proceedings. Technical Report 7, Faculty of Forestry, University of Toronto, Toronto, ON.

Bi, H. 2004. Stochastic frontier analysis of a classic self-thinning experiment. Austral Ecol. 29:408–417.

Bickerstaff, A. 1961. A variable quadrat regeneration survey method. For. Chron. 37:39–53.

Bickford, C. A. 1954. The place of individual-tree data in estimating growth. J. For. 52:423–426.

Bickford, C. A., F. S. Baker, and F. G. Wilson. 1957. Stocking, normality, and measurement of stand density. J. For. 55:99–104.

Biging, G. S., and M. Dobbertin. 1992. A comparison of distance-dependent competition measures for height and basal area growth of individual conifer trees. For. Sci. 38:695–720.

Birdsey, R. A. 2004. Data gaps for monitoring forest carbon in the United States: An inventory perspective. Environ. Manag. 33(Supplement 1):S1–S8.

Birth, E. E. 1977. Horizontal line sampling in upland hardwoods. J. For. 75:590–591.

Bitterlich, W. 1947. Die winkelzählmessung [Measurement of basal area per hectare by means of angle measurement]. Allg. Forst- Holzwirtsch. Ztg. 58:94–96.

Bitterlich, W. 1956. Fortschritte der Relaskopmessung [Progress with Relaskop measurement]. Holzkurier. 11:6–10.

Bitterlich, W. 1959. Sektorkluppern aus Leichmetall [Calipering forks made of light alloys]. Holzkurier. 14:15–17.

Bitterlich, W. 1976. Volume sampling using indirectly estimated critical heights. Commonw. For. Rev. 55:319–330.

Bitterlich, W. 1984. The relascope idea: Relative measurements in forestry. CAB International, Slough, England. 242p.

Blair, J. B., D. L. Rabine, and M. A. Hofton. 1999. The Laser Vegetation Imaging Sensor: A medium-altitude, digitisation-only, airborne laser altimeter for mapping vegetation and topography. ISPRS J. Photogramm. Remote Sens. 54:115–122.

Blaschke, T. 2010. Object based image analysis for remote sensing. ISPRS J. Photogramm. Remote Sens. 65:2–16.

BLM. 2009. Manual of surveying instructions (2009). Bureau of Land Management, US Department of Interior.

Böhl, J., and U.-B. Brändli. 2007. Deadwood volume assessment in the third Swiss National Forest Inventory: Methods and first results. Eur. J. For. Res. 126:449–457.

Bokalo, M., S. J. Titus, and D. P. Wiens. 1996. Sampling with partial replacement extended to include growth projections. For. Sci. 42:328–334.

Bonham, C. 1989. Measurements for terrestrial vegetation. John Wiley & Sons, Inc., New York. 338p.

Bonham, C. 2013. Measurements for terrestrial vegetation. Wiley/Blackwell, New York. 246p.

Bonnor, G. M. 1964a. A tree volume table for red pine by crown width and height. For. Chron. 40:339–346.

Bonnor, G. M. 1964b. The influence of stand density on the correlation of stem diameter with crown width and height for lodgepole pine. For. Chron. 40:347–349.

Bonnor, G. M. 1967. Estimation of ground canopy density from ground measurements. J. For. 65:544–547.

Borders, B. E., and W. D. Patterson. 1990. Projecting stand tables: A comparison of the Weibull diameter distribution method, a percentile-based projection method, and a basal area growth projection method. For. Sci. 36:413–424.

Botkin, D. B. 1993. Forest dynamics: An ecological model. First. Oxford University Press, Oxford, UK. 328p.

Bower, D. R., and W. W. Blocker. 1966. Accuracy of bands and tape for measuring diameter increments. J. For. 64:21–22.

Bowers, S., J. Reeb, and B. Parker. 2013. Tarif access tables: A comprehensive set. Extension Circular EC 1609, Oregon Agricultural Experiment Station, Oregon State University, Corvallis, OR. 26p.

Braathe, P., and T. Okstad. 1967. Trade of pulpwood based on weighing and dry matter samples. In: XIV IUFRO Congress, Section 41, International Union of Forestry Research Organization. 236–242.

Brack, C.1988. The RADHOP system. In: Modelling trees, stands, and forests. Bulletin 5, University of Melbourne, Melbourne, Australia. 509–526.

Bragg, D. C. 2008. An improved tree height measurement technique tested on mature southern pines. South. J. Appl. For. 32:38–43.

Bragg, D. C., L. E. Frelich, R. T. Leverett, W. Blozan, and D. J. Luthringer. 2011. The sine method: An alternative height measurement technique. Research Note SRS-RN-22, Southern Research Station, USDA Forest Service. 12p.

Braun, C. E. (Ed.). 2005. Techniques for wildlife investigations and management. Wildlife Society, Bethesda, MD. 974p.

Bravo-Oviedo, A., H. Pretzsch, C. Ammer, E. Andenmatten, A. Barbati, S. Barreiro, P. Brang, et al. 2014. European mixed forests: Definition and research perspectives. For. Syst. 23:518–533.

Breiman, L. 2001. Random forests. Mach. Learn. 45:5–32.

Brender, E. V. 1973. Silviculture of loblolly pine in the Georgia Piedmont. Report 33, Georgia Forest Research Council, Macon, GA. 74p.

Brewer, K. 1963. Ratio estimation and finite populations: Some results deducible from the assumption of an underlying stochastic process. Aust. J. Stat. 5:93–105.

Briegleb, P. A. 1943. Growth of ponderosa pine by Keen tree class. Research Note OSN-PNW-032, Old Series, Pacific Northwest Research Station, USDA Forest Service. 1–16.

Brissette, J. C., M. J. Ducey, and J. H. Gove. 2003. A field test of point relascope sampling of down coarse woody material in managed stands in the Acadian Forest. J. Torrey Bot. Soc. 130:79–88.

Brohman, R. J., and L. D. Bryant. 2005. Existing vegetation classification and mapping technical guide. Version 1.0. General Technical Report GTR-WO-67, Ecosystem Management Coordination Staff, USDA Forest Service, Washington, DC. 161p.

Brose, U., N. D. Martinez, and R. J. Williams. 2003. Estimating species richness: Sensitivity to sample coverage and insensitivity to spatial patterns. Ecology. 84:2364–2377.

Brown, G. S. 1967. Point density in stems per acre. New Zealand Forestry Research Notes 38, New Zealand Forestry Research Institute. 11p.

Brown, J. K. 1971. A planar intersect method for sampling fuel volume and surface area. For. Sci. 17:96–102.

Brown, J. K. 1974. Handbook for inventorying downed woody material. General Technical Report GTR-INT-16, Intermountain Forest and Range Experiment Station, USDA Forest Service. 24p.

Brown, J. K. 1976. Estimating shrub biomass from basal stem diameters. Can. J. For. Res. 6:153–158.

Brown, J. K., E. D. Reinhardt, and K. A. Kramer. 2003. Coarse woody debris: Managing benefits and fire hazard in the recovering forest. General Technical Report RMRS-GTR-105, Rocky Mountain Research Station, USDA Forest Service. 16p.

Brown, S. L. 1997. Estimating biomass and biomass change of tropical forests: A primer. FAO Forestry Paper 134, Forestry and Agricultural Organization, Rome, Italy. 55p.

Brown, S. L. 1999. Guidelines for inventorying and monitoring carbon offsets in forest–based projects. Winrock International Institute for Agricultural Development, Arlington, VA. 12p.

Brown, S. L. 2002. Measuring, monitoring and verification of carbon benefits for forest-based projects. Philos. Trans. R. Soc. Lond. A 360:1669–1683.

Bruce, D. 1926. A method of preparing timber yield tables. J. Agric. Res. 32:543–557.

Bruce, D. 1961. Prism cruising in the western United States and volume tables for use therewith. Mason, Bruce & Girard, Inc., Portland, OR.

Bruce, D. 1982. Butt log volume estimators. For. Sci. 28:489–503.

Bruce, D., and F. X. Schumacher. 1950. Forest mensuration. McGraw-Hill, New York. 483p.

Brünig, E. F. 1974. Das Risiko der forstlichen Funktionen-planung, dargestellt am Beispiel der Sturmgefahrdung. Allg. Forst- Jagd-Ztg. 145:60–67.

Buchman, R. G., and S. R. Shifley. 1983. Guide to evaluating forest growth projection systems. J. For. 81:232–234, 254.

Buckland, S. T., D. Anderson, K. P. Burnham, J. L. Laake, and L. Thomas. 2001. Introduction to distance sampling: Estimating abundance of biological populations. Revised. Oxford University Press, New York.

Buckland, S. T., D. Anderson, K. P. Burnham, J. L. Laake, and L. Thomas. 2004. Advanced distance sampling: Estimating abundance of biological populations. First. Oxford University Press, New York. 416p.

Buckman, R. E. 1962. Growth and yield of red pine in Minnesota. Technical Bulletin 1272, USDA Forest Service. 50p.

Buffon, Comte de, G. L. L. 1777. Essai d'arithmétique morale. In: Oeuvres Complètes de Buffon, vol. 10. Histoire Naturelle, Générale et Particulière, de L'Imprimerie Royale, Paris.

Bullock, J. 1996. Chapter 3: Plants. In: Sutherland, W. J. (Ed.) Ecological census techniques: A handbook. Cambridge University Press, New York. 111–138.

Bunting, P., and R. Lucas. 2006. The delineation of tree crowns in Australian mixed species forests using hyperspectral Compact Airborne Spectrographic Imager (CASI) data. Remote Sens. Environ. 101:230–248.

Burger, W., and M. J. Burge. 2008. Digital image processing: An algorithmic introduction using Java. Springer, New York. 564p.

Burk, T. E., and J. D. Newberry. 1984. A simple algorithm for moment-based recovery of Weibull distribution parameters. For. Sci. 30:329–332.

Burkhart, H. E. 1977. Cubic–foot volume of loblolly pine to any merchantable top limit. South. J. Appl. For. 1:7–9.

Burkhart, H. E., R. C. Parker, and R. G. Oderwald. 1972. Yields for natural stands of loblolly pine. FWS-2-72, Division of Forestry and Wildlife Resources, Virginia Polytechnic Institute and State University, Blacksburg, VA. 63p.

Burnham, K. P., and W. S. Overton. 1978. Estimation of the size of a closed population when capture probabilities vary among animals. Biometrika. 65:625–633.

Burnham, K. P., and W. S. Overton. 1979. Robust estimation of population size when capture probabilities vary among animals. Ecology. 60:927–936.

Büsgen, M., and E. Munch. 1929. The structure and life of forest trees (Thomson, T., Trans.). Chapman Hall, New York.

Bütler, R., L. Patty, R.-C. Le Bayon, C. Guenat, and B. Rodolphe Schlaepfer. 2007. Log decay of Picea abies in the Swiss Jura Mountains of central Europe. For. Ecol. Manag. 242:791–799.

Cailliez, F. 1980. Forest volume estimation and yield prediction. Volume 1 – Volume estimation. Forestry Paper 22/1, FAO, Rome, Italy. 98p.

Cain, S. A., and G. M. de Oliveira Castro. 1959. Manual of vegetation analysis. Harper and Row Publishers, New York. 325p.

Cajander, A. K. 1926. The theory of forest types. Acta For. Fenn. 29:1–108.

Calkins, H. A., and J. Yule. 1935. The Abney level handbook. USDA Forest Service.

Campbell, G. S. 1986. Extinction coefficients for radiation in plant canopies calculated using an ellipsoidal inclination angle distribution. Agric. For. Meteorol. 36:317–321.

Campbell, J. B., and R. H. Wynne. 2011. Introduction to remote sensing. 5th Ed. Guilford Press, New York. 667p.

Campbell, M., R. G. Congalton, J. Hartter, and M. J. Ducey. 2015. Optimal land cover mapping and change analysis in northeastern Oregon using Landsat imagery. Photogramm. Eng. Remote Sens. 81:37–47.

Canfield, R. H. 1941. Application of the line interception method in sampling range vegetation. J. For. 39:388–394.

Cannell, M. G. R., P. Rothery, and E. D. Ford. 1984. Competition within stands of *Picea sitchensis* and *Pinus contorta*. Ann. Bot. 53:349–362.

Cao, Q. V., and W. D. Pepper. 1986. Predicting inside bark diameter for shortleaf, loblolly, and long-leaf pines. South. J. Appl. For. 10:220–224.

Carmean, W. H. 1975. Forest site quality evaluation in the United States. Adv. Agron. 29:209–269.

Carr, B. 1992. Using laser technology for forestry and engineering applications. The Compiler. 10:5–16.

Cartus, O., M. Santoro, and J. Kellndorfer. 2012. Mapping forest aboveground biomass in the Northeastern United States with ALOS PALSAR dual-polarization L-band. Remote Sens. Environ. 124:466–478.

Castedo-Dorado, F., F. Crecente-Campo, P. Álvarez-Álvarez, and M. Barrio-Anta. 2009. Development of a stand density management diagram for radiata pine stands including assessment of stand stability. Forestry. 82:1–16.

Chamberlain, E. B., and H. A. Meyer. 1950. Bark volume in cordwood. TAPPI. 33:554–555.

Chamberlain, J. L., R. J. Bush, A. L. Hammett, and P. A. Araman. 2002. Eastern national forests: Managing for nontimber products. J. For. 100:8–14.

Chamberlain, J. L., G. Ness, C. J. Small, S. J. Bonner, and E. B. Hiebert. 2013. Modeling below-ground biomass to improve sustainable management of *Actaea racemosa*, a globally important medicinal forest product. For. Ecol. Manag. 293:1–8.

Chambers, J. Q., and S. E. Trumbore. 1999. An age-old problem. Trends Plant Sci. 4:385–386.

Chapman, D. G. 1961. Statistical problems in dynamics of exploited fisheries populations. In: Proceedings of the Fourth Berkley Symposium on Mathematical Statistics and Probability. University of California Press, Berkeley, CA. 153–168.

Chapman, H. H., and H. A. Meyer. 1949. Forest mensuration. McGraw-Hill, New York. 522p.

Chave, J., D. Coomes, S. Jansen, S. L. Lewis, N. G. Swenson, and A. E. Zanne. 2009. Towards a worldwide wood economics spectrum. Ecol. Lett. 12:351–366.

Chehock, C. R., and R. C. Walker. 1975. Sample weight scaling with 3–P sampling for multiproduct logging. Southeastern Area State and Private Forestry, USDA Forest Service. 27p.

Chen, J. M., P. M. Rich, S. T. Gower, J. M. Norman, and S. Plummer. 1997. Leaf area index of boreal forests: Theory, techniques, and measurements. J. Geophys. Res. Atmos. 102(D24):29429–29443.

Chin, H. C., and M. A. Quddus. 2003. Modeling count data with excess zeroes: An empirical application to traffic accidents. Sociol. Methods Res. 32:90–116.

Chisman, H. H., and F. X. Schumacher. 1940. On the tree-area ratio and certain of its applications. J. For. 38:311–317.

Chojnacky, D. C., J. C. Jenkins, and A. K. Holland. 2009. Improving North American forest biomass estimates from literature synthesis and meta-analysis of existing biomass equations. In: McWilliams, W., Moisen, G., and Czaplewski, R. (Comps.) Forest Inventory and Analysis (FIA) Symposium 2008. RMRS-P-56CD, Rocky Mountain Research Station, USDA Forest Service. 7.

Chojnacky, D. C., L. S. Heath, and J. C. Jenkins. 2014. Updated generalized biomass equations for North American tree species. Forestry. 87:129–151.

Chudnoff, M. 1984. Tropical timbers of the world. Agriculture Handbook 607, US Department of Agriculture. 466p.

Cieszewski, C., and R. L. Bailey. 2000. Generalized algebraic difference approach: Theory based derivation of dynamic site equations with polymorphism and variable asymptotes. For. Sci. 46:116–126.

Clark, A. I., and J. Schroeder. 1977. Biomass of yellow poplar in natural stands in western North Carolina. Research Paper RP-SE-165, Southeastern Forest Experiment Station, USDA Forest Service. 44p.

Clark, J. G. 1906. Measurement of sawlogs. For. Q. 4:79–93.

Clark, M. L., D. A. Roberts, and D. B. Clark. 2005. Hyperspectral discrimination of tropical rain forest tree species at leaf to crown scales. Remote Sens. Environ. 96:375–398.

Clark, N. A., R. H. Wynne, and D. L. Schmoldt. 2000. A review of past research on dendrometers. For. Sci. 46:570–576.

Cleveland, W. S. 1994. The elements of graphing data. Revised Edition. Chapman Hall, New York. 279p.

Clutter, J. L. 1963. Compatible growth and yield models for loblolly pine. For. Sci. 9:354–371.

Clutter, J. L., J. C. Fortson, L. V. Pienaar, G. H. Brister, and R. L. Bailey. 1983. Timber management. A quantitative approach. John Wiley & Sons, Inc., New York. 333p.

Cobb, D. F. 1988. Development of mixed western larch, lodgepole pine, Douglas-fir, and grand fir in eastern Washington. M.Sc. For. Thesis, University of Washington, Seattle, WA. 99p.

Cochran, W. G. 1977. Sampling techniques. John Wiley & Sons, Inc., New York. 448p.

Cody, J. B. 1976. Merchantable weight tables for New York State red pine plantations. Research Note 23, Applied Forest Research Institute, College of Environmental Studies and Forestry, Syracuse, NY. 4p.

Condés, S., and H. Sterba. 2005. Derivation of compatible crown width equations for some important tree species of Spain. For. Ecol. Manag. 217:203–218.

Congalton, R. G., and K. Green. 2009. Assessing the accuracy of remotely sensed data – Principles and practices. CRC Press, Taylor & Francis Group, Boca Raton, FL. 183p.

Cook, B. D., L. A. Corp, R. F. Nelson, E. M. Middleton, D. C. Morton, J. T. McCorkel, J. G. Masek, et al. 2013. NASA Goddard's LiDAR, hyperspectral and thermal (G-LiHT) airborne imager. Remote Sens. 5:4045–4066.

Cook, E. R., and I. A. Kairiukstis. 1990. Methods of dendrochronology—Applications in the environmental sciences. Kluwer Academic Publishers, New York. 394p.

Coops, N. C., T. Hilker, M. A. Wulder, B. St-Onge, G. Newnham, A. Siggins, and J. A. Trojmow. 2007. Estimating canopy structure of Douglas-fir forest stands from discrete-return LiDAR. Trees Struct. Funct. 21:295–310.

Cottam, G., J. T. Curtis, and B. W. Hale. 1953. Some sampling characteristics of a population of randomly dispersed individuals. Ecology. 34:741–757.

Creed, I. F., K. L. Webster, and D. L. Morrison. 2004. A comparison of techniques for measuring density and concentrations of carbon and nitrogen in coarse woody debris at different stages of decay. Can. J. For. Res. 34:744–753.

Cremer, K. W., C. J. Borough, F. H. McKinnell, and P. R. Carter. 1982. Effect of stocking and thinning on wind damage in plantations. N. Z. J. For. Sci. 12:244–268.

Crookston, N. L., and G. E. Dixon. 2005. The forest vegetation simulator: A review of its structure, content, and applications. Comput. Electron. Agric. 49:60–80.

Crookston, N. L., and A. O. Finley. 2008. yaImpute: An R package for kNN imputation. J. Stat. Softw. 23:1–16.

Curtis, J. T., and R. P. McIntosh. 1950. The interrelations of certain analytic and synthetic phytoso-ciological characters. Ecology. 31:434–455.

Curtis, R. O. 1970. Stand density measures: An interpretation. For. Sci. 16:403–414.

Curtis, R. O. 1971. A tree area power function and related stand density measures for Douglas-fir. For. Sci. 17:146–159.

Curtis, R. O., and D. D. Marshall. 2000. Why quadratic mean diameter? West. J. Appl. For. 15:137–139.

Curtis, R. O., and D. Marshall. 2005. Permanent-plot procedures for silvicultural and yield research. General Technical Report GTR-PNW-634, Pacific Northwest Research Station, USDA Forest Service. 86p.

Curtis, R. O., and D. L. Reukema. 1970. Crown development and site estimates in a Douglas-fir plantation spacing test. For. Sci. 16:287–301.

Dandois, J. P., and E. C. Ellis. 2010. Remote sensing of vegetation structure using computer vision. Remote Sens. 2:1157–1176.

Daniels, R. F. 1976. Simple competition indices and their correlation with annual loblolly pine tree growth. For. Sci. 22:454–456.

Daniels, R. F., H. E. Burkhart, and T. R. Clason. 1986. A comparison of competition measures for predicting growth of loblolly pine trees. Can. J. For. Res. 16:1230–1237.

Dassot, M., T. Constant, and M. Fournier. 2011. The use of terrestrial LiDAR technology in forest science: application fields, benefits and challenges. Ann. For. Sci. 68:959–974.

Daubenmire, R. F. 1945. An improved type of precision dendrometer. Ecology. 26:97–98.

Daubenmire, R. F. 1976. The use of vegetation in assessing the productivity of forest lands. Bot. Rev. 42:115–143.

Dawkins, H. C. 1957. Some results of stratified random sampling of tropical high-forest. Seventh British Commonwealth Forestry Conference. Adelaide, NZ, Aug. 26–29, 1957. Item 7 (iii), Holywell Press, Oxford.

de Liocourt, F. 1898. De l'amenagement des sapinières. [The arrangement of fir]. *Bull. Trimest. Société For. Franche-Comté Belfort Julliet.* 4:396–409.

de Vries, P. G. 1979. Line intersect sampling—Statistical theory, applications, and suggestions for extended use in ecological inventory. In: Cormack, R. M., Patil, G. P., and Robson, D. S. (Eds.) Sampling biological populations. International Cooperative Publishing House, Fairland, MD. 1–70.

de Vries, P. G. 1986. Sampling theory for forest inventory—A teach-yourself course. Springer, Berlin, Germany. 399p.

Dean, T. J., and V. C. Baldwin Jr. 1996. The relationship between Reineke's stand-density index and physical stem mechanics. For. Ecol. Manag. 81:25–34.

Dean, T. J., and J. N. Long. 1986. Variation in sapwood area-leaf area relations within two stands of lodgepole pine. For. Sci. 32:749–758.

Dean, T. J., J. N. Long, and F. W. Smith. 1988. Bias in leaf area—Sapwood area ratios and its impact on growth analysis in *Pinus contorta*. Trees. 2:104–109.

Decourt, N. 1956. Utilisation de la Photographie pour Mesurer las Surfaces Terrieres. Rev. For. Fr. 8:505–507.

Deeming, J. E., J. W. Lancaster, M. A. Fosberg, R. W. Furman, and M. J. Schroeder. 1972. National fire-danger rating system. Research Paper RP-RM-84, Rocky Mountain Forest and Range Experiment Station, USDA Forest Service. 165p.

Deitschman, G. H., and A. W. Green. 1965. Relations between western white pine site index and tree height of several associated species. Research Paper RP-INT-22, Intermountain Forest and Range Experiment Station, USDA Forest Service. 28p.

Delisle, G. P., P. M. Woodward, S. J. Titus, and A. F. Johnson. 1988. Sample size and variability of fuel weight estimates in natural stands of lodgepole pine. Can. J. For. Res. 18:649–652.

Dellenbaugh, M., M. J. Ducey, and J. C. Innes. 2007. Double sampling may improve the efficiency of litterfall estimates. Can. J. For. Res. 37:840–845.

Dennis, B. 1984. Distance methods for evaluating forest regeneration. In: Proceedings of the 1983 Society of American Foresters National Convention. Society of American Foresters, Portland, OR. 123–128.

Dick, A. R. 2012. Forest inventory using a camera: Concept, field implementation and instrument development. Unpublished M.Sc. For. Thesis, University of New Brunswick, Fredericton, NB, Canada. 67p.

Dickie, I. A., G. W. Yeates, M. G. St. John, S. Stevenson, J. T. Scott, M. C. Rillig, D. A. Peltzer, et al. 2011. Ecosystem service and biodiversity trade-offs in two woody successions. J. Appl. Ecol. 48:926–934.

Dickinson, Y. L., and E. K. Zenner. 2010. Allometric equations for the aboveground biomass of selected common eastern hardwood understory species. North. J. Appl. For. 27:160–165.

Diller, O. D., and L. D. Kellogg. 1940. Local volume tables for yellow poplar. Technical Note TN-CS-1, Central States Forest Experiment Station, USDA Forest Service.

Dobson, M. C. 2000. Forest information from synthetic aperture radar. J. For. 98:41–43.

Doebelin, E. O. 1966. Measurement systems: Applications and designs. McGraw-Hill, New York. 743p.

Donnelly, D. M., and R. L. Barger. 1977. Weight scaling for southwestern ponderosa pine. Research Paper RP-RM-181, Rocky Mountain Forest and Range Experiment Station, USDA Forest Service. 9p.

Doolittle, W. T., and J. P. Vimmerstedt. 1960. Site curves for natural stands of white pine in southern Appalachians. Research Note RN-SE-141, Southeastern Forest Experiment Station, USDA Forest Service. 2p.

Drew, T. J., and J. W. Flewelling. 1977. Some recent Japanese theories of yield-density relationships and their application to Monterey pine plantations. For. Sci. 23:517–534.

Drew, T. J., and J. W. Flewelling. 1979. Stand density management: An alternative approach and its application to Douglas-fir plantations. For. Sci. 25:518–532.

Dubayah, R. O., and J. B. Drake. 2000. Lidar remote sensing for forestry. J. For. 98:44–46.

Ducey, M. J. 2009a. Predicting crown size and shape from simple stand variables. J. Sustain. For. 28:5–21.

Ducey, M. J. 2009b. Sampling trees with probability nearly proportional to biomass. For. Ecol. Manag. 258:2110–2116.

Ducey, M. J. 2012. Evergreenness and wood density predict height–diameter scaling in trees of the northeastern United States. For. Ecol. Manag. 279:21–26.

Ducey, M. J., and R. Astrup. 2013. Adjusting for nondetection in forest inventories derived from terrestrial laser scanning. Can. J. Remote Sens. 39:410–425.

Ducey, M. J., and J. A. Kershaw Jr. 2011. Vertical point sampling with a camera. North. J. Appl. For. 28:61–65.

Ducey, M. J., and R. A. Knapp. 2010. A stand density index for complex mixed species forests in the northeastern United States. For. Ecol. Manag. 260:1613–1622.

Ducey, M. J., and B. C. Larson. 1997. Thinning decisions using stand density indices: The influence of uncertainty. West. J. Appl. For. 12:89–92.

Ducey, M. J., and B. C. Larson. 1999. Accounting for bias and uncertainty in nonlinear stand density indices. For. Sci. 45:452–457.

Ducey, M. J., and H. T. Valentine. 2008. Direct sampling for stand density index. West. J. Appl. For. 23:78–82.

Ducey, M. J., and M. S. Williams. 2011. Comparison of Hossfeld's method and two modern methods for volume estimation of standing trees. West. J. Appl. For. 26:19–23.

Ducey, M. J., J. H. Gove, G. Ståhl, and A. Ringvall. 2001. Clarification of the mirage method for boundary correction, with possible bias in plot and point sampling. For. Sci. 47:242–245.

Ducey, M. J., G. J. Jordan, J. H. Gove, and H. T. Valentine. 2002. A practical modification of horizontal line sampling for snag and cavity tree inventory. Can. J. For. Res. 32:1217–1224.

Ducey, M. J., J. H. Gove, and H. T. Valentine. 2004. A walkthrough solution to the boundary overlap problem. For. Sci. 50:427–435.

Ducey, M. J., M. S. Williams, J. H. Gove, and H. T. Valentine. 2008. Simultaneous unbiased estimates of multiple downed wood attributes in perpendicular distance sampling. Can. J. For. Res. 38:2044–2051.

Ducey, M. J., D. J. Zarin, S. S. Vasconcelos, and M. M. Araujo. 2009. Biomass equations for forest regrowth in the eastern Amazon using randomized branch sampling. Acta Amaz. 39:349–360.

Ducey, M. J., R. Astrup, S. Seifert, H. Pretzsch, B. C. Larson, and K. D. Coates. 2013a. Comparison of forest inventory and canopy attributes derived from two terrestrial LIDAR systems. Photogramm. Eng. Remote Sens. 79:245–258.

Ducey, M. J., M. S. Williams, J. H. Gove, S. Roberge, and R. S. Kenning. 2013b. Distance-limited perpendicular distance sampling for coarse woody debris: Theory and field results. Forestry. 86:119–128.

Duda, R. O., P. E. Hart, and D. G. Stork. 2001. Pattern classification. John Wiley & Sons, Inc., New York. 680p.

Duff, G. H., and N. J. Nolan. 1953. Growth and morphogenesis in the Canadian forest species. I. The control of cambial and apical activity in *Pinus resinosa* Ait. Can. J. Bot. 31:471–513.

Duff, G. H., and N. J. Nolan. 1957. Growth and morphogenesis in the Canadian forest species. II. Specific increments and their relation to the quantity and activity of growth in *Pinus resinosa* Ait. Can. J. Bot. 35:527–572.

Duff, G. H., and N. J. Nolan. 1958. Growth and morphogenesis in the Canadian forest species. III. The time scale of morphogenesis at the stem apex of *Pinus resinosa* Ait. Can. J. Bot. 36:687–706.

Eaton, R. J., and F. G. Sanchez. 2009. Quantitative and qualitative measures of decomposition: Is there a link? South. J. Appl. For. 33:137–141.

Eberhardt, L. L. 2003. What should we do about hypothesis testing? J. Wildl. Manag. 67:241–247.

Eddleman, L. E., E. E. Remmenga, and R. T. Ward. 1964. An evaluation of plot methods for alpine vegetation. Bull. Torrey Bot. Club. 91:439–450.

Eichenberger, J. K., G. R. Parker, and T. W. Beers. 1982. A method for ecological forest sampling. Research Bulletin 969, Agriculture Experiment Station, Purdue University, West Lafayette, IN.

Ek, A. R., and R. A. Monserud. 1974. FOREST: A computer model for simulating the growth and reproduction of mixed species forest stands. Report R2635, School of Natural Resources, College of Agriculture and Life Sciences, University of Wisconsin, Madison, WI. 72p.

Elias, M., and C. Potvin. 2003. Assessing inter- and intra-specific variation in trunk carbon concentration for 32 neotropical tree species. Can. J. For. Res. 33:1039–1045.

Ellis, B. D. 1966. Basic concepts of measurement. Cambridge University Press, New York. 219p.

Elzinga, C. L., D. W. Salzer, J. W. Willoughby, and J. P. Gibbs. 2001. Monitoring plant and animal populations: A handbook for field biologists. Blackwell Science, Malden, MA. 372p.

Enquist, B. J., and K. J. Niklas. 2002. Global allocation rules for patterns of biomass partitioning in seed plants. Science 295:1517–1520.

Eskelson, B. N. I., T. M. Barrett, and H. Temesgen. 2009. Imputing mean annual change to estimate current forest attributes. Silva Fenn. 43:649–658.

Evans, A. M., and M. J. Ducey. 2010. Carbon management and accounting for lying dead wood. Forest Guild, Climate Action Reserve, Santa Fe, NM. 75p.

Evans, F. C. 1952. The influences of size of quadrat on the distributional patterns of plant populations. Contrib. Lab. Vertebr. Biol. 54:1–15.

Falkowski, M. J., J. S. Evans, S. Martinuzzi, P. E. Gessler, and A. T. Hudak. 2009. Characterizing forest succession with lidar data: An evaluation for the Inland Northwest, USA. Remote Sens. Environ. 113:946–956.

FAO. 2006. Global forest resources assessment, 2005: Progress towards sustainable forest management. Forestry Paper 147, FAO, Rome, Italy. 320p.

Fastie, C. L. 2010. Estimating stand basal area from forest panoramas. In: Proceedings of the Fine International Conference on Gigapixel Imaging for Science. Carnegie Mellon University, Pittsburg, PA. 8.

Fehrmann, L., and C. Kleinn. 2006. General considerations about the use of allometric equations for biomass estimation on the example of Norway spruce in central Europe. For. Ecol. Manag. 236:412–421.

Fehrmann, L., T. G. Gregoire, and M. J. Ducey. 2012. Triangulation based inclusion probabilities: A design-unbiased sampling approach. Environ. Ecol. Stat. 19:107–123.

Fernandes, J. de F., A. L. T. de Souza, and M. O. Tanaka. 2014. Can the structure of a riparian forest remnant influence stream water quality? A tropical case study. Hydrobiologia 724:175–185.

Ferree, M. J. 1946. The pole caliper. J. For. 44:594–595.

FIA. 2011. Field methods for forest health (Phase 3) measurements. Version 5.1. Available online at: http://www.fia.fs.fed.us/library/field-guides-methods-proc/index.php; Last accessed August 31, 2016.

FIA. 2012. Forest inventory and analysis. National core field guide. Volume I: Field data collection procedures for Phase 2 plots. Version 6.0. USDA Forest Service. 427p.

FIA. 2013. Field guides, methods and procedures. Available online at: http://www.fia.fs.fed.us/library/field-guides-methods-proc/; Last accessed August 31, 2016.

Filho, A. F., S. A. Machado, and M. R. A. Carneiro. 2000. Testing accuracy of log volume calculation procedures against water displacement techniques (xylometer). Can. J. For. Res. 30:990–997.

Flewelling, J. W., and J. L. Strunk. 2013. The walk through and fro estimator for edge bias avoidance. For. Sci. 59:223–230.

Flewelling, J. W., R. L. Ernst, and L. M. Raynes. 2000. Use of three–point taper systems in timber cruising. In: Integrated tools for natural resources inventories in the 21st century. General Technical Report GTR-NC-212, Boise, Id, August 16–20, 1998, North Central Forest Experiment Station, USDA Forest Service. 364–371.

Flury, P. 1897. Berechnung der Holzmasse eines stehenden Bestandes mit dem Massenfaktor V/G [Calculation of the wood mass of a standing inventory with the mass factor V/G]. Schweiz. Forstl. Vers. 5:191–202.

Fogelberg, S. E. 1953. Volume charts based on absolute form class. Louisiana Tech Forestry Club of Louisiana Polytechnical Institute, Ruston, LA.

Fonseca, M. A. 2005. The measurement of roundwood: Methodologies and conversion ratios. CAB International, Boca Raton, FL. 288p.

Ford, E. D. 1975. Competition and stand structure in some even-aged plant monocultures. J. Ecol. 63:311–333.

Ford, E. D. 2000. Scientific method for ecological research. Cambridge University Press, New York. 564p.

Forslund, R. R. 1982. A geometrical tree volume model based on the location of the centre of gravity of the bole. Can. J. For. Res. 12:215–221.

Forward, D. F., and N. J. Nolan. 1961a. Growth and morphogenesis in the Canadian forest species. IV. Radial growth in branches and main axis of *Pinus resinosa* Ait. under conditions of open growth, suppression and release. Can. J. Bot. 39:385–409.

Forward, D. F., and N. J. Nolan. 1961b. Growth and morphogenesis in the Canadian forest species. V. Further studies of wood growth in branches and main axis of *Pinus resinosa* Ait. under conditions of open growth, suppression and release. Can. J. Bot. 39:411–436.

Foster, R. W. 1959. Relation between site indexes of eastern white pine and red maple. For. Sci. 5:279–291.

Francis, J. K. 1986. The relationship of bole diameters and crown widths of seven bottomland hardwood species. Research Note RN-SO-328, Southern Forest Experiment Station, USDA Forest Service. 3p.

Fraver, S., A. Ringvall, and B. G. Jonsson. 2007. Refining volume estimates of down woody debris. Can. J. For. Res. 37:627–633.

Freese, F. 1962. Elementary forest sampling. Agriculture Handbook 232, US Department of Agriculture. 91p.

Freese, F. 1973. A collection of log rules. General Technical Report GTR-FPL-1, Forest Products Laboratory, USDA Forest Service, Madison, WI. 65p.

Freese, F. 1974. Elementary statistical methods for foresters. Agriculture Handbook 317, US Department of Agriculture.

Fritts, H. C. 1976. Tree rings and climate. Academic Press, New York.

Fritts, H. C., and E. C. Fritts. 1955. A new dendrograph for recording radial changes of a tree. For. Sci. 1:271–276.

Fuchs, H., P. Magdon, C. Kleinn, and H. Flessa. 2009. Estimating aboveground carbon in a catchment of the Siberian forest tundra: Combining satellite imagery and field inventory. Remote Sens. Environ. 113:518–531.

Furnival, G. M. 1961. An index for comparing equations used in constructing volume tables. For. Sci. 7:337–341.

Furnival, G. M. 1979. Forest sampling—Past performance and future expectations. In: Forest resource inventories, Workshop Proceeding, Colorado State University, Fort Collins, CO, July 23–26, 1979. 320–326.

Gaiser, R. N., and R. W. Merz. 1951. Stand density as a factor in estimating white oak site index. J. For. 49:572–574.

Garay, L. 1961. An introduction to tarif volume tables. Review Paper 1, Forest Biometrics Research Group, University of Washington.

Garcia, O. 1998. Estimating top height with variable plot sizes. Can. J. For. Res. 28:1509–1517.

Garcia, O. 2006. Scale and spatial structure effects on tree size distributions: implications for growth and yield modelling. Can. J. For. Res. 36:2983–2993.

Garland, H. 1968. Using a poloroid camera to measure trucked hardwood pulpwood. Pulp Pap. Mag. Can. 68:86–87.

Gauch, H. G. 1982. Multivariate analysis in community ecology. Cambridge University Press, Cambridge, UK. 298p.

Gering, L. R., and D. M. May. 1997. Point-sampling of tree crowns using aerial photographs for forest inventory. South. J. Appl. For. 21:28–36.

Gerlach, F. 2000. Characteristics of Space Imaging's one-meter resolution satellite imagery products. Int. Arch. Photogramm. Remote Sens. 33 Part B1:128–135.

Gerrard, D. J. 1969. Competition quotient: A new measure of the competition affecting individual forest trees. Research Bulletin 20, Michigan Agricultural Experiment Station, Michigan State University. 32p.

Gevorkiantz, S. R., and L. P. Olsen. 1955. Composite volume tables for timber and their application in the Lake States. Technical Bulletin 1104, US Department of Agriculture. 51p.

Gingrich, S. F. 1964. Criteria for measuring stocking in forest stands. In: Proceedings of the 1963 Society of American Foresters National Convention, Denver, CO. Society of American Foresters. 198–201.

Gingrich, S. F. 1967. Measuring and evaluating stocking and stand density in upland hardwood forests in the Central States. For. Sci. 13:38–53.

Gingrich, S. F., and H. A. Meyer. 1955. Construction of an aerial stand volume table for upland oak. For. Sci. 1:140–147.

Girard, J. W. 1933. Volume tables for Mississippi bottomland hardwoods and southern pines. J. For. 31:34–41.

Glinski, D. S., K. J. Karnok, and R. N. Carrow. 1993. Comparison of reporting methods for root growth data from transparent-interface measurements. Crop. Sci. 33:310–314.

Glover, G. R., and J. N. Hool. 1979. A basal area ratio predictor of loblolly pine plantation mortality. For. Sci. 25:275–282.

Godman, R. M. 1949. The pole diameter tape. J. For. 47:568–569.

Goff, F. G., G. A. Dawson, and J. J. Rochow. 1982. Site examination for threatened and endangered plant species. Environ. Manage. 6:307–316.

Gonzalez, M., L. Augusto, A. Gallet-Budynek, J. Xue, N. Yauschew-Raguenesd, D. Guyon, P. Trichet, et al. 2013. Contribution of understory species to total ecosystem aboveground and belowground biomass in temperate *Pinus pinaster* Ait. forests. For. Ecol. Manag. 289:38–47.

Goodman, L. A. 1960. On the exact variance of products. J. Am. Stat. Assoc. 55:708–713.

Goreaud, F., B. Loussier, M. A. Ngo Bieng, and R. Allain. 2004. Simulating realistic spatial structure for forest stands : A mimetic point process. Interdisciplinary Spatial Statistics Workshop. December, 2-3, 2004. Paris, France. 22p.

Gotelli, N. J., and R. K. Colwell. 2001. Quantifying biodiversity: Procedures and pitfalls in the measurement and comparison of species richness. Ecol. Lett. 4:379–391.

Gove, J. H. 2003. Moment and maximum likelihood estimators for Weibull distributions under length- and area-biased sampling. Environ. Ecol. Stat. 10:455–467.

Gove, J. H., and S. E. Fairweather. 1989. Maximum-likelihood estimation of Weibull function parameters using a general interactive optimizer and grouped data. For. Ecol. Manag. 28:61–69.

Gove, J. H., and P. C. Van Deusen. 2011. On fixed-area plot sampling for downed coarse woody debris. Forestry. 84:109–117.

Gove, J. H., A. Ringvall, G. Ståhl, and M. J. Ducey. 1999. Point relascope sampling of downed coarse woody debris. Can. J. For. Res. 29:1718–1726.

Gove, J. H., M. J. Ducey, G. Ståhl, and A. Ringvall. 2001. Point relascope sampling: A new way to assess downed coarse woody debris. J. For. 99:4–11.

Gove, J. H., M. J. Ducey, and H. T. Valentine. 2002. Multistage point relascope and randomized branch sampling for downed coarse woody debris estimation. For. Ecol. Manag. 155:153–162.

Gove, J. H., M. S. Williams, G. Ståhl, and M. J. Ducey. 2005. Critical point relascope sampling for unbiased volume estimation of downed coarse woody debris. Forestry. 78:417–431.

Gove, J. H., M. J. Ducey, H. T. Valentine, and M. S. Williams. 2012. A distance limited method for sampling downed coarse woody debris. For. Ecol. Manag. 282:53–62.

Gove, J. H., M. J. Ducey, H. T. Valentine, and M. Williams. 2013. A comprehensive comparison of perpendicular distance sampling methods for sampling downed coarse woody debris. Forestry. 86:129–143.

Graves, H. S. 1906. Forest mensuration. John Wiley & Sons, Inc., New York. 458p.

Green, R. H. 1979. Sampling design and statistical methods for environmental biologists. John Wiley & Sons, Inc., New York. 272p.

Green, R. O., M. L. Eastwood, C. M. Sarture, T. G. Chrien, M. Aronsson, B. J. Chippendale, J. A. Faust, et al. 1998. Imaging spectroscopy and the airborne visible/infrared imaging spectrometer (AVIRIS). Remote Sens. Environ. 65:227–248.

Greenhill, G. 1881. Determination of greatest height consistent with stability that a vertical pole or mast can be made, and the greatest height to which a tree of given proportions can grow. Proc. Camb. Philos. Soc. 4:65–73.

Gregoire, T. G. 1982. The unbiasedness of the mirage correction procedure for boundary overlap. For. Sci. 28:504–508.

Gregoire, T. G. 1998. Design-based and model-based inference in survey sampling: Appreciating the difference. Can. J. For. Res. 28:1429–1447.

Gregoire, T. G., and N. Monkevich. 1994. The reflection method of line intercept sampling to eliminate boundary bias. Environ. Ecol. Stat. 1:219–226.

Gregoire, T. G., and C. T. Scott. 2003. Altered selection probabilities caused by avoiding the edge in field surveys. J. Agric. Biol. Environ. Stat. 8:36–47.

Gregoire, T. G., and H. T. Valentine. 1995. A sampling strategy to estimate the area and perimeter of irregularly shaped planar regions. For. Sci. 41:470–476.

Gregoire, T. G., and H. T. Valentine. 2003. Line intersect sampling: Ell-shaped transects and multiple intersections. Environ. Ecol. Stat. 10:263–279.

Gregoire, T. G., and H. T. Valentine. 2008. Sampling strategies for natural resources and the environment. Chapman Hall/CRC Press. 496p.

Gregoire, T. G., H. T. Valentine, and G. M. Furnival. 1986. Estimation of bole volume by importance sampling. Can. J. For. Res. 16:554–557.

Gregoire, T. G., H. T. Valentine, and G. M. Furnival. 1995. Sampling methods to estimate foliage and other characteristics of individual trees. Ecology. 76:1181–1194.

Gregoire, T. G., G. Ståhl, E. Næsset, T. Gobakken, R. Nelson, and S. Holm. 2011. Model-assisted estimation of biomass in a LiDAR sample survey in Hedmark County, Norway. Can. J. For. Res. 41:83–95.

Greig-Smith, P. 1957. Quantitative plant ecology. Academic Press, New York. 198p.

Grosenbaugh, L. R. 1948. Improved cubic volume estimates. J. For. 46:299–301.

Grosenbaugh, L. R. 1952a. Plotless timber estimates—New, fast, easy. J. For. 50:32–37.

Grosenbaugh, L. R. 1952b. Shortcuts for cruisers and scalers. Occasional Paper OSP-SO-126, Southern Forest Experiment Station, USDA Forest Service. 24p.

Grosenbaugh, L. R. 1954. New tree-measurement concepts: height accumulation, giant tree, taper and shape. Occasional Paper OSP-SO-134, Southern Forest Experiment Station, USDA Forest Service. 25p.

Grosenbaugh, L. R. 1955. Better diagnosis and prescription in southern forest management. Occasional Paper OSP-SO-145, Southern Forest Experiment Station, USDA Forest Service. 27p.

Grosenbaugh, L. R. 1958. Point-sampling and line-sampling: Probability theory, geometric implications, synthesis. Occasional Paper OSP-SO-160, Southern Forest Experiment Station, USDA Forest Service. 34p.

Grosenbaugh, L. R. 1960. Should continuity dominate forest inventories? In: Proceedings of a Short Course in Continuous Inventory Control in Forest Management. Center for Continuing Education, University of Georgia. 74–83.

Grosenbaugh, L. R. 1963a. Optical dendrometers for out-of-reach diameters: A conspectus and some new theory. For. Sci. Monogr. 4:48.

Grosenbaugh, L. R. 1963b. Some suggestions for better sample–tree measurement. In: Proceedings of the 1963 Society of American Foresters National Convention. Denver, CO. Society of American Foresters. 36–42.

Grosenbaugh, L. R. 1965. THREE-PEE SAMPLING THEORY and program "THRP" for computer generation of selection criteria. Research Paper RP-PSW-021, Pacific Southwest Forest Experiment Station, USDA Forest Service. 53p.

Grosenbaugh, L. R. 1967a. STX—Fortran-4 program for estimates of tree populations from 3P sample-tree-measurements. Research Paper RP-PSW-013, Pacific Southwest Forest Experiment Station, USDA Forest Service. 76p.

Grosenbaugh, L. R. 1967b. The gains from sample-tree selection with unequal probabilities. J. For. 65:203–206.

Grosenbaugh, L. R. 1971. STX 1-11-71 for dendrometry of multistage 3-P samples. Report FS-277, USDA Forest Service, Washington, DC. 63p.

Grosenbaugh, L. R. 1974. STX 3-3-73: Tree content and value estimation using various sample designs, dendrometry methods, and V-S-L conversion coefficients. Research Paper RP-SE-117, Southeastern Forest Experiment Station, USDA Forest Service. 112p.

Grosenbaugh, L. R. 1979. 3P sampling theory, examples, and rationale. Technical Note 331, Bureau of Land Management, US Department of Interior. 18p.

Grosenbaugh, L. R. 1980. Avoiding dendrometry bias when trees lean or taper. For. Sci. 26:203–215.

Gunn, J. S., M. J. Ducey, and A. A. Whitman. 2014. Late-successional and old-growth forest carbon temporal dynamics in the Northern Forest (Northeastern USA). For. Ecol. Manag. 312:40–46.

Guttenberg, S., D. Fassnacht, and W. C. Siegel. 1960. Weight–scaling southern pine sawlogs. Occasional Paper OSP-SO-129, Southern Forest Experiment Station, USDA Forest Service. 6p.

Haase, D. L. 2008. Understanding forest seedling quality: Measurements and interpretation. Tree Plant. Notes. 52:24–30.

Hagen, R. L. 1997. In praise of the null hypothesis statistical test. Am. Psychol. 52:15–24.

Hahn, J. T., C. D. MacLean, S. L. Arner, and W. A. Bechtold. 1995. Procedures to handle inventory cluster plots that straddle two or more conditions. For. Sci. Monogr. 31:12–25.

Hamilton, D. A. 1978. Specifying precision in natural resource inventories. In: Proceeding of the Integrated Inventories of Renewable Resources. General Technical Report GTR-RM-55, Rocky Mountain Research Station, USDA Forest Service. 276–281.

Hamilton, G. D. 1975. Forest mensuration handbook. Forestry Commission Booklet 39, Her Majestry's Stationery Office, London, UK. 274p.

Hann, D. W. 1983. Field procedures for measurement of felled trees. Southwest Oregon fundamental FIR growth and yield project. Department of Forest Management, Oregon State University, Corvallis, OR. 35p.

Hann, D. W., and R. K. McKinney. 1975. Stem surface area equations for four tree species of New Mexico and Arizona. Research Note RN-INT-190, Intermountain Forest and Range Experiment Station, USDA Forest Service. 7p.

Hansen, M. C., and T. R. Loveland. 2012. A review of large area monitoring of land cover change using Landsat data. Remote Sens. Environ. 122:66–74.

Hardy, S. S., and G. W. Weiland. 1964. Weight as a basis for the purchase of pulpwood in Maine. Technical Bulletin 14, Maine Agricultural Experiment Station, University of Maine, Orono, ME. 64p.

Harmon, M. E., and J. Sexton. 1996. Guidelines for measurements of woody detritus in forest ecosystems. LTER Publication 20, US LTER Network Office, University of Washington, Seattle, WA. 71p.

Harmon, M. E., J. F. Franklin, F. J. Swanson, P. Sollins, S. V. Gregory, J. D. Lattin, N. H. Anderson, et al. 1986. Ecology of coarse woody debris in temperate ecosystems. Adv. Ecol. Res. 15:133–302.

Harmon, M. E., C. W. Woodall, B. Fasth, and J. Sexton. 2008. Woody detritus density and density reduction factors for tree species in the United States: A synthesis. General Technical Report GTR-NRS-29, Northern Research Station, USDA Forest Service. 84p.

Harmon, M. E., C. W. Woodall, B. Fasth, J. Sexton, and M. Yatkov. 2011. Differences between standing and downed dead tree wood density reduction factors: A comparison across decay classes and tree species. Research Paper RP-NRS-15, Northern Research Station, USDA Forest Service. 40p.

Hart, H. M. J. 1928. Stamtal en dunning: Een oriënterend onderzoek naar de beste plantwijdte en dunningswijze voor den djati. [Number of trees and thinning : An exploratory study on the best planting width and thinning method for the teak] Meded. Van Het Proefstn. Voor Het Boschwezen. 21:1–219.

Hartig, T. 1863. Über die Zeit des Zuwachses der Baumwurzeln. [The growth of tree roots over time] Bot. Ztg. 21:288.

Hassan, H. A., C. Y. Mun, and N. Rahman (Eds.). 1996. Multiple resource inventory and monitoring of tropical forests. Asean Institute of Forest Management, Kuala Lumpur, Malaysia.

Haygreen, J. G., and J. L. Bowyer. 1996. Forest products and wood science. Third. Iowa State University Press, Ames, IA. 484p.

Hazard, J. W., and S. G. Pickford. 1986. Simulation studies on line intersect sampling of forest residue, Part II. For. Sci. 32:447–470.

Healey, S. P., Z. Yang, W. B. Cohen, and D. J. Pierce. 2006. Application of two regression-based methods to estimate the effects of partial harvest on forest structure using Landsat data. Remote Sens. Environ. 101:115–126.

Heath, L. S., M. Hansen, J. E. Smith, P. D. Miles, and B. W. Smith. 2009. Investigation into calculating tree biomass and carbon in the FIADB using a biomass expansion factor approach. In: McWilliams, W., Moisen, G., and Czaplewski, R. (Comps.). Forest Inventory and Analysis (FIA) Symposium 2008, Proceedings RMRS-P-6CD, Rocky Mountain Research Station, USDA Forest Service. 26p.

Heger, L. 1974. Longitudinal variation of specific gravity in stems of black spruce, balsam fir, and lodgepole pine. Can. J. For. Res. 4:321–326.

Hegyi, F. 1974. A simulation model for managing jack–pine stands. In: J. Fries (Ed.) Growth models for tree and stand simulation. Royal College of Forestry, Stockholm, Sweden. 74–90.

Helms, J. A. (Ed.) 1998. The dictionary of forestry. Society of American Foresters, Bethesda, MD. 210p.

Hendricksen, H. A. 1950. Højde–diameter diagram med logaritmisk diameter [Height–diameter diagram with logarithmic diameter]. Dan. Skovforen. Tidsskr. 35:193–202.

Henning, J. G., and P. J. Radtke. 2006a. Detailed stem measurements of standing trees from ground-based scanning lidar. For. Sci. 52:67–80.

Henning, J. G., and P. J. Radtke. 2006b. Ground-based laser imaging for assessing three-dimensional forest canopy structure. Photogramm. Eng. Remote Sens. 72(12):1349–1358.

Henry, J. D., and J. M. A. Swan. 1974. Reconstructing forest history from live and dead plant material—An approach to the study of forest succession in southwest New Hampshire. Ecology. 55:772–783.

Herrick, A. M. 1940. A defense of the Doyle rule. J. For. 38:563–567.

Hester, A. S., D. W. Hann, and D. R. Larsen. 1989. ORGANON: Southwest Oregon growth and yield model user manual. Version 2.0. Forest Research Laboratory, Oregon State University, Corvallis, OR. 59p.

Hicks, C. R., and K. V. Turner Jr. 1999. Fundamental concepts in the design of experiments. Oxford University Press, Oxford, UK. 565p.

Hilker, T., M. van Leeuwen, N. C. Coops, M. A. Wulder, G. Newnham, D. L. B. Jupp, and D. S. Culvenor. 2010. Comparing canopy metrics derived from terrestrial and airborne laser scanning in a Douglas-fir dominated forest stand. Trees Struct. Funct. 24:819–832.

Hinze, W. H. F., and N. O. Wessels. 2002. Stand stability in pines : An important silvicultural criterion for the evaluation of thinnings and the development of thinning regimes. South. Afr. For. J. 196:37–40.

Hirata, T. 1955. Height estimation through Bitterlich's method, vertical angle count sampling. Jpn. J. For. 37:479–480.

Honer, T. G. 1967. Standard volume tables and merchantable conversion factors for the commercial tree species of central and eastern Canada. Information Report FMR-X-5, Canadian Department of Forestry and Rural Development, Forest Management Research and Service Institute. 76p.

Honer, T. G., M. F. Ker, and I. S. Alemdag. 1983. Metric timber tables for the commercial tree species of central and eastern Canada. Information Report M-X-140, Canadian Forestry Service, Maritimes Forest Research Centre. 139p.

Huang, P., and H. Pretzsch. 2010. Using terrestrial laser scanner for estimating leaf areas of individual trees in a conifer forest. Trees Struct. Funct. 24:609–619.

Huang, S., D. Price, and S. J. Titus. 2000. Development of ecoregion-based height–diameter models for white spruce in boreal forests. For. Ecol. Manag. 129:125–141.

Hudak, A. T., N. L. Crookston, J. S. Evans, D. E. Hall, and M. J. Falkowski. 2008. Nearest neighbor imputation of species-level, plot-scale forest structure attributes from LiDAR data. Remote Sens. Environ. 112:2232–2245.

Huebner, C. D. 2007. Detection and monitoring of invasive exotic plants: A comparison of four sampling methods. Northeast. Nat. 14:183–206.

Hummel, F. C. 1953. Uses of the "volume/basal area line" for determining standing crop volumes. Report 1951/52, Forestry Commission, Her Majesty's Stationery Office. London, UK. 80p.

Hummel, F. C. 1955. The volume/basal area line; a study in forest mensuration. Bulletin 24, Forestry Commission, Her Majesty's Stationery Office, London, UK. 84p.

Hurlbert, S. H. 1971. The nonconcept of species diversity: A critique and alternative parameters. Ecology. 52:577–586.

Husch, B. 1947. A comparison between a ground and aerial–photogrammetric method of timber surveying. Unpubl. M.Sc. Thesis, New York State College of Forestry, Syracuse, NY.

Husch, B. 1956. Use of age at dbh as a variable in the site index concept. J. For. 54:340.

Husch, B. 1962. Tree weight relationships for white pine in southeastern New Hampshire. Technical Bulletin 106, Agriculture Experiment Station, University of New Hampshire, Durham, NH. 20p.

Husch, B. 1963. Forest mensuration and statistics. Ronald Press, New York. 474p.

Husch, B. 1971. Planning a forest inventory. Forest Products Studies 17, FAO, Rome, Italy. 120p.

Husch, B. 1980. How to determine what you can afford to spend on inventories. In: IUFRO Workshop on Arid Land Resource Inventories, La Paz, Mexico. General Technical Report WO-28, USDA Forest Service, Washington Office. 98–102.

Husch, B., and W. H. Lyford. 1956. White pine growth and soil relationship in southeastern New Hampshire. Technical Bulletin 95, Agriculture Experiment Station, University of New Hampshire, Durham, NH. 29p.

Husch, B., C. I. Miller, and T. W. Beers. 1983. Forest mensuration. John Wiley & Sons, Inc., New York.

Husch, B., T. W. Beers, and J. A. Kershaw Jr. 2003. Forest mensuration. John Wiley & Sons, Inc., Hoboken. 443p.

Hyder, D. N., R. E. Bement, E. E. Remmenga, and C. Terwilliger Jr. 1965. Frequency sampling of blue grama range. J. Range Manag. 18:90–94.

Hyink, D. M., and J. W. J. Moser. 1983. A generalized framework for projecting forest yield and stand structure using diameter distributions. For. Sci. 29:85–95.

Iles, K. 1979. Some techniques to generalize the use of variable plot and line intersect sampling. In W. E. Frayer (Ed.) Forest resource inventories workshop, Proceedings, Colorado State University, Ft. Collins, CO. 270–278.

Iles, K. 2003. A sampler of inventory topics. Kim Iles and Associates, Nanaimo, BC. 869p.

Iles, K. 2009. "Nearest-tree" estimations—A discussion of their geometry. Math. Comput. For. Nat. Resour. Sci. 1:47–51.

Iles, K., and M. Fall. 1988. Can an angle gauge really evaluate "borderline trees" accurately in variable plot sampling? Can. J. For. Res. 18:776–783.

Imprens, I. I., and J. M. Schalck. 1965. A very sensitive electric dendrograph for recording radial changes of a tree. Ecology. 46:183–184.

Innes, J. C., M. J. Ducey, J. H. Gove, W. B. Leak, and J. P. Barrett. 2005. Size–density metrics, leaf area, and productivity in eastern white pine. Can. J. For. Res. 35:2469–2478.

Irons, J. R., Dwyer, J. L., and Barsi, J. A. 2012. The next Landsat satellite: the Landsat data continuity mission. Remote Sens. Environ. 122:11–21.

IPCC. 1996. Revised 1996 IPCC guidelines for national greenhouse gas inventories: Volume 1. Reporting instructions. Available online at: http://www.ipcc-nggip.iges.or.jp/public/gl/invs1.html; Last accessed August 31, 2016.

IPCC. 2003. Good practice guidance for land use, land-use change, and forestry. IPCC National Greenhouse Gas Inventories Technical Programme, Kyoto, Japan. 549p.

IPCC. 2006. Forest lands. Intergovernmental Panel on Climate Change guidelines for national greenhouse gas inventories. Institute for Global Environmental Strategies, Hayama, Japan, 83p.

Irwin, K. G. 1960. The romance of weights and measures. Viking Juvenile, New York. 144p.

ISRO. 2011. PSLV-C16, RESOURCE-SAT2, YOUTHSAT, X-SAT [Indian Space Research Organization]. Available online at: http://www.isro.gov.in/pslvc16-brochure; Last accessed August 31, 2016.

IUFRO. 1959. The standardization of symbols in forest mensuration. International Union of Forestry Research Organization. 32p.

Jenkins, J. C., D. C. Chojnacky, L. S. Heath, and R. A. Birdsey. 2003. National scale biomass estimators for United States tree species. For. Sci. 49:12–35.

Jenkins, J. C., D. C. Chojnacky, L. S. Heath, and R. A. Birdsey. 2004. Comprehensive database of diameter-based biomass regressions for North American tree species. General Technical Report GTR-NE-319, Northeastern Forest Experiment Station, USDA Forest Service. 45p.

Jennings, S. B., N. D. Brown, and D. Sheil. 1999. Assessing forest canopies and understorey illumination: Canopy closure, canopy cover and other measures. Forestry. 72:59–74.

Jensen, J. R. 2004. Introductory digital image processing: A remote sensing perspective. Prentice Hall, Upper Saddle River, NJ. 544p.

Johnsen, K. H., L. Samuelson, R. O. Teskey, S. McNutty, and T. Fox. 2001. Process models as tools in forestry research and management. For. Sci. 47:2–8.

Johnson, D. H. 1999. The insignificance of statistical significance testing. J. Wildl. Manag. 63:763–772.

Johnson, E. W. 1972. Basic 3–P sampling. Forestry Department Series 5, Agriculture Experiment Station, Auburn University, Auburn, AL. 12p.

Johnson, E. W. 1973. Relationship between point density measurements and the subsequent growth of southern pines. Bulletin 447, Agriculture Experiment Station, Auburn University, Auburn, AL. 109p.

Johnson, E. W. 2001. Forest sampling desk reference. CRC Press, Boca Raton, FL. 1008p.

Johnson, F. A., J. B. Lowrie, and M. Gohlke. 1971. 3P sample log scaling. Research Note RN-PNW-162, Pacific Northwest Forest and Range Experiment Station, USDA Forest Service. 15p.

Johnson, K. M., and W. B. Ouimet. 2014. Rediscovering the lost archaeological landscape of southern New England using airborne light detection and ranging (LiDAR). J. Archaeol. Sci. 43:9–20.

Johnston, C. A. 1998. Geographic information systems in ecology. John Wiley & Sons, Inc., New York. 256p.

Jokela, E. J., and T. A. Martin. 2000. Effects of ontogeny and soil nutrient supply on production, allocation, and leaf area efficiency in loblolly and slash pine stands. Can. J. For. Res. 30:1511–1524.

Jonckheere, I., S. Fleck, K. Nackaerts, B. Muys, P. Coppin, M. Weiss, and F. Baret. 2004. Review of methods for in situ leaf area index determination: Part I. Theories, sensors and hemispherical photography. Agric. For. Meteorol. 121:19–35.

Jonson, T. 1912. Taxatoriska undersökningar öfer skogsträdens form. III. Form,–bestämninga stående träd [Forest mensurational investigations concerning forest tree form. Form determination of standing trees]. Sven. Skogsvårdsföreningens Tidskr. 10:235–275.

Jonsson, B., S. Holm, and H. Kallur. 1992. A forest inventory method based on density–adapted circular plot size. Scand. J. For. Res. 7:405–421.

Jonsson, B. G., N. Kruys, and T. Ranius. 2005. Ecology of species living on dead wood—Lessons for dead wood management. Silva Fenn. 39:289–309.

Jordan, G. J., M. J. Ducey, and J. H. Gove. 2004. Comparing line-intersect, fixed-area, and point relascope sampling for dead and downed coarse woody material in a managed northern hardwood forest. Can. J. For. Res. 34:1766–1775.

Jupp, D. L. B., D. S. Culvenor, J. L. Lovell, and G. Newnham. 2007. Evaluation and validation of canopy laser radar (LIDAR) systems for native and plantation forest inventory. Research Paper 020, CSIRO Marine and Atmospheric Research Center, Canberra, ACT. 150p.

Kahl, T., C. Wirth, M. Mund, G. Böhnisch, and E.-D. Schulze. 2009. Using drill resistance to quantify the density in coarse woody debris of Norway spruce. Eur. J. For. Res. 128:467–473.

Kaiser, L. 1983. Unbiased estimation in line-intercept sampling. Biometrics. 39:965–976.

Kangas, A., and M. Maltamo (Eds.). 2006. Forest inventory—Methodology and applications. Springer, The Hague. 362p.

Kasischke, E. S., J. M. Melack, and M. C. Dobson. 1997. The use of imaging radars for ecological applications—A review. Remote Sens. Environ. 59:141–156.

Keen, F. P. 1943. Ponderosa pine tree classes redefined. J. For. 41:249–253.

Kendall, R. H., and L. Sayn-Wittgenstein. 1959. An evaluation of the relaskop. Technical Note 77, Canadian Forestry Research Division, Department of Northern Affairs and Natural Resources. 26p.

Kenning, R. S. 2007. Field efficiency and bias of several methods for downed wood and snag inventory in western North American forests. Unpubl. M.Sc. For. Thesis, University of New Hampshire, Durham, NH. 104p.

Kenning, R. S., M. J. Ducey, J. C. Brissette, and J. H. Gove. 2005. Field efficiency and bias of snag inventory methods. Can. J. For. Res. 35:2900–2910.

Ker, M. F. 1984. Biomass equations for seven major Maritimes tree species. Information Report M-X-148, Canadian Forestry Service, Maritimes Forest Research Centre. 54p.

Kerns, B. K., and J. L. Ohmann. 2004. Evaluation and prediction of shrub cover in coastal Oregon forests (USA). Ecol. Indic. 4:83–98.

Kerr, G. 2014. The management of silver fir forests: de Liocourt (1898) revisited. Forestry. 87:29–38.

Kershaw, J. A., Jr. 2009. The Art of Kamera Sutra—Measuring trees and plots with a digital camera. Northeastern Mensurationists Organization Annual Meeting, Durham, NH, November 2–3, 2009.

Kershaw, J. A., Jr. 2010. Correcting for visibility bias in photo point samples. Northeastern Mensurationists Organization Annual Meeting, Stockbridge, MA, November 1–2, 2010.

Kershaw, J. A., Jr., and D. R. Larsen. 1992. A rapid technique for recording and measuring the leaf area of conifer needle samples. Tree Physiol. 11:411–417.

Kershaw, J. A., Jr., and D. A. Maguire. 1995. Crown structure in western hemlock, Douglas-fir, and grand fir in western Washington: Trends in branch-level mass and leaf area. Can. J. For. Res. 25:1897–1912.

Kershaw, J. A., Jr., E. W. Richards, and J. Larusic. 2007. A product ratio calculator for northeastern tree species. North. J. Appl. For. 24:307–311.

Kershaw, J. A., Jr., R. C. Morrissey, D. F. Jacobs, J. R. Seifert, and J. B. McCarter. 2008. Dominant height-based height-diameter equations for trees in southern Indiana. In: Jacobs, D. F., and Michler, C. H. (Eds.) Proceedings, 16th Central Hardwood Forest Conference. General Technical Report NRS-P-24, USDA Forest Service, Northern Research Station. 341–355.

Kershaw, J. A., Jr., E. W. Richards, J. B. McCarter, and S. Oborn. 2010. Spatially correlated forest stand structures: A simulation approach using copulas. Comput. Electron. Agric. 74:120–128.

Kershaw, K. A., and J. H. H. Looney. 1985. Quantitative and dynamic plant ecology. Edward Arnold, Baltimore, MD. 282p.

Ketterings, Q. M., R. Coe, M. van Noordwijk, Y. Ambagau, and C. A. Palm. 2001. Reducing uncertainty in the use of allometric biomass equations for predicting above-ground tree biomass in mixed secondary forests. For. Ecol. Manag. 146:199–209.

Kinerson, R. S. 1973. A transducer for investigations of diameter growth. For. Sci. 19:220–232.

Kissa, D. O., and D. Sheil. 2012. Visual detection based distance sampling offers efficient density estimation for distinctive low abundance tropical forest tree species in complex terrain. For. Ecol. Manag. 263:114–121.

Kitamura, M. 1962. On an estimate of the volume of trees in a stand by the sum of critical heights. Kai Nichi Rin Ko. 73:64–67.

Kittredge, J. 1944. Estimation of the amount of foliage of trees and stands. J. For. 42:905–912.

Kleinn, C., and F. Vilčko. 2006. Design-unbiased estimation for point-to-tree distance sampling. Can. J. For. Res. 36:1407–1414.

Kokkila, T., A. Mäkelä, and E. Nikinmaa. 2002. A method for generating stand structures using Gibbs marked point process. Silva Fenn. 36:265–277.

Köpf, E. U. 1976. Prediction of time consumption in logging. Publication 7, IUFRO Division 3, Forest Operations and Techniques, Royal College of Forestry, Garpenberg, Sweden.

Korhonen, L., and J. Heikkinen. 2009. Automated analysis of in situ canopy images for the estimation of forest canopy cover. For. Sci. 55:323–334.

Korhonen, L., K. T. Korhonen, M. Rautiainen, and P. Stenberg. 2006. Estimation of forest canopy cover: A comparison of field measurement techniques. Silva Fenn. 40:577–588.

Kozak, A. 1988. A variable-exponent taper equation. Can. J. For. Res. 18:1363–1368.

Kozak, A. 2004. My last words on taper equations. For. Chron. 80:507–515.

Kozlowski, T. T., P. J. Kramer, and S. G. Pallardy. 1991. The physiological ecology of woody plants. Academic Press, San Diego, CA. 678p.

Kraft, G. 1884. Beiträge zur Lehre von den Durchforstungen, [Contributions to the theory of thinning] Schlagstellungen und Lichtungshieben. Klindworth, Hanover, Germany. 147p.

Krajicek, J. E., K. A. Brinkman, and S. F. Gingrich. 1961. Crown competition—A measure of density. For. Sci. 7:35–42.

Krebs, C. J. 1989. Ecological methodology. First. Harper and Row Publishers, New York. 624p.

Kutner, M., C. Nachtsheim, and J. Neter. 2004. Applied linear regression models. Fourth. McGraw-Hill, New York. 701p.

Kuusela, K. 1979. Sampling of tree stock by angle gauge in proportion to tree characteristics. Commun. Inst. For. Fenn. 95:1–16.

Kwak, D.-A., W.-K. Lee, H.-K. Cho, S.-H. Lee, Y. Son, M. Kafatos, and S.-R. Kim. 2010. Estimating stem volume and biomass of *Pinus koraiensis* using LiDAR data. J. Plant Res. 123:421–432.

Läärä, E. 2009. Statistics: Reasoning on uncertainty, and the insignificance of testing null. Ann. Zool. Fenn. 46:138–157.

Lahiri, D. G. 1951. A method of sample selection providing unbiased ratio estimates. Bull. Inst. Int. Stat. 33:133–140.

Lam, T. Y., and C. Kleinn. 2008. Estimation of tree species richness from large area forest inventory data: Evaluation and comparison of jackknife estimators. For. Ecol. Manag. 255:1002–1010.

Lambert, M.-C., C.-H. Ung, and F. Raulier. 2005. Canadian national tree aboveground biomass equations. Can. J. For. Res. 35:1996–2018.

Lamlom, S. H., and R. A. Savidge. 2003. A reassessment of carbon content in wood: Variation within and between 41 North American species. Biomass Bioenergy. 25:381–388.

Lappi, J., H. Smolander, and A. Kotisaari. 1983. Height relascope for regeneration surveys. Silva Fenn. 17(1):77–82.

Larjavaara, M., and H. C. Muller-Landau. 2010. Comparison of decay classification, knife test, and two penetrometers for estimating wood density of coarse woody debris. Can. J. For. Res. 40:2313–2321.

Larjavaara, M., and H. C. Muller-Landau. 2011. Cross-section mass: An improved basis for woody debris necromass inventory. Silva Fenn. 49:291–298.

Larsen, D. R., and J. A. Kershaw Jr. 1990. The measurement of leaf area. In: J. Lassoie, and T. Hinkley (Eds.) Techniques in forest tree ecophysiology. CRC Press, Boca Raton, FL. 465–475.

Larson, P. R. 1963. Stem form development of forest trees. For. Sci. Monogr. 5:1–42.

Leak, W. B. 1976. Relation of tolerant species to habitat in the White Mountains of New Hampshire. Research Paper RP-NE-351, Northeastern Forest Experiment Station, USDA Forest Service. 10p.

Leak, W. B. 1978. Relationship of species and site index to habitat in the White Mountains of New Hampshire. Research Paper RP-NE-397, Northeastern Forest Experiment Station, USDA Forest Service. 9p.

Leak, W. B. 1982. Habitat mapping and interpretation in New England. Research Paper RP-NE-496, Northeastern Forest Experiment Station, USDA Forest Service. 28p.

Leckie, D. G., F. A. Gougeon, S. Tinis, T. Nelson, C. N. Burnett, and D. Paradine. 2005. Automated tree recognition in old growth conifer stands with high resolution digital imagery. Remote Sens. Environ. 94:311–326.

Lefsky, M. A. 2010. A global forest canopy height map from the Moderate Resolution Imaging Spectroradiometer and the Geoscience Laser Altimeter System. Geophys. Res. Lett. 37:L15401.

Lefsky, M. A., W. B. Cohen, D. J. Harding, G. G. Parker, S. A. Acker, and S. T. Gower. 2002. Lidar remote sensing of above-ground biomass in three biomes. Glob. Ecol. Biogeogr. 11:393–399.

Leick, A. 2003. GPS satellite surveying. John Wiley & Sons, Inc., Hoboken. 464p.

Lemmon, P. E. 1956. A spherical densiometer for estimating forest overstory density. For. Sci. 2:314–320.

Lemmon, P. E. 1957. A new instrument for measuring over-story density. J. For. 55:667–669.

Lessard, V., D. D. Reed, and N. Monkevich. 1994. Comparing N-tree distance sampling with point and plot sampling in northern Michigan forest types. North. J. Appl. For. 11:12–16.

Levy, E. B., and E. A. Madden. 1933. The point method of pasture analysis. N. Z. J. Agric. 46:267–279.

Lexen, B. 1941. The application of sampling to log scaling. J. For. 39:624–631.

Li, B. H., J. Zhang, J. Ye, X. G. Wang, and Z. Q. Hao. 2008. Ying Yong Sheng Tai Xue Bao Chin [Seasonal dynamics and spatial distribution patterns of herbs diversity in broadleaved Korean pine (*Pinus koraiensis*) mixed forest in Changbai Mountains]. J. Appl. Ecol. 19:467–473.

Lightner, G., H. T. Schreuder, B. Bollenbacher, and K. McMenus. 2001. Integrated inventory and monitoring. In: S. J. Barras (Ed.) Proceedings: National Silvicultural Workshop, P-RMRS-019, Kalispell, MT, October 5–7, 1999, Rocky Mountain Research Station, USDA Forest Service. 78–83.

Liknes, G. C., R. S. Morin, and C. D. Canham. 2013. Trend analyses and projections using national forest inventory data. Math. Comput. For. Nat. Resour. Sci. 5:112–114.

Lim, K., P. Treitz, M. A. Wulder, B. St-Onge, and M. Flood. 2003. LiDAR remote sensing of forest structure. Prog. Phys. Geogr. 27:88–106.

Liming, F. G. 1957. Homemade dendrometers. J. For. 55:575–577.

Lindberg, E., J. Holmgren, K. Olofsson, J. Wallerman, and H. Olsson. 2010. Estimation of tree lists from airborne laser scanning by combining single-tree and area-based methods. Int. J. Remote Sens. 31:1175–1192.

Lindsay, S. R., G. R. Wood, and R. C. Wollons. 1996. Stand table modelling through the Weibull distribution and usage of skewness information. For. Ecol. Manag. 81:19–23.

Loetsch, F., and K. A. Haller. 1964. Forest inventory. Vol. I. BLV Verlagsgesellschaft, Munich, Germany. 436p.

Loetsch, F., F. Zöhrer, and K. A. Haller. 1973. Forest inventory. Vol. II. BLV Verlagsgesellschaft, Munich, Germany. 469p.

Loewenstein, E. F. 1996. An analysis of the size– and age–structure of an uneven–aged oak forest. Unpubl. PhD Dissertation, University of Missouri, Columbia, MO. 167p.

Loewenstein, E. F., P. S. Johnson, and H. E. Garrett. 2000. Age and diameter structure of a managed uneven-aged oak forest. Can. J. For. Res. 30:1060–1070.

Long, J. N. 1985. A practical approach to density management. For. Chron. 61:23–27.

Long, J. N., and T. W. Daniel. 1990. Assessment of growing stock in uneven-aged stands. West. J. Appl. For. 5:93–96.

Long, J. N., and F. W. Smith. 1988. Leaf area—Sapwood area relations of lodgepole pine as influenced by stand density and site index. Can. J. For. Res. 18:247–250.

Long, J. N., and F. W. Smith. 1992. Volume increment in *Pinus contorta* var. *latifolia*: The influence of stand development and crown dynamics. For. Ecol. Manag. 53:53–64.

Longley, P. A., M. F. Goodchild, D. J. Maguire, and D. W. Rhind. 2011. Geographic information systems and science. John Wiley & Sons, Inc., Hoboken. 560p.

Lorimer, C. G. 1985. Methodological considerations in the analysis of forest disturbance history. Can. J. For. Res. 15:200–213.

Lovell, J. L., D. L. B. Jupp, D. S. Culvenor, and N. C. Coops. 2003. Using airborne and ground-based ranging lidar to measure canopy structure in Australian forests. Can. J. Remote Sens. 29:607–622.

Lu, D. 2006. The potential and challenge of remote sensing-based biomass estimation. Int. J. Remote Sens. 27:1297–1328.

Luhmann, T., S. Robson, S. Kyle, and I. Hartley. 2007. Close range photogrammetry: Principles, techniques and applications. John Wiley & Sons, Inc., Hoboken. 528p.

Luhmann, T., S. Robson, S. Kyle, and J. Boehm. 2013. Close range photogrammetry: 3D imaging techniques. Walter De Gruyter, Berlin, Germany. 684p.

Lynch, T. B., and J. H. Gove. 2013. An antithetic variate to facilitate upper-stem height measurements for critical height sampling with importance sampling. Can. J. For. Res. 43:1151–1161.

Lynch, T. B., and R. Rusydi. 1999. Distance sampling for forest inventory in Indonesian teak plantations. For. Ecol. Manag. 113:215–221.

Lyr, H., and G. Hoffmann. 1967. Growth rates and growth periodicity of tree roots. Int. Rev. For. Res. 2:181–236.

Maass, A. 1939. Tallens formbedömd av diametern 2.3. meter fran mårken [Stem form of pine as determined by diameter 2.3 meters above ground]. Sven. Skogsvårdsföreningens Tidskr. 37:120–140.

MacArthur, R. H., and H. S. Horn. 1969. Foliage profile by vertical measurements. Ecology. 50:802–804.

MacArthur, R. H., and E. O. Wilson. 1967. The theory of island biogeography. Princeton University Press, Princeton, NJ. 224p.

MacDicken, K. G. 1997. A guide to monitoring carbon storage in forestry and agro forestry projects. Winrock International Institute for Agricultural Development, Arlington, VA. 87p.

MacKinney, A. L., and L. E. Chaiken. 1939. Volume, yield and growth of loblolly pine in the Mid–Atlantic Region. Technical Paper 33, Appalachian Forest Experiment Station, USDA Forest Service. 30p.

MacKinney, A. L., F. X. Schumacher, and L. E. Chaiken. 1937. Construction of yield tables for nonnormal loblolly pine stands. J. Agric. Res. 54:531–545.

MacLean, M. G., M. J. Campbell, D. S. Maynard, M. J. Ducey, and R. G. Congalton. 2013. Requirements for labelling forest polygons in an object-based image analysis classification. Int. J. Remote Sens. 34:2531–2547.

MacLean, R. G., M. J. Ducey, and C. M. Hoover. 2014. A comparison of carbon stock estimates and projections for the northeastern United States. For. Sci. 60:206–213.

MacLeod, M. N. 1919. Mapping from air photographs. Geogr. J. 53:382–396.

MacLeod, W. K., and A. W. Blyth. 1955. Yield of even–aged, fully–stocked spruce–poplar stands in northern Alberta. Technical Note 18, Department of Northern Affairs and National Resources. 33p.

Magnussen, S. 1999. Effect of plot size on top height in Douglas-fir. West. J. Appl. For. 14:17–27.

Magnussen, S. 2015. Arguments for a model-dependent inference? Forestry. 88:317–325.

Magnussen, S., C. Kleinn, and N. Picard. 2008. Two new density estimators for distance sampling. Eur. J. For. Res. 127:213–224.

Magnussen, S., B. Smith, C. Kleinn, and L. F. Sun. 2010. An urn model for species richness estimation in quadrat sampling from fixed-area populations. Forestry. 83:293–306.

Magnussen, S., L. Fehrmann, and W. J. Platt. 2012. An adaptive composite density estimator for k-tree sampling. Eur. J. For. Res. 131:307–320.

Maguire, D. A., and J. L. F. Batista. 1996. Sapwood taper models and implied sapwood volume and foliage profiles for coastal Douglas-fir. Can. J. For. Res. 26:849–863.

Maguire, D. A., and D. W. Hann. 1989. The relationship between gross crown dimensions and sapwood area at crown base in Douglas-fir. Can. J. For. Res. 19:557–565.

Maguire, D. A., and D. W. Hann. 1990. Bark thickness and bark volume in southwestern Oregon Douglas-fir. West. J. Appl. For. 5:5–8.

Magurran, A. E. 2003. Measuring biological diversity. John Wiley & Sons, Inc., Hoboken. 264p.

Mäkelä, A. 1992. Process-oriented growth and yield models: Recent advances and future prospects. In: T. Preuhsler (Ed.) Research on growth and yield with emphasis on mixed stands. Proceedings of S4.01 Mensuration, Growth and Yield, Bayerische Forsrliche Versuchsund Foreschungsanstalt, Freising, Germany. 85–96.

Mäkelä, A., and H. T. Valentine. 2006. Crown ratio influences allometric scaling in trees. Ecology. 87:2967–2972.

Mäkipää, R., and T. Linkosalo. 2011. A non-destructive field method for measuring wood density of decaying logs. Silva Fenn. 45:1135–1142.

Mandallaz, D. 2008. Sampling techniques for forest inventories. Chapman Hall, Boca Raton, FL. 272p.

Marden, R. M., D. C. Lothner, and E. Kallio. 1975. Wood and bark percentage and moisture contents of Minnesota pulpwood species. Research Paper RP-NC-114, North Central Forest Experiment Station, USDA Forest Service. 9p.

Marquis, D. A., and T. W. Beers. 1969. A further definition of some forest growth components. J. For. 67:493.

Marshall, D. D., K. Iles, and J. F. Bell. 2004. Using a large-angle gauge to select trees for measurement in variable plot sampling. Can. J. For. Res. 34:840–845.

Marshall, J. D., and R. H. Waring. 1986. Comparison of methods of estimating leaf-area index in old-growth Douglas-fir. Ecology. 67:975–979.

Marshall, P. L., G. Davis, and V. M. LeMay. 2000. Using line intersect sampling for coarse woody debris. Forest Research Technical Report TR-003, Vancouver Forest Region, BC Ministry of Forests. 34p.

Martin, A. J. 1984. Testing volume equation accuracy with water displacement techniques. For. Sci. 30:41–50.

Martin, G. L. 1982. A method for estimating ingrowth on permanent horizontal sample points. For. Sci. 28:110–114.

Martin, G. L., and A. R. Ek. 1984. A comparison of competition measures and growth models for predicting plantation red pine diameter and height growth. For. Sci. 30:731–743.

Martin, M. E., and J. D. Aber. 1997. High spectral resolution remote sensing of forest canopy lignin, nitrogen, and ecosystem processes. Ecol. Appl. 7:431–443.

Martin, M. E., S. D. Newman, J. D. Aber, and R. G. Congalton. 1998. Determining forest species composition using high spectral resolution remote sensing data. Remote Sens. Environ. 65:249–254.

Martínez-Ramos, M., and E. R. Alvarez-Buylla. 1998. How old are tropical rain forest trees? Trends Plant Sci. 3:400–405.

Matérn, B. 1956. On the geometry of the cross-section of a stem. Medd. Från Statens Skogforsk. Inst. 46:1–28.

Matérn, B. 1990. On the shape of the cross-section of a tree stem: An empirical study of the geometry of mensurational methods. Report 28, Swedish University of Agriculture, Section of Forest Biometry, Umeå, Sweden. 46p.

Matney, T. G., and A. D. Sullivan. 1982. Variable top volume and height predictors for slash pine trees. For. Sci. 28:274–282.

Mawson, J. C. 1982. Diameter growth estimation on small forests. J. For. 80:217–219.

Max, T. A., and H. E. Burkhart. 1976. Segmented polynomial regression applied to taper equations. For. Sci. 22:283–289.

Max, T. A., H. T. Schreuder, J. W. Hazard, J. Tepley, and J. Alegria. 1996. The region 6 vegetation and monitoring system. Research Paper RP-PNW-493, Pacific Northwest Forest and Range Experiment Station, USDA Forest Service. 22p.

Maynard, D. S., M. J. Ducey, R. G. Congalton, and J. Hartter. 2013. Modeling forest canopy structure and density by combining point quadrat sampling and survival analysis. For. Sci. 59:681–692.

Maynard, D. S., M. J. Ducey, R. G. Congalton, J. A. Kershaw Jr., and J. Hartter. 2014. Vertical point sampling with a digital camera: Slope correction and field evaluation. Comput. Electron. Agric. 100:131–138.

McCarter, J. B., and J. N. Long. 1986. A lodgepole pine density management diagram. West. J. Appl. For. 1:6–11.

McClure, J. P., and R. L. Czaplewski. 1986. Compatible taper equation for loblolly pine. Can. J. For. Res. 16:1272–1277.

McCormac, J. C., W. Sarasua, and W. Davis. 2012. Surveying. John Wiley & Sons, Inc., Hoboken. 400p.

McGarrigle, E., J. A. Kershaw Jr., M. B. Lavigne, A. R. Weiskittel, and M. J. Ducey. 2011. Predicting the number of trees in small diameter classes using predictions from a two-parameter Weibull distribution. Forestry. 84:431–439.

McLintock, T. F., and C. A. Bickford. 1957. A proposed site index for red spruce in the Northeast. Station Paper SP-NE-93, Northeastern Forest Experiment Station, USDA Forest Service. 30p.

McRoberts, R. E. 2001. Imputation and model-based updating techniques for annual forest inventories. For. Sci. 47:322–330.

McRoberts, R. E., and E. O. Tomppo. 2007. Remote sensing support for national forest inventories. Remote Sens. Environ. 110:412–419.

McTague, J.-P. 1988. Estimation of stand density with probability proportional to size from aerial photography. West. J. Appl. For. 3:89–92.

McTague, J.-P., and R. L. Bailey. 1985. Critical height sampling for stand volume estimation. For. Sci. 31:899–911.

McTague, J.-P., and D. R. Patton. 1989. Stand density index and its application in describing wildlife habitat. Wildl. Soc. Bull. 17:58–62.

Mesavage, C. 1965. Three–P sampling and dendrometry for better timber estimating. South. Lumberm. 211:107–109.

Mesavage, C. 1967. Random integer dispenser. Research Note RN-S0-49, Southern Forest Experiment Station, USDA Forest Service. 3p.

Mesavage, C. 1971. STX timber estimating with 3P sampling and dendrometry. Handbook 415, US Department of Agriculture. 135p.

Mesavage, C., and J. W. Girard. 1946. Tables for estimating board-foot volume of timber. USDA Forest Service, Washington, DC. 94p.

Mesavage, C., and L. R. Grosenbaugh. 1956. Efficiency of several cruising designs on small tracts in north Arkansas. J. For. 54:569–576.

Meyer, H. A. 1953. Forest mensuration. Penn's Valley Publishers, State College, PA. 357p.

Meyer, V., S. S. Saatchi, J. Chave, J. Dalling, S. Bohlman, G. A. Fricker, C. Robinson, and et al. 2013. Detecting tropical forest biomass dynamics from repeated airborne Lidar measurements. Biogeosciences. 10:5421–5438.

Meyer, W. H. 1934. Growth in selectively cut ponderosa pine forests of the Pacific Northwest. Technical Bulletin 407, US Department of Agriculture, Washington, DC. 59p.

Miles, P. D., and W. B. Smith. 2009. Specific gravity and other properties of wood and bark for 156 tree species found in North America. Research Note RN-NRS-38, Northern Research Station, USDA Forest Service. 35p.

Miller, C. I. 1959. Comparison of Newton's, Smalian's, and Huber's formulas. Unpublished Manuscript, Department of Forestry and Conservation, Purdue University.

Miller, C. I., and T. W. Beers. 1975. Thin prisms as angle gauges in forest inventory. Research Bulletin 929, Agriculture Experiment Station, Purdue University, West Lafayette, IN. 8p.

Miner, C. L., N. R. Walters, and M. L. Belli. 1988. A guide to the TWIGS program for the North Central United States. General Technical Report GTR-NC-125, North Central Forest Experiment Station, USDA Forest Service. 105p.

Minor, C. O. 1951. Stem-crown diameter relations in southern pine. J. For. 49:490–493.

Mitchell, K. J. 1975. Dynamics and simulated yield of Douglas-fir. For. Sci. Monogr. 17:1–39.

Miyata, E. S., H. M. Steinhilb, and S. A. Winsauer. 1981. Using work sampling to analyze logging operations. Research Paper RP-NC-213, North Central Forest Experiment Station, USDA Forest Service. 8p.

Monleon, V. J. 2009. An assessment of the impact of FIA's default assumptions on the estimates of coarse woody debris volume and biomass. In: McWilliams, W., Moisen, G., Czaplewski, R. (Comps.) Forest Inventory and Analysis (FIA) Symposium 2008. Proceedings RMRS-P-56CD, Rocky Mountain Research Station, USDA Forest Service. 8p.

Monserud, R. A. 2003. Evaluating forest models in a sustainable forest management context. For. Biometry Model. Inf. Sci. 1:35–47.

Monserud, R. A., and J. D. Marshall. 1999. Allometric crown relations in three northern Idaho conifer species. Can. J. For. Res. 29:521–535.

Moody, A., and C. E. Woodcock. 1995. The influence of scale and the spatial characteristics of landscapes on land-cover mapping using remote sensing. Landsc. Ecol. 10:363–379.

Moore, J. A., C. A. Budelsky, and R. C. Schlesinger. 1973. A new index representing individual tree competitive status. Can. J. For. Res. 3:495–500.

Moore, M. M., and D. A. Deiter. 1992. Stand density index as a predictor of forage production in northern Arizona pine forests. J. Range Manag. 45:267–271.

Moore, P. G. 1954. Spacing in plant populations. Ecology 35:222–227.

Morrison, M. L., B. G. Marcot, and R. W. Mannan. 1992. Wildlife–habitat relationships: Concepts and applications. University of Wisconsin Press, Madison, WI. 343p.

Morsdorf, F., E. Meier, B. Kötz, K. I. Itten, M. Dobbertin, and B. Allgöwer. 2004. LIDAR-based geometric reconstruction of boreal type forest stands at single tree level for forest and wildland fire management. Remote Sens. Environ. 92:353–362.

Moser, J. W. J. 1976. Specification of density for the inverse J-shaped diameter distribution. For. Sci. 22:177–180.

Moser, J. W. J., and O. F. Hall. 1969. Deriving growth and yield functions for uneven-aged forest stands. For. Sci. 15:183–188.

Motulsky, H., and A. Christopoulos. 2004. Fitting models to biological data using linear and nonlinear regression: A practical guide to curve fitting. Oxford University Press, New York. 352p.

Muhairwe, C. K. 2000. Bark thickness equations for five commercial tree species in regrowth forests of northern New South Wales. Aust. For. 63:34–43.

Mullahy, J. 1986. Specification and testing of some modified count data models. J. Econom. 33:341–365.

Muller-Landau, H. C., R. S. Condit, J. Chave, S. C. Thomas, S. A. Bohlman, S. Bunyavejchewin, S. Davies, et al. 2006. Testing metabolic ecology theory for allometric scaling of tree size, growth and mortality in tropical forests. Ecol. Lett. 9:575–588.

Myers, C. C., and T. W. Beers. 1968. Point sampling for forest growth estimation. J. For. 66:927–929.

Næsset, E. 1999. Relationship between relative wood density of *Picea abies* logs and simple classification systems of decayed coarse woody debris. Scand. J. For. Res. 14:454–461.

Næsset, E. 2002. Predicting forest stand characteristics with airborne scanning laser using a practical two-stage procedure and field data. Remote Sens. Environ. 80:88–99.

NASA. MODIS: Moderate Resolution Imaging Spectroradiometer. National Atmospheric and Space Administration. Available online at: http://modis.gsfc.nasa.gov/about/; Last accessed August 31, 2016.

Nash, A. J. 1948. The Nash scale for measuring tree crown widths. For. Chron. 24:117–120.

Newberry, J. D., and L. V. Pienaar. 1978. Dominant height growth models and site index curves for site–prepared slash pine plantations in the lower coastal plain of Georgia and North Florida. Research Paper 4, Plantation Management Research Cooperative, University of Georgia, Athens, GA.

Niebel, B. W. 1992. Motion and time study. McGraw-Hill, New York. 880p.

Niinimäki, S., O. Tahvonen, A. Mäkelä, and T. Linkosalo. 2013. On the economics of Norway spruce stands and carbon storage. Can. J. For. Res. 43:637–648.

NOAA. 2013. NGDC geomagnetic calculators. Available online at: https://www.ngdc.noaa.gov/geomag/WMM/DoDWMM.shtml; Last accessed September 1, 2016.

Noone, C. S., and J. F. Bell. 1980. An evaluation of eight intertree competition indices. Research Note 66, Forest Research Laboratory, Oregon State University, Corvallis, OR.

Nordén, B., M. Ryberg, F. Götmark, and B. Olausson. 2004. Relative importance of coarse and fine woody debris for the diversity of wood-inhabiting fungi in temperate broadleaf forests. Biol. Conserv. 117(1):1–10.

Norman, E. L., and J. W. Curlin. 1968. A linear programming model for forest production control at the AEC Oak Ridge Reservation. Report ORNL-4349, Oak Ridge National Laboratory. 48p.

Nylinder, P. 1958. Variations in weight of barked spruce pulpwood. Uppsats 15, Institutioner Virkeslära K. Skogshögsk, Swedish University of Agricultural Sciences, Stockholm, Sweden. 23p.

Nylinder, P. 1967. Weight measurement of pulpwood. Rapport R57, Institutioner Virkeslära K. Skogshögsk, Swedish University of Agricultural Sciences, Stockholm, Sweden.

Nyyssönen, A. 1955. On the estimation of the growing stock from aerial photographs. Commun. Inst. For. Fenn. 46:1–57.

Oderwald, R. G. 1981. Point and plot sampling—The relationship. J. For. 79:377–378.

Oderwald, R. G., and B. A. Boucher. 1997. Where in the world and what? An introduction to global positioning systems. Kendall/Hunt Publishing Company, Dubuque, IA.

O'Hara, K. L., and N. I. Valappil. 1995. Sapwood–leaf area prediction equations for multi-aged ponderosa pine stands in western Montana and central Oregon. Can. J. For. Res. 25:1553–1557.

Ohmann, J. L., and M. J. Gregory. 2002. Predictive mapping of forest composition and structure with direct gradient analysis and nearest-neighbor imputation in coastal Oregon, U.S.A. Can. J. For. Res. 32:725–741.

Oliver, C. D. 1982. Stand development—Its uses and methods of study. In: Maans, J. E. (Ed.) Forest succession and stand development research in the Pacific Northwest. Forest Research Laboratory, Oregon State University, Corvallis, OR. 100–112.

Oliver, C. D., and B. C. Larson. 1996. Forest stand dynamics. John Wiley & Sons, Inc., New York. 544p.

Oliver, C. D., and E. P. Stephens. 1977. Reconstruction of a mixed-species forest in central New England. Ecology. 58:562–572.

Olson, C. M., and R. E. Martin. 1981. Estimating biomass of shrubs and forbs in central Washington Douglas-fir stands. Research Note PNW-RN-380, Pacific Northwest Forest and Range Experiment Station, USDA Forest Service. 6p.

Olsthoorn, A. F. M., H. H. Bartelink, J. J. Gardiner, H. Pretzsch, H. J. Hekhuis, and A. Franc. 1999. Management of mixed-species forest: Silviculture and economics. IBN Scientific Contributions, Wageningen, The Netherlands. 389p.

Ondok, J. P. 1984. Simulation of stand geometry in photosynthetic models based on hemispherical photographs. Photosynthetica. 18:231–239.

Opie, J. E. 1968. Predictability of individual tree growth using various definitions of competing basal area. For. Sci. 14:314–323.

Ore, O. 1988. Number theory and its history. Reprint Edition. Dover Publications, New York. 400p.

O'Regan, W. G., and L. G. Arvanitis. 1966. Cost-effectiveness in forest sampling. For. Sci. 12:406–414.

Oren, R., R. H. Waring, S. G. Stafford, and J. W. Barrett. 1987. Twenty-four years of ponderosa pine growth in relation to canopy leaf area and understory competition. For. Sci. 33:538–547.

Ormerod, D. W. 1973. A simple bole model. For. Chron. 49:136–138.

Osborne, J. G. 1942. Sampling errors of systematic and random surveys of cover-type areas. J. Am. Stat. Assoc. 37:256–264.

Osborne, J. G., and F. X. Schumacher. 1935. The construction of normal-yield and stand tables for even-aged timber stands. J. Agric. Res. 51:547–564.

Özçelik, R., H. V. Wiant, and J. R. Brooks. 2008. Accuracy using xylometry of log volume estimates for two tree species in Turkey. Scand. J. For. Res. 23:272–277.

Paine, D. P., and J. D. Kiser. 2012. Aerial photography and image interpretation. John Wiley & Sons, Inc., Hoboken. 648p.

Palace, M., M. Keller, G. P. Asner, S. Hagen, and B. Braswell. 2008. Amazon forest structure from IKONOS satellite data and the automated characterization of forest canopy properties. Biotropica. 40:141–150.

Paletto, A., and V. Tosi. 2010. Deadwood density variation with decay class in seven tree species of the Italian Alps. Scand. J. For. Res. 25:164–173.

Pardé, J. 1955. Le mouvement forestier à l'étranger. Un dendromètre pratique: le Dendromètre Blume-Leiss [The forest movement abroad. A practical dendrometer: The dendrometer Blume-Leiss]. Rev. For. Fr. 7:207–210.

Parresol, B. R. 1999. Assessing tree and stand biomass: A review with examples and critical comparisons. For. Sci. 45:573–593.

Parresol, B. R. 2001. Additivity of nonlinear biomass equations. Can. J. For. Res. 31:865–878.

Parsons, H. H. 1930. Aerial forest surveying. For. Chron. 6:56–60.

Paton, D., J. Nuñez, D. Bao, and A. Muñoz. 2002. Forage biomass of 22 shrub species from Monfragüe Natural Park (SW Spain) assessed by log–log regression models. J. Arid Environ. 52:223–231.

Payandeh, B., and A. R. Ek. 1986. Distance methods and density estimators. Can. J. For. Res. 16:918–924.

Pearson, T. R. H., S. L. Brown, and R. A. Birdsey. 2007. Measurement guidelines for the sequestration of forest carbon. General Technical Report GTR-NRS-18, Northern Research Station, US Department of Agriculture. 42p.

Peet, F. G., D. J. Morrison, and K. W. Pellow. 1997. Using a hand-held electronic laser-based survey instrument for stem mapping. Can. J. For. Res. 27:2104–2108.

Peet, R. K. 1974. The measurement of species diversity. Annu. Rev. Ecol. Syst. 5:285–307.

Pesonen, A., O. Leino, M. Maltamo, and A. Kangas. 2009. Comparison of field sampling methods for assessing coarse woody debris and use of airborne laser scanning as auxiliary information. For. Ecol. Manag. 257:1532–1541.

Peterson, D. L., M. A. Spanner, S. W. Running, and K. B. Teuber. 1987. Relationship of thematic mapper simulator data to leaf area index of temperate coniferous forests. Remote Sens. Environ. 22:323–341.

Petterson, H. 1955. Yield of coniferous forests. Meddelelse Statens Skogsförsöksanst 45:1B.

Phipps, R. L., and G. E. Gilbert. 1960. An electric dendrograph. Ecology. 41:389–390.

Pickford, S. G., and J. W. Hazard. 1978. Simulation studies on line intersect sampling of forest residue. For. Sci. 24:469–483.

Pielou, E. C. 1966. The measurement of diversity in different types of biological collections. J. Theor. Biol. 13:131–144.

Pielou, E. C. 1969. An introduction to mathematical ecology. John Wiley & Sons, Inc., New York. 286p.

Pielou, E. C. 1977. Mathematical ecology. John Wiley & Sons, Inc., New York.

Pienaar, L. V., and K. J. Turnbull. 1973. The Chapman-Richards generalization of von Bertalanffy's growth model for basal area growth and yield in even—Aged stands. For. Sci. 19:2–22.

Pierce, L. L., and S. W. Running. 1988. Rapid estimation of coniferous forest leaf area index using a portable integrating radiometer. Ecology. 69:1762–1767.

Pinheiro, J. C., and D. M. Bates. 2000. Mixed effects models in S and S-Plus. Springer, New York. 528p.

Pirotti, F. 2011. Analysis of full-waveform LiDAR data for forestry applications: A review of investigations and methods. IForest J. Biogeosciences For. 4:100–106.

Pitt, M. D., and F. E. Schwab. 1988. Quantitative determination of shrub biomass and production: A problem analysis. Land Management Report 54, B.C. Ministry of Forests, Research Branch. 68p.

Pojar, J., K. Klinka, and D. V. Meidinger. 1987. Biogeoclimatic ecosystem classification in British Columbia. For. Ecol. Manag. 22:119–154.

Popescu, S. C., R. H. Wynne, and R. F. Nelson. 2003. Measuring individual tree crown diameter with lidar and assessing its influence on estimating forest volume and biomass. Can. J. Remote Sens. 29:564–577.

Poulos, H. M. 2009. Mapping fuels in the Chihuahuan Desert borderlands using remote sensing, geographic information systems, and biophysical modeling. Can. J. For. Res. 39:1917–1927.

Pressler, M. R. 1860. Aus der Holzzuwachs Lehre. [The theory of timber growth] *Allg. Forst- Jagd Ztg.* 173–191.

Prodan, M. 1965. Holzmesslehre. Sauerlaender's Verlag, Frankfurt, Germany.

Prodan, M. 1968a. Forest biometrics (Gardiner, S.H., Trans.). Pergamon Press, New York.

Prodan, M. 1968b. Punkstichprobe für die Forsteinrichtung [Random point sampling for forest inventory]. Forst- Holzwirtsch. 23:225–226.

Prodan, M., R. Peters, F. Cox, and P. Real. 1997. Mensura forestal. Deutsche Gesellschaft für Zusammenarbeit (GTZ) FmbH: Instituto Interamericano de Cooperación para la Agricultura (IICA), San José, Costa Rica.

Pyle, C., and M. M. Brown. 1999. Heterogeneity of wood decay classes within hardwood logs. For. Ecol. Manag. 114:253–259.

QGIS. 2013. QGIS: A free and open source geographic information system. Available online at: http://www.qgis.org/en/site/; Last accessed September 1, 2016.

R Development Core Team. 2013. R: A language and environment for statistical computing. R Foundation for Statistical Computing, Vienna, Austria. Available online at: https://cran.r-project.org/; Last accessed September 1, 2016.

Radtke, P. J., and P. V. Bolstad. 2001. Laser point-quadrat sampling for estimating foliage-height profiles in broad-leaved forests. Can. J. For. Res. 31:410–418.

Raspopov, I. M. 1955. K metodike izucenija proekcii kron derevjev [A method of studying the crown projection of trees]. Bot. Ztg. 40:825–827.

Raunkiaer, C. 1934. The life forms of plants and statistical plant geography. Oxford University Press, Oxford, UK. 632p.

Ravindranath, N. H., and M. Ostwald. 2008. Carbon inventory methods: Handbook for greenhouse gas inventory, carbon mitigation and roundwood production projects. Springer, New York. 304p.

Rebain, S. 2010. The fire and fuels extension to the forest vegetation simulator: Updated model documentation (revised March 20, 2012). Forest Management Service Center, USDA Forest Service.

Reed, D. D., and G. D. Mroz. 1997. Resource assessment in forested landscapes. John Wiley & Sons, Inc., New York. 386p.

Reineke, L. H. 1927. A modification of Bruce's method of preparing timber yield table. J. Agric. Res. 35:843–856.

Reineke, L. H. 1932. A precision dendrometer. J. For. 30:692–699.

Reineke, L. H. 1933. Perfecting a stand-density index for even-aged forests. J. Agric. Res. 46:627–638.

Rennie, J. C. 1976. Point-3P sampling: A useful timber inventory design. For. Chron. 52:145–146.

Rényi, A. 1961. On measures of information and entropy. In: Proceedings of Fourth Berkeley Symposium on Mathematical Statistics and Probability. University of California Press. 547–561.

Repo, T., J. Laukkanen, and R. Silvennoinen. 2005. Measurement of the tree root growth using electrical impedance spectroscopy. Silva Fenn. 39:159–166.

Resa, F. 1877. Ueber die Periode der Wurzelbildung. [The timing of rooting] Carthaus, Bonn. 37p.

Reynolds, M. R., T. E. Burk, and W.-C. Huang. 1988. Goodness-of-fit tests and model selection procedures for diameter distribution models. For. Sci. 34:373–399.

Rheinhardt, R., M. Brinson, G. Meyer, and K. Miller. 2012. Integrating forest biomass and distance from channel to develop an indicator of riparian condition. Ecol. Indic. 23:46–55.

Ribe, R. G. 1989. The aesthetics of forestry: What has empirical preference research taught us? Environ. Manag. 13:55–74.

Rice, B., A. R. Weiskittel, and R. G. Wagner. 2014. Efficiency of alternative forest inventory methods in partially harvested stands. Eur. J. For. Res. 133:261–272.

Rich, P. M. 1990. Characterizing plant canopies with hemispherical photographs. Remote Sens. Rev. 5:13–29.

Richards, F. J. 1959. A flexible growth function for empirical use. J. Exp. Bot. 10:290–301.

Riemann Hershey, R., and W. A. Befort. 1995. Aerial photo guide to New England forest cover types. General Technical Report GTR-NE-195, Northeastern Forest Experiment Station, USDA Forest Service. 70p.

Rijal, B., A. R. Weiskittel, and J. A. Kershaw Jr. 2012. Development of regional height to diameter equations for 15 tree species in the North American Acadian Region. Forestry. 85:379–390.

Ring, J. 1963. The laser in astronomy. New Sci. 18:344–345.

Ringvall, A., and G. Ståhl. 1999a. Field aspects of line intersect sampling for assessing coarse woody debris. For. Ecol. Manag. 119:163–170.

Ringvall, A., and G. Ståhl. 1999b. On the field performance of transect relascope sampling for assessing downed coarse woody debris. Scand. J. For. Res. 14:552–557.

Ringvall, A., G. Ståhl, V. Teichmann, J. H. Gove, and M. J. Ducey. 2001. Two-phase approaches to point and transect relascope sampling of downed logs. Can. J. For. Res. 31:971–977.

Ritson, P., and S. Sochacki. 2003. Measurement and prediction of biomass and carbon content of *Pinus pinaster* trees in farm forestry plantation, south-western Australia. For. Ecol. Manag. 175:103–117.

Ritter, T., and J. Saborowski. 2012. Point transect sampling of deadwood: A comparison with well-established sampling techniques for the estimation of volume and carbon storage in managed forests. Eur. J. For. Res. 131:1845–1856.

Ritter, T., A. Nothdurft, and J. Saborowski. 2013. Correcting the nondetection bias of angle count sampling. Can. J. For. Res. 43:344–354.

Rivoire, M., and G. Le Moguedec. 2012. A generalized self-thinning relationship for multi-species and mixed-size forests. Ann. For. Sci. 69:207–219.

Roach, B. A. 1977. A stocking guide for Allegheny hardwoods and its use in controlling intermediate cuttings. Research Paper RP-NE-373, Northeastern Forest Experiment Station, USDA Forest Service. 30p.

Roach, B. A., and S. F. Gingrich. 1962. Timber management guide for upland central hardwoods. Central States Forest Experiment Station, USDA Forest Service. 33p.

Roach, B. A., and S. F. Gingrich. 1968. Even-aged silviculture for upland central hardwoods. Agriculture Handbook 355, US Department of Agriculture. 39p.

Robbins, C. R. 1929. Air survey and forestry. Emp. For. J. 8:205–228.

Rose, C. E., S. W. Martin, K. D. Wannemuehler, and B. D. Plikaytis. 2006. On the use of zero-inflated and hurdle models for modeling vaccine adverse event count data. J. Biopharm. Stat. 16:463–481.

Rosenzweig, M. L. 1995. Species diversity in space and time. Cambridge University Press, New York. 436p.

Ross, J. W. 1981. The radiation regime and architecture of plant stands. Dr. W. Junk, The Hague. 420p.

Ross, S. M. 1980. Introduction to probability models. Academic Press, Boston, MA. 782p.

Row, C., and C. Fasick. 1966. Weight scaling tables by electronic computer. For. Prod. J. 16:41–45.

Row, C., and S. Guttenberg. 1966. Determining weight–volume relationships for sawlogs. For. Prod. J. 16:39–47.

Rozendaal, D. M. A., and P. A. Zuidema. 2011. Dendroecology in the tropics: A review. Trees. 25:3–16.

Rubilar, R. A., H. L. Allen, and D. L. Kelting. 2005. Comparison of biomass and nutrient content equations for successive rotations of loblolly pine plantations on an Upper Coastal Plain site. Biomass Bioenergy. 28:548–564.

Running, S. W., D. L. Peterson, M. A. Spanner, and K. B. Teuber. 1986. Remote sensing of coniferous forest leaf area. Ecology. 67:273–276.

Rustagi, K. P., and R. S. Loveless Jr. 1991. Compatible variable-form volume and stem-profile equations for Douglas-fir. Can. J. For. Res. 21:143–151.

Rytter, L. 2002. Nutrient content in stems of hybrid aspen as affected by tree age and tree size, and nutrient removal with harvest. Biomass Bioenergy. 23:13–26.

Saatchi, S., K. Halligan, D. G. Despain, and R. L. Crabtree. 2007. Estimation of forest fuel load from radar remote sensing. IEEE Trans. Geosci. Remote Sens. 45:1726–1740.

Sandström, F., H. Petersson, N. Kruys, and G. Ståhl. 2007. Biomass conversion factors (density and carbon concentration) by decay classes for dead wood of *Pinus sylvestris*, *Picea abies* and *Betula* spp. in boreal forests of Sweden. For. Ecol. Manag. 243:19–27.

Satoo, T., and H. A. I. Madgwick. 1982. Forest biomass. Martinus Nijhoff/Dr. W. Junk Publishers, The Hague. 152p.

Schaeffer, R. L., W. Mendenhall, and L. Ott. 1990. Elementary survey sampling. PWS-Kent Publishing, Boston, MA. 390p.

Schiffel, A. 1899. Form und Inhalt der Fichte [Form and volume of spruce]. Mittheilungen aus dem forstlichen Versuchswesen Österreichs. 139p.

Schmid-Haas, P. 1969. Stichproben am Waldrand [Sampling at the edge of the forest]. Mitteilungen Eidgenöss. Anst. Für Forstl. Vers. 43:234–303.

Schnell, S., J. Wikman, and G. Ståhl. 2013. Relascope sampling for crown ratio estimation. Can. J. For. Res. 43:459–468.

Schniewind, A. P. 1962. Horizontal specific gravity variation in tree stems in relation to their support function. For. Sci. 8:111–118.

Schnur, G. L. 1937. Yield, stand, and volume tables for even-aged upland oak forests. Technical Bulletin 560, US Department of Agriculture. 87p.

Schreuder, H. T. 2004. Sampling using a fixed number of trees per plot. Research Note RMRS-RN-17, Rocky Mountain Research Station, USDA Forest Service. 4p.

Schreuder, H. T., T. G. Gregoire, and G. B. Wood. 1993. Sampling methods for multiresource forest inventory. John Wiley & Sons, Inc., New York. 446p.

Schumacher, F. X. 1939. A new growth curve and its application to timber-yield studies. J. For. 37:819–820.

Schumacher, F. X., and F. D. S. Hall. 1933. Logarithmic expression of timber-tree volume. J. Agric. Res. 47:719–734.

Scott, C. T. 1979. Midcycle updating: Some practical suggestions. In: Proceedings of the Forest Resource Inventories Workshop, July 23–26, 1979. Fort Collins, CO. IUFRO. 362–370.

Shannon, C. E. 1948. A mathematical theory of communication. Bell Syst. Tech. J. 27:379–423, 623–656.

Sharma, M., and J. Parton. 2009. Modeling stand density effects on taper for jack pine and black spruce plantations using dimensional analysis. For. Sci. 55:268–282.

Sharma, M., and S. Y. Zhang. 2004. Height-diameter models using stand characteristics for *Pinus banksiana* and *Picea mariana*. Scand. J. For. Res. 19:442–451.

Sheil, D., M. J. Ducey, K. Sidiyasa, and I. Samsoedin. 2003. A new type of sample unit for the efficient assessment of diverse tree communities in complex forest landscapes. J. Trop. For. Sci. 15:117–135.

Shepperd, W. D. 1973. An instrument for measuring tree crown width. Research Note RN-RM-229, Rocky Mountain Forest and Range Experiment Station, USDA Forest Service. 3p.

Shinozaki, K., K. Yoda, K. Hozumi, and T. Kira. 1964. A quantitative analysis of plant form; 1. The pipe model theory. Jpn. J. Ecol. 14:97–105.

Shiryayev, A. N. 1984. Probability. Springer-Verlag, New York. 624p.

Shiue, C.-J. 1960. Systematic sampling with multiple random starts. For. Sci. 6:42–50.

Shiver, B. D., and B. E. Borders. 1996. Sampling techniques for forest resource inventory. John Wiley & Sons, Inc., New York. 368p.

Sievänen, R., and T. E. Burk. 1994. Fitting process-based models with stand growth data: Problems and experiences. For. Ecol. Manag. 69:145–156.

Sikkink, P. G., and R. E. Keane. 2008. A comparison of five sampling techniques to estimate surface fuel loading in montane forests. Int. J. Wildland Fire. 17:363–379.

Simpson, E. H. 1949. Measurement of diversity. Nature. 163:688.

Skog, J., and J. Oja. 2007. Improved log sorting combining X-ray and 3D scanning—a preliminary study. In: COST E 53 Conference—Quality Control for Wood and Wood Products, October 15 –17, 3007. Warsaw, Poland. 133–140.

Skole, D., and C. Tucker. 1993. Tropical deforestation and habitat fragmentation in the Amazon: Satellite data from 1978 to 1988. Science. 260:1905–1910.

Skovsgaard, J. P., and J. K. Vanclay. 2008. Forest site productivity: A review of the evolution of dendrometric concepts for even-aged stands. Forestry. 81:13–31.

Skovsgaard, J. P., V. K. Johannsen, and J. K. Vanclay. 1998. Accuracy and precision of two laser dendrometers. Forestry. 71:131–139.

Smith, D. M., B. C. Larson, M. J. Kelty, and P. M. S. Ashton. 1997. The practice of silviculture: Applied forest ecology. John Wiley & Sons, Inc., New York. 560p.

Smith, J. E., L. S. Heath, and J. C. Jenkins. 2003. Forest volume-to-biomass models and estimates of mass for live and standing dead trees of U.S. forests. General Technical Report GTR-NE-298, Northeastern Research Station, USDA Forest Service. 57p.

Smith, J. E., L. S. Heath, K. E. Skog, and R. A. Birdsey. 2006. Methods for calculating forest ecosystem and harvested carbon with standard estimates for forest types of the United States. General Technical Report GTR-NE-343, Northern Research Station, USDA Forest Service. 216p.

Smith, J. H. G., and J. W. Ker. 1959. Empirical yield equations for young forest growth. Br. Columbian Lumberm. 43:22–25.

Smith, J. H. G., J. W. Ker, and J. Csizmazia. 1961. Economics of reforestation of Douglas fir, western hemlock, and western red cedar in the Vancouver Forest District. Forestry Bulletin 3, University of British Columbia. 144p.

Smith, J. R. 1970. Optical distance measurement. Crosby Lockwood and Sons, London. 124p.

Smith, M.-L., S. V. Ollinger, M. E. Martin, J. D. Aber, R. A. Hallett, and C. L. Goodale. 2002. Direct estimation of aboveground forest productivity through hyperspectral remote sensing of canopy nitrogen. Ecol. Appl. 12:1286–1302.

Smith, W. B., and G. J. Brand. 1983. Allometric biomass equations for 98 species of herbs, shrubs, and small trees. Research Note RN-NC-299, North Central Forest Experiment Station, USDA Forest Service. 8p.

Snowdon, P., D. Eamus, P. Gibbons, H. Keith, J. Raison, and M. Kirschbaum. 2000. Synthesis of allometrics, review of root biomass and design of future woody biomass sampling strategies. National Carbon Accounting System Technical Report 17, Australian Greenhouse Office, Canberra, Australia. 133p.

Solberg, S., E. Næsset, and O. M. Bollandsas. 2006. Single tree segmentation using airborne laser scanner data in a structurally heterogeneous spruce forest. Photogramm. Eng. Remote Sens. 72:1369–1378.

Somers, G. L., and S. K. Nepal. 1994. Linking individual-tree and stand-level growth models. For. Ecol. Manag. 69:233–243.

Somogyi, Z., E. Cienciala, R. Mäkipää, P. Muukkonen, A. Lehtonen, and P. Weiss. 2009. Indirect methods of large-scale forest biomass estimation. Eur. J. For. Res. 126:197–207.

Spurr, S. H. 1948a. Aerial photographs in forestry. Ronald Press, New York. 333p.

Spurr, S. H. 1948b. Aerial photography. Unasylva. 4(2). Available online at: http://www.fao.org/docrep/x5345e/x5345e00.htm; Last accessed September 1, 2016.

Spurr, S. H. 1952. Forest inventory. Ronald Press, New York. 476p.

Spurr, S. H. 1962. A measure of point density. For. Sci. 8:85–96.

Srivastava, V. K., and D. E. A. Giles. 1987. Seemingly unrelated regression equations models. Marcel Dekkar, New York. 374p.

Stage, A. R. 1962. A field test of point-sample cruising. Research Paper RP-INT-67, Intermountain Forest and Range Experiment Station, USDA Forest Service. 17p.

Stage, A. R. 1973. Prognosis model for stand development. Research Paper RP-INT-137, Intermountain Forest and Range Experiment Station, USDA Forest Service. 32p.

Stage, A. R. 1976. An expression for the effect of aspect, slope, and habitat type on tree growth. For. Sci. 22:457–460.

Stage, A. R., and D. E. Ferguson. 1984. Linking regeneration surveys to future yields. In: Proceedings of the 1983 Society of American Foresters National Convention. Society of American Foresters. 133–137.

Ståhl, G. 1997. Transect relascope sampling for assessing coarse woody debris: The case of a $\pi/2$ relascope angle. Scand. J. For. Res. 12:375–381.

Ståhl, G. 1998. Transect relascope sampling—A method for the quantification of coarse woody debris. For. Sci. 44:58–63.

Ståhl, G., A. Ringvall, and J. Fridman. 2001. Assessment of coarse woody debris—A methodological overview. Ecol. Bull. 49:57–70.

Ståhl, G., A. Ringvall, J. H. Gove, and M. J. Ducey. 2002. Correction for slope in point and transect relascope sampling of downed coarse woody debris. For. Sci. 48:85–92.

Ståhl, G., J. H. Gove, M. S. Williams, and M. J. Ducey. 2010. Critical length sampling: A method to estimate the volume of downed coarse woody debris. Eur. J. For. Res. 129:993–1000.

Ståhl, G., S. Holm, T. G. Gregoire, T. Gobakken, E. Næsset, and R. Nelson. 2011. Model-based inference for biomass estimation in a LiDAR sample survey in Hedmark County, Norway. Can. J. For. Res. 41:96–107.

Stamatellos, G. S. 1995. Comparison of Point and Point-3P sampling for forest volume estimation with cost analysis. For. Ecol. Manag. 74:75–79.

Staples, C., S. Ahmed, and R. M. Ewers. 2012. Sensitivity of GIS patterns to data resolution: A case study of forest fragmentation in New Zealand. N. Z. J. Ecol. 36:203–209.

Staudhammer, C., and V. M. LeMay. 2000. Height prediction equations using diameter and stand density measures. For. Chron. 76:303–309.

Steber, G. D., and J. C. Space. 1972. New inventory system sweeping the south. J. For. 70:76–79.

Stein, W. I. 1984a. Fixed plot methods for evaluating forest regeneration. In: Proceedings of the 1983 Society of American Foresters National Convention. Society of American Foresters. 129–135.

Stein, W. I. 1984b. Regeneration surveys: An overview. In: Proceedings of the 1983 Society of American Foresters National Convention. Society of American Foresters. 111–116.

Steininger, M. K. 2000. Satellite estimation of tropical secondary forest above-ground biomass: Data from Brazil and Bolivia. Int. J. Remote Sens. 21:1139–1157.

Stenberg, P., L. Korhonen, and M. Rautiainen. 2008. A relascope for measuring canopy cover. Can. J. For. Res. 38:2545–2550.

Stephens, E. P. 1955. The historical-developmental method of determining forest trends. Unpubl. PhD Dissertation, Harvard University, Cambridge, MA. 228p.

Sterba, H. 1975. Assmanns theorie der Grundflächenhaltung und die "Competition-Density-Rule" der Japaner Kira, Ando, und Tadaki. [Assmann's theory of the basal area and the "Competition-density Rule" of the Japanese Kira, Ando, and Tadaki] Zentralbl. Gesamte Forstwes. 92:46–62.

Sterba, H., and R. A. Monserud. 1993. The maximum density concept applied to uneven-aged mixed-species stands. For. Sci. 39:432–452.

Stevens, C. L. 1931. Root growth of white pine (*Pinus strobus* L.). Bulletin 32, School of Forestry, Yale University. 62p.

Stevens, S. S. 1946. On the theory of scales of measurement. Science. 103:677–680.

Stewart, B., C. Cieszewski, and M. Zasada. 2004. Use of a camera as an angle-gauge in angle-count sampling. In: Second International Conference on Forest Measurements and Quantitative Methods and Management. Warnell School of Forestry and Natural Resources, University of Georgia, Athens, GA. 375–380.

Stoffels, A., and J. Van Soest. 1953. Principiele vraagstukken bije proefperken. 3, Hoogteregressie [The main problem in sample plots. 3, Height regression]. Ned. Bosb.-Tijdschr. 66:834–837.

Stokland, J. N., J. Siitonen, and B. G. Jonsson. 2012. Biodiversity in dead wood. Cambridge University Press, New York. 521p.

Strahler, A. H., D. L. B. Jupp, C. E. Woodcock, C. B. Schaaf, T. Yao, F. Zhao, X. Yang, et al. 2008. Retrieval of forest structural parameters using a ground-based lidar instrument (Echidna®). Can. J. Remote Sens. 34(S2):S426–S440.

Strand, L. 1957. "Relaskopisk" høyde- og kubikmassebestemmelse. ["Relascope" height and cubic volume determination] Nor. Skogbr. 3:535–538.

Strickler, G. S., and F. W. Stearns. 1962. Determination of plant density. In: Range research methods: A symposium. Miscellaneous Publication 940, USDA Forest Service. 30–40.

Stubblefield, G. W. 1978. Reconstruction of red alder/Douglas-fir/western hemlock/western redcedar mixed stand and its biological and silvicultural implications. Unpubl. M.Sc. For. Thesis, University of Washington, Seattle, WA. 139p.

Sullivan, A. D., and J. L. Clutter. 1972. A simultaneous growth and yield model for loblolly pine. For. Sci. 18:76–86.

Sutherland, W. J. (Ed.). 1996. Ecological census techniques—A handbook. Cambridge University Press, New York. 336p.

Swamy, S. L., S. Puri, and A. K. Singh. 2003. Growth, biomass, carbon storage and nutrient distribution in Gmelina arborea Roxb. stands on red lateritic soils in central India. Bioresour. Technol. 90:109–126.

Swamy, S. L., S. K. Kushwaha, and S. Puri. 2004. Tree growth, biomass, allometry and nutrient distribution in Gmelina arborea stands grown in red lateritic soils of Central India. Biomass Bioenergy. 26:306–317.

Swan, D. A. 1959. Weight scaling in the Northeast. Technical Release 59-R5, American Pulpwood Association.

Swank, W. T., and H. T. Schreuder. 1974. Comparison of three methods of estimating surface area and biomass for a forest of young eastern white pine. For. Sci. 20:91–100.

Tallent-Halsell, N. G. (Ed.). 1994. Forest health monitoring field methods guide. EPA/620/R-94/027, US Environmental Protection Agency, Washington, DC. 225p.

Taras, M. A. 1956. Buying pulpwood by weight as compared with volume measure. Station Paper (Old Series) SP-SE-074, Southeastern Forest Experiment Station, USDA Forest Service. 13p.

Tardif, G. 1965. Some considerations concerning the establishment of optimum plot size in forest survey. For. Chron. 41:93–102.

Taylor, A. R., H. Y. H. Chen, and L. VanDamme. 2009. A review of forest succession models and their suitability for forest management planning. For. Sci. 55:23–36.

Temesgen, H., and J. M. Ver Hoef. 2015. Evaluation of the spatial linear model, random forest and gradient nearest-neighbour methods for imputing potential productivity and biomass of the Pacific Northwest forests. Forestry. 88:131–142.

Temesgen, H., V. M. LeMay, and S. J. Mitchell. 2005. Tree crown ratio models for multi-species and multi-layered stands of southeastern British Columbia. For. Chron. 81:133–141.

Tesch, S. D. 1980. The evolution of forest yield determination and site classification. For. Ecol. Manag. 3:169–182.

Thomas, J. W. (Ed.). 1979. Wildlife habitats in managed forests : The Blue Mountains of Oregon and Washington. Agriculture Handbook 553, USDA Forest Service. 512p.

Thomas, S. C., and A. R. Martin. 2012. Carbon content of tree tissues: A synthesis. Forests. 3:332–352.

Thompson, S. K. 2012. Sampling. Wiley/Blackwell, New York. 472p.

Timson, F. G. 1974. Weight and volume variation in truckloads of logs hauled in the Central Appalachians. Research Paper RP-NE-300, Northeastern Forest Experiment Station, USDA Forest Service. 9p.

Tokola, T. 2006. Chapter 18: Europe. In: A. Kangas and M. Maltamo (Eds.) Forest inventory—Methodology and applications, Springer, Dordrecht, The Netherlands. 295–308.

Tolunay, D. 2009. Carbon concentrations of tree components, forest floor and understorey in young Pinus sylvestris stands in north-western Turkey. Scand. J. For. Res. 24:394–402.

Tomé, M., and H. E. Burkhart. 1989. Distance-dependent competition measures for predicting growth of individual trees. For. Sci. 35:816–831.

Tomppo, E. 2006. Chapter 11: The Finnish national forest inventory. In: A. Kangas and M. Maltamo (Eds.) Forest inventory—Methodology and applications. Springer, Dordrecht, The Netherlands. 179–194.

Tomppo, E., and M. Katila. 1991. Satellite image-based national forest inventory of Finland. In: Proceedings of IGARSS'91, Remote Sensing: Global Monitoring for Earth Management, June 3–6, 1991 Espoo, Finland. 1141–1144.

Toumey, J. W., and C. F. Korstian. 1947. Foundations of silviculture upon an ecological basis. John Wiley & Sons, Inc., New York. 468p.

Treitz, P., and P. J. Howarth. 1999. Hyperspectral remote sensing for estimating biophysical parameters of forest ecosystems. Prog. Phys. Geogr. 23:359–390.

Trincado, G., and C. Leal. 2006. Local and generalized height-diameter equations for radiata pine (*Pinus radiata*). Bosque. 27:23–34.

Trofymow, J. A., H. J. Barclay, and K. M. McCullough. 1991. Annual rates and elemental concentrations of litter fall in thinned and fertilized Douglas-fir. Can. J. For. Res. 21:1601–1615.

Trorey, L. G. 1932. A mathematical expression for the construction of diameter height curves based on site. For. Chron. 18:3–14.

Tucker, C. J. 1979. Red and photographic infrared linear combinations for monitoring vegetation. Remote Sens. Environ. 8:127–150.

Tufte, E. R. 1990. Envisioning information. Graphics Press, Cheshire, CT. 126p.

Turnbull, K. J., and G. E. Hoyer. 1965. Construction and analysis of comprehensive tree-volume tarif tables. Washington Department of Natural Resources Management, Olympia, WA. 63p.

Turnbull, K. J., G. R. Little, and G. E. Hoyer. 1963. Comprehensive tree-volume tarif tables. Washington Department of Natural Resources Management, Olympia, WA. 127p.

Turner, D. P., S. A. Acker, J. E. Means, and S. L. Garman. 2000. Assessing alternative allometric algorithms for estimating leaf area of Douglas-fir trees and stands. For. Ecol. Manag. 126:61–76.

USDA. 1991. National Forest cubic scaling handbook. Forest Service Handbook 2409.11a, USDA Forest Service, Washington, DC. 157p.

USDA. 1992. Forest Service resource inventories: An overview. Forest Inventory, Economics, and Recreation Research, USDA Forest Service, Washington, DC. 39p.

USDA. 2013. National Agricultural Imagery Program (NAIP): Information sheet. United States Department of Agriculture. Available online at: https://catalog.data.gov/dataset/the-national-agriculture-imagery-program-naip-information-sheet; Last accessed August 31, 2016.

USMA. 2007. Guide to the use of the metric system. US Metric Association, Colorado State University, Fort Collins, CO.

Valentine, H. T., L. M. Tritton, and G. M. Furnival. 1984. Subsampling trees for biomass, volume, or mineral content. For. Sci. 30:673–681.

Valentine, H. T., D. A. Herman, J. H. Gove, and D. Y. Hollinger. 2000. Initializing a model stand for process-based projection. Tree Physiol. 20:393–398.

Valentine, H. T., J. H. Gove, and T. G. Gregoire. 2001. Monte Carlo approaches to sampling forested tracts with lines or points. Can. J. For. Res. 31:1410–1424.

Valentine, H. T., M. J. Ducey, J. H. Gove, A. Lanz, and D. L. R. Affleck. 2006. Corrections for cluster-plot slop. For. Sci. 52:55–66.

Valentine, H. T., J. H. Gove, M. J. Ducey, T. G. Gregoire, and M. S. Williams. 2008. Chapter 6: Estimating the carbon in coarse woody debris with perpendicular distance sampling. In: C. Hoover (Ed.) Field measurements for forest carbon monitoring: A landscape-scale approach, Springer, New York. 73–90.

Valentine, H. T., A. Mäkelä, E. J. Green, R. L. Amateis, H. Makinen, and M. J. Ducey. 2012a. Models relating stem growth to crown length dynamics: Application to loblolly pine and Norway spruce. Trees Struct. Funct. 26:469–478.

Valentine, H. T., A. Mäkelä, E. J. Green, R. L. Amateis, H. Mäkinen, and M. J. Ducey. 2012b. Models relating stem growth to crown length dynamics: Application to loblolly pine and Norway spruce. Trees Struct. Funct. 26:469–478.

Van Den Meersschaut, D., and K. Vandekerkhove. 2000. Development of a stand–scale forest biodiversity index based on the State forest inventory. In: Integrated tools for natural resources inventories in the 21st century. Proceedings, General Technical Report GTR-NC-212, North Central Forest Experiment Station, USDA Forest Service. 340–347.

Van Deusen, P. C. 1987. Combining taper functions and critical height sampling for unbiased stand volume estimation. Can. J. For. Res. 17:1416–1420.

Van Deusen, P. C., and J. H. Gove. 2011. Sampling coarse woody debris along spoked transects. Forestry. 84:93–98.

Van Deusen, P. C., and W. J. Meerschaert. 1986. On critical-height sampling. Can. J. For. Res. 16:1310–1313.

Van Dyne, G. M., W. G. Vogel, and H. G. Fisser. 1964. Influence of small plot size and shape on range herbage production estimates. Ecology. 44:746–759.

Van Hooser, D. D. 1972. Evaluation of two-stage 3P sampling for forest surveys. Research Paper RP-SO-77, Southern Forest Experiment Station, USDA Forest Service. 9p.

Van Hooser, D. D. 1973. Field evaluation of two-stage 3P sampling. Research Paper RP-SO-86, Southern Forest Experiment Station, USDA Forest Service. 15p.

Van Lear, D. H., J. B. Waide, and M. J. Teuke. 1984. Biomass and nutrient content of a 41-year-old loblolly pine (*Pinus taeda* L.) plantation on a poor site in South Carolina. For. Sci. 30:395–404.

van Leeuwen, M., and M. Nieuwenhuis. 2010. Retrieval of forest structural parameters using LiDAR remote sensing. Eur. J. For. Res. 129:749–770.

van Leeuwen, M., T. Hiker, N. C. Coops, G. Frazer, M. A. Wulder, G. J. Newnham, and D. S. Culvenor. 2011. Assessment of standing wood and fiber quality using ground and airborne laser scanning: A review. For. Ecol. Manag. 261:1467–1478.

Van Pelt, R., S. C. Sillett, and N. M. Nadkami. 2004. Chapter 3: Quantifying and visualizing canopy structure in tall forests: methods and a case study. In: M. Lowman, and B. Rinker (Eds.) Forest canopies. Elsevier, Boston, MA. 49–72.

Van Wagner, C. E. 1968. The line intersect method in forest fuel sampling. For. Sci. 14:20–26.

Vanclay, J. K. 1994. Modelling forest growth and yield: Applications to mixed tropical forests. CAB International, Wallingford, UK. 312p.

Vanderwel, M. C., H. C. Thorpe, J. L. Shuter, J. Caspersen, and S. C. Thomas. 2008. Contrasting downed woody debris dynamics in managed and unmanaged northern hardwood stands. Can. J. For. Res. 38:2850–2861.

Vogelmann, J. E., B. Tolk, and Z. Zhu. 2009. Monitoring forest changes in the southwestern United States using multitemporal Landsat data. Remote Sens. Environ. 113:1739–1748.

von Bertalanffy, L. 1957. Quantitative laws in metabolism and growth. Q. Rev. Biol. 32:217–231.

Vose, J. M., and H. L. Allen. 1988. Leaf area, stemwood growth, and nutrition relationships in loblolly pine. For. Sci. 34:547–563.

Wagner, R. G., K. M. Little, B. Richardson, and K. McNabb. 2006. The role of vegetation management for enhancing productivity of the world's forests. Forestry. 79:57–79.

Wahlenberg, W. G. 1941. Methods of forecasting timber growth in irregular stands. Technical Bulletin 796, US Department of Agriculture. 55p.

Wang, Y. P., and P. G. Jarvis. 1990. Influence of crown structural properties on PAR absorption, photosynthesis, and transpiration in Sitka spruce: application of a model (MAESTRO). Tree Physiol. 7:297–316.

Wang, Y. S., and D. R. Miller. 1987. Calibration of the hemispherical photographic technique to measure leaf area index distributions in hardwood forests. For. Sci. 33:210–216.

Ware, K. D. 1967. Sampling properties of three pee estimates. In: *Proceedings of the 1967 Society of American Foresters—Canadian Institute of Forestry Joint Convention*. October, 1967. Ottawa, Ontario, Canada. Society of American Foresters.

Waring, R. H., K. Newman, and J. Bell. 1981. Efficiency of tree crowns and stemwood production at different canopy leaf densities. Forestry. 54:129–137.

Waring, R. H., P. E. Schroeder, and R. Oren. 1982. Application of the pipe model theory to predict canopy leaf area. Can. J. For. Res. 12:556–560.

Warren, W. G., and P. F. Olsen. 1964. A line intersect technique for assessing logging waste. For. Sci. 10:267–276.

Weggler, K., M. Dobbertin, E. Jüngling, E. Kaufmann, and E. Thürig. 2012. Dead wood volume to dead wood carbon: The issue of conversion factors. Eur. J. For. Res. 131:1423–1438.

Weiskittel, A. R., P. Gould, and H. Temesgen. 2009a. Sources of variation in the self-thinning boundary line for three species with varying levels of shade tolerance. For. Sci. 55:84–93.

Weiskittel, A. R., J. A. Kershaw Jr., P. V. Hofmeyer, and R. S. Seymour. 2009b. Species differences in total and vertical distribution of branch- and tree-level leaf area for the five primary conifer species in Maine, USA. For. Ecol. Manag. 258:1695–1703.

Weiskittel, A. R., D. W. Hann, J. A. Kershaw Jr., and J. K. Vanclay. 2011. Forest growth and yield modeling. Wiley/Blackwell, New York. 430p.

Wenger, K. F. (Ed.). 1984. Forestry handbook. John Wiley & Sons, Inc., New York. 1360p.

Wesley, R. 1956. Measuring the height of trees. Park Adm. 21:80–84.

West, G. B., J. H. Brown, and J. Enquist. 1999. A general model for the structure and allometry of plant vascular systems. Nature. 400:664–667.

West, P. W. 2013. Precision of inventory using different edge overlap methods. Can. J. For. Res. 43:1081–1083.

Westfall, J. A. 2012. A comparison of above-ground dry-biomass estimators for trees in the northeastern United States. North. J. Appl. For. 29:26–34.

Westfall, J. A., and C. W. Woodall. 2007. Measurement repeatability of a large-scale inventory of forest fuels. For. Ecol. Manag. 253:171–176.

Westoby, J. 1987. The purpose of forests. Basil Blackwell, Oxford, UK. 343p.

Wharton, E. H., and D. M. Griffith. 1998. Estimating total forest biomass in Maine, 1995. Research Bulletin RB-NE-142, Northeastern Forest Experiment Station, USDA Forest Service. 50p.

Whitlock, G. F. A., Lt.-Colonel Newcombe, Lt.-Colonel Salmon, M. Brock, T. Holdich, M. Hinks, G. Hardy, et al. 1919. Mapping from air photographs: Discussion. Geogr. J. 53:396–403.

Whittaker, R. H. 1965. Dominance and diversity in land plant communities. Science. 147:250–260.

Whittaker, R. H. (Ed.). 1978. Classification of plant communities. Springer-Verlag, The Hague. 408p.

Whittaker, R. H., and G. M. Woodwell. 1968. Dimension and production relations of trees and shrubs in the Brookhaven Forest, New York. J. Ecol. 56:1–25.

Wiant, H. V., G. B. Wood, and R. R. Forslund. 1991. Comparison of centroid and paracone estimates of tree volume. Can. J. For. Res. 21:714–717.

Wiant, H. V., G. B. Wood, and G. M. Furnival. 1992a. Estimating log volume using the centroid position. For. Sci. 38:187–191.

Wiant, H. V., G. B. Wood, and T. G. Gregoire. 1992b. Practical guide for estimating the volume of a standing sample tree using either importance or centroid sampling. For. Ecol. Manag. 49:333–339.

Wiemann, M. C., and G. B. Williamson. 2013. Biomass determination using wood specific gravity from increment cores. General Technical Report FPL-GTR-225, Forest Products Laboratory, USDA Forest Service, Madison, WI. 9p.

Williams, M. S. 2001a. New approach to areal sampling in ecological surveys. For. Ecol. Manag. 154:11–22.

Williams, M. S. 2001b. Nonuniform random sampling: an alternative method of variance reduction for forest surveys. Can. J. For. Res. 31:2080–2088.

Williams, M. S., and J. H. Gove. 2003. Perpendicular distance sampling: An alternative method for sampling downed coarse woody debris. Can. J. For. Res. 33:1564–1579.

Williams, M. S., M. J. Ducey, and J. H. Gove. 2005. Assessing surface area of coarse woody debris with line intersect and perpendicular distance sampling. Can. J. For. Res. 35:949–960.

Wilson, F. G. 1946. Numerical expression of stocking in terms of height. J. For. 44:758–761.

Wilson, F. G. 1951. Control of growing stock in even-aged stands of conifers. J. For. 49:692–695.

Wilson, F. G. 1979. Thinning as an orderly discipline: A graphic spacing schedule for red pine. J. For. 77:483–486.

Wilson, J. W. 1959a. Analysis of the spatial distribution of foliage by two-dimensional point quadrats. New Phytol. 58:92–99.

Wilson, J. W. 1959b. Inclined point quadrats. New Phytol. 59:1–8.

Wilson, J. W. 1965. Stand structure and light penetration. I. Analysis by point quadrats. J. Appl. Ecol. 2:383–390.

Wilson, R. L. 1989. Elementary forest surveying and mapping. Oregon State University Bookstore, Corvallis, OR. 181p.

Winner, L., and C. B. Stott. 1954. Forest control by continuous inventory. CFI Notes 2, North Central Forest Experiment Station, USDA Forest Service. 17p.

Winthers, E., D. Fallon, J. Haglund, T. DeMeo, G. Nowacki, D. Tart, M. Ferwerda, et al. 2005. Terrestrial Ecological Unit inventory technical guide. General Technical Report GTR-WO-68, Ecosystem Management Coordination Staff, USDA Forest Service, Washington, DC. 245p.

Woldendorp, G., R. J. Keenan, S. Barry, and R. D. Spencer. 2004. Analysis of sampling methods for coarse woody debris. For. Ecol. Manag. 198:133–148.

Wonn, H. T., and K. L. O'Hara. 2001. Height:diameter ratios and stability relationships for four northern Rocky Mountain tree species. West. J. Appl. For. 16:87–94.

Wood, G. B., and H. V. Wiant. 1992. Comparison of point-3P and modified point-list sampling for inventory of mature native hardwood forest in southeastern New South Wales. Can. J. For. Res. 22:725–728.

Wood, G. B., H. V. Wiant, R. J. Loy, and J. A. Miles. 1990. Centroid sampling: A variant of importance sampling for estimating the volume of sample trees of radiata pine. For. Ecol. Manag. 36:233–243.

Woodall, C. W., and M. S. Williams. 2005. Sampling protocol, estimation, and analysis procedures for the down woody materials indicator of the FIA program. General Technical Report GTR-NC-256, North Central Research Station, USDA Forest Service. 47p.

Woodall, C. W., P. D. Miles, and J. S. Vissage. 2005. Determining maximum stand density index in mixed species stands for strategic-scale stocking assessments. For. Ecol. Manag. 216:367–377.

Woodall, C. W., L. S. Heath, and J. E. Smith. 2008. National inventories of down and dead woody material forest carbon stocks in the United States: Challenges and opportunities. For. Ecol. Manag. 256:221–228.

Woodall, C. W., J. Rondeux, P. J. Verkerk, and G. Ståhl. 2009. Estimating dead wood during national forest inventories: A review of inventory methodologies and suggestions for harmonization. Environ. Manage. 44:624–631.

Woodall, C. W., L. S. Heath, G. M. Domke, and M. C. Nichols. 2011. Methods and equations for estimating aboveground volume, biomass, and carbon for trees in the U.S. forest inventory, 2010. General Technical Report GTR-NRS-88, Northern Research Station, USDA Forest Service. 30p.

Worbes, M., and W. J. Junk. 1989. Dating tropical trees by means of ^{14}C from bomb tests. Ecology. 70:503–507.

Wulder, M. A., C. W. Bater, N. C. Coops, T. Hilker, and J. C. White. 2008. The role of LiDAR in sustainable forest management. For. Chron. 84:807–826.

Wulder, M. A., J. G. Masek, W. B. Cohen, T. R. Loveland, and C. E. Woodcock. 2012. Opening the archive: How free data has enabled the science and monitoring promise of Landsat. Remote Sens. Environ. 122:2–10.

Wykoff, W. R., N. L. Crookston, and A. R. Stage. 1982. User's guide to the Stand Prognosis model. General Technical Report GTR-INT-1333, Intermountain Forest and Range Experiment Station, USDA Forest Service. 112p.

Yerkes, V. P. 1966. Weight and cubic–foot relationships for Black Hills ponderosa pine sawlogs. Research Note RN-RM-78, Rocky Mountain Forest and Range Experiment Station, USDA Forest Service. 4p.

Yocom, H. A. 1970. Vernier scales for diameter tapes. J. For. 68:725.

Young, H. E., L. Strand, and R. Altenberger. 1964. Preliminary fresh and dry weight tables for seven tree species in Maine. Technical Bulletin 12, Agriculture Experiment Station, University of Maine, Orono, ME. 76p.

Young, H. E., W. C. Robbins, and S. Wilson. 1967. Errors in volume determination of primary forest products. In: Proceedings of 14th IUFRO Congress, Munich, September 4–9, 1967. Part VI, Sect. 25. 546–562.

Zanakis, S. H. 1979. A simulation study of some simple estimators for the three–parameter Weibull distribution. J. Stat. Comput. Simul. 9:101–116.

Zar, J. H. 2009. Biostatistical analysis. Pearson, New York. 960p.

Zarin, D. J., E. A. Davidson, E. Brondizio, I. C. G. Vieira, T. Sá, T. Feldpausch, E. A. G. Schuur, et al. 2005. Legacy of fire slows carbon accumulation in Amazonian forest regrowth. Front. Ecol. Environ. 3:365–369.

Zarin, D. J., M. J. Ducey, J. M. Tucker, and W. A. Salas. 2001. Potential biomass accumulation in Amazonian regrowth forests. Ecosystems. 4:658–668.

Zarnoch, S. J., and T. R. Dell. 1985. An evaluation of percentile and maximum likelihood estimators of Weibull parameters. For. Sci. 31:260–268.

Zeide, B. 1980. Plot size optimization. For. Sci. 26:251–257.

Zeide, B. 2004. Intrinsic units in growth modeling. Ecol. Model. 175:249–259.

Zenger, A. 1964. Systematic sampling in forestry. Biometrics. 20:553–565.

Zhang, L., H. Bi, J. H. Gove, and L. S. Heath. 2011. A comparison of alternative methods for estimating the self-thinning boundary line. Can. J. For. Res. 35:1507–1514.

Zheng, D., L. S. Heath, and M. J. Ducey. 2007. Forest biomass estimated from MODIS and FIA data in the Lake States: MN, WI and MI, USA. Forestry. 80:265–278.

Zheng, D., L. S. Heath, M. J. Ducey, and J. E. Smith. 2009. Quantifying scaling effects on satellite-derived forest area estimates for the conterminous USA. Int. J. Remote Sens. 30:3097–3114.

Zianis, D., P. Muukkonen, R. Mäkipää, and M. Mencuccini. 2005. Biomass and stem volume equations for tree species in Europe. Silva Fenn. Monogr. 4:63.

Zöhrer, F. 1978. Fundamentale Stichprobenkonzepte der Forstinventur. I. [Fundamental sampling concepts of forest inventory. I.] Allg. Forstztg. 89:180–185.

Zöhrer, F. 1979. Fundamentale Stichprobenkonzepte der Forstinventur. II. [Fundamental sampling concepts of forest inventory. II.] Allg. Forstztg. 90:112–115.

Zon, R. 1910. The forest resources of the world. Bulletin 83, USDA Forest Service. 91p.

Zuur, A., E. N. Ieno, N. Walker, A. A. Saveliev, and G. M. Smith. 2009. Mixed effects models and extensions in ecology with R. Springer, New York. 574p.

INDEX

Page numbers with *f* refers figures; those t refers to tables

Forest Mensuration, Fifth Edition. John A. Kershaw, Jr., Mark J. Ducey, Thomas W. Beers and Bertram Husch.
© 2017 John Wiley & Sons, Ltd. Published 2017 by John Wiley & Sons, Ltd.